›hysica status solidi (a)

›pplied research

Volume 111 Number 1 January 1989

AKADEMIE-VERLAG BERLIN

SSN 0031-896 phys. stat. sol. (a), Berlin **111** (1989) 1, 1–394, K1–K132

International Classification System for Physics*)

60. Condensed matter: structure, mechanical and thermal properties

    61. Structure of liquids and solids; crystallography
    (see also 68.20. Solid surface structure, 71. Electron states)

    62. Mechanical and acoustical properties of condensed matter
    (see also 61.70. Defects in crystals, 68.30. Surfaces and interfaces)

    63. Lattice dynamics and crystal statistics
    (see also 65. Thermal properties, 66.70. Thermal conduction, 68.30. Dynamics of surface and interface vibrations, 78.30. Infrared and Raman spectra)

    64. Equations of state, phase equilibria, and phase transitions

    65. Thermal properties of condensed matter
    (see also 63. Lattice dynamics; for thermodynamic properties of quantum fluids, see 67.40; for thermal properties of solid helium, see 67.80)

    66. Transport properties of condensed matter (nonelectronic)

    67. Quantum fluids and solids; liquid and solid helium

    68. Surfaces and interfaces; thin films and whiskers
    (for impact phenomena, see 79; for crystal growth, see 61.50)

70. Condensed matter: electronic structure; electrical, magnetic, and optical properties

    71. Electronic states
    (see also 63. Lattice dynamics, 73. Electronic structure and electrical properties of surfaces, interfaces, and thin films)

    72. Electronic transport in condensed matter
    (for surfaces, interfaces, and thin films, see 73)

    73. Electronic and electrical properties of surfaces, interfaces, and thin films

    74. Superconductivity

    75. Magnetic properties and materials

    76. Magnetic resonances and relaxation in condensed matter; Mössbauer effect

    77. Dielectric properties and materials
    (for conductivity phenomena, see 72.20 and 72.80)

    78. Optical properties and condensed-matter spectroscopy and other interactions of matter with particles and radiation
    (for phonon spectra, see 63)

    79. Electron and ion emission by liquids and solids; impact phenomena

85. Devices**)

---

*) Excerpt; reproduced with permission of International Council for Scientific and Technical Information (ICSTI).

**) Outside the ICSTI Classification for Physics.

(The Substance Classification is given on cover three)

Important Information

The Proceedings of the International Conference on Optical Nonlinearity and Bistability in Semiconductors, Berlin, August 1988, published in the December issue (Vol. 150, No. 2) of physica status solidi (b), can be ordered also separately for the usual price per issue.

Your order should be sent to

Akademie-Verlag Berlin
Vertriebsabteilung
Habersaathstr. 31
Berlin
DDR-1040

physica status solidi (a)

applied research

Volume 111 · Number 1 · Pages 1 to 394 and K1 to K132

January 16, 1989

AKADEMIE-VERLAG · BERLIN

Subscriptions and orders for single copies should be directed
in the GDR:

to the Postzeitungsvertrieb or
to the Akademie-Verlag Berlin, Leipziger Str. 3–4, DDR-1086 Berlin;

in the other socialist countries:

to a book-shop for foreign language literature or
to the competent news-distributing agency;

in the FRG and BERLIN (WEST):

to a book-shop or
to the wholesale distributing agency Kunst und Wissen, Erich Bieber, OHG, Wilhelmstr. 4–6, P. O. Box 102844, D-7000 Stuttgart 10;

in the other Western European countries:

to Kunst und Wissen, Erich Bieber GmbH, General Wille-Str. 4, CH-8002 Zürich;

in USA and CANADA:

to VCH Publishers, Inc., 303 NW 12th Avenue, Deerfield Beach, FL 33442–1788, USA;

in other countries:

to the international book- and journal-selling trade,
to Buchexport, Volkseigener Außenhandelsbetrieb der Deutschen Demokratischen Republik. Postfach 160, DDR-7010 Leipzig, or
to the Akademie-Verlag Berlin, Leipziger Str. 3–4, DDR-1086 Berlin.

Editorial Note:

"physica status solidi (a)" undertakes that an original paper *accepted for publication* before the 23rd of any month will be published within 50 days of this date unless the author requests a postponement. In special cases there may be some delay between *receipt* and *acceptance* of a paper due to the review and, if necessary, revision of the paper.

Schriftleiter und verantwortlich für den Inhalt:
Professor Dr. E. Gutsche, Leipziger Straße 3–4, Postfach 1233, DDR-1086 Berlin.

Verlag:
Akademie-Verlag Berlin, Leipziger Str, 3–4, DDR-1086 Berlin; Fernruf: 2236221 und 2236229; Telex-Nr.: 114420; Bank: Staatsbank der DDR, Berlin, Kto.-Nr.: 6836-26-20712.

Chefredakteur:
Dr. H.-J. Hänsch.

Anschrift der Redaktion:
Leipziger Straße 3–4, Postfach 1233, DDR-1086 Berlin, Fernruf: 2236279.

Veröffentlicht unter der Lizenznummer 1620 des Presseamtes beim Vorsitzenden des Ministerrates der Deutschen Demokratischen Republik.

Gesamtherstellung:
VEB Druckerei „Thomas Müntzer", DDR-5820 Bad Langensalza.

Erscheinungsweise:
Die Zeitschrift „physica status solidi (a)" erscheint jeweils am 16. eines jeden Monats. Jährlich erscheinen 6 Bände zu je 2 Heften.

Bezugspreis:
Je Band 220,– DM zuzüglich Versandspesen; je Jahr 1320,– DM. Der gültige Jahresbezugspreis für die DDR ist der Postzeitungsliste zu entnehmen.

Bestellnummer dieses Bandes: 1085/111.

Printed in the German Democratic Republic.
AN (EDV) 20735
06500

phys. stat. sol. (a) **111**, No. 1 (1989)

Contents

Lattice properties

Surfaces, interfaces, thin films; lower-dimensional systems

Extended electronic states

Localized electronic states

Electronic transport; superconductivity

Magnetic properties; resonances

Dielectric and optical properties

Device-related phenomena

physica status solidi (a) is indexed in Current Contents/Physical, Chemical & Earth Sciences.

Systematic List

Subject classification:	Corresponding papers begin on the following pages (pages given in italics refer to the principle subject classification):

Contents of Volume 111 Continued on Page 397

International Classification System for Physics*)

60. Condensed matter: structure, mechanical and thermal properties

61. Structure of liquids and solids; crystallography
    (see also 68.20 Solid surface structures, 71. Electron states)
    61.10. X-ray determination of structures
        (for specific determinations, see 61.55 to 61.80)
    61.12. Neutron determination of structures
        (for specific determinations, see 61.55 to 61.80)
    61.14. Electron determination of structures
        (for specific determinations, see 61.55 to 61.80)
    61.16. Other determination of structures
        (for specific determinations, see 61.55 to 61.80)
    61.20. Classical, semiclassical, and quantum theories of liquid structure
        (for electronic states, see 71; for liquid helium, see 67)
    61.25. Studies of specific liquid structures
    61.30. Liquid crystals
    61.40. Amorphous and polymeric materials
    61.50. Crystalline state (including molecular motions in solids)
        (for magnetic structure and spin systems, see 75.25)
    61.55. Specific structures of elements and alloys
    61.60. Specific structures of inorganic compounds
    61.65. Specific structures of organic compounds
    61.70. Defects in crystals
        (see also 61.80. Radiation damage, 62. Mechanical and acoustical properties, 71.55. Impurities and defect levels, 76.30. EPR of color centers and other defects, 78.50. Impurity and defect absorption in solids)
    61.80. Radiation damage and other irradiation effects
        (for techniques of structure determination, see 61.10 to 61.16; for electron and ion impact phenomena, see 79.20)
    61.90. Other topics in structure of liquids and solids

62. Mechanical and acoustical properties of condensed matter
    (see also 61.70. Defects in crystals, 68.30. Surfaces and interfaces)
    62.10. Mechanical properties of liquids
        (for viscosity of liquids, see also 66.20)
    62.20. Mechanical properties of solids (related to microscopic structure)
    62.30. Mechanical and elastic waves
    62.40. Anelasticity, internal friction, and mechanical resonances
        For thermomechanical effects, see 65.70
        For magnetomechanical effects, see 75.80
        For piezoelectric effects, see 77.60
        For elasto-optical effects, see 78.20
    62.50. High-pressure and shock-wave effects in solids
    62.60. Acoustical properties of liquids
        For lattice dynamics, phonons, see 63
        For second sound in quantum fluids, see 67.40
    62.65. Acoustical properties of solids
        For magnetoacoustic effects in metals, see 75.80
        For acoustoelectric effects, see 72.50
        For acousto-optical effects, see 78.20
    62.80. Ultrasonic relaxation
        (see also 74.30. Ultrasonic attenuation in superconductors)
    62.90. Other topics in mechanical and acoustical properties of condensed matter

---

*) Excerpt, reproduced with permission of International Council for Scientific and Technical Information (ICSTI).

63. Lattice dynamics and crystal statistics
(see also 65. Thermal properties, 66.70. Thermal conduction, 68.30. Dynamics of surface and interface vibrations, 78.30. Infrared and Raman spectra)
63.10. General theory
63.20. Phonons and vibrations in crystal lattices
63.50. Vibrational states in disordered systems
63.70. Statistical mechanics of lattice vibrations
(see also 65. Thermal properties of condensed matter, and 66.70. Thermal conduction)
63.75. Statistical mechanics of displacive phase-transitions
For order–disorder and statistical mechanics of model systems, see 64.60
For crystallographic aspects of polymorphic and order–disorder transformations, see 61.50
63.90. Other topics in lattice dynamics and crystal statistics

64. Equations of state, phase equilibria, and phase transitions
64.10. General theory of equations of state and phase equilibria
64.30. Equations of state of specific substances
(see also 65.70. Thermal expansion)
64.60. General studies of phase transitions
(for magnetic, superconducting, and quantum fluid critical phenomena, see 75, 74, and 67, respectively)
64.70. Phase equilibria, phase transitions, and critical points of specific substances
64.75. Solubility, segregation, and mixing
64.80. Other phase properties of systems
64.90. Other topics in equations of state, phase equilibria, and phase transitions

65. Thermal properties of condensed matter
(see also 63. Lattice dynamics; for thermodynamic properties of quantum fluids, see 67.40; for thermal properties of solid helium, see 67.80)
65.20. Heat capacities of liquids
65.40. Heat capacities of solids
(for specific heat of superconductors, see 74.30; for specific heat of magnetic systems, see 75.40)
65.50. Thermodynamic properties and entropy
65.70. Thermal expansion and thermomechanical effects
(see also 64.30. Equations of state)
For thermal conduction in nonmetallic liquids, see 66.60; for nonmetallic solids, see 66.70
For electronic thermal conduction, see 72.10 and 72.20
For thermal conductivity of superconductors, see 74.30
For pyroelectric and electrocaloric effects, see 77.70
65.90. Other topics in thermal properties of condensed matter

66. Transport properties of condensed matter (nonelectronic)
66.10. Diffusion and ionic conduction in liquids
66.20. Diffusive momentum transport
(for viscosity of liquids, see also 62.10)
66.30. Diffusion in solids
66.60. Thermal conduction in nonmetallic liquids
(for thermal conduction in liquid metals, see 72.15)
66.70. Nonelectronic thermal conduction and heat-pulse propagation in nonmetallic solids
(for thermal conduction in solid metals, see 72.15; for statistical mechanics of lattice vibrations, see 63.70)
66.90. Other topics in nonelectronic transport properties

Author Index

67. Quantum fluids and solids; liquid and solid helium
    67.20. Quantum effects on the structure and dynamics of nondegenerate fluids
    67.40. Boson degeneracy and superfluidity of helium-4
    67.50. Fermi fluids; liquid helium-3
    67.60. Mixed systems; liquid helium-3, -4 mixtures
    67.70. Films (including physical adsorption)
    67.80. Solid helium and related quantum crystals
    67.90. Other topics in quantum fluids
    (e.g., neutron-star matter)

68. Surfaces and interfaces; thin films and whiskers
    (for impact phenomena, see 79; for crystal growth, see 61.50)
    68.10. Fluid surfaces and fluid–fluid interfaces
    68.15. Liquid thin films
    68.20. Solid surface structures
    68.25. Mechanical and acoustical properties of solid surfaces and interfaces
    68.30. Dynamics of solid surfaces and interface vibrations
    68.40. Surface energy of solids; thermodynamic properties
    68.45. Solid–fluid interface processes
    68.48. Solid–solid interfaces (including bicrystals)
    (for grain boundaries, see 61.70)
    68.55. Thin film growth, structure, and epitaxy
    68.60. Physical properties of thin films: nonelectronic
    68.70. Whiskers and dendrites: growth, structure, and nonelectronic properties
    68.90. Other topics in the structure and nonelectronic properties of surfaces and thin films

70. Condensed matter: electronic structure; electrical, magnetic, and optical properties

71. Electronic states
    (see also 63. Lattice dynamics, 73. Electronic structure and electrical properties of surfaces, interfaces, and thin films)
    71.10. General theories and computational techniques
    71.20. Electronic density of states determinations
    (including energy states of liquid semiconductors)
    (see also 65.40. Electronic heat capacity)
    71.25. Nonlocalized single-particle electronic states
    71.30. Metal–insulator transitions
    71.35. Excitons and related phenomena (including electron–hole drops)
    71.36. Polaritons
    (including photon–photon and phonon–magnon interactions)
    71.38. Polarons and electron–phonon interactions
    (see also 63.20. phonon–electron interactions in lattices)
    71.45. Collective effects
    71.50. Localized single-particle electronic states (excluding impurities)
    71.55. Impurity and defect levels
    71.60. Positron states
    (see also 78.70. Positron annihilation)
    71.70. Level splitting and interactions
    (see also 75.10. — in magnetic phenomena, 75.30. Exchange and superexchange interactions, 73.20. Electronic surface states)
    71.90. Other topics in electron states

72. Electronic transport in condensed matter
    (for surfaces, interfaces, and thin films, see 73)
    72.10. Theory of electronic transport; scattering mechanism
    72.15. Electronic conduction in metals and alloys
    72.20. Conductivity phenomena in semiconductors and insulators
        (for nonelectronic conduction, see 66.70)
    72.30. High-frequency effects; plasma effects
    72.40. Photoconduction and photovoltaic effects; photodielectric effects
    72.50. Acoustoelectric effects
    72.55. Magnetoacoustic effects
    72.60. Mixed conductivity and conductivity transitions
    72.70. Noise processes and phenomena
    72.80. Conductivity of specific semiconductors and insulators
    72.90. Other topics in electronic transport in condensed matter

73. Electronic structure and electrical properties of surfaces, interfaces, and thin films
    73.20. Electronic surface states
        (for emission and impact phenomena, see 79)
    73.25. Surface conductivity
    73.30. Surface double layers, Schottky barriers, and work functions
    73.40. Interfaces
    73.60. Electronic properties of thin films
    73.90. Other topics in electrical properties of surfaces, interfaces, and thin films

74. Superconductivity
    74.10. Occurrence, critical temperature
    74.20. Theory
    74.30. General properties
    74.40. Fluctuations and critical effects
    74.50. Tunneling phenomena, Josephson effect, and proximity effects
    74.55. Type-I superconductivity
    74.60. Type-II superconductivity
    74.70. Superconducting materials
    74.90. Other topics in superconductivity

75. Magnetic properties and materials
    75.10. General theory and models of magnetic ordering
        (see also 71.25. Nonlocalized single-particle electronic states, 71.70. Level splitting and interactions)
    75.20. Diamagnetism and paramagnetism
    75.25. Spin arrangements in magnetically ordered materials
        (neutron studies, etc.)
    75.30. Magnetically ordered materials, other intrinsic properties
        (for critical point effects, see 75.40)
    75.40. Critical-point effects, specific heats, short-range order
        (see also 65.40. Heat capacities)
    75.50. Studies of specific magnetic materials
    75.60. Domain effects, magnetization curves, and hysteresis
    75.70. Magnetic films and plates
    75.80. Magnetomechanical and magnetoelectric effects, magnetostriction
    For galvanomagnetic effects, see 72.15 and 72.20
    For magneto-optical effects, see 78.20
    75.90. Other topics in magnetic properties and materials

76. Magnetic resonances and relaxation in condensed matter; Mössbauer effect
    - 76.20. General theory of resonances and relaxation
    - 76.30. Electron paramagnetic resonance and relaxation
    - 76.40. Diamagnetic and cyclotron resonances
    - 76.50. Ferromagnetic, antiferromagnetic, and ferrimagnetic resonances; spin-wave resonance (see also 75.30. Spin waves)
    - 76.60. Nuclear magnetic resonance and relaxation
    - 76.70. Magnetic double resonances and cross effects
    - 76.80. Mössbauer effect; other γ-ray spectroscopy
    - 76.90. Other topics in magnetic resonances and relaxation

77. Dielectric properties and materials
    (for conductivity phenomena, see 72.20 and 72.80)
    - 77.20. Permittivity
    - 77.30. Polarization and depolarization effects
    - 77.40. Dielectric loss and relaxation
    - 77.50. Dielectric breakdown and space-charge effects
    - 77.55. Dielectric thin films
    - 77.60. Piezoelectricity and electrostriction (for piezo-optical effects, see 78.20)
    - 77.70. Pyroelectric and electrocaloric effects
    - 77.80. Ferroelectricity and antiferroelectricity
    - 77.85. Electrical resonances
    - 77.90. Other topics in dielectric properties and materials

78. Optical properties and condensed-matter spectroscopy and other interactions of matter with particles and radiation
    (for phonon spectra, see 63)
    - 78.20. Optical properties and materials
    - 78.30. Infrared and Raman spectra and scattering
    - 78.35. Brillouin and Rayleigh scattering
    - 78.40. Visible and ultraviolet spectra
    - 78.45. Stimulated emission
    - 78.50. Impurity and defect absorption in solids
    - 78.55. Photoluminescence
    - 78.60. Other luminescence and radiative recombination

      For photoconduction and photovoltaic effects, see 72.40
    - 78.65. Optical properties of thin films
    - 78.70. Other interactions of matter with particles and radiation
    - 78.90. Other topics in optical properties of condensed matter and other interactions of matter with particles and radiation

79. Electron and ion emission by liquids and solids; impact phenomena
    - 79.20. Impact phenomena, including electron spectra and sputtering
    - 79.40. Thermionic emission
    - 79.60. Photoemission and photoelectron spectra
    - 79.70. Field emission and field ionization
    - 79.75. Exoelectron emission
    - 79.80. Resonance tunneling
    - 79.90. Other topics in emission and impact phenomena in condensed matter

85. Devices*)

---

*) Outside the ICSTI Classification for Physics.

Substance Classification*)

S1. Transition metals and their alloys and compounds (for oxides, see S10).
- S1.1. Fe
- S1.2. Other 3d metals
- S1.3. Noble metals
- S1.4. Lanthanides
- S1.5. Actinides
- S1.61. Carbides, silicides, germanides
- S1.62. Nitrides, phosphides, arsenides
- S1.63. Sulfides, selenides, tellurides

S2. Alkali metals and their alloys and compounds (for alkali halides, see S9.11)

S3. Alkaline earth metals and their alloys and compounds (for alkaline earth halides, see S9.12)

S4. Other metals and their alloys

S5. Nonmetallic elements
- S5.11. Silicon
- S5.12. Germanium
- S5.2. Solidified Gases

S6. Group IV compounds (for transition metal compounds, see S1.61)

S7. Group V compounds (for transition metal compounds, see S1.62)
- S7.1. III–V compounds
- S7.11. Phosphides
- S7.12. Arsenides
- S7.13. Antimonides
- S7.15. Mixed crystals
- S7.16. Higher compounds

S8. Chalcogenides (for transition metal compounds, see S1.63)
- S8.1. II–VI compounds
- S8.11. Sulfides
- S8.12. Selenides
- S8.13. Tellurides
- S8.15. Mixed crystals
- S8.16. Higher compounds

S9. Halides
- S9.11. Alkali halides
- S9.12. Alkaline earth halides
- S9.15. Mixed crystals
- S9.16. Higher compounds

S10. Oxides
- S10.1. Simple oxides
- S10.15. Mixed crystals

S11. Other higher compounds
- S11.1. Salts of oxy-acids
- S11.2. Spinels, garnets, ferrites

S12. Organic compounds
- S12.1. Quasi-one-dimensional systems

---

*) Outside the ICSTI Classification for Physics.

Review Article

phys. stat. sol. (a) **111**, 17 (1989)

Subject classification: 68.55; 73.40; 85; S8

School of Energy Studies, Department of Physics, Shivaji University, Kolhapur[1])

Electrodeposition of Thin Film Semiconductors

By

C. D. Lokhande and S. H. Pawar

Contents

1. Introduction

In recent years, electrodeposited thin film semiconductors are becoming popular in the fields of solar cells, optoelectronic devices, solar selective coating, etc. For instance, requirements of a photovoltaic cell for large scale terrestrial use, namely, high efficiency, stability, and low cost are met with electrodeposited thin film semiconductor solar cells. With all-electrodeposited thin films, n-CdS–p-$CuInSe_2$ and n-CdS–p-HgCdTe cells, greater than 9% solar conversion efficiencies have been reported [1, 2]. These cells are quite reproducible. The solar selective coatings based on oxides of Ni, Cr,

[1]) Kolhapur 416004 (M.S.), India.

Pb, etc. have been widely used [3, 4]. The electrodeposition of semiconductors, or one of the components in the cases of binary or ternary semiconductors, could be very useful in the process of device formation.

The preparation of thin film semiconductors with the electrodeposition technique has some advantages over other physical and chemical deposition techniques: it is easy and economical, as semiconductors with no or small waste of materials could be prepared. Because of the purification, which often results from electrodeposition, it is not required to have very pure starting materials, as is the case for other methods. The semiconductor properties like n- or p-type conductivity, band-gap variation, control of stoichiometry, doping, etc. could be controlled with a reasonable accuracy, as the parameters like deposition current (charge) and plating time can be controlled to the accuracy of μC and ms, respectively [5 to 7].

Electrodeposition techniques are successfully employed for preparation of elemental semiconductors like Si, Ge, Se, Te, etc., binary semiconductors like GaAs, CdS, CdSe, CdTe, InSe, Bi_2S_3, etc., intermetallic compounds like InSb, PbTe, CuTe, HgTe, $NiTe_2$, etc., and ternary compounds like $CuInS_2$, $CuInSe_2$, $CuInTe_2$, $AgInSe_2$, HgCdTe, etc. Electrodeposition offers the following attractive features for the preparation of thin film solar cells [8]:

1. It is an isothermal process, mainly controlled by electrical parameters such as electrode voltage and current density, which are easily adjusted to control film thickness, morphology, composition, etc.

2. Electrodeposition or electroless deposition usually have low operating temperatures. Apart from the obvious advantages in terms of energy saving, the low deposition temperature yields sharper junctions without deleterious effects of interdiffusion, contamination, and dopant redistribution, which is typical of high temperature processes.

3. It is particularly suited to the fabrication of heterojunction solar cells, simply by changing the deposition electrolyte, an n-type film can be deposited onto a p-type substrate.

The interesting feature of electrodeposition is that the plating could be employed as one of the steps in the preparation of binary or ternary semiconductors. For instance, the combination of electrodeposition with electrochemical and/or chemical deposition and/or anodization can be developed to get desired quality material. An example is the preparation of CdSe; it is an n-type II–VI compound semiconductor with a band gap of 1.6 eV. It is a direct band-gap semiconductor and is used as a photoelectrode material in solar cells and in optoelectronic devices. It has been prepared by the following ways [9 to 11]:

1. Direct cathodic electrodeposition from both aqueous and non-aqueous baths.

2. Cadmium metal or electrodeposited cadmium films on other metallic surfaces have been anodized to result in CdSe in a selenium containing solution.

3. Electrodeposited cadmium has been dipped in a selenium containing solution to form CdSe.

4. Electrodeposited cadmium has been heated at high temperature in H_2Se atmosphere to form CdSe.

Such combinations also work for the preparation of ternary semiconductors like $CuInS_2$, $CuInSe_2$, $CuInTe_2$, HgCdTe, etc. from their respective alloys. Therefore, even metal or alloy deposition has great importance for their potential applications of forming semiconductors.

In solar selective coatings, mostly metal oxides or sulfides have been used. They have a desirable band gap and other optical properties related to absorption and thermal emissivity. Such oxides or sulfides are prepared by metallic deposition followed by anodization or modification of the electrodeposition bath itself. The desired morphology for obtaining the above optical properties is easily derived from the electrodeposition technique.

The flexibility in the electrodeposition technique is attractive. A variety of electrodeposition techniques, either singly or in combination with others have been reported to produce better quality semiconductors. The following techniques are commonly used for electrodeposition of a metal, alloy, or compound semiconductors: dc, pulsed, dc + pulsed, electroless, anodization, etc.

In this review article, a survey of thin film electrodeposited semiconductors is given. The theoretical aspects for the electrodeposition of elemental and binary (compound) semiconductors are discussed by considering the Nernst equation and Pourbaix (potential–pH) diagrams. The elemental, binary, and ternary semiconductors with their electrodeposition and physical and chemical properties, are given. The solid state solar cells and photoelectrochemical (PEC) cells with electrodeposited semiconductors are described and their solar conversion efficiencies are reported.

2. Theoretical Considerations

2.1 Elemental semiconductors

The electrodeposition of semiconductors is based on the Nernst equation

$$E_{\mathrm{m}} = E_{\mathrm{m}}^{0} + \frac{RT}{nF} \ln\left(\frac{a_{\mathrm{m}}^{m+}}{a_{\mathrm{m}}}\right), \tag{1}$$

where E_{m} is the potential of the metal electrode M in a given solution, E_{m}^{0} is the standard potential for the reduction to form M, R is the gas constant, T the absolute temperature, N the number of electrons required for the reduction, F the Faraday constant, and a_{m}^{m+} and a_{m} are the activities of M^{m+} metal ions in the electrolyte and of M in the deposit, respectively.

Electrodeposition of M can occur at potentials more negative than the equilibrium potential due to the neglect of interactions of the solute ion M^{m+} with the solvent, or with complexing ligands, in the above equation. Such potentials are known as overpotentials. A Pourbaix diagram (potential–pH) has been used as a guide to fix the preliminary electrodeposition conditions, particularly, the pH-value of the electrolyte, as the activity is pH dependent [12]. Thermodynamically reversible potentials can be established only if the following conditions are fulfilled [13]:

1. All the processes that give rise to the electrodeposition of M are fast compared to other competing reactions.
2. All soluble intermediate species are present in the solution at equilibrium concentration.
3. The element M is pure and of well-defined phase.
4. The deposited crystal grains are in their equilibrium form and state.
5. The crystals are not too small.

The above conditions are rarely met in practice and hence Kröger [14] has introduced the concept of "quasi-rest potentials" to help to explain zero current conditions that otherwise only approximate the true thermodynamic reversibility. These are many kinetic factors that influence the deposition of elemental semiconductors, as the rate of the electron transfer reaction, dissociation kinetics of solvated or complexed ions, etc.

2.2 *Alloy and compound semiconductors*

The co-deposition of binary or ternary alloy or compound semiconductors is quite difficult because the conditions favourable for deposition of one of the metals may differ from those necessary for the other metals or constituents. Further, the electrodeposition gets complicated for alloys or compounds if one wishes to have a desired stoichiometry, a critical condition necessary for semiconductors. The governing factors in the co-deposition of binary or ternary compounds at any given temperature and current density are [13 to 18] (i) the electrode potentials of individual ions in the electrolyte, (ii) cathodic polarization caused by the difference in deposition potentials, (iii) relative ion concentrations in the electrolyte, (iv) the solution potential that is the dissolving tendency of the deposited compound, and (v) the hydrogen overpotential on the deposited cathode surface. Further, these parameters change due to the change in electrolyte temperature and electrolysis current density. Even though the direction of the effect of each variable can often be predicted, it is not possible to estimate the magnitude of such changes, especially if two or more conditions are simultaneously varied.

The role of such parameters is discussed by Brenner [17]. The condition for the simultaneous deposition of two different metals at the cathode can be written as

$$E_M + \eta^M = E_N + \eta^N , \tag{2}$$

where E_M and E_N are the equilibrium potentials for the metals M and N, respectively, and η^M and η^N are their respective overvoltages. Activities of metals in the compound or alloy are determined by their concentrations and the thermodynamic stability of the deposit. The reversible potential of a metal alloyed with a component should be more positive than that of the pure metal.

When the standard potentials of metals are far apart and the metals at the cathode do not interact with one another according to (2), the deposition potential may be brought together by adjusting the concentration of noble metal ions or by changing the activity of discharging ions by a complex formation or inhibiting the rate of reduction of the nobler metal by introducing suitable surfactants. The co-deposition requires that the electrode potential for the two ions should be close (<200 mV) to each other.

Interaction between the components in a deposit usually shifts the deposition potentials of a deposit to the values that are positive relative to the deposition potential of the less noble metal. It is necessary to consider the interaction energy of the alloy components as well as the change in the point of zero charge due to the shift of static potentials and therefore resulting in a change in the overall kinetics. There is a wide diversity of alloy phase structures and therefore, electrochemical conditions during crystallization are (i) eutectic mixtures of pure metals, (ii) homogeneous solid solutions, (iii) intermetallic compounds, and (iv) alloys are different.

The dependence of the deposition potential on the pH-value of an electrolyte can be understood with the help of the Pourbaix diagram by considering the deposition of Cu, In, and Se as an example [19]. In Fig. 1 the corrosion, immunity, and the passivation domains of copper, indium, and selenium are shown separately. For obtaining corrosion free, pure electrodeposits, the condition should be so adjusted that it lies in their respective immunity domains. For simultaneous or co-deposition of Cu, In, and Se, it is shown in Fig. 2 by the black area ABC. This indicates that the pH-value of the electrolysis solution should lie in the corresponding region ABC. If the experiment is performed under galvanostatic conditions, the suitable current density will be that which will develop the potential lying in the region ABC.

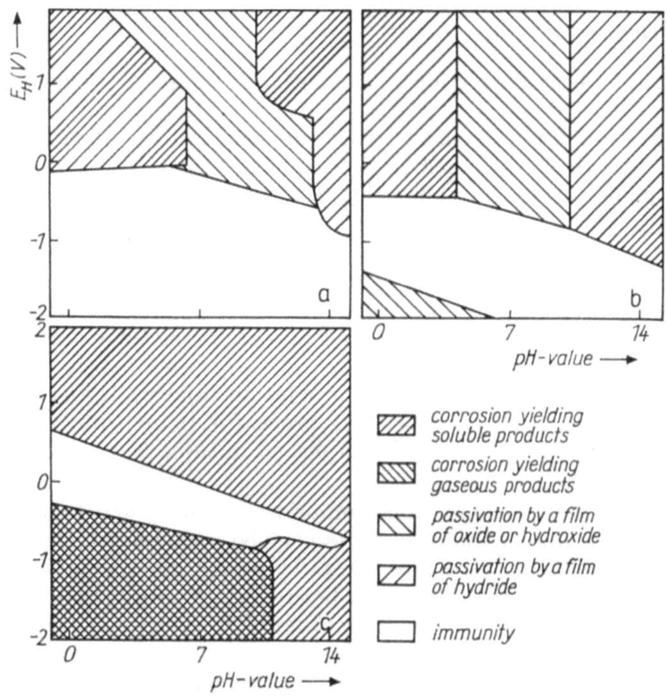

Fig. 1. Pourbaix diagram showing corrosion, passivation, and immunity domains of a) Cu, b) In, c) Se

According to the Nernst equation, the electrode potential for Se and Cu being more positive than that for indium, deposition of Se and Cu is expected to precede the indium deposition. For simultaneous deposition of Cu, In, and Se, the concentration and pH-value of an electrolyte should be adjusted such that the electrode potentials of all the individual deposits may come closer to each other. A relatively higher concentration of In shifts its potential in the positive direction approaching closer that of Cu and Se. The presence of Cu in the electrolyte solution enhances the deposition of Se with Cu.

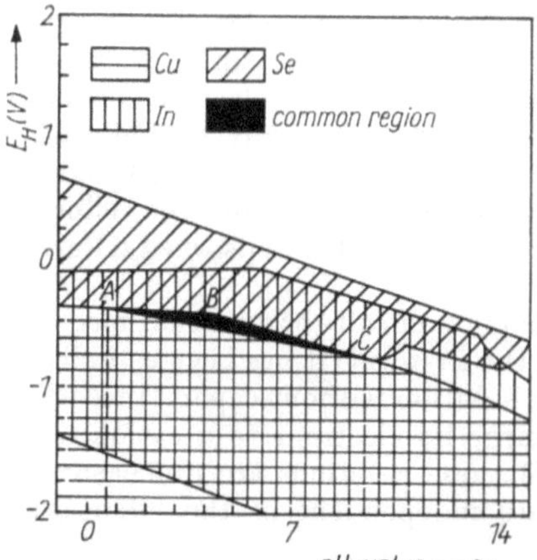

Fig. 2. Overlapping of immunity domains of Cu, In, Se

2.3 Mechanism of electrodeposition

The steps in the mechanism of electrodeposition of elemental or compound semiconductors are as follows [20]:

1. The aquo- or complexed metal ion is transferred or deposited as an adion (still partly bound) to a surface site. Such sites include the plane surface, edges, corners, crevices, or holes with the plane surface providing the primary sites.
2. The adion diffuses across the surface until it meets a growing edge or step where further dehydration occurs.
3. Continued transfer or diffusion steps may occur into a kink or vacancy or coordinate with other adions, accompanied by more dehydration until it is finally coordinated with other ions (and electrons) and becomes a part of the metal and being incorporated into the lattice.
4. Deposition of metal ions results in depletion in the solution adjacent to the surface. These ions must be replenished if the deposition process is to continue. This replenishment or mass transport of ions can be accomplished in three ways as ionic migration, convection, and diffusion.

Electrodeposition of compound semiconductors comprises placing two or more elements in a solution as cations. One of these elements may be a semimetal such as Te, As, or Sb and the other element may be any substance which has a deposition potential less noble than that of the semimetal when both elements are in solution. The formation of such compound involves the following order of necessary occurrences [21]:

1. Transfer of solvated cations to the cathode with a resulting polarization according to commonly accepted principles.
2. Neutralization of a nobler ion.
3. Conversion of the cathode surface to the layer of adsorbed anions.
4. Complete depolarization of the cathode with respect to the less noble ion.
5. Reduction of the less noble ion at the rate of first-order electron conductance.
6. Polarization of the cathode as the neutral character of surface is restored.
7. Building of the crystal lattices and macrocrystals of a compound of the two or more participating elements through repetition of the polarization and depolarization steps.

Electrodeposition of CdTe is one of the examples of the above class. Cadmium and tellurium in nitric acid are present as Cd^{2+} and Te^{4+} ions surrounded by equivalent dielectric shells of the solvent and are about 0.8 V apart in deposition potential in the solvent. Therefore, at the onset of current flow, tellurium being nobler, deposits preferentially on a cathode. This tellurium is adsorbed as a positive ion on a cathodic surface and is dehydrated and reduced to the neutral state by the electrons provided by an external energy source. Accordingly, the cathode is initially coated with a monomolecular layer of tellurium. Once a tellurium surface is established and the cathode film exhausted of tellurium cations, the electrons passing into the tellurium surface from a source are captured to form valence electrons of the adsorbed tellurium ion.

An ion on a cathode may be treated either thermodynamically or quantum-mechanically and the result in either case is the same. In the first case it has been established that the electrons can be discharged directly into solution without capacitance effect and in the second case that an anion will discharge its electrons to a cation if the energy level is lower in the cation. The rates of these reactions are unmeasurably faster than ordinary electrode processes and are independent of the "deposition potential".

A choice between these two different cathodic deposition mechanisms can be made after a study of the current efficiency of deposition and of the secondary products formed in the solution.

The cathode having preferentially stripped the contageous electrolyte layer of tellurium, now deposits cadmium at an extremely high rate with electrons transferring to the cation through the anion. The combination of a cadmium atom and a tellurium atom is a neutral state and the anionic character of the cathode surface is terminated whereupon the cathode is polarized and normal preferential deposition of tellurium is resumed. The repetition of the series of events creates the cadmium telluride.

The factors which determine the deposition process include [20] (i) the electrical double layer (1 nm thick) and adsorption of ions at the surface (some 0.2 to 0.3 nm away): At any electrode immersed in an electrolyte, a double layer of charges is set up in the metal and the solution ions adjacent to the surface. At solid electrode surfaces, which are usually heterogeneous, the character and constitution of this double layer may exhibit local variations, resulting in variations in the kinetics of the deposition process. This could affect the electro-crystallization processes involved in the overall growth process; (ii) the energy and geometry of solvated ions: especially those involving complex ions. All metal ions are associated with either the solvent (water) or, complexed with other solution constituents by electrostatic of co-ordinated covalent bonding. A desolvation energy is required in transferring the metal ion out of solution to the growing crystal lattice.

3. Electrodeposition of Thin Film Semiconductors

3.1 Elemental semiconductors

3.1.1 Silicon

A deposition of silicon from an aqueous solution is not possible due to its larger reduction potential in an aqueous solution than hydrogen. Hence two types of deposition baths, namely, a room temperature operating non-aqueous and a high temperature molten salt baths have been developed [22, 23].

Amorphous silicon has been electroplated from non-aqueous baths using $SiHCl_3$ as a silicon source [24]. A typical bath composition was 10 M $SiHCl_3$ in propylene carbonate containing 0.1 M tetrabutylammonium chloride as the supporting electrolyte. Deposits were obtained potentiostatically at around -2.5 V versus Pt at temperatures 35 to 140 °C under argon gas. A variety of materials including Pt, Ti, Ti–Al alloy, n-Si, tin oxide, etc. have been used as substrates. The deposition of a silicon layer on electrically conducting low-cost substrates such as steel, brass, nickel, tungsten, graphite could be a very promising approach to reduce the material processing cost of solar cells. Unfortunately, the majority of these substrates are not suitable for silicon electrodeposition since they form silicides at the operating temperature of 750 °C in high temperature molten salt bath [25].

Dense and well adherent silicon was successfully deposited on an inexpensive graphite substrate from the K_2SiF_6/LiF–KF solution at (745 ± 5) °C [26, 27]. The solution concentration range was 8 to 14 mol%. The cross-sectional views through the deposits indicated a nodular or dendritic growth up to several mm on top of an underlying coherent layer of 0.2 to 0.5 mm thick silicon. The polycrystalline silicon has columnar structure with normal grain size of 250 μm [28].

The proposed reaction for the deposition of silicon is

$$Si + Si^{4+} \rightarrow 2\,Si^{2+}\,. \qquad (3)$$

Deposition and subsequent dissolution of silicon due to the above reaction was responsible for the current variations during electrodeposition. Initially deposited silicon was assumed to be partially redissolved because of its interaction with tetravalent Si ions in the solution according to (3) to give divalent silicon ions. In consequence of this secondary reaction, the deposits for short periods were spongy. When the divalent silicon ion concentration increased, the equilibrium value of (3) should shift to the left side, which results in coherent, dense, and crystalline deposits.

Similarly a blue thin film of amorphous silicon has been deposited on a nickel cathode by electrolysis of a solution of tetraethylorthosilicate in acetic acid [29]. The maximum thickness was 0.5 μm. A dc current density of 0.2 to 20 mA/cm² at room temperature was applied.

Lee and Kröger have electrodeposited silicon from solutions of K_2SiF_6 in acetone with HF as a supporting electrolyte [30]. The anode was platinum and the cathode stainless steel or nickel. They found that as the amount of HF in the bath was increased, a significant amount of carbon was incorporated in the deposit. Doping with boron was carried out by adding triethylborate to the bath, however, the resistivity was unaffected by doping.

3.1.2 Germanium

Elwell has reviewed the electrodeposition of germanium in detail [22]. Generally, GeO_2 has been used as a solute. Three methods have been used to electrodeposit germanium; a) deposition from molten salt, b) deposition from aqueous alkaline solutions, and c) deposition from organic solutions. Deposits from aqueous solutions stopped as soon as the cathode was completely covered with germanium.

Electrodeposition of germanium has been carried out from a non-aqueous bath, containing germanium tetrachloride in propylene glycol by Szekely [31]. The optimum conditions were 7% of $GeCl_4$ in propylene glycol, temperature 59 °C, and cathode current density 0.4 A/m². The anode was a graphite rod. The bath efficiency was 75% in the 0.2 to 0.4 mA/cm² current density range. The microstructure was dependent on the germanium concentration in the solution: with higher Ge concentration neddle-like structures were obtained.

3.1.3 Selenium

Thin films (0.3 μm thickness) of selenium have been electrodeposited from an aqueous solution by Shin et al. [32]. A purified 0.5 M SeO_2 (99%) solution was used for the cathodic electrodeposition. Single crystals of tellurium and tungsten were used as the substrates. The electrodeposition was carried out at 90 °C with the current density 0.3 mA/cm². The films deposited at such a low current density were quite smooth and consisted of fine grains with an average dimension of 0.5 μm and adhered well to the substrates.

3.1.4 Tellurium

Rosamilia and Miller [32a] have deposited Te films on Au disks from 1 mM Na_2TeO_3–0.1 M $HClO_4$ solution by cycling the potential between 0 and −0.5 V (SCE) for 3 min at 20 mV/s. Such films have been used further for the formation of $CdSe_xTe_{1-x}$ films.

3.2 Binary semiconductors

3.2.1 Cadmium sulfide

Electrodepositions of CdS have been carried out from the aqueous and non-aqueous solution baths and by the anodization of cadmium in sulfide solution. The CdS films were cathodically deposited from an aqueous solution containing cadmium and thiosulfate ions. The electrode reaction is

$$Cd^{2+} + S_2O_3^{2-} + 2\,e^- \rightarrow CdS + SO_3^{2-} . \quad (4)$$

This process on the Pt electrode has been studied by Power et al. with electrochemical techniques [33]. Fatus et al. [34] used a variety of substrates such as platinum, molybdenum, and aluminum. The plating solution was 20 mM $CdSO_4$–100 mM $Na_2S_2O_3$. The pH-value was adjusted to 2.3 by H_2SO_4. The deposition was carried out at 25 °C. The potential applied was −1 V (SCE) and the electrodeposition time was 72 h. The films on molybdenum electrodes showed a maximum thickness of 0.1 μm and were sulphur rich in composition. The films on aluminum substrates were thick enough to form devices. The microstructure indicated that the CdS films on aluminum substrate consisted of three zones, each with different composition. The thermal treatment in vacuum (200 °C, 1 h) resulted in a higher degree of homogeneity.

An ac electrodeposition technique has been applied by Fatus et al. [35] for CdS film deposition from an aqueous solution containing $CdSO_4$, $(NH_4)_2SO_4$, $Na_2S_2O_3$, NaCl, and glycerol at 25 °C on indium doped tin oxide substrates. During the electrodeposition two electrodes were alternately connected to positive and negative potentials between +8 V and −8 V with half periods corresponding to the positive potential longer than corresponding to negative potential. The films are Cd rich, homogeneous, and crack free. X-ray diffraction showed the formation of β-CdS. Thermal treatment in nitrogen gas at 350 °C for 1 h decreased the resistivity to 100 Ω cm and the band-gap energy to 2.43 eV.

Baranski and Faqcett [36] have used a new approach to deposit thin films of semiconducting metal chalcogenides on various conducting substrates. They used an electrolyte constituted of the corresponding metal salt and elemental chalcogen, dissolved in a suitable non-aqueous solvent such as DMF (dimethyl formamide), DMSO (dimethyl sulfoxide), and ethylene glycol. The nascent metal formed at the cathode rapidly reacts with the dissolved chalcogen to form the required metal chalcogen. The electrode reaction for CdS deposition is

$$Cd^{2+} + S + 2\,e^- \rightarrow CdS . \quad (5)$$

The different substrates like Pt, Au, stainless steel, Ni, Zn, Mo, Al, and indium–tin oxide glasses have been used successfully for CdS electrodeposition. The mechanism of electrodeposition of CdS on Hg, Au, and Pt electrodes has been studied in DMSO solution by performing cyclic voltametry at stationary electrodes and pulse polarography at dropping Hg electrodes in solutions containing Cd^{2+} ions and elemental sulphur. The formation of the initial layer of CdS involves reaction of Cd^{2+} ions with chemisorbed sulphur or with a pre-existing layer of metal sulfide [37, 38].

The CdS films were deposited from DMSO solution containing 0.055 M $CdCl_2$–0.19 M S on the Pt foils and stainless steel substrates [39 to 41]. The voltage applied was 5 V ($Cd/CdCl_2$) and the deposition temperature was 110 °C. The films grown for longer time than 250 s, at 2.5 mA/cm² current density, had significant cracking and defects. Such cracking occurs during deposition. The cracks developed in CdS films grown in non-aqueous solutions were attributed to the presence of tensile stress in CdS layers resulting from piezoelectric effects. This is because the electric field applied for deposition may result in the displacement of deposited ions along the axis which has no centre of symmetry. Another reason for cracks may be the different thermal expansion coefficients of CdS and the substrates. The quality of the deposit was independent of sulphur and $CdCl_2$ concentrations; moreover, addition of up to 10% water to the solution had no significant effect on the deposit quality. Deposit characteristics depend on temperature. Highly non-stoichiometric deposits were obtained below 90 °C [41].

XRD showed a high degree of orientation along the ⟨111⟩ planes of β-CdS. Carrier concentrations of 10^{14} cm^{-3} and carrier mobilities of approximately 50 cm^2 V^{-1} s^{-1} have been reported [40].

CdS films grown from diethylene glycol (DEG) on stainless steel [41] and Mo [42] substrates have been reported. In the latter case, DEG–water mixtures containing two different cadmium salts, namely cadmium chloride and cadmium perchlorate, have been used. The depositions were carried out at 120 °C at 1 mA/cm^2 current density. The CdS films prepared with cadmium perchlorate were not cracked and completely coherent after annealing. These films did not exhibit a completely columnar structure and were not well orientated crystallographically. The annealed films (200 °C, 2 h) showed higher mobility and electron density than the CdS films prepared with $CdCl_2$ solution. The resistivity was of the order of 0.6 to 6 Ω cm. CdS films deposited from $CdCl_2$ showed some cracks, and the films were highly oriented along the [002] direction.

The CdS films deposited from DEG based bath showed lower resistivity than the films from DMSO based bath. In order to overcome the problem of solubility in DMSO and DMG, triphenystibine sulfide has been used as a source of sulphur [43]. The reduced valency of the sulphur allows for room temperature electrodeposition of CdS. By using temperatures above 80 °C, the deposition proceeds electrolessly. The deposited CdS films are stoichiometric, as confirmed by Rutherford backscattering spectroscopy. Electrodeposition was successfully achieved on Mo, Ti, Pt, graphite, and indium–tin oxide (ITO) substrates. The optimized parameters were as follows: 1.0 M lithium perchlorate, 0.2 mM triphenystibine sulfide, 0.2 M cadmium perchlorate, current density 50 to 100 μA/cm^2. The film thickness was 1.0 μm at the deposition time of 30 to 120 min. The cathode current efficiency was 80%. Annealing did not alter the density and conductivity of CdS films significantly. The films showed no cracking. The freshly deposited films were very soft and adhered well to the substrate. The band gap of the as-deposited CdS was 2.63 eV and after annealing (200 °C, 30 min, N_2) it increased to 2.67 eV. However, the problem faced with annealing and higher thickness, were cracks developed on the films.

The formation of thin anodic films of CdS has been studied by few workers [44 to 47]. The electrolytic solution used for anodization was 1 M $NaHCO_3$–0.1 M Na_2S. The films formed with such a technique, were uniform but with very fine grains and some cracks [41]. CdS films up to 5 nm were formed under these experimental conditions. It was seen that the films formed initially under galvanostatic conditions grew according to the high field assisted migration of ions. The exchange current density was first order with respect to the concentration of Na_2S. This dependence could be accounted for with a process at the anodic film–solution interface, as the rate determining step. S^{2-} ions were the species participating in the rate determining step across the inner Helmholtz layer.

Mixed films of CdS_xSe_{1-x} have been prepared by Baranski et al. [48] and Loutfy and Ng [49] from a non-aqueous organic bath and by Mahapatra and Roy [5] from an aqueous bath. Baranski et al. [48] have used the DMSO electrolyte solution containing $CdCl_2$, elemental sulphur, and elemental selenium. The substrates were high purity Pt sputtered on glass. The temperature of the bath was 110 °C. The CdS_xSe_{1-x} films were very smooth, uniform, and exhibited extremely good adhesion to the Pt substrate. The composition of the mixed films altered with the concentration of selenium. Crackless and amorphous films were obtained with a DMSO bath saturated with selenium. The films were highly electrically resistive and showed excellent mechanical stability. Loutfy and Ng [49] have used the ethylene glycol electrolyte containing $CdCl_2$, S, and SeO_2 for the electrodeposition of CdS_xSe_{1-x} films. The substrates were

ITO glasses. The composition of the bath was 6.7 g/l $CdCl_2$, 3.3 g/l elemental sulphur, and SeO_2. The deposition was carried out at 160 °C under a constant current density. The films were annealed (360°, air, 30 min) to improve stability, performance, and adhesion of the CdS_xSe_{1-x} films. Mahapatra and Roy [5] have prepared CdS_xSe_{1-x} films by partial replacement of sulfide ions from CdS with selenide ions. The electrodeposited CdS electrodes were kept in 0.22 M aqueous sodium selenosulphate solution at 78 to 80 °C for about 90 min. The electrodes were rinsed and heated in nitrogen at 200 °C for about 1 h.

3.2.2 Cadmium selenide

Electrodeposition of CdSe has been reported from acidic and basic baths. In the case of an acidic bath, H_2SeO_3 has been used as a selenium source. Kazacos and Miller [9] have proposed two types of deposition mechanisms for the cathodic reduction of H_2SeO_3, depending on the nature of the metal substrate. From group I substrates, which include Au, Pt, and carbon, there is a competition between the formation of H_2Se and Se that passivates the electrode. The reactions that take place are

$$H_2SeO_3 + 6\ H^+ + 6\ e^- \rightarrow H_2Se + 3\ H_2O\ , \tag{6}$$

followed by the chemical reduction

$$H_2SeO_3 + 2\ H_2Se \rightarrow 3\ Se + 3\ H_2O\ . \tag{7}$$

The rate of chemical reaction is dependent on the concentration of H_2SeO_3, so that if this concentration is large enough, reaction (6) is fast and the overall process appears to be

$$H_2SeO_3 + 4\ H^+ + 4\ e^- \rightarrow Se + 3\ H_2O\ . \tag{8}$$

The group II substrates contain active metals like Hg, Ag, and Cu; for these metals the mechanism is reported to be

$$n\mathrm{M} + H_2SeO_3 + 4\ H^+ + 4\ e^- \rightarrow \mathrm{M}_n\mathrm{Se} + 3\ H_2O\ , \tag{9}$$

where M is the metal.

As the flux of the metal through the electronically conducting selenide film decreases, reactions (6) and (7) take over and Se could be found on the surface of these electrodes.

Houston et al. [50] prepared CdSe films from an acidic bath. The bath consisted of 0.1 M $CdCl_2$–0.5 M HCl and millimolar concentration of SeO_2. The solution was held at (40 ± 1) °C and stirred during deposition. The substrates were Ti. The CdSe films were annealed at 430 °C (2 h, air). The electrodeposition potential was −0.44 to −0.5 V (SCE).

The films, at optimized potential were gray or black in appearance, adhered well to the substrate and had a good coverage of about 3 to 5 μm thickness. Similarly Tomkiewicz et al. [51] deposited CdSe films from 0.03 M SeO_2–0.1 M $CdSO_4$–0.5 M H_2SO_4 solution at a constant current density of 6 mA/cm^2 on Ti substrates. The plating duration was 30 min. The films were annealed in Ar at 750 °C for 40 min. The microstructure consisted of cauliflower-like structure morphology which remained unchanged after annealing. The Cd:Se ratio was between 1.2 and 1.3.

Alkaline selenosulfite baths have been used to solve the problems associated with the acidic bath; namely co-deposition of free elemental selenium along with CdSe from H_2SeO_3 solution. Cocivera et al. [52] complexed the cadmium ions using nitrilotriacetate (NTA) and mixed with selenosulfate solution. The Cd complex was used to stabilize the bath. The pH-value of the solution was maintained at 9.0. The substrates were Ti and deposition was carried out at room temperature. A constant potential of -0.3 V was applied. The films of thicknesses 1.6 to 1.9 μm were obtained in 1 h deposition time. SEM showed some cracks on the films, however, cauliflower-like appearance of the microstructure was not obtained. Kazacos [53] had prepared CdSe, $CdSe_{0.65}Te_{0.35}$, and CdTe thin films onto titanium substrates by the reduction of Se(O) and Te(O) species in solutions of KCN containing a high concentration of Cd^{2+} ions to eliminate the problems of co-deposition of elemental Se and Te. The complexing agent was EDTA.

Hodes et al. [54] have used both acidic and basic baths for the electrodeposition of $CdSe_{0.65}Te_{0.35}$ films. For the acidic deposition bath, the composition of the bath was 0.2 M $CdSO_4$–20 mM SeO_2 and 10 mM TeO_2 in 2 M H_2SO_4. In most cases, triton × 100 was added as a surface active agent. The deposition was carried out under potentiostatic conditions at -650 mV (SCE) on to Ti substrates in the temperature range between room temperature and 50 °C without stirring. The deposition current densities were ≈ 10 mA/cm² for about 30 min.

For the alkaline deposition bath, an electrolyte was prepared containing CdO (11.5 g/l), NaOH (7.5 g/l), NaCN (40 g/l), SeO_2 (2.5 g/l), TeO_2 (up to 1.5 g/l), and 0.25% triton × 100. The plating was carried out at -1.65 V (SCE) at 10 mA/cm² current density at room temperature with stirring. Triton × 100 gave a much more regular surface; for the acidic deposition, spherical particles with homogeneous composition were obtained, however, the particle size was several times smaller than that of layers plated from the acidic bath. The films were annealed at 600 °C in Ar gas atmosphere. As-deposited $CdSe_{0.55}Te_{0.35}$ layers were cubic in structure regardless of the anion present in the solution, but changed to hexagonal on annealing.

Bhardwaj et al. [10] outlined an electrochemical conversion method for the preparation of CdSe film by electrodeposition of a cadmium film on nickel substrate and dipping it into an acidic solution of SeO_2. The cadmium was deposited from 1 M acidic chloride solution. The films of cadmium with different thickness ranging from 0.5 to 2.5 μm were dipped into a 1 M acidic solution of SeO_2 for times ranging from 10 min to 3 h. The resultant CdSe films are reported to be 0.8 to 3.0 μm thick. The films were annealed at 500 °C in air for best PEC results.

3.2.3 *Cadmium telluride*

Electrodeposition of CdTe has been carried out from aqueous and organic baths on a variety of substrates like stainless steel, ITO, Mo, Ti, etc. Panicker et al. [55] carried out CdTe deposition from 1.0 M $CdSO_4$ and 10^{-5} to 10^{-6} M TeO_2 solution on nickel and Sb doped tin oxide (10 to 20 Ω/□) substrates. The experiments were carried out at different temperatures between 25 and 100 °C. The bath was stirred with a magnetic stirrer with different rotation speed. The deposition potential was between -0.2 and -0.5 V (SCE). The deposits were amorphous when deposited at room temperature and semicrystalline with grain size 50 to 100 nm, when deposited in the temperature range 25 to 90 °C. The annealing at 350 °C in argon gas increased the crystalline size to ≧ 500 nm. The material was n-type with deposition potential ≧ 0.3 V (SCE) and p-type with ≦ 0.3 V (SCE). The films doped with donors (In) or acceptors (Cu) were n- and p-type, respectively, for all deposition conditions.

Llabres and Delmas [56], see also [18], electrodeposited CdTe from 0.5 to 1 M $CdSO_4$–saturated TeO_2 solution under potentiostatic conditions. The cell temperature was (82 ± 2) °C and pH was 2.0 to 3.5. In this pH-range, the films grown were uniform and free of macroscopic defects or pinholes. The substrates were stainless steel, ITO, Ti, etc. The potential was -630 to -675 mV (SCE). There was no evidence for co-deposition of Te or Cd. The CdTe crystal structure was hexagonal, single phase with ⟨111⟩ direction. The average grain size was 5 μm on stainless steel substrates, however, the films grown on ITO or Ti substrates had grain size smaller than 1.0 μm. Auger electron microscopy showed uniform composition all over the films.

Bhattacharaya and Rajeshwar [57] electrodeposited CdTe on Ti and Nesatron glass substrates. The as-deposited films were amorphous, heating at temperatures <350 °C produced cubic crystal structure, and 600 to 650 °C argon annealing produced hexagonal crystal structure. The Cd:Se ratio was 45:55. Different types of morphologies were obtained with Ti and Nesatron glass substrates.

Llabres [58] obtained hexagonal structure when the CdTe films were electrodeposited at potentials −620 to −660 mV (SCE). The average grain size was 5 μm. The deposition temperature was (82 ± 3) °C. Gerritsen [59] prepared CdTe films on tellurium rods using 0.1 M $CdSO_4$ solution and applying 0.63 V voltage. The electrodeposition mechanism of CdTe from an acidic aqueous solutions containing Cd^{2+} and $HTeO_2^{2+}$ ions has been discussed by Sella et al. [60]. The formation of CdTe is controlled by diffusion and adsorption of $HTeO_2^{2+}$. Voltametric studies were carried out to support this mechanism [60].

Bhattacharaya et al. [61] used an "electroless" technique to deposit CdTe films. The method is based on short-circuiting the thin film substrate to an easily oxidizable redox component (e.g., Al, Cd) in the electrolyte bath. The solution contained 100 mM CdOAc and 100 mM TeO_2 at pH = 2. The temperature was 40 to 60 °C. The substrates were suspended in the solution with a Cu alligator clip suitably clamped such that intimate contact of the substrate with the solution was facilitated. The aluminum contained in the clip screw set up the short-circuit cell; viz. the dissociation reaction $Al \rightarrow Al^{3+}(aq.) + 3\,e$. The films prepared by such a technique are better in grain size (0.15 to 0.2 μm) and morphology when deposited at high temperature. EDAX showed the presence of free tellurium. The composition of the films was Cd:Te = 55:45. The films were crystalline and p-type in nature.

Employment of tri-n-butylphosphine telluride (BPT) and Cd ions in propylene carbonate allows for electroreduction of CdTe rather than alloy formation [62]. CdTe was electrodeposited on CdS substrates at 100 °C under nitrogen atmosphere without stirring for 30 min. The applied potential was −1100 mV (Ag/AgCl). The deposition solution consisted of 20 ml propylene carbonate containing 0.004 M Cd perchlorate, 0.018 M BPT, and 0.120 M lithium perchlorate. The CdTe film thickness was 1 μm and films showed no cracks before or after annealing at 400 °C.

From a similar bath, CdTe was electrodeposited on Ti substrates [63]. Smooth, gray films were obtained with 4 to 5 μm thickness. The Cd:Te ratio changed from 0.63 to 1.1, depending upon the potential and solution composition. As-deposited films were amorphous and became polycrystalline (cubic) upon annealing at 400 °C. These films could be n- or p-type and were converted to n-type after annealing.

3.2.4 Copper sulfide

Thin films of $Cu_{1+x}S$ and $Cu_{2-x}S$ have been electrodeposited at temperatures from 21 to 120 °C from propylene glycol solutions of $CuNa_2EDTA$, elemental sulfur, and KCl [64]. The films were grown on a variety of substrates including ITO, graphite, nickel, tin, and titanium. Surface preparation was found to be critical for the uniformity and adherence of the deposits. The voltage applied was −0.7 to −0.5 V (Ag/AgCl) and the current density was 3 to 16 μA/cm² at room temperature and was changed from 30 to 150 μA/cm² at 100 °C. The solution was stirred with different speeds. The stoichiometry of the films grown at room temperature was controlled through the electrolyte stirring rate, which controls the rate of sulfur diffusion toward the cathode. Heterojunction-like double layer structures of $Cu_{2-x}S$ on $Cu_{1+x}S$ and Cu_{1+x} on $Cu_{2-x}S$ were electrodeposited by changing the stirring rate during the deposition. Similarly, a continuous composition change and hence a band-gap change could be obtained.

The $Cu_{2-x}S$ based films were grown to dark gray and the $Cu_{1+x}S$ films were blue-green to black in appearance. These films were amorphous and showed nodule-covered surfaces. The films were cracked on ITO substrates. The films deposited at 90 to 120 °C were more grainy, but nonuniform in composition. The most commonly deposited phase of $Cu_{2-x}S$ is digenite. Polycrystalline $Cu_{2-x}S$ exhibited a spongy-like or columnar surface. The X-ray microanalysis data indicated a continuous change of the Cu/S ratio from 2.33 to 1.06. Further the optical absorbance edge minimum shifted continuously from 835 to 575 nm as the sulfur content increased. The use of a pulse plated scheme allowed nearly complete reaction of copper and sulfur to form $Cu_{1+x}S$. The on and off periods were 1 to 20 and 20 to 60 s, respectively.

3.2.5 Copper selenide

Electrodeposition of copper selenide has been carried out from aqueous and non-aqueous baths. Lokhande [65] carried out electrodeposition of CuSe from a solution containing 100 mM CuCl–100 mM SeO_2. The pH-value of the solution was 2.4. The substrates were Mo and Ti. The applied potential was −400 mV (SCE). The layers were gray in appearance, 2 to 3 μm thick, and well adhesive to the substrate. The surface contained needle-like structures with the same composition as that of the layer. These layers are polycrystalline in nature and consisted of different phases as $Cu_{2-x}Se$, Se, Cu, etc. The layer contained excess Se; heating at 400 °C resulted into $Cu_{1.70}Se$. In [66] Cu_xSe layers were electrodeposited using selenosulfate as a source of Se, and a cupric salt complexed with NTA. Densely packed coherent layers of composition Cu_xSe, where x varied from 1.2 to 1.8, depending on deposition conditions, were obtained on Mo substrates. The electrodeposited layers were of Berzelianite structure with a strong ⟨111⟩ orientation.

To improve the quality of the Cu selenide depositions on metal surfaces and to decrease the quantity of H_2 evolved, a non-aqueous solution of the following composition was used: $CuCl_2$–SeO_2–LiCl–HCO_2H [67]. The cathodes were Cu, NiPt, or Mo, while the anode was a Pt rod. The composition and structure of the electrodeposits depend on the conditions of deposition and the ratio of the components (Cu:Se) which varied from 3:1 to 6:1. The total concentration of components was constant at 70 mM. At a current density of 10 mA/cm² and a Cu:Se ratio of solution components as 5.4, the composition of the deposit was near to Cu_2Se with a rhombic structure.

3.2.6 Tungsten selenide

Thin films of WSe_2 have been deposited from an aqueous electrolyte containing H_2WO_4–SeO_2. The pH-value was adjusted to 9.2 with the addition of ammonia solution. The depositions were carried out at current density of 6 mA/cm² at 60 °C [68]. The first deposit was always selenium-rich which was followed by increasing tungsten deposition with time, and hence different compositions were incorporated into the deposit. The SeO_2 concentration was found to be the rate determining step. The presence of a selenium layer followed by tungsten deposition is suggestive of the mechanism that Se somehow induces the deposition of W and hence is called "induced codeposition".

Stoichiometric WSe_2 films were obtained for about 20 to 22 min deposition time. The presence of W and WSe_2 was detected by XRD and EDAX analyses. The WSe_2 films were polycrystalline in nature. SEM showed that the microstructure was step-ladder-like in appearance, however, after few days cracks were developed in the films which were known as hydrogen induced delayed cracking. Electrodeposited WSe_2 films have been characterized optically, electrically, and electrochemically [69]. The optical absorption studies showed that the material is an indirect band-gap semiconductor ($E_g = 1.0$ eV). The temperature dependent resistivity showed that the trap density at room temperature was of the order of 10^{15} cm^{-3}.

3.2.7 *Molybdenum selenide*

The $MoSe_2$ films have been electrodeposited on Ti substrates from an ammoniacal solution of H_2MoO_4 and SeO_2 under galvanostatic conditions [70]. The pH-value was adjusted to 9.7 and deposition was carried at 2.4 mA/cm^2 current density. The initial layers of the films were selenium, followed by an increasing amount of Mo in the deposit and therefore, combinations of all molybdenum selenide compounds were obtained. Some intermediate layer had a composition equivalent to $MoSe_2$. The amount of SeO_2 solution had a greater effect on the deposition current density than ammonium salt and hence SeO_2 was the rate determining step in $MoSe_2$ deposition. The electrodeposits were Mo rich with electrolytes having low SeO_2 concentration. An induced codeposition for the $MoSe_2$ film, involving first selenium and then molybdenum deposition has been suggested.

The as-deposited layers were polycrystalline in nature with preferred orientations. Scanning electron microscopy indicated that the films are non-uniform and showed the presence of excessive voids or pits or dimples. The films after some time developed cracks which were due to the hydrogen induced delayed cracking. Electrodeposited $MoSe_2$ films further have been characterized by optical, electrical, and electrochemical techniques [71]. The optical absorption spectra showed the material to be an indirect band-gap semiconductor with a band gap equal to 1.14 eV. The electrical conductivity at different temperatures showed an impurity conduction and the presence of deep level traps.

3.2.8 *Antimony sulfide*

The Sb_2S_3 films have been prepared by anodization of antimony [72, 73]. Polycrystalline antimony samples (99.998%) were cathodically polarized to remove any reducible surface films prior to the experiment. The solutions were prepared with 0.1 M Na_2S–1.0 M $NaHCO_3$. The potential applied was −0.51 V (NHE). The pH-value of the solution was 9.0. The films were n-type with donor concentration 10^{20} cm^{-3}. An indirect band edge was found at 1.77 eV and a direct transition observed at energies above 2.28 eV.

The growth and reduction of anodic Sb_2S_3 films has been investigated by using potentiostatic, potentiodynamic, and galvanostatic techniques. The sulfidization mechanism and the efficiency of charge storage in the sulfide film were dependent on the electrode potential and current density. The first anodic process possibly involves the adsorption of hydrosulfide on the surface, followed by the nucleation of an Sb_2S_3 phase and its lateral spread across the electrode. The initial stage of film thickening has been interpreted as a high field assisted ionic migration process. The films were non-stoichiometric, which might be due to the differences in the cationic and anionic conductivities.

3.2.9 *Bismuth sulfide*

Thin films of Bi_2S_3 have been prepared by electrodeposition from non-aqueous solutions [74] and by anodization of bismuth metal [75, 46]. Nearly stoichiometric Bi_2S_3 films have been obtained from an electrolyte containing 0.032 M $BiCl_3$–S–0.93 M NH_4Cl in diethylene glycol. The substrates were high purity Pt, Mo, Ni foils, and platinum sputtered glass and the counter-electrode was graphite. The electrolyte was stirred and depositions were carried out at current density of 2.0 mA/cm^2 at different temperatures. The efficiency of the bath was 90% [74]. Depending on the deposition temperature, two types of Bi_2S_3 films were obtained. Below 98 °C the deposit was glossy and metallic blue in colour with a mirror-like surface and such deposits were amorphous in nature. Above 102 °C, the deposits were darker and non-glossy. The crystallites were strongly oriented in (001) direction with an orthorhombic crystal structure. The surface was slightly rough with a structure characteristic of a collection of crystallites to have diameters of the order of 1 to 2 μm. The composition in both phases was the same. The structural properties were independent of the nature of the substrate. The resistivity was of the order of 10^{-3} Ω cm.

In the anodization process, a 10 mm sheet or cast rods of bismuth were anodized in 1.0 M (Na_2S–NaOH) solution using platinum or carbon as counter-electrodes. The films showed n-type behaviour in the PEC cells [46]. Peter [75] showed that the electrodeposited Bi_2S_3 had two transitions, an indirect one at 1.25 eV and a direct one at 1.7 eV.

3.2.10 *Tin selenide*

$Sn_{1-x}Se$ films have been electrodeposited from aqueous solutions containing $SnCl_2$ and SeO_2 or $HSeO_3$ and non-aqueous solution containing $SnCl_2$ and molecularly dissolved selenium powder [76]. As-deposited films ranged from amorphous to polycrystalline in structure and exhibited indirect band gaps from 0.85 to 0.95 eV. The pH-value of the solution was adjusted to 3.0 and electrodeposition was carried out using ITO coated glass as cathode, and tin or graphite as anode. The temperature range studied was 20 to 90 °C. There was a little difference between the material data from films grown by the above two techniques. After annealing, the band gap was raised from 0.95 to 1.39 eV.

3.2.11 *Indium selenide*

The indium selenide layers were deposited from a solution containing 100 mM $InCl_3$–100 mM SeO_2. The pH-value of the solution was adjusted to 1.2 by using diluted HCl. The layers were deposited on Mo, Ti, and ITO substrates [65]. The plating potential was −1000 mV versus SCE. The current density was 20 to 30 mA/cm^2 during the deposition. Different compounds of In–Se were deposited by changing the deposition potentials.

The as-deposited layers are well adhesive to the substrates, 2 to 3 μm thick, and reddish in colour. XRD showed that the as-deposited In–Se layers were polycrystalline and contained different phases like In_6Se_7, In_2Se_3, Se, and In. The layers on Ti substrates were thin. The layers contained excess selenium, which was taken out by heating the InSe films at high temperature. Heating at 300 °C resulted in In_2Se_3, while at 400 °C in InSe.

3.2.12 *Gallium arsenide*

Chandra and Khare [65a] have reported on the electrodeposition of GaAs from an aqueous chloride bath. The bath contains 10 g/l of $GaCl_3$ and As_2O_3 solutions. The current density was 2.5 mA/cm^2 and the deposition time 7 min. The deposition was carried out at 22 °C. The stoichiometry of the films was $Ga_{1.04}As_{0.96}$.

The EDAX studies showed that the films were non-uniform in stoichiometry. The films were polycrystalline and the band gap of GaAs films was 1.5 eV. The resistivity of the films was of the order of 10^3 Ω cm and the trap density was 1.4×10^{14} cm^{-3}.

3.3 Ternary semiconductors

3.3.1 Copper–indium disulfide

Both types of n- or p-$CuInS_2$ films have been electrodeposited from aqueous and non-aqueous solutions. Bhattacharaya et al. [77] and Hodes et al. [78, 6] have prepared $CuInS_2$ films from aqueous solutions of CuCl, $InCl_3$, and thiourea. The composition of the bath was 10 mM $InCl_3$–6 mM CuCI, 0.2 vol% triethanolamine, and 0.25 vol% ammonia–0.125 M thiourea. The pH-value was adjusted to 2 by diluted HCl. Deposition was carried out for 10 to 15 min at potentials of −1.5 to −2.0 V (SCE) and current densities of typically 6 to 8 mA/cm^2. All electrodepositions were carried out in stirred bath and at room temperature.

The as-deposited layers did not show an XRD pattern, after annealing the characteristic chalcopyrite peaks of $CuInS_2$ appeared invariably mixed with In_2O_3. Large variations in the stoichiometry were found from sample to sample. The deficiency of sulphur was common to all films. Needle-like indium sulfide growths appeared on the surface. A dependence of morphology on the composition was observed.

Hodes et al. [79] deposited $CuInS_2$ films from a non-aqueous bath using an electrolyte consisting of $InCl_3$, $CuCl_2$, and elemental sulphur dissolved in DMSO at 135 °C. The elemental analyses showed that $CuInS_2$ could be obtained. The morphology of the layers was not better than that of the layers deposited from aqueous solutions, and consisted of isolated particles. Both n- and p-type $CuInS_2$ films were obtained by adjusting the Cu:In ratio in the bath composition. When the ratio Cu:In < 1, the films were p-type and when the ratio Cu:In > 1, the films were n-type. Similarly $CuIn_5S_8$ films were also prepared by modifying the bath composition.

Similarly, the $CuInS_2$ films were prepared by annealing Cu–In alloys in sulphur containing atmosphere [78, 79]. The Cu–In alloys were prepared from an aqueous bath containing millimolar concentrations of $CuSO_4$, $In_2(SO_4)_3$, and H_2SO_4 at pH $= 2$ or from a bromide bath containing HBr glycollic acid, CuBr, and $InBr_3$. The deposition potential was −1.0 V (SCE). The alloys reacted with H_2S:Ar at 550 °C for 30 min. $CuInS_2$ films prepared in this way were more homogeneous than those prepared using the ternary deposition bath.

3.3.2 Copper–indium diselenide

The electrodeposition of $CuInSe_2$ has been carried out using aqueous acidic and basic and non-aqueous baths. Bhattacharaya [80] prepared $CuInSe_2$ films under potentiostatic conditions on F-doped conducting glasses. The plating solution contained 0.018 M $InCl_3$–0.18 M CuCl–0.025 M SeO_2–0.006 vol% triethanolamine and 0.007 vol% ammonia. The pH-value of the solution was adjusted to ≈ 1 by adding diluted HCl. The deposition was carried out under constant stirring and at room temperature. The voltage was maintained at −700 mV (SCE) and the deposition current density was ≈ 15 mA/cm^2. The as-deposited $CuInSe_2$ films were heated in argon gas at 600 °C. The as-deposited layers did not show an XRD pattern indicating their amorphous or very fine-grained polycrystalline nature. After annealing the characteristic peaks of $CuInSe_2$ appeared along with In_2O_3 peaks.

Singh et al. [19] deposited $CuInSe_2$ films under galvanostatic conditions from an aqueous solution containing $CuCl_2$, $InCl_3$, and SeO_2. The electrodeposition was carried out at 22 °C on Ti substrates. The pH-value was ≈ 1 and the current density was 5 mA/cm². The film thickness was 2 to 3 μm. The band gap was found to be 1.08 eV by using the optical absorption technique. The resistivity of the $CuInSe_2$ films was ≈ 10^3 Ω cm. This was attributed to a high trap density of ≈ 10^{14} cm^{-3} due to the defective bulk nature of the film. The films were polycrystalline.

$CuInSe_2$ layers on Mo, ITO, and Ti substrates were deposited from an aqueous acidic bath containing 50 mM each of $InCl_3$, SeO_2, and CuCl. The deposition potential was −800 mV (SCE) and the current density was 9 to 10 mA/cm². The as-deposited layers were black in colour, thin, rough, and less adhesive to the substrates. The layers were Cu-rich and inhomogeneous in composition. XRD showed that these layers were polycrystalline and contained mixed phases of $CuInSe_2$, $Cu_{2-x}Se$, and In–Se. The layers were heated at 500 °C in Ar atmosphere. XRD showed the presence of $CuInSe_2$ and $Cu_{2-x}Se$ phases. The layers became homogeneous with an average grain size of 2 to 3 μm.

Hodes et al. [81] carried out electrodeposition of $CuInSe_2$ from an alkaline selenosulfate bath containing In^{3+} and Cu^{1+}. NTA was used as a complexing agent for indium. The pH-value of the solution was 9 to 10. The electrodeposition was carried out under potentiostatic conditions at room temperature. $CuInSe_2$ in reasonably controllable proportions was obtained; however, it was difficult to get films thicker than several hundred nm when the deposition was carried out for longer time, increased roughness and large scale cracking were observed.

Lokhande [81a] used the pulse plating technique to deposit $CuInSe_2$ films from an aqueous acidic bath. The bath consisted of 50 mM each of CuCl, $InCl_3$, and SeO_2. The following conditions were found suitable to get stoichiometric $CuInSe_2$ films: pulse height −840 mV (SCE), duty cycle 33% (5 mS on, 10 mS off), deposition temperature 25 °C, the substrates were Mo and Ti. The as-deposited films were polycrystalline. The morphology of Cu-rich and In-rich $CuInSe_2$ films was different; the Cu-rich films consisted of isolated round grains, 3 to 4 μm in diameter, while for In-rich films, the grains were typically 2 to 3 μm in diameter and interconnected. Stoichiometric films were smooth with fine grains. The films on Mo cracked for shorter deposition times. The films were heated at 400 °C for 20 to 60 min in argon gas. The crystallinity of the $CuInSe_2$ films improved after annealing as evidenced by XRD.

The $CuInSe_2$ films using electrophoretic techniques were deposited by Shin et al. [82]. Single-phase polycrystalline $CuInSe_2$ ingots were ground to a fine powder. A suspension consisting of $CuInSe_2$ with suitable organic solvents, e.g., acetone, alcohol, tetramethyl ammonium hydroxide, was prepared. The substrate was molybdenum. The deposition temperature was 20 to 60 °C. The as-deposited films were non-uniform and easily detachable from the substrate. The films were annealed at 850 °C for 15 min. Two different types of morphologies were obtained: one with bright area consisting of grains 2 to 3 μm in size and another with dark area of fine grains.

$CuInSe_2$ films were prepared by the sequential electrodeposition [65] of InSe over previously electroplated CuSe layers and vice versa, or by annealing a Cu–In alloy in H_2Se or H_2 + Se atmosphere at high temperature [83].

The $CuInSe_2$ films prepared by sequential CuSe over InSe or InSe over CuSe techniques are polycrystalline, well adhesive to the substrate, uniform and pinhole free. These films were heated at high temperature in argon gas. XRD showed the presence of different $Cu_{2-x}Se$ phases in both the methods. The films were often Cu-rich.

Electrodeposition of Cu–In alloys was carried out using sulfate [83], cynide [84], and sulfamate [85] baths. The morphology was dependent on many factors; a good morphology was obtained with cynide bath. Lokhande [83] used the pulse plating technique to deposit CuIn alloys from sulfate bath. Noufi et al. [86] prepared Cu–In alloys from non-aqueous methanol bath containing cupric and indium perchlorates and ethanolamine.

Before chalcogenization the layers were heated at 140 °C and then at high temperatures ranging from 500 to 600 °C for different time periods in H_2Se to form $CuInSe_2$ films [79]. The films formed with such a techniques are polycrystalline, 4 to 6 μm in average grain size, well packed, and had the chalcopyrite structure. Stoichiometric $CuInSe_2$ films were obtained above 500 °C on both Ti and Mo substrates.

3.3.3 Copper–indium ditelluride

The $CuInTe_2$ films have been deposited by Bhattacharaya and Rajashwar [87] and Lokhande [88] from acidic baths. Bhattacharya and Rajashwar [87] used a 0.5 M $InCl_3$–0.5 M CuCl–0.1 M TeO_2 solution for the electrodeposition. The deposition potential was −1.0 V (SCE) at 12 mA/cm² current density. The substrates were Mo, Ti, and ITO glass. After annealing, XRD showed characteristics peaks of chalcopyrite $CuInTe_2$, with several small unidentified peaks. The composition of the films was irreproducible.

Lokhande [88] obtained $CuInTe_2$ films from a bath containing 0.1 M $InCl_3$–0.05 M CuCl–saturated TeO_2 solutions. The pH-value was adjusted to ≈ 1 using diluted HCl. The plating potential was −800 mV (SCE) at 3 to 4 mA/cm². The substrates were Mo and Ti. These films were annealed at 400 °C in Ar for 30 min. The as-deposited films were rough, easily detachable from the substrates, and Se-rich. XRD showed the polycrystalline nature of the films. After annealing the films became p-type with fine grains and improved crystalline structure. The composition of such films was uniform and in stoichiometric ratio.

4. Applications of Electrodeposited Semiconductors in Solar Cells

Electrodeposited semiconductors have found applications in solar selective coatings, solid state solar cells, Schottky barrier devices, and photoelectrochemical (PEC) cells. Below are given some examples of solar cells, where electrodeposited semiconductors have been utilized fully or partly in device formation.

4.1 n-CdS–p-CdTe solar cells

Thin film n-CdS–p-CdTe solar cells have been amongst the most promising candidates for low-cost terrestrial applications ever since the demonstration of over 5% efficiency with this structure. It has been only recently that some fundamental problems such as high resistivity and low carrier lifetime, etc. associated with the thin CdTe films have been overcome by better control of the deposition processes. Among the many deposition processes, three yielded all-thin-film solar cells with more than 8% efficiency. They are close-spaced sublimation, screen printing, and electrodeposition [89].

Recently, Basol et al. [90] have reported n-CdS–p-CdTe all-thin-film electroplated devices with 9% conversion efficiency. An n-CdS layer was electrodeposited onto indium–tin oxide (ITO) coated glass substrates (10 Ω/□) to a thickness of 60 to 100 nm. This was followed by the deposition of a 1 to 1.5 μm thick CdTe film on the CdS layer. SEM analyses of the CdS films indicated a fine grain structure where the grain size was comparable to the film thickness. The CdTe films were polycrystalline with columnar grains around 0.5 to 1 μm. The material had a strong ⟨111⟩ preferred orientation as indicated by X-ray diffraction. The as-deposited CdTe layer was n-type. The Cd:Te ratio through the film was determined by Auger depth profile and found to be very close to unity [91]. Such n-type CdTe layers were highly resistive.

The production of a CdS–CdTe heterojunction involved a post-heat treatment step that converted the as-deposited n-CdTe thin film layers into p-type [90, 91]. Since annealing did not affect the type of the CdS layer, at the end of this step, a rectifying junction was formed between the CdS and the converted CdTe layer. Devices were then completed by making evaporated gold contacts to the p-CdTe layer [90].

4.2 n-CdS-p-CuInSe$_2$ solar cells

The p-CuInSe$_2$ layers were prepared utilizing the electrodeposition technique [1]. The soda lime glass substrates were metallized by sputtering molybdenum. Copper and indium metal layers were sequentially deposited on the Mo coated glass substrates. The Cu:In ratio was maintained in the range of 0.98 to 1.0. This range is important for obtaining single phase chalcopyrite material with desired electrooptical properties. Copper was plated from $CuSO_4$ solution for 7.5 to 15 s at 40 mA/cm^2 current density. The resulting thickness was 100 to 250 nm. Indium metal was electroplated from an indium sulfamate bath with a thickness of 220 to 440 nm. The stacked glass–Mo–Cu–In layers were placed in a tube furnace and heated at 250 to 450 °C in H_2Se–Ar gas mixture. The resulting films are CuInSe$_2$ with $\approx$ 1.5 μm thickness.

The device was formed with evaporated n-CdS on p-CuInSe$_2$ film. The CdS deposition was carried out in two steps: first a layer of high resistivity CdS was evaporated (0.6 μm) and then In-doped CdS (1.2 μm) was evaporated. The cell was heat-treated in air for 15 min at 155 °C. The device area was 1 cm^2 and the following results were obtained under 100 mW/cm^2 illumination intensity: J_{sc} = 32 mA/cm^2, U_{oc} = 410 mV, FF = 0.54, and efficiency 7.27%.

4.3 n-CdS-p-HgCdTe solar cells

Basol and Tseng [2] et al. have proposed and demonstrated the use of Cd-rich HgCdTe in the place of p-CdTe in electroplated heterojunction solar cells. The resistivity of electroplated and converted (to p-type) HgCdTe can typically be around 8×10^{-3} to 8×10^{-4} Ω cm, depending upon its stoichiometry. Because of the lower resistivities attainable in p-HgCdTe films, good ohmic contacts to this material are easier to obtain.

Various concentrations of Hg^{2+} ions were introduced into the electrolyte using $HgCl_2$. HgCdTe deposition was carried out onto glass–ITO–CdS substrates at 90 °C using two anodes, one carbon and one tellurium. The CdS layer was electroplated on ITO (80 nm). The as-deposited HgCdTe films were 1.2 to 1.3 μm thick and were n-type. The n-CdS–p-HgCdTe junction was formed by heat treatment of the CdS–HgCdTe structure in air for 8 min at 400 °C. This treatment converts n-HgCdTe into p-type and forms an active junction between p-HgCdTe and the n-CdS layer. The following results have been reported: light intensity 100 mW/cm^2, J_{sc} = 27.03 mA/cm^2, U_{oc} = 0.62 V, FF = 0.633, and η = 10.6%.

4.4 n-CdS-p-AgInSe$_2$ solar cells

Thin polycrystalline AgInSe$_2$ films were prepared by electrodeposition from aqueous solutions of $AgNO_3$, $In(NO_3)_3 \cdot 3\,H_2O$, and SeO_2 [93]. The as-deposited films mainly consisted of AgInSe$_2$ and excess free selenium. Single-phase chalcopyrite films were obtained by annealing the as-deposited films in N_2 atmosphere in the temperature range from 500 to 600 °C. Thin film n-CdS–p-$AgInSe_2$ solar cells with $\eta \approx 2\%$ are fabricated using annealed electrodeposited $AgInSe_2$ and evaporated CdS films. The following results were reported: area 2.5×2.5 cm², illumination 100 mW/cm², $J_{sc} = 14.2$ mA/cm², $U_{oc} = 0.29$ V, FF = 0.48, $\eta = 2\%$.

4.5 Photoelectrochemical solar cells

The electrodeposited semiconductors like CdS, $CuInSe_2$, etc. have shown efficiencies >5% in PEC cells. Here, a brief discussion will be made only for the cells with 5% efficiency.

CdSe was prepared by the electrodeposition technique and a PEC cell was formed with polysulfide (1 M Na_2S–1 M S–1 M NaOH) electrolyte using brass as a counter-electrode after giving routine treatments to the CdSe electrode [94]. The performance parameters under illumination intensity of 100 mW/cm² from a W-halogen lamp, were $J_{sc} = 19$ mA/cm², $U_{oc} = 700$ mV, FF = 0.4, and $\eta = 6.7\%$.

On the same line, $CdSe_{0.65}Te_{0.35}$ films were electrodeposited from a SeO_2–TeO_2 bath by adjusting the bath composition and after proper treatments they were used in polysulfide based PEC cells [95]. The power conversion efficiency was 5.1% at an illumination intensity of 80 mW/cm². The performance parameters were $J_{sc} = 17$ mA/cm², $U_{oc} = 600$ mV, FF = 0.55, and $\eta = 5.1\%$.

In another PEC cell where an efficiency of 5.7% was obtained, the CdSe films were prepared by first depositing cadmium on to a nickel substrate and then dipping the resultant cadmium plated nickel into an acidic SeO_2 solution for 30 min to 3 h [10]. These films were annealed at 500 °C and were used in polysulfide based PEC cells. The performance parameters were $J_{sc} = 9.4$ mA/cm², $U_{oc} = 0.65$, FF = 0.65, and $\eta = 5.7\%$.

5. Conclusions

The preparation of thin semiconductors with the electrodeposition technique possesses many advantages: it is an easy and economic process; there is no need for purified materials; one has to control the stoichiometry, etc. Electrodeposition offers flexibility in choosing different steps in forming compound semiconductors. The physical and chemical properties of such semiconductors are comparable with the semiconductors prepared by other methods.

Such films have been used in devices as solar cells, optoelectronic devices, solar selective coatings, etc. In solar cells, junction formation with all-electrodeposited n- and p-type semiconductors has resulted in cells with solar conversion efficiency >9%.

References

[1] V. K. Kapur, B. B. Basol, and E. S. Tseng, 18th IEEE P.V. Spec. Conf., Oct. 1985, Las Vegas (USA), IEEE, in the press.
[2] B. M. Basol and E. S. Tseng, Appl. Phys. Letters 48, 946 (1986).
[3] M. M. Koltun, Selective Optical Surfaces for Solar Energy Converters, Allerton Press, Inc., New York 1981.
[4] O. P. Agnihotri and B. K. Gupta, Solar Selective Surfaces, John Wiley and Sons, New York 1980.

[5] P. K. MAHAPATRA and C. B. ROY, Electrochim. Acta **29**, 1435 (1984).
[6] G. HODES and D. CAHEN, Solar Cells **16**, 245 (1986).
[7] R. D. SRIVASTAVA and R. C. MUKERJEE, J. appl. Electrochem. **6**, 321 (1976).
[8] J. LLABREM and V. DELMAS, 6th EEC Photovoltaic Solar Energy Conf., London April 15 to 19, 1985.
[9] M. S. KAZACOS and B. MILLER, J. Electrochem. Soc. **127**, 869 (1980).
[10] R. C. BHARDWAJ, C. M. JADHAV, and M. M. TAQUI KHAN, Solar Cells **13**, 293 (1984).
[11] B. MILLER and A. HELLER, Nature **262**, 680 (1976).
[12] M. POURBAIX, Lectures on Electrochemical Corrosion, W. Roger Stachie/Plenum, New York 1973.
[13] G. F. FULOP and R. M. TAYLOR, Annu. Rev. Mater. Sci. **15**, 197 (1985).
[14] F. A. KRÖGER, J. Electrochem. Soc. **125**, 2028 (1978).
[15] R. WEIL, Plating and Surface Finishing **12**, 46 (1982).
[16] K. M. GORUBUNOVA and Y. M. POLUKAROV, in: Advances in Electrochemistry and Electrochemical Engineering, Vol. 5. Ed. C. W. TOBIAS and P. DELAPAY, Wiley/Interscience, New York 1967.
[17] A. BRENNER, Electrodeposition of Alloys: Principle and Practice, Academic Press, New York 1963.
[18] B. E. CONWAY, J. O. M. BOCKRIS, E. YEAGER, S. U. M. KHAN, and R. E. WHITE (Ed.), Comprehensive Treatise of Electrochemistry, Vol. 7, Plenum Press, New York 1983.
[19] R. P. SINGH, S. L. SINGH, and S. CHANDRA, J. Phys. C **19**, 1299 (1980).
[20] R. F. BUNSHAH, Deposition Technologies for Films and Coating: Development and Applications, D. Reidel Publ. Co., Dordrecht 1986.
[21] F. F. AMMERMAN, A. ARBOR, and D. J. SCHIMELHETTE, U.S. Patent 3, 419, 484 (1968).
[22] D. ELWELL, J. Crystal Growth **52**, 741 (1981).
[23] D. ELWELL and R. S. FEIGELSON, Solar Energy Mater. **6**, 123 (1982).
[24] A. K. AGARWAL and A. E. AUSTIN, J. Electrochem. Soc. **128**, 2292 (1981).
[25] R. BOEN and J. BOUTEILLON, J. appl. Electrochem. **13**, 277 (1983).
[26] G. M. RAO, D. ELWELL, and R. S. FEIGELSON, J. Electrochem. Soc. **128**, 1708 (1981).
[27] D. ELWELL, R. S. FEIGELSON, and G. M. RAO, J. Electrochem. Soc. **127**, 1940 (1980).
[28] G. M. RAO, D. ELWELL, and R. S. FEIGELSON, Solar Energy Mater. **7**, 15 (1982).
[29] T. R. RAMAMOHAN and F. A. KRÖGER, Electrochim. Acta **27**, 371 (1982).
[30] C. H. LEE and F. A. KRÖGER, J. Electrochem. Soc. **129**, 936 (1982).
[31] G. SZEKELY, J. Electrochem. Soc. **98**, 318 (1951).
[32] L. SHIN, C. X. QIU, S. K. ZHANG, and J. K. CHEN, Mater. Letters **3**, 226 (1985).
[32a] J. M. ROSAMILIA and B. MILLER, J. electroanal. Chem. **215**, 249 (1986).
[33] G. P. POWER, R. D. PEGGS, and A. J. PARKER, Electrochim. Acta **26**, 681 (1981).
[34] B. FATUS, P. HERRASTI, R. DUO, F. ARJONA, and E. G. CAMARERO, J. Electrochem. Soc. **131**, 2243 (1984).
[35] B. FATUS, P. HERRASTI, F. ARJONA, and E. G. CAMARERO, 169th Electrochem. Soc. Meeting, Boston, May 4 to 9, 1986, Abstr. 327, Vol. 86-1.
[36] A. S. BARANSKI and W. R. FAWCETT, J. Electrochem. Soc. **127**, 766 (1980).
[37] A. S. BARANSKI, W. R. FAWCETT, and A. C. MCDONALD, J. Electrochem. Soc. **131**, 271 (1984).
[38] F. MONDON, J. Electrochem. Soc. **132**, 319 (1985).
[39] A. S. BARANSKI, W. R. FAWCETT, A. C. MCDONALD, and R. M. DE NOBRIGA, J. Electrochem. Soc. **128**, 963 (1981).
[40] A. S. BARANSKI and W. R. FAWCETT, J. Electrochem. Soc. **127**, 766 (1980).
[41] E. FATAS, P. HERRASTI, T. GARCIA, F. ARJONA, and E. G. CAMARERO, Mater. Chem. Phys. **13**, 497 (1985).
[42] A. S. BARANSKI, M. S. BENNETT, and W. R. FAWCETT, J. appl. Phys. **54**, 6390 (1983).
[43] M. COCIVERA and S. PREUSSER, see [35], Abstr. 336.
[44] L. S. R. YEH, P. G. HUDSON, and A. DAMJANOVIC, J. appl. Electrochem. **12**, 133 (1982).
[45] L. S. R. YEH, P. G. HUDSON, and D. DAMJANOVIC, J. appl. Electrochem. **12**, 1982 (1982).
[46] B. MILLER and A. HELLER, Nature **262**, 680 (1976).
[47] T. K. BANDYOPADHYAY and S. R. CHOUDHARI, phys. stat. sol. (a) **84**, 651 (1984).

[48] A. S. BARANSKI, W. R. FAWCETT, K. GATNER, and A. C. MCDONALD, J. R. MACDONALD, and M. SELEN, J. Electrochem. Soc. **130**, 579 (1983).
[49] R. O. LOUTFY and D. S. NG, Solar Energy Mater. **11**, 319 (1984).
[50] G. J. HOUSTON, J. F. MCCANN, and D. HENEMANN, J. Electrochem. Soc. **129**, 37 (1982).
[51] M. TOMKIEWICZ, I. LING, and W. S. PARSON, J. Electrochem. Soc. **129**, 2017 (1982).
[52] M. COCIVERA, A. DARKOWSKI, and B. LOVE, J. Electrochem. Soc. **131**, 2514 (1984).
[53] M. S. KAZACOS, J. electroanal. Chem. **148**, 233 (1983).
[54] G. HODES, J. MANASSEN, S. NEAGU, D. CAHEN, and Y. MIROVSKY, Thin Solid Films **90**, 433 (1982).
[55] M. P. R. PANICKER, M. KNASTER, and F. A. KRÖGER, J. Electrochem. Soc. **125**, 506 (1978).
[56] J. LLABRES and V. DELMAS, J. Electrochem. Soc. **131**, 464 (1984).
[57] R. N. BHATTACHARAYA and K. RAJESHWAR, J. Electrochem. Soc. **131**, 2032 (1984).
[58] J. LLABRES, J. Electrochem. Soc. **131**, 465 (1984).
[59] H. J. GERRITSEN, J. Electrochem. Soc. **131**, 137 (1984).
[60] C. SELLA, P. BONCOOPS, and J. VEDEL, J. Electrochem. Soc. **133**, 2043 (1986).
[61] R. N. BHATTACHARAYA, K. RAJESHWAR, and R. NOUFI, J. Electrochem. Soc. **131**, 939 (1984).
[62] A. DARKOWSKI, T. V. WIRDHEIM, and M. COCIVERA, see [35], Abstr. 324.
[63] A. DARKOWSKI and M. COCIVERA, J. Electrochem. Soc. **132**, 2768 (1985).
[64] R. D. ENGELKEN and H. E. MCCLOUD, J. Electrochem. Soc. **132**, 567 (1985).
[65] C. D. LOKHANDE, Bull. Electrochem. **3**, 219 (1987).
[65a] S. CHANDRA and N. KHARE, Semicond. Sci. Technol. **2**, 214 (1987).
[66] G. HODES and D. CAHEN, Tech. Rep. May to Aug. 1985, Proj. No. IL-5-04132-1 to SERI (USA).
[67] A. J. ALEKPEROU, Z. A. ALIVAROVA, and K. A. ASKEROVA, Azerb. Khim. Zh. **5**, 104 (1981) (in Russian).
[68] S. CHANDRA, O. N. SRIVASTAVA, and S. N. SAHU, phys. stat. sol. (a) 88, 497 (1985).
[69] S. CHANDRA and S. N. SAHU, phys. stat. sol. (a) **89**, 321 (1985).
[70] S. CHANDRA and S. N. SAHU, J. Phys D **17**, 2115 (1984).
[71] S. CHANDRA, D. P. SINGH, P. C. SRIVASTAVA, and S. N. SAHU, J. Phys. D **17**, 2125 (1984).
[72] A. VICHBECK and N. HACKERMANN, J. Electrochem. Soc. **131**, 1315 (1984).
[73] A. VICHBECK and N. HACKERMANN, in: Photoelectrochemistry: Fundamental Processes and Measurement Techniques, Ed. W. L. WALLANCE, A. J. NOZIK, S. K. DEB, and R. H. WILSON, The Electrochem. Soc., Pennington (N.J.) 1982 (p. 478).
[74] A. S. BARANSKI, W. R. FAWCETT, and C. M. GILBERT, J. Electrochem. Soc. **130**, 2423 (1983).
[75] L. M. PETER, J. electroanal. Chem. **98**, 49 (1979).
[76] R. D. ENGELKEN, A. K. BERRY, T. P. VANDOREN, and A. SHAHNAZARY, J. Electrochem. Soc. **133**, 581 (1986).
[77] R. N. BHATTACHARAYA, D. CAHEN, and G. HODES, Solar Energy Mater. **10**, 41 (1984).
[78] G. HODES, T. ENGELHARD, D. CAHEN, L. L. KAREVMERSKI, and C. R. HERRINGTON, Thin Solid Films **128**, 93 (1985).
[79] G. HODES, G. ALEXANDROWITZ, and D. CAHEN, Tech. Rep. Nov. 1984 to Feb. 1985, Proj. No. IL-5-04132-1 to SERI (USA).
[80] R. N. BHATTACHARAYA, J. Electrochem. Soc. **130**, 2040 (1983).
[81] G. HODES, C. D. LOKHANDE, and D. CAHEN, Tech. Rep. Nov. 1985 to Jan. 1986, Proj. No. IL-5-04132-1 to SERI (USA).
[81a] C. D. LOKHANDE, J. Electrochem. Soc. **134**, 1727 (1987).
[82] I. SHIN, S. K. ZHANG, and C. X. QIA, Solar Cells **16**, 283 (1986).
[83] C. D. LOKHANDE, Bull. Electrochem. **4**, 131 (1988).
[84] G. HODES, D. CAHEN, and C. D. LOKHANDE, Tech. Rep. Oct. 1985 to Nov. 1986, Proj. No. IL-5-04132-1 to SERI (USA).
[85] C. D. LOKHANDE and G. HODES, Solar Cells **21**, 215 (1987).
[86] R. NOUFI, R. POWELL, R. AXTON, Y. CHEN, and T. DATTA, Annu. Rep. Solid State Photovoltaic Research Branch 1984, SERI/PR-212-2601 DE 85008812, April 1985.
[87] R. N. BHATTACHARAYA and K. RAJESHWAR, Solar Cells **16**, 237 (1986).
[88] C. D. LOKHANDE, J. Phys. C **26**, 1213 (1987).
[89] B. M. BASOL, J. appl. Phys. **55**, 601 (1984).

[90] B. M. BASOL, R. L. ROD, E. S. TSENS, D. S. LO, S. S. OU, and O. M. STANFSUDD, Proc. 5th Europ. Conf. Photovoltaic Energy, 1983, D. Reidel Publ. Co., 1984 (p. 888).
[91] B. M. BASOL, O. M. STANFSUDD, R. C. ROD, and E. S. W. TSENG, Proc. 4th Europ. Conf. Photovoltaic Energy, 1982, D. Reidel Publ. Co., 1983 (p. 715).
[92] S. S. QU, A. BINDAL, O. M. STAFSUDD, K. L. WANG, and B. M. BASOL, J. appl. Phys. 55, 1020 (1984).
[93] D. RAVIENDRA and J. K. SHARMA, phys. stat. sol. (a) 88, 365 (1985).
[94] R. A. BOUDREAU and R. D. RAUH, Solar Energy Mater. 7, 385 (1982).
[95] G. HODES, J. MANESSEN, S. MEOGU, D. CAHEN, and Y. MIROSVSKY, Thin Solid Films 90, 433 (1982).

(Received September 15, 1987; in revised form April 7, 1988)

Original Papers

phys. stat. sol. (a) **111**, 41 (1989)

Subject classification: 61.10

Faculty of Physics, Erevan State University[1])

Method of Determining Supersmall Dilatations in a Crystal Lattice[2])

By

V. S. Haroutyunyan and P. H. Bezirganyan

A method to measure supersmall dilatation lattice distortions in the range $\Delta d/d \approx 10^{-10}$ to 10^{-8} by a bicrystal wedge-shaped gap X-ray interferometer is predicted.

Предсказан способ определения сверхмалых дилатационных нарушений $\Delta d/d \approx$ $\approx 10^{-10}$ до 10^{-8} кристаллической решетки при помощи двухкристального рентгеновского интерферометра с клиновидным зазором.

1. Introduction

In [1 to 3] Bonse and Hart used X-ray moiré interference patterns obtained in a three-crystal interferometer (see Fig. 1a) for the precision measurement of small lattice dilatations and distortions. For these measurements Lang and Miuscov [4] and Chikawa [5] used similar interference patterns obtained in bicrystal systems (see Fig. 1b). With the aid of the above-mentioned methods dilatations $\Delta d/d \approx 10^{-8}$ to 10^{-5} (d is the interplanar spacing of reflecting planes) can be found. Dilatations $\Delta d/d \approx 10^{-8}$ can be measured using the scanning technique. The measurement of smaller distortions ($\Delta d/d < 10^{-8}$) by the methods described in [1 to 5] is impossible, since for these small deviations of the parameter $\Delta d/d$ the moiré fringe period should be too large to be observed (see Table 1).

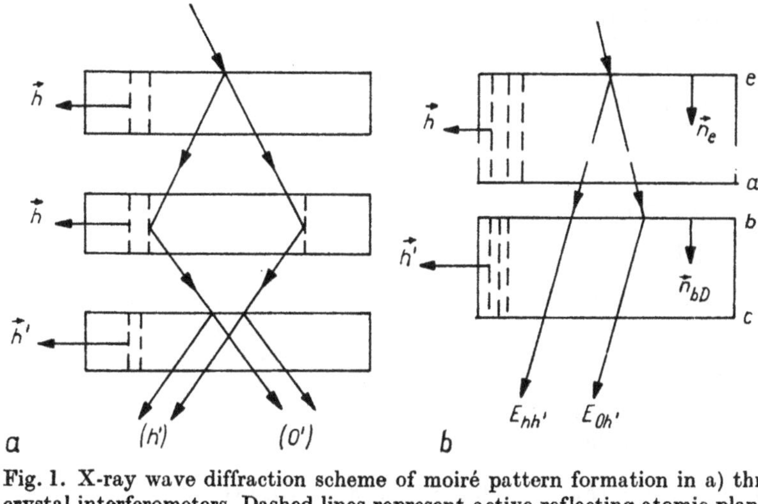

Fig. 1. X-ray wave diffraction scheme of moiré pattern formation in a) three-crystal and b) bicrystal interferometers. Dashed lines represent active reflecting atomic planes

[1]) Ul. Mravyana 1, SU-375049 Erevan, USSR.

[2]) Dedicated to Prof. U. Bonse on the occasion of his 60th birthday.

Table 1

$\Delta d/d$	σ_D (μm)	σ (μm)
10^{-6}	192	183
10^{-8}	192×10^2	3151
10^{-9}	192×10^3	3698
10^{-10}	192×10^4	3763

In the present work we propose a method to determine supersmall dilatations $\Delta d/d \approx 10^{-10}$ to 10^{-8} in any block of a bicrystal X-ray interferometer with wedge-shaped gap.

2. Description of the Method and Interference Pattern Period Determination

In this method we take advantage of the results obtained in [6], as well as the scheme represented in Fig. 2. Our investigation is based on the following assumptions:

1. the incident X-ray wave is plane and monochromatic;
2. both entrance (e) and exit (a) surfaces of the first crystal are parallel to the exit surface (c) of the second one, whereas the entrance (b) surface of the second crystal is tilted;
3. the rib of the wedge gap is perpendicular to the plane of incidence;
4. in the first block of the interferometer X-ray diffraction occurs in compliance with the symmetrical Laue reflection scheme;
5. the diffraction vectors for the acting Bragg reflection in both blocks of the interferometer slightly differ from one another only by their modulus ($\Delta d/d \leqq 10^{-8}$);
6. anomalous Borrmann transmission of X-rays takes place in both crystals (assuming sufficiently thick crystals).

The rectangular coordinate system XYZ has the orientation shown in Fig. 2. The interference of the weakly absorbing waves $E_{0h'}$ and $E_{hh'}$ is considered. $E_{0h'}$ and $E_{hh'}$ are the waves diffracted in the first block in the directions of (o) and (h), respectively, whereas in the second crystal both of them are diffracted in the direction of the reflection in the first one. The interference field region in the second crystal (dashed region in Fig. 2) will be considerable, if the gap between the crystals would be sufficiently narrow. The amplitudes of the interfering waves can be represented in the following way:

$$E_{0h'} = A_{0h'} \exp(i2\pi \boldsymbol{h'r}) \exp(i\varphi_{0h'}) , \tag{1}$$

$$E_{hh'} = A_{hh'} \exp(i2\pi \boldsymbol{hr}) \exp(i\varphi_{hh'}) , \tag{2}$$

where $A_{0h'}$, $A_{hh'}$ and $\varphi_{0h'}$, $\varphi_{hh'}$ are amplitude factors and phases of the corresponding waves, respectively (the dependence of these values on the X-ray incidence angle and inclination angle of the second block entrance surface in an obvious form is obtained in [6]), $\boldsymbol{h}$ and $\boldsymbol{h'}$ are diffraction vectors in blocks 1 and 2, respectively, $\boldsymbol{r}$ is the position vector of the observation point. For the intensity distribution of the interference pattern formed in $\boldsymbol{h'}$-direction the following expression can be found:

$$I_{h'} = |E_{0h'} + E_{hh'}|^2 = |E_{0h'}|^2 + |E_{hh'}|^2 + 2\,|E_{0h'}|\,|E_{hh'}| \cos\psi , \tag{3}$$

where

$$\psi = \varphi_{0h'} - \varphi_{hh'} + 2\pi(\boldsymbol{h'} - \boldsymbol{h})\,\boldsymbol{r} . \tag{4}$$

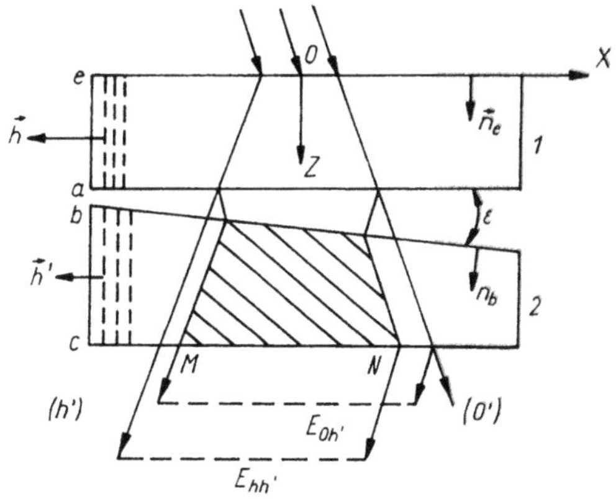

Fig. 2. Diffraction scheme of X-ray plane wave in a bicrystal interferometer with narrow wedge-shaped gap

The equations $|\boldsymbol{h}| = 1/d$ and $|\boldsymbol{h}'| = 1/d'$ (d and d' are the spacings of reflecting planes in blocks 1 and 2, respectively) as well as the phase difference $\varphi_{0h'} - \varphi_{hh'}$ obtained in [6] being taken into account, (4) comes to the form

$$\psi \approx [2\pi(\vartheta_B - \vartheta)\,\varepsilon/d]\,X + 2\pi(\Delta d/d^2)\,X\,, \tag{5}$$

where

$$\Delta d = d' - d\,,$$

ϑ_B being the exact Bragg angle in the first crystal, ϑ the X-ray incidence angle, ε the inclination angle of the second block entrance surface with respect to the $0X$-axis, X the observation point coordinate.

We assume that in (5) $\varepsilon \approx 1°$, since the diffraction curves in crystals 1 and 2 considerably overlap at these values of the wedge-shaped gap angle [6]. When obtaining the first term in (5) the approximation $\vartheta_B = \vartheta'_B$ (ϑ'_B is the exact Bragg angle in the second crystal) is valid at $\Delta d/d \lesssim 10^{-8}$.

Here we find the expression for the interference pattern period σ. From the condiion $\psi(X_k) = 2\pi k$, where $\psi(X)$ is given by (5), k is an integer, X_k correspond to the intensity maxima, follows:

$$\sigma = \left|\frac{\sigma_\varepsilon \sigma_D}{\sigma_\varepsilon + \sigma_D}\right| = \left|\sigma_D \Big/ 1 + \frac{\varepsilon(\vartheta_B - \vartheta)\,d}{\Delta d}\right|, \tag{6}$$

where

$$\sigma_\varepsilon = d/(\vartheta_B - \vartheta)\,\varepsilon\,, \qquad \sigma_D = d^2/\Delta d\,.$$

$|\sigma_\varepsilon|$ is the period of the interference fringes arising in a wedge-shaped gap system at $\Delta d = 0$ [6], $|\sigma_D|$ is the period of dilatation moiré fringes at $\varepsilon = 0$ [1, 7] (the diffraction schemes given in Fig. 1 correspond to this case). In (6) the values $\vartheta_B - \vartheta$ (the incidence angle deviation from ϑ_B), Δd, and ε may be either positive or negative. If the entrance surface of the second crystal has a clockwise inclination with respect to the X-axis, the value of ε is positive, and ε is supposed negative in the opposite case [6]. From (5) follows that straight fringes perpendicular to the plane of incidence are formed (these fringes have the same orientation as the dilatation moiré fringes).

3. Numerical Results

The comparison of the period of interference fringes σ with the period of moiré fringes σ_D at different $\Delta d/d$ ratios ($\Delta d > 0$ being assumed) can be made using the relevant data from Table 1. Corresponding curves are shown in Fig. 3. The calculations are

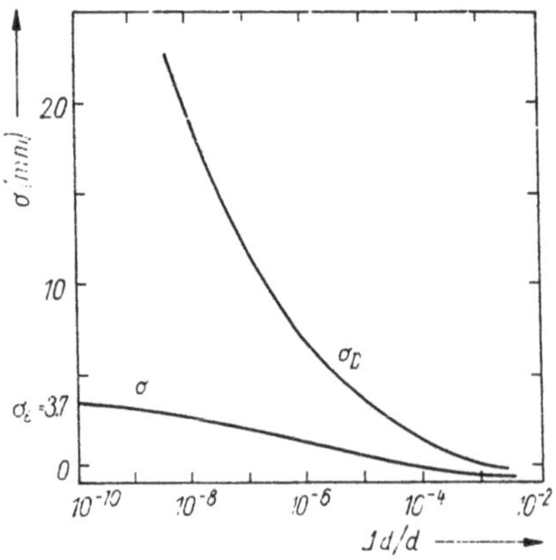

Fig. 3. Comparison of the period of interference fringes σ with the period of moiré fringes σ_D at different $\Delta d/d$ ratios ($\Delta d > 0$)

carried out using (6) and $\sigma_D = d^2/\Delta d$ for an Si interferometer, MoK_{α_1} radiation (220) reflecting planes, and parameters $\vartheta_B - \vartheta = 0.2''$, $\varepsilon = 3°$.

The results given in Table 1 show that for $\Delta d/d < 10^{-8}$ the period σ_D is too large to be detected (even using scanning techniques), whereas for $\Delta d/d \approx 10^{-10}$ to 10^{-8} the period σ gives us the opportunity to detect these small dilatations. From (6) follows, that for $\Delta d/d \to 0$ $\sigma \to |\sigma_\varepsilon|$ ($|\sigma_\varepsilon| = 3770\ \mu m$), therefore for $\Delta d/d < 10^{-10}$ the changes of the period $\Delta\sigma < 7\ \mu m$, which is comparable with the available accuracy of the measurements.

4. Reciprocal Space Diagram

In Fig. 4 the reciprocal space diagram is represented, qualitatively interpreting the results given in Table 1. This diagram shows the formation principle of dilatation moiré fringes [1 to 5] and fringes arising in a bicrystal X-ray interferometer with wedge-shaped gap and small dilatation in any block. If the approximation $\chi_h = \chi_{h'}$ is made (χ_h and $\chi_{h'}$ are the h-th order Fourier coefficients of the polarizability of the first and second crystals, respectively), then the dispersion surface for the first crystal coincides with the dispersion surface for the second one. In the case represented in Fig. 4 $\Delta d = d' - d > 0$, $\vartheta_B - \vartheta > 0$, (e) is the entrance surface of block 1, $\boldsymbol{n}_e$ in the unit normal of surface (e), $\overrightarrow{OH} = \boldsymbol{h}$, $\overrightarrow{O'H'} = \boldsymbol{h}'$, L and L′ are Laue points of blocks 1 and 2, respectively; S_0, S_h and S_0', S_h' are incidence and reflection spheres in blocks 1 and 2, respectively, Q is the Lorentz point, DS the weakly absorbing branch of the dispersion surface corresponding to the σ-polarisation state, $\overrightarrow{TO} = \boldsymbol{K}_e$ the wave vector of the initial wave (incident on block 1), $\overrightarrow{T'O'} \equiv \boldsymbol{K}_0 = \boldsymbol{K}_e$ and $\overrightarrow{RH} = \overrightarrow{R'H'} = \boldsymbol{K}_h$ are wave vectors of transmitted and reflected vacuum waves, respectively, in the gap (these waves are incident on block 2). Obviously, for small values of the parameter $\Delta d/d$ the following conditions are fulfilled (see Fig. 4):

$$|\boldsymbol{k}_{0h'}^{D} - \boldsymbol{k}_{hh'}^{D}| = |\overrightarrow{P_{1D}P_{2D}}| = |\Delta \boldsymbol{k}_D| \ll |\boldsymbol{k}_{0h'} - \boldsymbol{k}_{hh'}| = |\overrightarrow{P_1P_2}| = |\Delta \boldsymbol{k}|$$

and

$$\sigma = 2\pi/|\Delta \boldsymbol{k}| \ll \sigma_D = 2\pi/|\Delta \boldsymbol{k}_D|\,.$$

5. Conclusions

If the ratio $L/\sigma \geqq 2$ (the values for the period σ are given in Table 1), where L is the linear dimension of the interference pattern along the X-axis ($L = \overline{MN}$ in Fig. 2), then the method described above allows one to determine dilatation lattice distor-

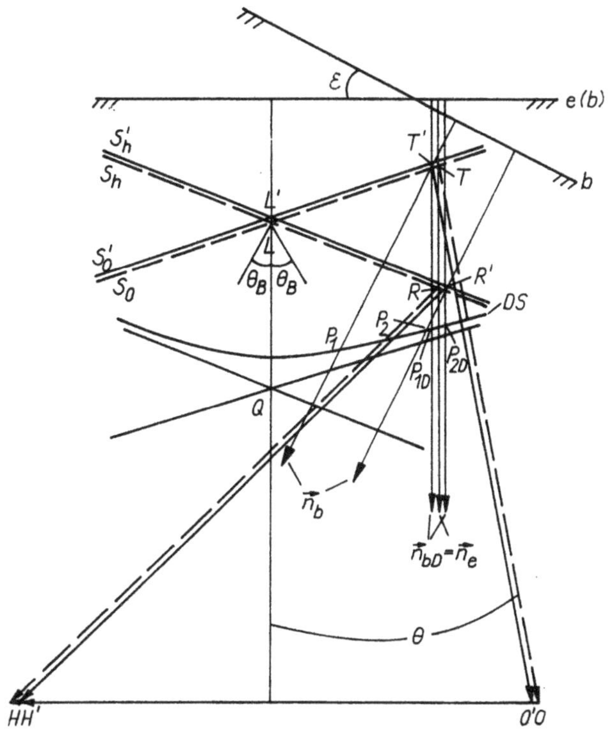

Fig. 4. Reciprocal space diagram. Dilatation moiré case, 1. corresponding to the scheme given in Fig. 1b ($\boldsymbol{n}_{bD}$ is the unit normal of the surface (b) ($\boldsymbol{n}_{bD} = \boldsymbol{n}_e$), P_{1D} and P_{2D} are wave points excited in block 2, $\overrightarrow{P_{1D}H'} = \boldsymbol{k}_{0h'}^{D}$ and $\overrightarrow{P_{2D}H'} = \boldsymbol{k}_{hh'}^{D}$ are wave vectors of interfering waves $E_{0h'}$ and $E_{hh'}$ (in second block) in $\boldsymbol{h}'$-direction, respectively.) 2. corresponding to the scheme given in Fig. 2 ($\varepsilon > 0$. $\boldsymbol{n}_b$ is the unit normal of the surface (b) ($\boldsymbol{n}_b \neq \boldsymbol{n}_e$), P_1 and P_2 are excited wave points, $\overrightarrow{P_1H'} = \boldsymbol{k}_{0h'}$ and $\overrightarrow{P_2H'} = \boldsymbol{k}_{hh'}$)

tions $\Delta d/d \approx 10^{-10}$ to 10^{-8} without scanning techniques (to obtain an initial X-ray plane wave with wide wave front a collimator crystal utilizing the asymmetric Bragg diffraction can be used [8]), while at $L/\sigma < 2$ scanning along the X-axis is needed.

Thus, from the theoretical results obtained in the present paper follows that the use of a bicrystal wedge-shaped gap X-ray interferometer makes it possible to determine supersmall dilatations in any of the crystals.

References

[1] U. Bonse and M. Hart, Appl. Phys. Letters **6a**, 155 (1965).
[2] U. Bonse and M. Hart, Appl. Phys. Letters **7b**, 99 (1965).
[3] U. Bonse and M. Hart, Z. Phys. **190**, 455 (1966).
[4] A. R. Lang and V. F. Miuscov, Appl. Phys. Letters **7**, 214 (1965).
[5] J. Chikawa, Appl. Phys. Letters **7**, 193 (1965).
[6] P. H. Bezirganyan and V. S. Haroutyunyan, phys. stat. sol. (a) **99**, 37 (1987).
[7] R. Gevers, Phil. Mag. **7**, 1681 (1962).
[8] H. Hashizume, H. Ishida, and K. Kohra, phys. stat. sol. (a) **12**, 453 (1972).

(Received May 31, 1988)

phys. stat. sol. (a) **111**, **47** **(1989)**

Subject classification: 61.16 and 61.70; S1.2

Physics Department, Busan Sanub University[1]) *(a)*
and Department of Physics, University of Alberta, Edmonton[2]) *(b)*

An Electron Microscope Method for Determining the Nature of Stacking Faults in F.C.C. Materials which Does not Rely on Effects of Absorption

By

H. S. KIM (a) and S. S. SHEININ (b)

Previous methods for determining whether stacking faults in f.c.c. materials are extrinsic or intrinsic make use of an asymmetry in dark field images which is due to the effects of absorption. As a result, the techniques are more difficult to apply in thinner regions of a crystal and in crystals with low absorption, such as silicon. A method for determining the nature of stacking faults in f.c.c. materials is presented which does not depend on absorption effects. The method makes use of an asymmetry in bright field images which is obtained when non-symmetrical Laue diffraction conditions prevail and/or the stacking fault is inclined so that the column approximation is not valid. These images can provide sufficient information to apply either the method of Hashimoto et al. or the method of Gevers et al. with the rules for the direction of the reciprocal lattice vector reversed.

Frühere Methoden zur Bestimmung, ob ein Stapelfehler in k.f.z.-Materialien intrinsisch oder extrinsisch ist, machen Gebrauch von einer Asymmetrie in Dunkelfelddiagrammen, die durch Absorptionseinflüsse verursacht wird. Als Ergebnis läßt sich die Technik in dünneren Bereichen eines Kristalls und in Kristallen mit niedriger Absorption, wie Silizium, schwerer anwenden. Es wird eine Methode zur Bestimmung der Art der Stapelfehler in k.f.z.-Materialien angegeben, die nicht von Absorptionseffekten abhängt. Die Methode macht Gebrauch von einer Asymmetrie in den Hellfelddiagrammen, die auftritt, wenn nicht-symmetrische Laue-Beugungsbedingungen überwiegen und/oder der Stapelfehler geneigt ist, so daß die Spaltennäherung nicht gültig ist. Diese Bilder können genügend Informationen liefern, um entweder die Methode von Hashimoto et al. oder die Methode von Gevers et al. anzuwenden, mit umgekehrten Regeln für die Richtung des reziproken Gittervektors.

1. Introduction

An important problem in electron microscope observations of stacking faults in f.c.c. materials is a determination of the nature of the fault. Two well established methods for determining whether a stacking fault in an f.c.c. material is intrinsic or extrinsic have been given in the literature [1, 2]. Both methods involve the use of asymmetrical dark field images of stacking faults obtained under strong beam diffraction conditions. The asymmetry in these images is due to the effects of anomalous absorption and, therefore, this asymmetry can only be observed in thicker regions of a specimen. It is evident, therefore, that the methods available in the literature for determining the nature of a fault in f.c.c. materials will be more difficult to apply in thinner regions of a crystal or in crystals which have low absorption coefficients such as silicon [3]. It is, therefore, desirable to develop a method for determining the nature of faults in f.c.c. materials which does not depend on effects of absorption. The purpose of this paper is to present such a method.

[1]) Busan 608, Korea.

[2]) Edmonton, Alberta T6G 3J1, Canada.

2. Basis of the Method

The method to be presented in this paper is based on results obtained by the authors which show an asymmetry in bright field images of stacking faults [4]. This asymmetry is not associated with effects of absorption but rather is due to the inclination of the entrance surface of the specimen and/or the stacking fault plane to the incident beam. (Inclination of the entrance surface results in non-symmetrical Laue diffraction conditions while inclination of the stacking fault plane results in the column approximation in the standard theory of image contrast not being valid.) The asymmetry in bright field images referred to above is illustrated in Fig. 1 a, b which show theoretical images of a stacking fault in a wedge crystal obtained at positive and negative values of α, respectively. (It should be noted that in the discussion of bright field images to follow, the reciprocal lattice vector used to define α will correspond to the most strongly excited of the low-order reflections, $\boldsymbol{g}$ or $-\boldsymbol{g}$, in a systematic set.) The asymmetry in the bright field images in Fig. 1 can be easily seen if attention is focussed on the

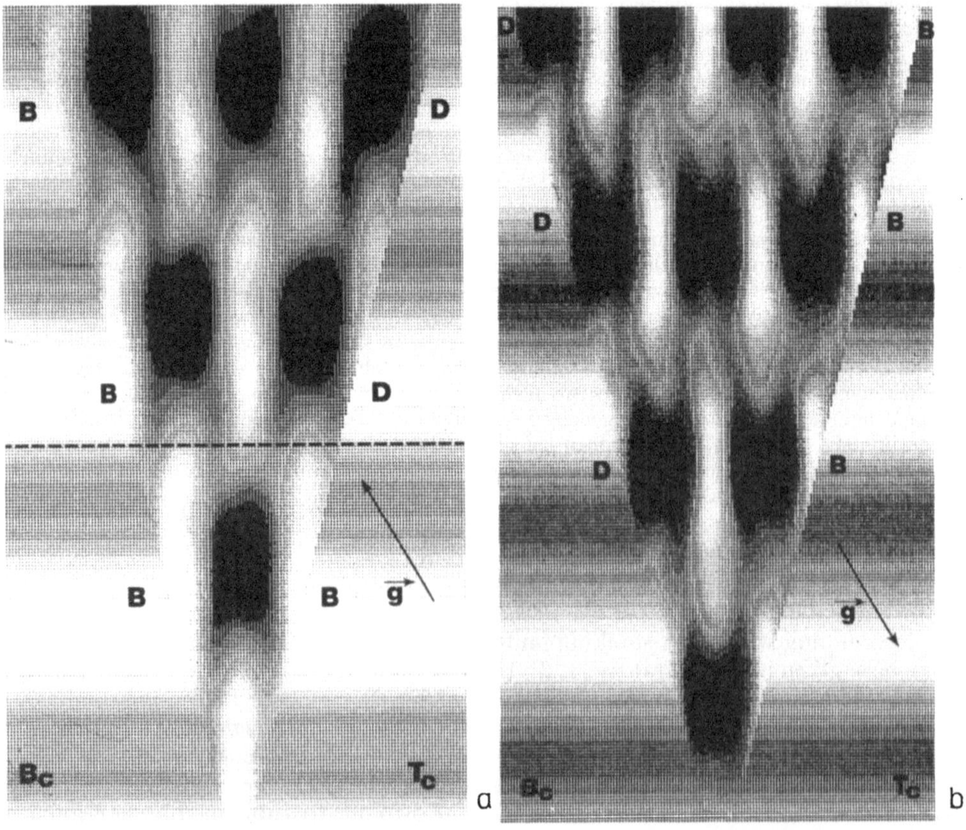

Fig. 1. Theoretical bright field images of a stacking fault in a wedge crystal of f.c.c. cobalt for a) a positive value of α and b) a negative value of α. T_c and B_c indicate the side of the image corresponding to the intersection of the fault with the top and bottom surfaces of the specimen, respectively. D and B denote a dark and bright edge fringe, respectively, while the dashed line indicates the approximate value of the crystal thickness at which the transition from a symmetrical to an asymmetrical image takes place. The diffraction conditions used to obtain these images are given in Section 3 and are the same as for the experimental images in Fig. 2

high contrast regions such as at positions labelled D and B, indicating dark and bright edge fringes, respectively. It is also worthwhile noting that the same general contrast features were found for the other low-order systematic sets of reflections.

It is next of interest to see how the asymmetry referred to above can be used to determine the nature of a fault in f.c.c. materials. This can be done in one of two ways. In the first, the method of Gevers et al. [2] can be adapted to determine the nature of the fault from an asymmetrical bright field image. The arguments used are very similar to those used in [2] and, therefore, will not be repeated here. The results obtained give a rule which is exactly opposite to the rule given by Gevers et al. [2] for strong beam dark field images. Using the same classification of reflections as in [2], the rule for asymmetrical bright field images can be stated as follows for class B reflections such as {111}, {220}, and {400}: when the reciprocal lattice vector associated with the most strongly excited low-order reflection in the systematic set is placed at the centre of the stacking fault image, it will point to a bright fringe if the fault is intrinsic and to a dark fringe if the fault is extrinsic. The rule is reversed for class A reflections such as {200}, {222}, and {440}.

The second way in which the nature of a fault can be determined from asymmetrical bright field images is based on the method of Hashimoto et al. [1]. In this method both the sign of α and the sense of inclination of the fault must be determined. The manner in which the sign of α can be determined from an asymmetrical bright field image can be seen from Fig. 1 a, b which show that the high contrast regions are shifted from positions in the image corresponding to crystal thicknesses of $n\xi$ (where n is an integer) by an amount equal to about $+1/3\xi$ for the positive value of α in Fig. 1a and $-1/3\xi$ for the negative value of α in Fig. 1b (ξ is the effective extinction distance at the deviation parameter concerned). This permits the sign of α to be easily determined from an experimental image. The sense of inclination of the fault can be determined by noting from Fig. 1a that for positive α, the edge fringe corresponding to the intersection of the fault with the bottom surface is bright while Fig. 1b shows that for negative α, the opposite is true. It should be noted that, in the method described above, both the sign of α and the sense of inclination of the fault can be determined from a single asymmetrical bright field image whereas the method of Hashimoto et al. [1] requires that both a bright and a dark field image be recorded.

3. Comparison of Theoretical and Experimental Results

In order to test the applicability of the method described in Section 2, experimental bright field images of stacking faults in wedge crystals of cobalt were studied. The orientation of the crystal for both the experimental image in Fig. 2 and the theoretical images in Fig. 1 was such that a [123] zone axis was nearly parallel to the electron beam giving a strongly excited $(-1, -1, 1)$ systematic set. This gave values of the deviation parameters, S_g in Fig. 1a and S_{-g} in Fig. 1b of about $2.7 \times 10^{-2}\ \text{nm}^{-1}$, corresponding to an intersection of the Ewald sphere lying approximately half-way between the first- and second-order reflections. Other diffraction conditions used to obtain these images were the same as those used to obtain the experimental image in Fig. 2. The values of θ_s and θ_f used in the calculations were 60° and 52°, respectively, where θ_s and θ_f are the angles of the normals to the entrance surface and the stacking fault plane measured with respect to the zone axis defining the orientation of the crystal. (Details of the procedure used to calculate the theoretical images are described in [4].) A comparison of theoretical results with experiment shows that there is ex-

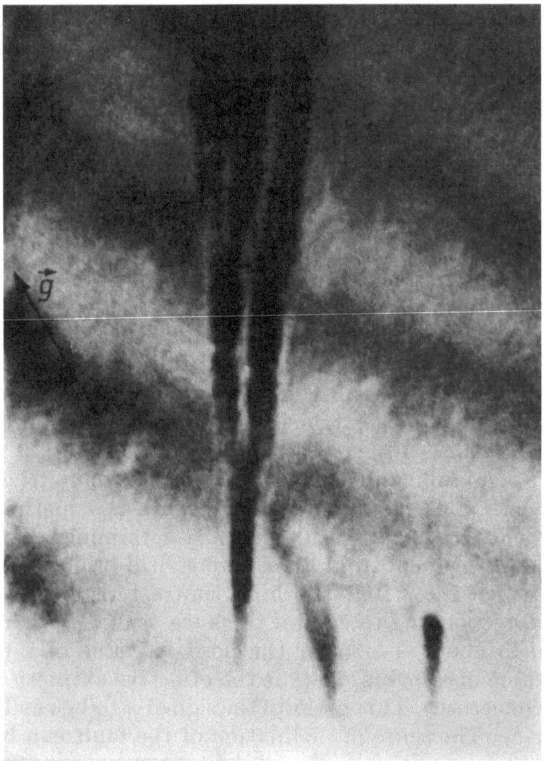

Fig. 2. Experimental bright field image of a stacking fault in f.c.c. cobalt. The diffraction conditions used to obtain this image are given in Section 3

cellent agreement between the experimental image in Fig. 2 and the theoretical image in Fig. 1a.

In order to determine the nature of the fault, the origin of the reciprocal lattice vector was placed in the asymmetrical region of Fig. 2 and at the centre of the image. As can be seen, this vector points to a bright fringe. Since (−1, −1, 1) is a class B reflection, the rules given in Section 2 indicate that the fault is intrinsic. It is also worth noting that the faults in Fig. 3 of [4] are also intrinsic. This can be seen from the fact that the (002) reflection is a class A reflection and $\boldsymbol{g}$ points to the dark edge fringe. This result is also confirmed by the theoretical images in Fig. 4 of [4].

The information required to apply the method of Hashimoto et al. [1] can also be determined from the experimental image in Fig. 2. The sign of α can be determined by noting the position of the first dark patch or low intensity region in the image which is greater than $n\xi$ (in agreement with the theoretical image in Fig. 1a), thus indicating a positive value of α. The sense of inclination of the fault can be determined by noting that up to a crystal thickness of about 2.5ξ ($\xi = 19.1$ nm), the image is symmetrical with both edge fringes being bright. Beyond this thickness, the image becomes asymmetrical with the fringe on the left being bright and the fringe on the right being dark. As a result the bright edge fringe or the left-hand side of the image corresponds to the intersection of the fault with the bottom surface of the crystal. Finally, it should be noted that the results obtained from an analysis of asymmetrical bright field images, as described above, were consistent with determinations made using the conventional methods [1] and [2].

4. Discussion

An important aspect of the experimental method presented in this paper for determining the nature of a fault is the proper choice of diffraction conditions which yield the asymmetry shown in Fig. 1 a and b. The first point to note is that the crystal should first be oriented so that non-systematic reflections are weak and a strong systematic set is obtained. Once this has been done, the procedure recommended by the authors to obtain an asymmetrical bright field image is to first set the lowest-order reflection, $\boldsymbol{g}$, in the systematic set at its Bragg condition. The specimen should then be gradually tilted toward the Bragg condition of the reflection $2\boldsymbol{g}$ until the required asymmetry in the image is observed on the screen. This will often occur at an orientation corresponding to an intersection of the Ewald sphere with the reciprocal lattice row about half-way between $\boldsymbol{g}$ and $2\boldsymbol{g}$. If the asymmetry is difficult to detect, then the authors recommend that the specimen should be tilted through the symmetry orientation until the reflection $-\boldsymbol{g}$ becomes the most strongly excited reflection. The same procedure referred to above for the reflection $\boldsymbol{g}$ should then be implemented. In most cases this will overcome the difficulty. However, if the asymmetry in the bright field image is still too weak to be easily observed, then it may be necessary to tilt the crystal to an orientation which gives increased values of θ_s and/or θ_f. The authors also recommend that faults in wedge crystals be examined since the contrast effects associated with non-symmetrical Laue diffraction conditions and the non-column case are more easily seen in this case.

Finally, it can be seen from the above discussion that the procedure required to obtain asymmetrical bright field images may be more involved than in the conventional methods [1, 2]. As a result, the authors recommend that the method presented in this paper be considered as one which is complementary to the conventional methods, to be used primarily in situations where absorption effects are low and the conventional methods are difficult to apply.

Acknowledgements

The authors are grateful for the financial support of the Korea Science and Engineering Foundation and the Natural Sciences and Engineering Council, Canada.

References

[1] H. Hashimoto, A. Howie, and M. J. Whelan, Proc. Roy. Soc. **A269**, 80 (1962).
[2] R. Gevers, A. Art, and S. Amelinckx, phys. stat. sol. **3**, 1563 (1963).
[3] A. K. Head, Austral. J. Phys. **22**, 569 (1969).
[4] H. S. Kim and S. S. Sheinin, phys. stat. sol. (a) **90**, 231 (1985).

(Received May 24, 1988; in revised form September 12, 1988)

phys. stat. sol. (a) **111**, 53 (1989)

Subject classification: 61.60; 76.80; S10

Institute of Physics Jagellonian University, Cracow[1])

Investigation of ^{57}Fe-Doped ϑ-$Al_xV_2O_5$ Bronzes

By

J. J. BARA and B. F. BOGACZ

ϑ-$^{57}Fe_{0.01}Al_{x-0.01}V_2O_5$ ($0.33 \leqq x \leqq 0.58$) vanadium oxide bronzes are synthesized by solid state reaction and investigated at 78 and 295 K using the Mössbauer effect method. The high-spin ferric iron isomorphously substitutes for the aluminium. Four different types of the Fe-site surroundings are indicated by the Mössbauer absorption spectra. Their relative populations are both, composition and temperature dependent.

ϑ-$^{57}Fe_{0.01}Al_{x-0,01}V_2O_5$ ($0,33 \leqq x \leqq 0,58$) Vanadiumoxidebronze wird durch Festkörperreaktionen synthetisiert und mit Mößbauerspektroskopie untersucht. High-spin-Eisen ersetzt isomorph Aluminium. Vier verschiedene Gitterumgebungen des Eisens können durch die Absorptions-Mößbauerspektren identifiziert werden. Ihre relativen Anteile am Mößbauerspektrum hängen sowohl von der Substanzzusammensetzung als auch von der Temperatur ab.

1. Introduction

As found by Galy et al. [1] the monoclinic vanadium oxide bronzes formed with iron (ϑ-$Fe_xV_2O_5$) are isostructural with those formed with aluminium (ϑ-$Al_xV_2O_5$). Their structure consists of the distorted octahedra VO_6 which by sharing edges form two kinds of single and double ribbons in the *ab* plane. The two kinds of ribbons, running along the *b*-axis and sharing their corners, are alternatively placed along the *c*-axis. As a result of this the so-called "tunnels" (Fig. 1), parallel to the *b*-axis, are formed in the contact area between the two kinds of ribbons. The tunnels are constituted of a new set of oxygen octahedra whose centres may be occupied by the donor (Fe^{3+}, Al^{3+}) cations. There are 2/3 holes per formula unit accessible for the donor cations, but only a fraction of them ($3x/2$) is occupied.

As reported by Galy et al. [1] the lower limit of the homogeneity range is $x = 0.33$ for both ϑ-$Fe_xV_2O_5$ and ϑ-$Al_xV_2O_5$. Indeed, in order to stabilize the crystal structure at least 50% of the accesible donor sites must be occupied. Due to different atomic radii of Fe and Al the upper limit of the homogeneity range [1] for ϑ-$Fe_xV_2O_5$ ($x = 0.38$) is smaller than the one for ϑ-$Al_xV_2O_5$ ($x = 0.58$). Both ϑ-$Fe_xV_2O_5$ and ϑ-$Al_xV_2O_5$ are known as Curie-Weiss paramagnets and hopping semiconductors. The magnetic susceptibility data [2, 3] have shown conclusively that each donor atom transfers its three valence electrons to three V^{5+} cations thereby creating three $V^{4+}(3d^1)$ paramagnetic cations. The results of electric conductivity measurements support the hypothesis [4] according to which the charge transport in ϑ-type vanadium oxide bronzes occurs via a thermally activated electron hopping process between adjacent vanadium sites: $V^{4+}(V^{5+}) \overset{e^-}{\leftrightarrow} V^{5+}(V^{4+})$.

The purpose of this paper is to report the results of investigations performed at 78 and 295 K for the ϑ-$^{57}Fe_{0.01}Al_{x-0.01}V_2O_5$ series ($0.33 \leqq x \leqq 0.58$) by the Mössbauer effect method. The results of our earlier studies carried out on the ϑ-$Fe_xV_2O_5$ and ϑ-$(Fe, Al)_{0.33}V_2O_5$ systems have been published elsewhere [5].

[1]) Reymonta 4, PL-30-059, Cracow, Poland.

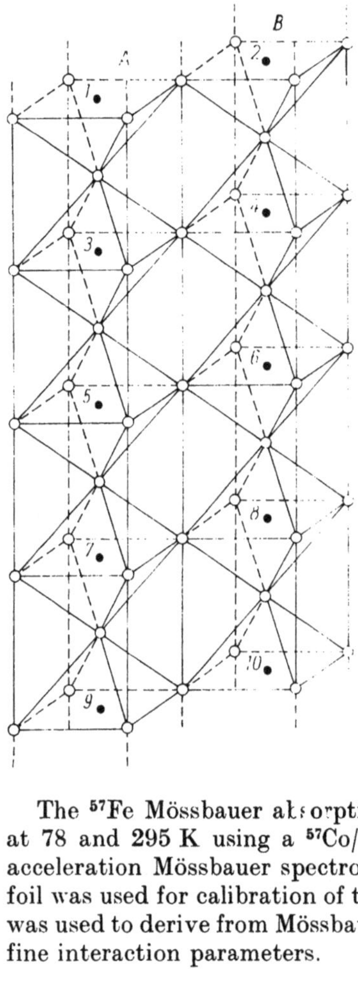

Fig. 1. The tunnels formed by oxygen octahedra in the crystal structure of the ϑ-phase vanadium oxide bronzes. ○ oxygen atoms, ● holes accessible for ^{57}Fe or Al^{3+} cations

2. Experimental

The compounds of the stoichiometric formula ϑ-$^{57}Fe_{0.01}Al_{x-0.01}V_2O_5$ (x=0.33, 0.35, 0.40, 0.45, 0.50, 0.55, and 0.58) were produced by mixing ϑ-$Al_xV_2O_5$ with ϑ-$^{57}Fe_{0.33}V_2O_5$ in appropriate proportions, prolonged grinding in an agate mortar, and then heating in vacuum at 893 K for 100 h. The product was regrinded and again heated under the same conditions. It was proved by controlling experiments that further repeating of the same procedure did not cause any measurable differences in ^{57}Fe Mössbauer absorption spectra. The described method of synthesis yielded fine polycrystalline preparations gradually changing in colour from dark-blue (small x value) to lead-grey (large x value). Microscopic observations revealed no inhomogeneities of the samples. The X-ray diffraction analysis showed no other phase but the ϑ-type phase.

The ^{57}Fe Mössbauer absorption spectra were recorded for powdered samples both at 78 and 295 K using a ^{57}Co/Cr source kept at room temperature and a constant acceleration Mössbauer spectrometer of the Polon type. A high purity metallic iron foil was used for calibration of the velocity scale. The least squares computer program was used to derive from Mössbauer absorption spectra the discrete values of the hyperfine interaction parameters.

3. Results and Discussion

All samples studied were checked for homogeneity by the X-ray powder method. The results of the X-ray analysis performed for the four ϑ-$Al_xV_2O_5$ bronzes are in good agreement with those reported by Galy et al. [1] for this system. It is worth to note that the lattice parameters a, b, c, and β of the investigated samples do not change by more than 1 pm and 0.5°, respectively, over the whole homogeneity range.

The Fe^{3+} cations isomorphously substitute for the Al^{3+} cations in the ϑ-$Al_xV_2O_5$ oxide bronzes. Therefore, the ^{57}Fe isotope was used in our Mössbauer effect investigations. It served as a microscopic probe sensitive to its nearest surroundings. Some Mössbauer absorption spectra recorded at 78 and 295 K for the ^{57}Fe-doped ϑ-$Al_xV_2O_5$ system are shown in Fig. 2. Each spectrum was resolved into four symmetric quadrupole doublets of equal linewidth. Unfortunately, we have not succeeded in assigning them to four non-equivalent types of ^{57}Fe-site surroundings. In the ϑ-$Al_xV_2O_5$ system not all crystal sites accessible for donor cations are occupied. For $x = 0.33$ as much as 50% of the holes remain unoccupied while for $x = 0.58$ this number is reduced to

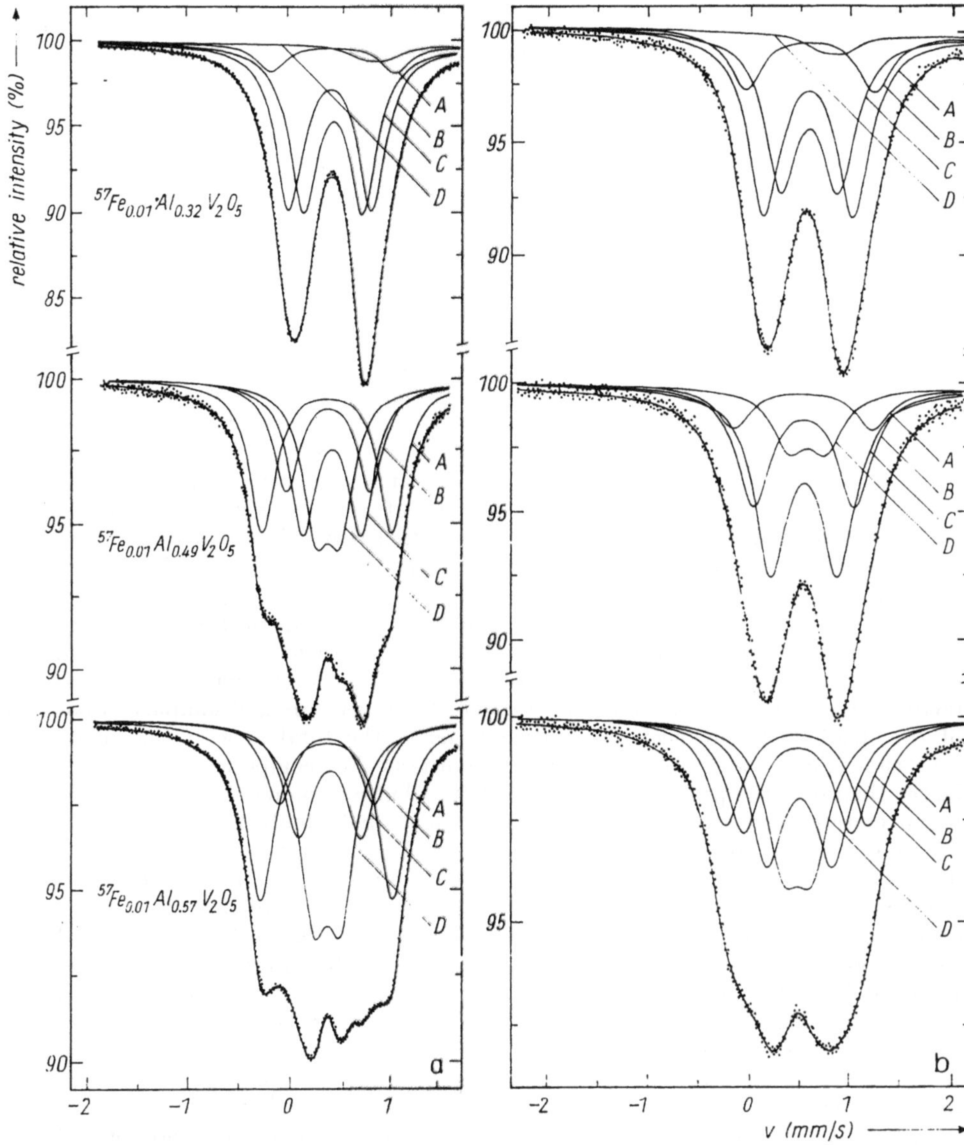

Fig. 2. The typical ^{57}Fe Mössbauer absorption spectra recorded a) at room temperature (295 K) and b) at liquid nitrogen temperature (78 K) for selected samples of ^{57}Fe-doped ϑ-$Al_xV_2O_5$. The spectra were resolved into four symmetric quadrupole doublets

13%. In the simplest case ($x = 0.33$) two different arrangements of donor cations in the oxygen tunnels (Fig. 1), one so-called "zigzag" (1, 4, 5, 8, 9, . . .) and the other "ladder" (1, 2, 5, 6, 9, 10, . . .), are in principle possible. Each arrangement leads to its own unique type of position for all donor cations inside the tunnels. However, the deconvolution of the Mössbauer absorption spectrum into four quadrupole doublets

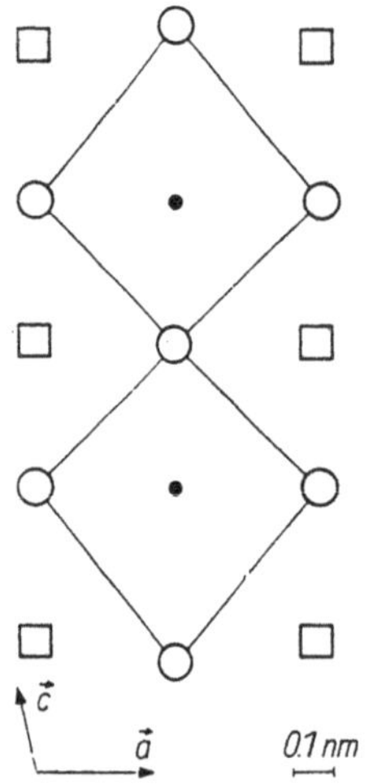

Fig. 3. Nearest neighbours to the lattice holes on the *ac* plane. ○ oxygen, □ vanadium, ● holes accessible for Fe^{3+} and Al^{3+} cations

suggests the existence in the $x = 0.33$ sample of four different surroundings associated with donor sites. Thus, the distribution of the donor cations amoung accessible crystal sites is still an open question, even for the $x = 0.33$ composition. The standard X-ray diffraction technique cannot solve this problem univocally. Most probable the V^{4+} and V^{5+} cations constitute four different surroundings of a donor site. Each pair of donor sites (1–2, 3–4, 5–6, ...; Fig. 1) is surrounded in the *ac* plane by six vanadium cations (Fig. 3) which are located in the centres of the oxygen octahedron-sharing edges with the tunnel edges. Due to the electrostatic balance the trivalent donor cations (Al^{3+}, Fe^{3+}) are surrounded preferentially by the V^{4+} cations while the V^{5+} cations are concentrated around the empty donor holes. The relative populations of various vanadium surroundings of a donor site should be both temperature and composition (x) dependent, since the thermally activated charge transport goes via adjacent V^{4+} and V^{5+} cations and the concentration of the V^{5+} cations is strongly reduced with the increase in donor concentration. Variations with temperature and sample composition of the quadrupole subspectrum contributions to the total area of the Mössbauer spectrum (Fig. 5) seem to support this hypothesis. For compositions with x values larger than 0.33 the problem of the donor distribution among the accessible sites is much more complicated. However, no new quadrupole doublets are needed for the deconvolution of Mössbauer absorption spectra. The donor cations locally distort the oxygen tetrahedra what may also have some influence on the Mössbauer spectra.

It should be pointed out that not any magnetic ordering of the ^{57}Fe cations was observed neither at room nor at liquid nitrogen temperatures. This is in accordance with previous magnetic susceptibility [2, 3, 6] and Mössbauer effect [4] investigations. In the ϑ-$Al_xV_2O_5$ vanadium oxide bronzes only one kind of the paramagnetic cation (V^{4+}) is present, while in the ϑ-$Fe_xV_2O_5$ and ϑ-$(Fe, Al)_xV_2O_5$ systems both V^{4+} and Fe^{3+} cations possess magnetic moments. The V^{4+} and Fe^{3+} subsystems may be treated as decoupled. A long-range magnetic ordering was not observed in previous investigations neither for ϑ-$Al_xV_2O_5$ nor for ϑ-$Fe_xV_2O_5$. However, in both V^{4+} and Fe^{3+} subsystems antiferromagnetic interactions between magnetic moments were found [3, 6] below approximately 100 K. The low temperature magnetic susceptibility data of both the ϑ-$Al_xV_2O_5$ and ϑ-$(Fe, Al)_{0.33}V_2O_5$ bronzes have been interpreted [6] by a simple model containing V^{4+} dimers with $S = 0$ and paramagnetic V^{4+} cations with $S = 1/2$. The contribution of dimers to the total susceptibility of the ϑ-$Al_xV_2O_5$ bronzes was found to be proportional to x. It varies between about 52% for $x = 0.33$ and about 68% for $x = 0.45$. It cannot be excluded that formation of the V^{4+} dimers has some influence on the low temperature Mössbauer absorption spectra.

The hyperfine interaction parameters derived from the Mössbauer absorption spectra are shown in Fig. 4 and 5. The isomer shift and quadrupole splitting diagrams (Fig. 4) indicate the ^{57}Fe cations in the Fe^{3+} high spin state. This supports the hypothesis according to which the valence electrons of the donor atoms (Fe, Al) are trans-

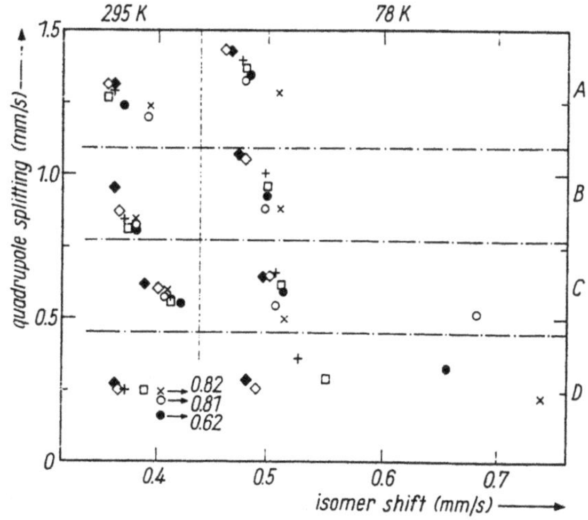

Fig. 4. The correlation between the room temperature isomer shift and quadrupole splitting values of investigated vanadium oxide bronzes. The liquid-nitrogen temperature data are inserted for comparison. The number written near the arrow indicates the value of isomer shift to which the point should be shifted. × $x = 0.33$, ○ 0.35, ● 0.40, □ 0.45, + 0.50, ◇ 0.55, ◆ 0.58

ferred to vanadium atoms in the ϑ-type vanadium oxide bronzes. As can be seen in Fig. 4 there is a correlation between the values of the isomer shift and quadrupole splitting. The quadrupole doublets with larger isomer shifts have smaller quadrupole splittings. The values of quadrupole splitting derived from spectra recorded at 78 K are slightly larger than those derived from spectra recorded for the same samples at 295 K. This indicates the lattice origin of the electric field gradients and is consistent with the ferric high spin state of the iron ions.

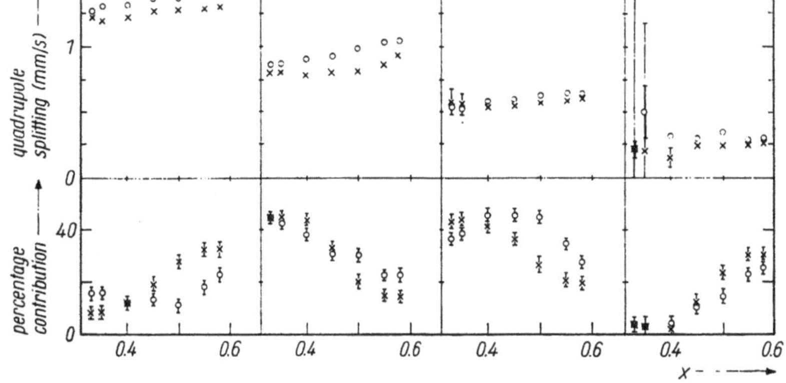

Fig. 5. The isomer shift, the quadrupole splitting, and the relative intensity of the quadrupole subspectrum (A, B, C, and D) as functions of sample composition (x). Experimental errors larger than the size of the symbols used are marked. ○ 78, × 295 K

The isomer shift of each quadrupole doublet slowly decreases (Fig. 5) with increase in the aluminium content in the ^{57}Fe-doped ϑ-$Al_xV_2O_5$ bronzes. This behaviour indicates the small increase in the S-electron density inside the ^{57}Fe nuclei caused by the increase in x. The opposite dependence is observed for the quadrupole splittings of the A, B, and C doublets. The quadrupole splitting of each doublet slowly decreases when the aluminium content increases. No conclusive information can be derived for $x = 0.33$, 0.35, and 0.40 samples from the values of the isomer shift and quadrupole splitting of doublet D due to its small contribution to the total Mössbauer absorption.

The percentage contribution of each quadrupole doublet to the total nuclear absorption of γ-rays is both concentration (x) and temperature dependent (Fig. 5). The intensities of doublets A and D increase with the increase both in temperature and in aluminium concentration. The opposite dependences are observed for the other doublets (B and C). These variations are pronounced for the samples with $x > 0.40$. It is worth pointing out that the sum of the relative intensities of doublets A and C and similarly of doublets B and D is, in first approximation, neither temperature nor composition dependent. The appropriate values determined at 78 K (295 K) are 55% (56%) for doublets A + C and consequently 45% (44%) for doublets B + D. This observation suggests that some kind of exchange process, which is induced by temperature or by sample composition, takes place in the investigated samples. In each pair the population of one type of donor site is increased with respect to the other one. The observed dependences may be attributed to some modifications in the distribution of the V^{4+} and V^{5+} cations, or to a redistribution of aluminium cations among accessible crystal sites. The X-ray diffraction patterns have indicated some changes in the intensities of their lines when the temperature was changed from 78 to 295 K. Further investigations carried out at higher temperatures are in progress.

Acknowledgements

The authors would like to express their gratitude to Dr. A. Polaczek and Dr. M. Pekała for preparation of the samples, the X-ray analysis and helpful discussions. The financial support of the 01.09 Central Research Programme is acknowledged.

References

[1] J. Galy, A. Casalot, J. Darriet, and P. Hagenmuller, Bull. Soc. Cim. Franc. **1**, 227 (1967).
J. Galy, M. Pouchard, A. Casalot, and P. Hagenmuller, Bull. Soc. Franc. Miner. Crist. **90**, 544 (1967).
P. Hagenmuller, J. Galy, M. Pouchard, and A. Casalot, Mater. Res. Bull. **1**, 95 (1966).
[2] M. Pekała and A. Polaczek, Polon. J. Chem. **52**, 353 (1978).
[3] M. Pekała and A. Polaczek, phys. stat. sol. (a) **58**, 533 (1980).
[4] S. Krupicka, Fizika ferritov, Vol. 2, Izd. Mir, Moscow 1976 (p. 422).
[5] J. J. Bara, B. F. Bogacz, M. Pekała, and A. Polaczek, J. Solid State Chem. **58**, 143 (1985).
[6] D. Hüsert, J. A. Mydosh, M. Pekała, and A. Polaczek, J. Phys. C **18**, 3797 (1985).

(Received July 15, 1988)

phys. stat. sol. (a) **111**, 59 (1989)

Subject classification: 68.48; 61.16; 61.80; S7.12

Sektion Physik der Friedrich-Schiller-Universität Jena[1]) *(a) and Institut für Festkörperphysik und Elektronenmikroskopie der Akademie der Wissenschaften der DDR, Halle (Saale)*[2]) *(b)*

Investigation of Low-Temperature Epitaxial Regrowth of Ion-Implanted Amorphous GaAs

By

J. HEROLD (a), H. BARTSCH (b), W. WESCH (a), and G. GÖTZ (a)

A detailed description of optical reflectivity technique for monitoring laser-induced solid phase epitaxial regrowth in real time is given. An example illustrates the use of this technique for the investigation of interface structures. By cooling the sample the epitaxial regrowth of the crystalline/amorphous interface is stopped before the interface reaches the surface. These stages are additionally investigated by TEM- and RBS-studies which illustrate the dependence of the crystalline quality on the regrowth process.

Es wird eine genaue Beschreibung der optischen Echtzeitreflexionstechnik zur Aufzeichnung von laserinduzierter Festphasenepitaxie gegeben. Die Anwendung dieser Methode zur Untersuchung der Struktur der Phasengrenze wird an einem Beispiel gezeigt. Bevor die kristallin/amorphe Phasengrenze die Oberfläche erreicht, wird die Epitaxie durch Kühlung gestoppt. Diese Zustände werden zusätzlich durch TEM- und RBS-Messungen untersucht. Sie zeigen die Kristallqualität in Abhängigkeit vom Wachstumsprozeß.

1. Introduction

In the past numerous studies have been carried out concerning the regrowth of ion-implanted amorphous GaAs layers in the temperature interval up to 700 K [1 to 6], but only a few had been directed to the investigation of the dynamics of the regrowth process. A useful method for doing this is the direct measurement using time-resolved reflectivity [7]. This technique has been extensively utilized for studying the kinetics of the solid-phase epitaxy in Si [8, 9]. Results obtained for silicon have shown that solid phase recrystallization of ion-implanted layers is a simple epitaxial process resulting in a complete crystal recovery and a complete activation of the dopants [8].

The annealing of damage in ion implanted GaAs, on the other hand, is a much more complex process consisting of three distinct annealing stages [10]. A sharp amorphous-to-crystalline transition corresponding to an epitaxial regrowth occurs between 450 and 700 K. A second stage in the temperature interval up to 1100 K is characterized by the removal of extended defects remaining as a result of the low-temperature recrystallization. The third annealing stage, the electrical activation of dopants, occurs at temperatures higher than about 1200 K.

In the following a detailed description of the use of time-resolved reflectivity measurements for the investigation of the low-temperature regrowth of ion-implanted GaAs layers is given and the optical constants necessary for the analysis of the experimental results are presented. The results of time-resolved reflectivity measurement

[1]) Max-Wien-Platz 1, DDR-6900 Jena, GDR.

[2]) Weinberg 2, DDR-4050 Halle (Saale), GDR.

are compared with those obtained with Rutherford backscattering (RBS) and transmission electron microscopy (TEM) investigations.

2. Method

Chromium — doped semi-insulating ⟨100⟩ — GaAs substrates were implanted with 5×10^{14} P^+ ions cm^{-2} (energy $E = 85$ keV) at the temperature $T_i = 80$ K. Under these implantation conditions an amorphous layer reaching to the sample surface is created. The recrystallization process was induced by heating the sample with a cw argon laser.

The time-resolved reflectivity method is based on the differences between the complex reflective indices $\tilde{n} = n - ik$ (n refractive index, k absorption index) of amorphous ($\tilde{n}_a$) and of virgin material ($\tilde{n}_c$) as well as on the variation of the reflectivity as a consequence of the movement of the interface during the recrystallization process [11]. For carrying out theoretical simulations of the dependence of the reflectivity on the thickness of the amorphous layer the knowledge of $\tilde{n}_a$ and $\tilde{n}_c$ and of the dependence of the refractive index of the recrystallized material ($\tilde{n}_{c'}$) on the annealing temperature T_A is necessary. This dependence of $\tilde{n}_{c'}$ on T_A is a consequence of the variation of the defect concentration in the recrystallized layers with T_A.

With the help of ellipsometric measurements the complex refractive index of chromium-doped virgin GaAs at the wavelength $\lambda = 0.6328$ µm was determined to $\tilde{n}_c = 3.79 - 0.22i$, a value close to those approximated from [12, 13]. Furthermore, the unknown complex refractive indices of amorphous and recrystallized GaAs $\tilde{n}_a$ and $\tilde{n}_{c'}$ were measured for P^+-implanted GaAs using angular dependent reflectivity measurements. For amorphous GaAs $\tilde{n}_a = 4.24 - 1.09i$ at $\lambda = 0.6328$ µm was achieved. In all cases the surface oxide layers were removed by chemical etching immediately before the measurements. By means of reflectivity measurements it was checked that at room temperature oxide layers influencing the reflectivity are not observed during the time of measurement. The complex refractive indices $\tilde{n}_{c'} = n_{c'} - ik_{c'}$ of amorphous GaAs recrystallized in the temperature interval $T_A = 470$ to 670 K are represented in Fig. 1 as a function of the recrystallization temperature T_A (values measured at $T = 300$ K). As can be seen, the deviation of $n_{c'}$ and $k_{c'}$ from the values for crystalline GaAs decreases with increasing T_A.

In order to determine the recrystallization temperature directly, the reflectivity of recrystallized GaAs as a function of sample temperature must be known. According

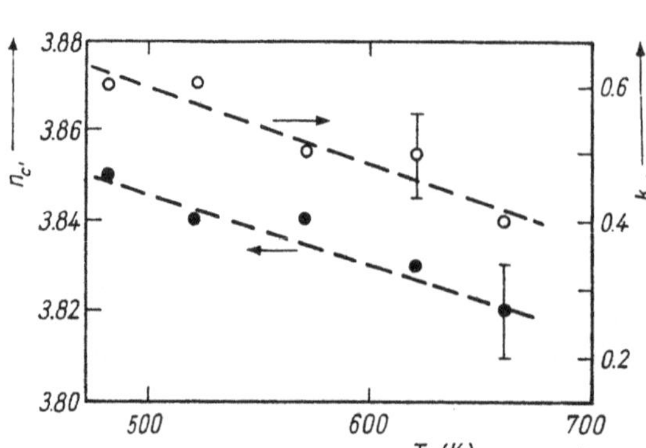

Fig. 1. The components $n_{c'}$ and $k_{c'}$ of the complex refractive index of recrystallized GaAs as a function of the recrystallization temperature T_A obtained for P^+-implanted material (85 keV $P^+ \rightarrow$ ⟨100⟩ GaAs, 5×10^{14} cm^{-2}, $T_i = 80$ K)

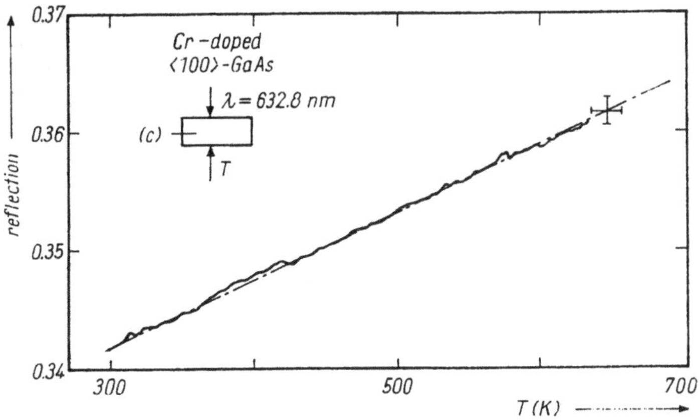

Fig. 2. Reflection of crystalline chromium-doped GaAs as a function of the sample temperature T at the wavelength $\lambda = 0.6328\ \mu m$. The dash-dotted curve represents the approximation to the experimental results

to [9] the recrystallization temperature can be calculated from the decrease of the reflection when the heating laser is switched off and the sample temperature decreases. As a supposition, the reflectivity must be measured as a function of the annealing temperature. In a first step, a crystalline sample was mounted on a heating target. The temperature was measured by a thermocouple and the reflection was registered using a time-resolved reflectivity measurement apparatus. The results are illustrated in Fig. 2. By approximating a linear dependence of the reflection R_c of crystalline GaAs on the sample temperature in the temperature interval between 300 and 700 K, $\Delta R_c/\Delta T = (0.0055 \pm 0.0002)$ %/K is obtained. In a second step, amorphous GaAs layers produced by different implantation conditions, and crystalline GaAs were laser irradiated and the change of the reflection was measured after switching off the heating laser at different laser powers, i.e. for different heating temperatures. Fig. 3 shows the change of the reflectivity $\Delta R_{c'}$ for recrystallized GaAs as a function of the

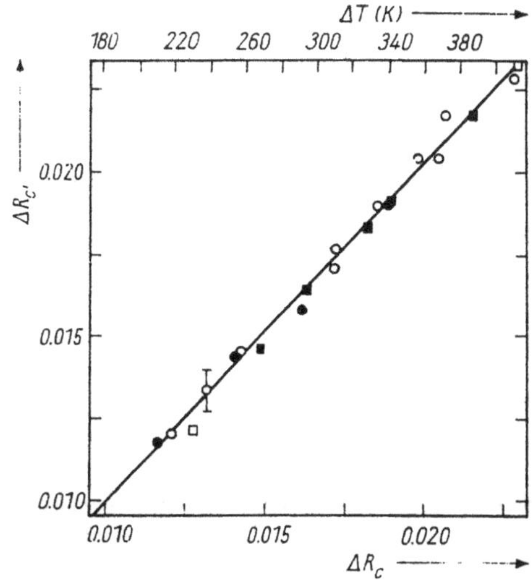

Fig. 3. Comparison of the reflection rise of recrystallized GaAs $\Delta R_{c'}$ with the reflection rise of crystalline GaAs ΔR_c caused by a temperature rise ΔT (upper scale). ○ 90 keV, Ar^+ (1×10^{15} cm^{-2}); □ 100 keV, N^+ (1×10^{15} cm^{-2}); ● 300 keV, Se^+ (1×10^{15} cm^{-2}); ■ 85 keV, P^+ (5×10^{14} cm^{-2}). Open symbols $T_i = 80$ K; full symbols $T_i = 300$ K

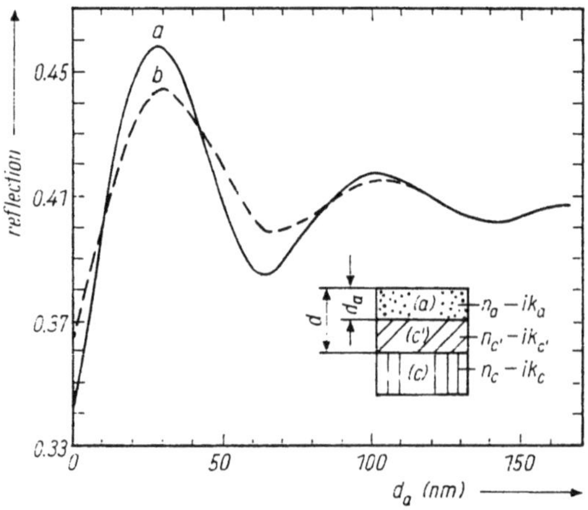

Fig. 4. Calculated reflectivity as a function of the thickness of the amorphous layer ($\lambda = 632.8$ nm): (a) the complex refractive index of recrystallized GaAs $\tilde{n}_{c'}$ is equal to the value of the crystal (perfect recrystallization), (b) the complex refractive index of recrystallized GaAs $\tilde{n}_{c'}$ is different to the value of the crystal ($n_{c'} = 3.85$, $k_{c'} = 0.6$, $d = 113$ nm). The value $\tilde{n}_{c'}$ with the greatest deviation from the crystalline one $\tilde{n}_c$ (see Fig. 1 at $T_A = 490$ K) is used (most mperfect recrystallization)

change of the reflectivity ΔR_c for crystalline material. With the measured values of ΔR_c and the value $\Delta R_c/\Delta T$ for crystalline GaAs the change of the temperature ΔT was calculated, which is shown in the upper scale of the diagram. The results show that in the temperature region 500 to 700 K a certain temperature rise ΔT causes identical variations of the reflectivity in crystalline and recrystallized GaAs, i.e. $\Delta R_{c'}/\Delta T = = \Delta R_c/\Delta T$.

Fig. 4 shows reflectivity curves for the SPE process as a function of the thickness of the amorphous layer calculated according to [8, 9] with the use of the experimentally determined complex refractive indices. The calculations were carried out for a smooth interface and with the optical data for room temperature. For curve a) a complex refractive index of the recrystallized GaAs equal to that of virgin GaAs was assumed (perfect recrystallization). Curve b) is obtained assuming a complex refractive index $\tilde{n}_{c'}$ of the recrystallized material different to that of the virgin GaAs, which corresponds to the experimental results shown in Fig. 1. In the example in Fig. 4 the value $\tilde{n}_{c'}$ achieved for the lowest recrystallization temperature $T_A = 490$ K for P^+-implant-

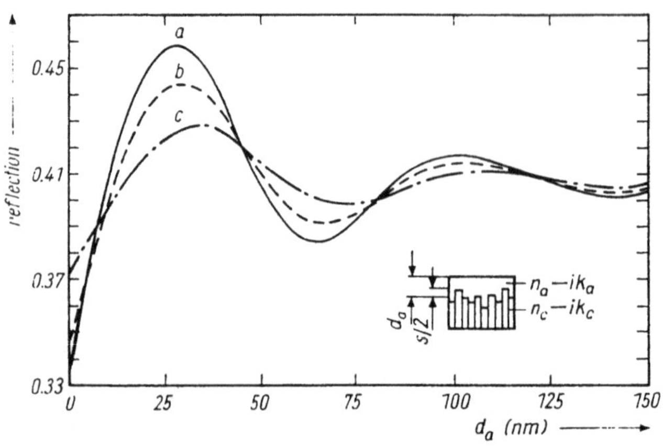

Fig. 5. Reflection as a function of the thickness of the amorphous layer for a system consisting of an amorphous layer with a rough interface on a crystalline substrate. The calculations correspond to different values of the roughness (see inset). (a) $s = 0$, (b) 30, (c) 50 nm

ed GaAs (most imperfect recrystallization) was used. It can be seen that the oscillation amplitude is smaller in the case of imperfect recrystallization. In the framework of the parameters discussed in this paper the two curves represent the limiting cases of perfect and most imperfect recrystallization.

In the preceding discussion it was assumed that SPE is characterized by a uniform movement of a planar interface. However, in GaAs it was found that the interface is nonplanar [2, 7, 14]. Such a nonplanarity will cause deviations of the reflectivity pattern compared to that for the ideal case. Calculated reflectivity patterns for a nonplanar interface are illustrated in Fig. 5. For the calculation a simple model of the nonplanar interface is used assuming that the interface consists of steps with step heights randomly distributed around the thickness of the amorphous layer d_a in the limits of $(d_a \pm s/2)$, where s characterizes the extent of the nonplanarity (see inset in Fig. 5). A supposition for the application of this model is that the width of the steps is large compared to the thickness d_a of the amorphous layer and to the wavelength λ of the laser light. By means of TEM cross section measurements it was shown that this is fulfilled in the cases investigated here. Fig. 5 clearly illustrates that with increasing step height the amplitude of the reflection curves decreases, and the position of the interference extrema shifts towards higher values of the thickness d_a of the amorphous layer.

Using the optical data presented in the following the time-resolved reflectivity measurement is applied to the characterization of the recrystallization process as well as to the investigation of the modification of the planarity of the interface.

3. Results

Fig. 6 shows the yield of backscattered ions for P^+-implanted GaAs after implantation and after annealing at 505 K for different recrystallization times. An annealing time of 3.5 s leads to a shift of the interface crystalline–amorphous (left edge of the damage peak) from a depth of (147 ± 4) nm to (113 ± 4) nm indicating an SPE process with a regrowth rate of 8.3 nm/s. Further annealing at 2.5 s mainly leads to a reduction of the height of the damage peak near the interface. After a 12 s-annealing the yield

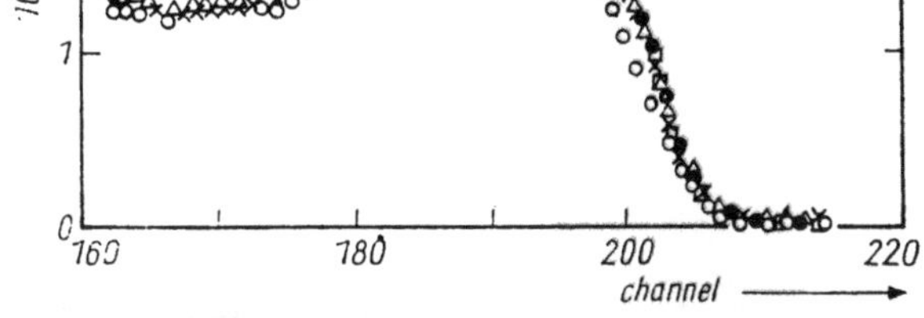

Fig. 6. Channeling spectra from samples implanted with 85 keV P^+ to a dose of 5×10^{14} cm^{-2} after implantation (●) and after the recrystallization time of t_A 3.5 (×), 6 (△), and 12 s (○) at a temperature $T_A = 505$ K. □ random. $T_i = 80$ K

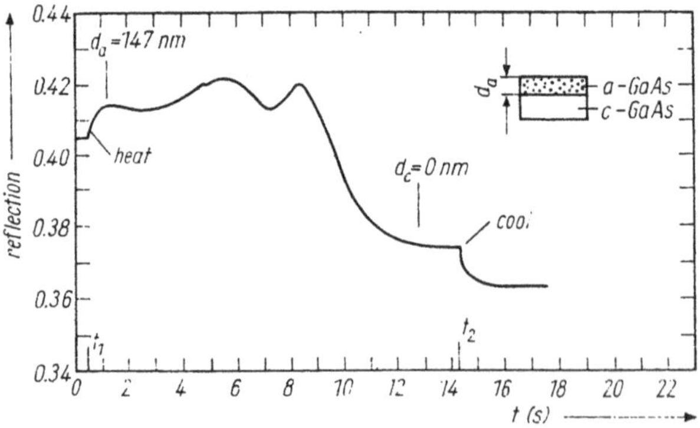

Fig. 7. Reflection as a function of the time observed during the recrystallization of P$^+$-implanted GaAs (5×10^{14} cm^{-2}) at a recrystallization temperature of 505 K. T_i = 80 K

of backscattered ions does not reach the random level indicating that the amorphous layer is recrystallized. It is evident that a good regrowth only happened in a layer of about 34 nm of the initial amorphous material leading to a remarkable decrease of the yield of backscattered ions. In the layer remained more imperfect regrowth occurred indicated by a relatively high backscattering yield.

In Fig. 7 the corresponding variation of the reflection is depicted as a function of the time. At the beginning of heating the sample by the laser (time t_1) an increase of

a

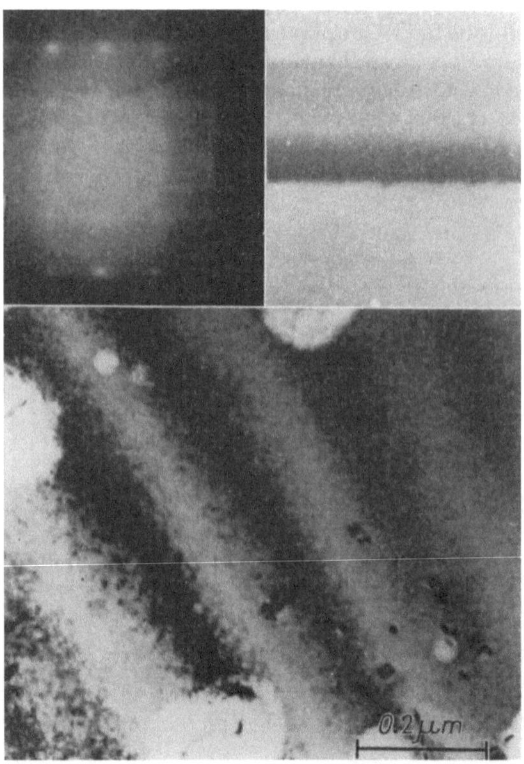

c

b

Fig. 8. a) TED pattern, b) TEM bright field micrograph, and c) dark field cross-section view of P$^+$-implanted GaAs at T_i = 80 K

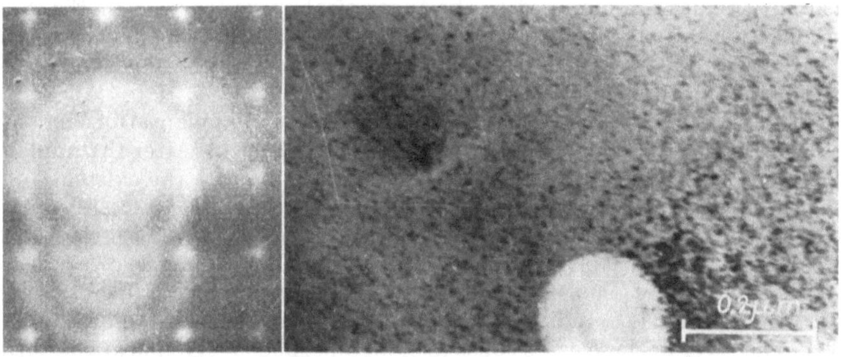

Fig. 9. a) TED pattern and b) TEM bright field micrograph of P^+-implanted GaAs and recrystallized at 505 K for 3.5 s

the reflection is observed, followed by oscillations with the time, which are typical for SPE processes. At the time t_2 the sample is cooled down from the recrystallization temperature to the ground temperature, and the reflection decreases. This decrease is used for the calculation of the process temperature. The number of the extrema of the reflection curve and the value of the reflection of the initial amorphous layer at room temperature compared with the theoretical calculation shown in Fig. 4 (curve a) indicate that the growth has indeed started at a depth of about 147 nm, which was

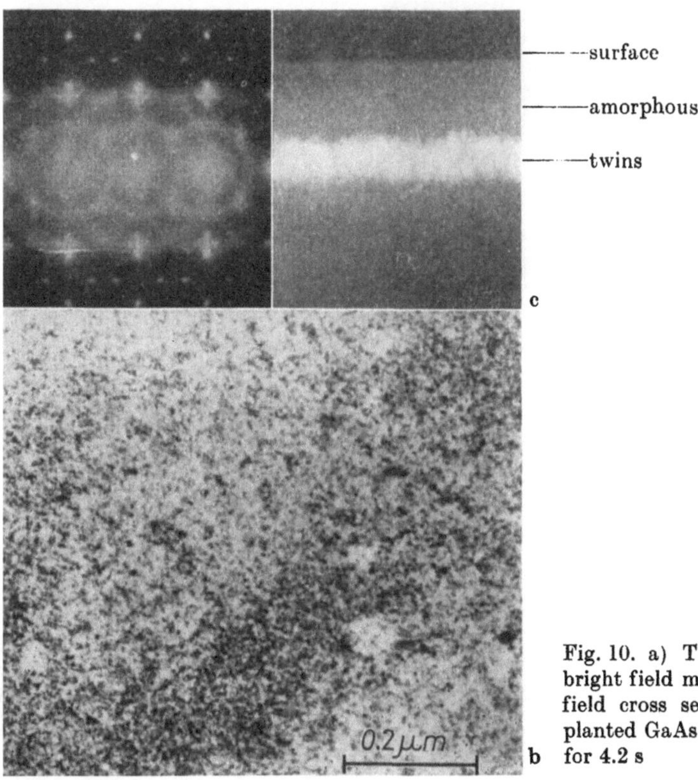

Fig. 10. a) TED pattern, b) TEM bright field micrograph, and c) dark field cross section view of P^+-implanted GaAs and annealed at 505 K for 4.2 s

also measured with RBS. Compared to the theoretical curve (curve a in Fig. 4) the measured reflection exhibits a markedly smaller amplitude indicating deviations from the ideal case of recrystallization with a planar interface. Therefore, an imperfect regrowth can be expected.

The results of TEM investigations carried out after the implantation and after annealing at different annealing times are shown in Fig. 8 to 12. After the implantation the diffraction fringes in the TED pattern (Fig. 8a) indicate the existence of an amorphous layer reaching to a depth of about (140 ± 5) nm (TEM cross section view in Fig. 8c). It is evident that the interface crystalline–amorphous is rough, and the extent of the rough region is in the order of about 10 nm. With increasing annealing time the thickness of the amorphous layer decreases (compare TEM cross-section pictures in Fig. 10c and 11c) indicating a proceeding partial annealing of the amorphous layer. The layer in these cases consists of an amorphous region near the surface and the underlying recrystallized layer. For a recrystallization time of 4.2 s the thickness varies between 90 and 120 nm (Fig. 10c), for a time of 6 s between 52 and 95 nm (Fig. 11c). The increase of the nonplanarity during the recrystallization is clearly illustrated. A complete recrystallization of the amorphous layer is obtained for the annealing time of 12 s (Fig. 12c).

The satellite spots in the TED pattern (Fig. 9a and 10a) loo king like crosses indicate, that the recrystallized layers mainly contain microtwins. Such twins obviously grow already during the beginning of the regrowth (Fig. 9a). In no case an indication for the formation of polycrystalline regions was found. From the TEM

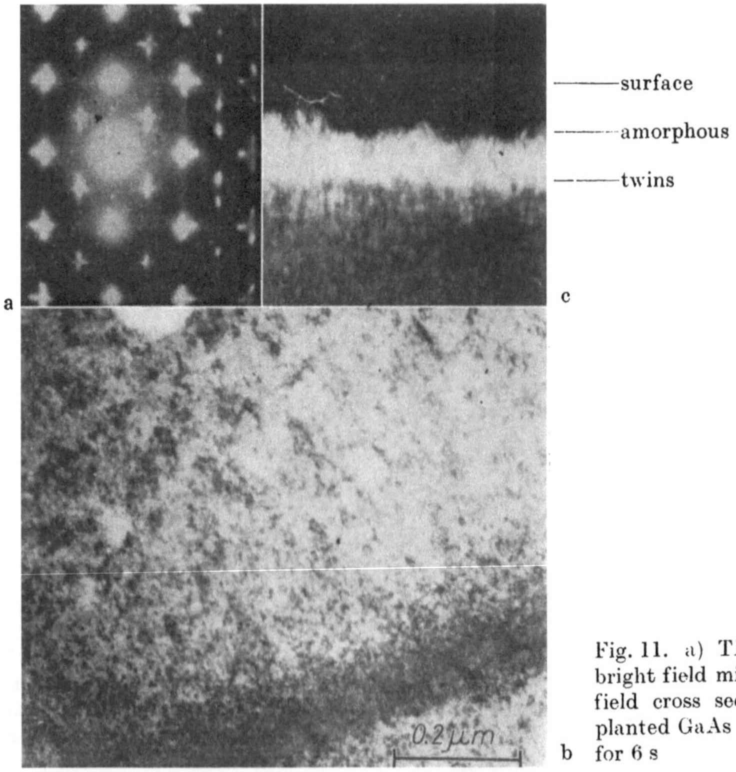

Fig. 11. a) TED pattern, b) TEM bright field micrograph, and c) dark field cross section view of P^+-implanted GaAs and annealed at 505 K for 6 s

bright-field micrographs (Fig. 9b to 12b) it is obvious that the density of the defects (black spots in the pictures) decreases, whereas their size increases during the regrowth process.

4. Discussion

By means of time-resolved reflectivity measurements it was shown that the movement of the crystalline–amorphous interface during low-temperature recrystallization of amorphous GaAs layers is connected with an increasing roughness of the interface. To estimate the evolution of the nonplanarity of the epitaxially growing interface, a three-layer approximation (succession of amorphous, recrystallized and virgin layers) as illustrated in Fig. 4 is used. The roughness of the interface between amorphous and recrystallized layer is taken into account by assuming randomly distributed columns with different heights to exist in the recrystallized layer (see inset in Fig. 5). The complex refractive index of the recrystallized layer is taken from Fig. 1 (recrystallization temperature 505 K). The temperature dependences of the reflectivity of recrystallized and amorphous GaAs are assumed to be equal. Fig. 13 shows the reflectivity as a function of the thickness d_a of the amorphous layer for various values of the maximum extent $s/2$ of the interface. For comparison experimental values of the reflectivity determined from the interference extrema in the dependences of the reflectivity on the heating time (see Fig. 7) are represented in the figure. The results clearly indicate the increase of the roughness of the interface with proceeding position of the interface (increasing heating time).

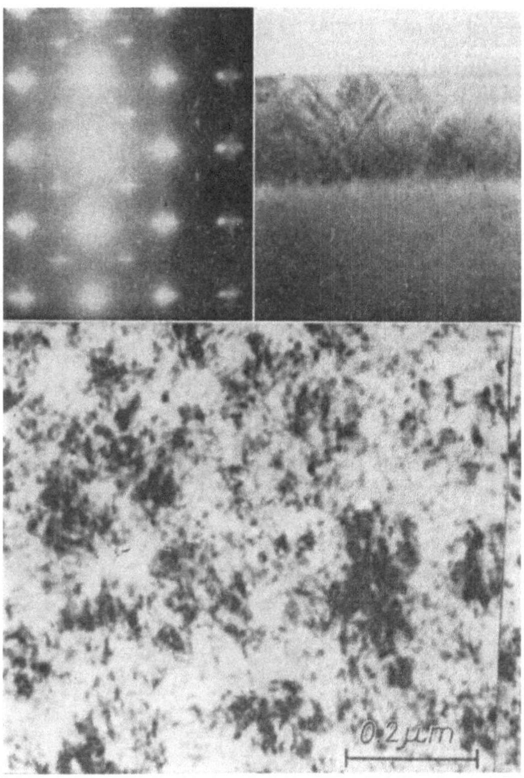

Fig. 12. a) TED pattern, b) TEM bright field micrograph, and c) cross section bright field view of P^+-implanted GaAs and annealed at 505 K for 12 s

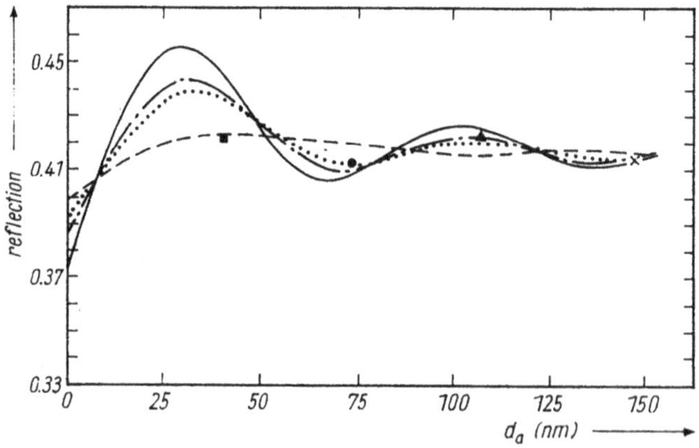

Fig. 13. Calculated curves of reflection as a function of the depth d_a at 505 K. These curves are compared with the experimental values of reflection at the different extrema of the SPE oscillation shown in Fig. 7. Theory: ——— $s/2 = 0$, ·—·—· 18, ·········· 24, – – – 37 nm; experiment: × $t_A = 0$, ▲ 4.2, ● 6.0, ■ 7.3 s. $T_i = 80$ K

The roughness of the interface in some cases can also be determined from the energy spectra of backscattered ions. For the standard measurement geometry (angle of incidence $\varphi_1 = 0°$, angle of emergence $\varphi_2 = 10°$, scattering angle $\theta = 170°$) used in our experiments the energy spread of the interface (left edge of the damage peak) δE is given by [15]

$$(\delta E)^2 = (\delta E_{ir})^2 + (\delta E_d)^2 + (\delta E_c)^2 + (\delta E_s)^2$$

with δE_{ir} the energy spread due to the interface roughness, δE_d the detector energy resolution, δE_c the energy spread caused by the components Ga and As, and δE_s the energy straggling in the target. With the use of the corresponding energy-to-depth conversion the roughness of the interface δz_{ir} can be calculated according to

$$\delta z_{ir} = \delta E_{ir}/(\delta E/\delta z)$$

with $(\delta E/\delta z)$ the energy loss per unit depth. Because the interface in the case discussed has stopped after moving to the surface over a distance of only 34 nm (see Fig. 6) and further increase of the annealing time resulted only in a decrease of the backscattering yield, this method can only be applied for recrystallization times smaller than 4.2 s. The values of the roughness determined by TRR, RBS, and TEM cross section investigations are listed in Table 1.

Table 1

Roughness of the growth interface for various recrystallization times t_A and interface positions d_a in a P^+-implanted GaAs layer recrystallized at 505 K

t_a (s)	d_a (nm)	TRR s (nm)	RBS δz_{ir} (nm)	TEM Δz (nm)
0	147	0 ± 10	10 ± 10	10 ± 5
3.5	118	—	21 ± 10	—
4.2	105	36 ± 3	43 ± 10	30 ± 5
6.0	75	48 ± 3	—	43 ± 5
7.3	40	74 ± 3	—	—

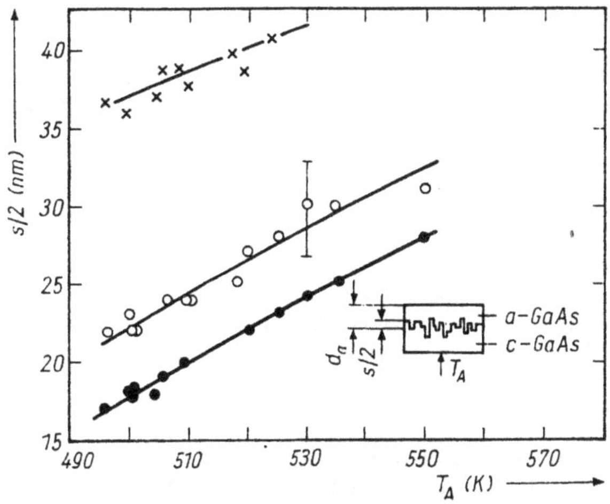

Fig. 14. Extent of nonplanarity $s/2$ as a function of the recrystallization temperature T_A for different interface-positions, calculated like those shown in Fig. 13. × $d_a = 35$, ○ 70, ● 105 nm

Within the error limits the values of the roughness determined with the different methods are in good agreement. However, the absolute variation of the reflection with the roughness decreases with increasing thickness d_a of the amorphous layer, as can be seen in Fig. 13. Therefore, it is concluded that for layer thicknesses greater than about 150 nm TEM and RBS investigations give more accurate information about the roughness than TRR measurements. Fig. 14 shows the values of the nonplanarity $s/2$ as a function of the recrystallization temperature T_A for different depths of the interface. It can be seen that the roughness not only increases with proceeding recrystallization, but also with increasing recrystallization temperature causing a further decrease of the amplitude of the oscillation at recording time-resolved curves of SPE processes.

For temperatures higher than 750 K the oscillations vanish. Concerning the defective growth during recrystallization of GaAs several authors [3, 4, 14, 16] have argued that this is a consequence of local nonstoichiometric disruption caused by the implantation, and incorporated dopant concentrations above the solid solubility at temperatures below 750 K. Starting with these assumptions our experimental results can be explained in the following way. The already initially rough interface proceeds towards the surface. (111)-facets existing in the layer provide favoured sites for the accumulation of dopants with concentrations above the solid solubility and for the generation of local nonstoichiometry [4]. These particles become accumulated, and the fact that two equivalent configurations of atoms can be attached in the growing (111)-interface make the formation of defects possible [4]. Especially the formation of microtwins is imaginable. Their density rises very quickly during the beginning of the recrystallization. Because of the great density in the early stage of recrystallization doubly and multiply twinned structures are formed with different growth directions. From other experiments it is known that the velocity of the interface in the (111)-direction is half of that in the (100)-direction. If we consider an interface structure containing only these two directions, the increase of the nonplanarity can be explained by the differences of the velocities of the two directions. Because the favoured interface directions grow fast enough the size of twins oriented along these directions increases during the regrowth.

References

[1] J. S. WILLIAMS, F. M. ADAMS, and K. ROSSITER, Thin Solid Films **93**, 139 (1982).
[2] R. S. BHATTACHARYA, A. K. RAI, P. P. PRONOKO, J. NARAYAN, S. C. LING, and S. R. WILSON, J. Phys. Chem. Solids **44**, 61 (1983).
[3] J. S. WILLIAMS and M. W. AUSTIN, Appl. Phys. Letters **36**, 994 (1980).
[4] D. K. SADANA, Nuclear Instruments and Methods **B7/8**, 375 (1985).
[5] SOOK-JL KWUN, M. H. LEE, L. L. LION, W. G. SPITZER, H. L. DUNLAP, and K. V. VAIDYANATHAN, J. appl. Phys. **57**, 1022 (1985).
[6] R. S. BHATTACHARYA and P. P. PRONKO, Appl. Phys. Letters **40**, 502 (1982).
[7] C. LICOPPE, Y. I. NISSIM, and C. MERIADEC, J. appl. Phys. **58**, 3094 (1985).
[8] L. CSEPREGI, E. F. KENNEDY, J. W. MAYER, T. W. SIGMON, J. appl. Phys. **49**, 3907 (1978).
[9] G. L. OLSON, S. A. KOKOROWSKI, R. A. MCFARLANE, and L. D. HESS, Appl. Phys. Letters **37**, 1019 (1980).
[10] J. S. WILLIAMS, in: Laser Processing of Semiconductors, Ed. J. M. POATE and J. W. MAYER, Academic Press, New York 1982 (p. 285).
[11] G. L. OLSON, S. A. KOKOROWSKI, J. A. ROTH, R. S. TURLEY, and L. D. HESS, Laser Induced Crystal Growth Measurements by Time-Resolved Optical Reflectivity, MRS Symposium, Boston 1981.
[12] J. S. BLAKEMORE, J. appl. Phys. **53**, 123 (1982).
[13] J. R. MEYER, M. R. KRUER, and F. J. BARTOLI, J. appl. Phys. **51**, 5513 (1980).
[14] C. LICOPPE, Y. I. NISSIM, C. MERIADEC, and P. KRANZ, J. appl. Phys. **60**, 1352 (1986).
[15] G. GÖTZ and K. GÄRTNER (Ed.), High Energy Ion Beam Analysis of Solids, Akademie-Verlag, Berlin, (1988).
[16] L. CRISTEL and J. GIBBONS, J. appl. Phys. **52**, 5050 (1981).

(Received March 24, 1988; in revised form November 2, 1988)

phys. stat. sol. (a) **111**, 71 (1989)

Subject classification: 61.16; 61.70; S1.2; S1.3

Institut für Metallphysik der Universität Göttingen[1])

Calculation of TEM Contrast of Small Coherent Precipitates[2])

By

F. HAIDER

A new simple method to compute the image contrast of small coherent precipitates without strain contrast is proposed. The method is based on a perturbational expansion of the extra scattering due to the precipitate. Two approaches are possible — z-dependent and z-independent perturbation theory, which each have special advantages. Application of both methods to suited alloy systems is demonstrated.

Eine einfache Methode zur Berechnung des elektronenmikroskopischen Bildkontrasts kleiner kohärenter Ausscheidungen ohne Verzerrungskontrast wird vorgeschlagen. Das Verfahren beruht auf einer Störungsrechnung im Zusatzpotential der Ausscheidung. Es sind zwei Ansätze möglich — z-abhängige und z-unabhängige Störungsrechnung, wobei beide Verfahren spezifische Vorteile zeigen. Für beide Verfahren wird die Anwendung auf geeignete Legierungssysteme gezeigt.

1. Introduction

In the case of very small coherent precipitates ($D < 2$ nm), the strain contrast often used in conventional TEM becomes very weak, especially if the lattice misfit between matrix and precipitate is small. Furthermore, the size determination for particles with a strain field is rather incorrect. Hence we tried other contrast mechanisms, that is superstructure contrast and mass contrast to image very small precipitates, which represent the early stages of decomposition reactions by nucleation and growth [1].

To image these small particles it is appropriate to work under high resolution conditions in the TEM, i.e. with a many-beam image, since only this guarantees a spatial resolution down to some tenth of a nanometer. This implies in the case of mass contrast that one has to use high defocus values to transmit the large spatial frequencies which belong to the usual interatomic distances in metals.

For a complete interpretation of the TEM images it is necessary to compare them with computed images, because the observed size and shape of the precipitates is not always real. In particular the visibility of a particle may strongly depend on its position in the foil.

The usual algorithms for the calculation of the outgoing wave function are either rather unwieldy for large defects (e.g. multislice methods and descendents [2 to 4]) and give no insight in the analytical dependence of image contrast on specimen parameters, or they are unrealistically simplifying for high resolution conditions [5]. Therefore, we propose a method which is both simple enough to do all calculations on a small personal computer as well as exact enough to give results comparable to multislice calculations at least for small precipitates.

The method depends on a perturbation series for the potential in the Howie-Whelan equations, where we treat the ideal crystal as unperturbed system and the additional electrostatic potential of the precipitate as a perturbation.

---

[1]) Hospitalstr. 3/5, D-3400 Göttingen, FRG.

[2]) Göttinger Dissertation D7.

Both z-dependent and z-independent formulations are possible, in analogy to time-dependent and time-independent perturbation theory in quantum mechanics, because the Howie-Whelan equations are formally analogous to the Schrödinger equation. Both give similar results, but have certain advantages each.

2. Ideal Crystal

The scattering of high energy electrons by a crystalline specimen is described by the Howie-Whelan equations [6],

$$\frac{\mathrm{d}\psi_g}{\mathrm{d}z} = i\pi \left(s_g \psi_g + \sum_{g'} v_{g-g'} \psi_{g'}\right), \tag{1}$$

where the incoming plane wave propagates parallel to the z-axis with wave vector $\boldsymbol{k}_0 \cdot \psi_{g'}$ denotes the excitation of the diffracted beam in the direction $(\boldsymbol{k}_0 + \boldsymbol{g})$ in reciprocal space. s_g is the excitation error, defined by

$$s_g = \frac{\boldsymbol{k}_0^2 - (\boldsymbol{k}_0 + \boldsymbol{g})^2}{2(\boldsymbol{k}_0 + \boldsymbol{g})_z},$$

v_g the scattering amplitude. Equation (1) may be rewritten in matrix notation writing Ψ, the vector of diffraction amplitudes ψ

$$\frac{\mathrm{d}}{\mathrm{d}z} \boldsymbol{\psi} = i\pi (S + V) \boldsymbol{\psi};$$

$$\boldsymbol{\Psi}_i = \psi_{g_i}, \tag{2}$$

$$S_{ik} = s_{g_i} \delta_{ik},$$

$$V_{ik} = v_{g_i - g_k}.$$

The formal solution of (2) reads

$$\boldsymbol{\Psi}(z) = \exp i\pi (S + V) z \boldsymbol{\Psi}(0), \tag{3}$$

$$\boldsymbol{\Psi}(0) = \begin{pmatrix} 1 \\ 0 \\ 0 \\ \vdots \end{pmatrix} := \boldsymbol{e}_1, \tag{4}$$

and may be constructed by a unitary transformation

$$\frac{\mathrm{d}}{\mathrm{d}z} (U^t \boldsymbol{\Psi}) = i\pi \big(U^t (S + V) U\big) (U^t \boldsymbol{\Psi}), \tag{2'}$$

where

$$\big(U^t (S + V) U_{ik}\big) = \lambda_i \delta_{ik}$$

is diagonal.

The column vectors of U, the eigenvectors of $(S + V)$ are the well-known Bloch waves $\boldsymbol{\varphi}_i$, which propagate freely through the ideal crystal. The boundary condition (4) yields the initial excitation c_i of each Bloch wave as the scalar product

$$c_i = \big(\boldsymbol{\varphi}_i, \boldsymbol{\Psi}(0)\big) \equiv \boldsymbol{\varphi}_{i,1}.$$

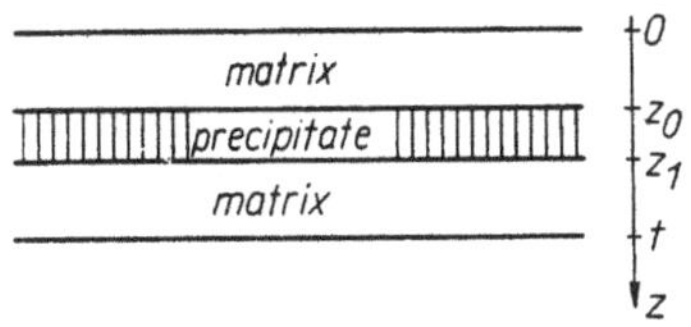

Fig. 1. The coordinates $d = z_1 - z_0$

3. Defects

To calculate the image of large lattice defects, one may apply the column approximation, if structural details of the lattice are not important. Then in order to compute the image of a precipitate it suffices to calculate the outgoing wave function in reciprocal space $\boldsymbol{\Psi} = \boldsymbol{\Psi}(\tau, d(\boldsymbol{x}))$ for stackings matrix — precipitate — matrix (Fig. 1) for a set of precipitate thicknesses, combine them for different columns $\boldsymbol{x}$ and Fourier-transform this combination into direct space

$$\psi(\boldsymbol{x}, \tau) = \sum_i \boldsymbol{\Psi}_i(\tau, d(\boldsymbol{x}))\, e^{2\pi i g_i x}\,.$$

Then this wave function must be folded with the transfer function of the electron microscope in the usual way by Fourier transformation — multiplication — back transformation.

3.1 *Perturbational formulation*

The Bloch-wave scattering due to a thin layer of precipitated material with the same structure as the matrix or a superstructure can easily be treated by a perturbation theory: supposed there is no change in degeneracy (i.e. in structure) one can apply z-independent perturbation theory with appropriate boundary conditions at the upper and lower surface of each layer. Otherwise it is easier to use the z-dependent formulation, where the perturbation potential is a step function in z-direction.

3.1.1 *z-independent formulation*

Denoting with index M matrix and with index P precipitate Bloch waves, $\boldsymbol{\Psi}(z)$ equals

$$\begin{aligned}
z \leqq 0&: \quad \boldsymbol{e}_1\,,\\
0 < z \leqq z_0&: \quad \sum_n \boldsymbol{\varphi}_n^{\mathrm{M}}(\boldsymbol{\varphi}_n^{\mathrm{M}}, \boldsymbol{e}_1)\, e^{i\lambda_n^{\mathrm{M}} z}\,,\\
z_0 < z \leqq z_1&: \quad \sum_m \boldsymbol{\varphi}_m^{\mathrm{P}}(\boldsymbol{\varphi}_m^{\mathrm{P}}, \boldsymbol{\Psi}(z_0))\, e^{i\lambda_m^{\mathrm{P}}(z-z_0)}\,,\\
z_1 < z \leqq t&: \quad \sum_l \boldsymbol{\varphi}_l^{\mathrm{M}}(\boldsymbol{\varphi}_l^{\mathrm{M}}, \boldsymbol{\Psi}(z_1))\, e^{i\lambda_l^{\mathrm{M}}(z-z_1)}\,.
\end{aligned} \tag{5}$$

First-order perturbation theory yields

$$\begin{aligned}
\boldsymbol{\varphi}_m^{\mathrm{P}} &\approx \boldsymbol{\varphi}_m^{\mathrm{M}} + \sum_{m'} \frac{V_{mm'}}{\lambda_m - \lambda_{m'}}\, \boldsymbol{\varphi}_{m'}\,,\\
\lambda_m^{\mathrm{P}} &\approx \lambda_m^{\mathrm{M}} + V_{mm'}\,,
\end{aligned} \tag{6}$$

where

$$V_{mm'} = \pi(\boldsymbol{\varphi}_{m'}^{\mathrm{M}}, (V^{\mathrm{P}} - V^{\mathrm{M}})\, \boldsymbol{\varphi}_m^{\mathrm{M}})\,.$$

Up to terms quadratic in V this gives

$$\boldsymbol{\Psi}(t, z_0, d) = \sum_l \varphi_l \, e^{i(\lambda_l t + V_{ll} d)} \left[\varphi_{l,1} + \sum_{\substack{n \\ n \neq l}} \varphi_{n,l} \frac{V_{ln}}{\lambda_l - \lambda_n} e^{i(\lambda_n - \lambda_l) z_0} \left(1 - e^{i(\lambda_n - \lambda_l + V_{nn} - V_{ll}) d}\right) \right] \quad (7)$$

($\varphi_{l,k}$ denotes the k-th component of Bloch wave φ_l).

3.1.2 z-dependent formulation

A somewhat more general approach is the case of z-dependent perturbation theory, which allows for z-dependent potentials, but also facilitates to treat degeneracy, i.e. cases where for $n \neq m$ $\lambda_n = \lambda_m$. This seems to happen if the symmetry in the matrix is different from that in the precipitate (superstructure).
Expanding the solution (3) in terms of Bloch waves, but with z-dependent excitation,

$$\boldsymbol{\Psi}(z) = \sum_n c_n(z) \, e^{i\lambda_n z} \varphi_n$$

leads to a differential equation for the excitations $c_n(z)$,

$$\frac{d}{dz} c_n(z) = i \sum_m e^{i(\lambda_m - \lambda_n) z} \, V_{mn}(z) \, c_m(z) ,$$

with first-order perturbative solution

$$c_m(z) \approx i \int_0^z \sum_m e^{i(\lambda_m - \lambda_n) z'} \, V_{mn}(z') \, c_m(0) \, dz' .$$

Here the perturbation potential is a step function,

$$V(z) = \pi(V^{\mathrm{P}} - V^{\mathrm{M}}) \, \Theta(z_0, z_1) ,$$

$$\Theta(z_0, z_1) = \begin{cases} 1 & \text{if} \quad z_0 < z \leqq z_1 , \\ 0 & \text{else} , \end{cases}$$

so

$$c_n(z) = \sum_{\substack{m \\ \lambda_m \neq \lambda_n}} \frac{V_{mn}}{\lambda_m - \lambda_n} \left[e^{i(\lambda_m - \lambda_n) z_1} - e^{i(\lambda_m - \lambda_n) z_0}\right] c_m(0) - i \sum_{\substack{m \\ \lambda_m = \lambda_n}} V_{mn} \, d c_m(0) . \quad (8)$$

The possible Bloch waves divide into two classes: those with components representing allowed reflections and those representing forbidden reflections. In the matrix these classes are orthogonal, the forbidden reflections are not excited (we neglect the case where reflections are only kinematically forbidden but dynamically allowed). In the precipitate Bloch waves are no longer strictly orthogonal but overlap, so the superstructure reflections are excited.

Applying this decomposition to the above solution leads to

$$\boldsymbol{\Psi}(\tau, z_0, d) = \begin{pmatrix} \boldsymbol{\Psi}^{(1)} \\ \hline \boldsymbol{\Psi}^{(2)} \end{pmatrix} ,$$

$$\boldsymbol{\Psi}^{(1)}(t, z_0, d) = \sum_l e^{i\lambda_l t} \varphi_l^{(1)} \varphi_{l,1}^{(1)} (1 + i V_{ll} d) ,$$

$$\boldsymbol{\Psi}^{(2)}(t, z_0, d) = \sum_l e^{i\lambda_l t} \varphi_l^{(2)} \left[\sum_{\substack{n \\ \lambda_n \neq \lambda_l}} \varphi_{n,1}^{(1)} \frac{V_{ln}}{\lambda_l - \lambda_n} e^{i(\lambda_m - \lambda_l) z_0} \left(1 - e^{i(\lambda_n - \lambda_l) d}\right) + \right.$$
$$\left. + i \sum_{\substack{n \\ \lambda_n = \lambda_l}} \varphi_{n,1}^{(1)} V_{ln} d \right] , \quad (9)$$

if one neglects the weak overlap between matrix Bloch waves. This result is similar to that above, except that the intra-Bloch wave scattering is taken into account only up to first order, and, more important, it is easy to consider degeneracy.

3.2 Visibility of small precipitates

To demonstrate the ease of qualitative statements with the above theory we calculate the position dependence of the contrast of a small particle.

For illumination along a low index zone axis usually only two Bloch waves are excited. The wave function then reads (z-independent form assumed)

$$\Psi(t, z_0, d) = \varphi_1(\varphi_{1,1} + i\varepsilon\varphi_{2,1})\, e^{i(\lambda_1 t + V_{11} d)} + \varphi_2(\varphi_{2,1} + i\varepsilon\varphi_{1,1})\, e^{i(\lambda_2 t + V_{22} d)}; \qquad (10)$$

$$\varepsilon(z_0, d) = -i\, \frac{V_{12}}{\lambda_1 - \lambda_2}\, e^{i(\lambda_1 - \lambda_2) z_0} (1 - e^{i(\lambda_1 - \lambda_2 + V_{11} - V_{22}) d})\,.$$

For small d and $(\lambda_1 - \lambda_2)\, t = 2\pi n$, $n \in N$ (a foil thickness between two extinction contours), the visibility of a precipitate may be estimated

$$|\Psi_i|^2 \approx \begin{cases} 1 - 4\varphi_{1,1}^2 \varphi_{2,1}^2 \operatorname{Im} \varepsilon\,, \\ 2\varphi_{1,1}\varphi_{1,i}\varphi_{2,1}\varphi_{2,i}(1 - (\varphi_{1,1}\varphi_{2,1}(\varphi_{1,i}^2 + \varphi_{2,i}^2) + \varphi_{1,i}\varphi_{2,i}) \operatorname{Im} \varepsilon)\,; \end{cases}$$

$$\operatorname{Im} \varepsilon \approx \frac{V_{12}}{\lambda_2 - \lambda_1} (\lambda_1 - \lambda_2 + V_{11} - V_{22})\, d \sin (\lambda_1 - \lambda_2)\, z_0\,,$$

what means that the visibility is highest if the particle is situated in the middle of the foil.

4. Numerical Results

The above formalism was applied to two alloys: NiAl, where an ordered $L1_2$-phase Ni_3Al precipitates, and AlAg, where Ag-rich clusters, so-called Guinier-Preston zones form during the early stages. The latter case was used to test the z-independent formulation, because the main contrast originates from the change in atomic form factor between matrix and precipitate, whereas for NiAl with ordered precipitates with only weak enrichment of Al, but superstructure, the application of z-dependent theory is appropriate.

Images are calculated for spherical precipitates of radius R. Microscopic aberrations were applied in the usual way by folding the outgoing wave function with a transfer function. The constants used are tabulated in Table 1.

4.1 Results for NiAl

Here precipitates are visible in a many-beam image, if the image is formed by interference of the zero beam with the {001}-superstructure reflections. If the foil normal

Table 1

The aberration constants used

spherical aberration	1.2 mm
focus fluctuations	10.0 nm
beam divergence	1.0 mrad

Table 2
The Bloch-wave characteristics for Ni–10 at% Al

foil normal	⟨110⟩			⟨100⟩			
numbers of Bloch waves coupling to the zero beam	2	4		1		9	
eigenvalues (nm^{-1})	−0.82	−0.07		−0.48		−0.01	
extinction length (nm)	8.395			13.403			
numbers of Bloch waves coupling to the {001}-beam	1	8	10	2	3	13	14
eigenvalues (nm^{-1})	−0.82	0.015	0.016	0.48		0.106	

is ⟨110⟩, two {001}-reflections are excited, if it is ⟨100⟩, there are four. Table 2 shows the Bloch wave characteristics for the matrix for both cases. The overlap matrix elements are tabulated in Table 3.

Fig. 2 and 3 show calculated images of precipitates of different size in two different positions in the foil. In the ⟨110⟩-direction the particles have only poor contrast, whereas along ⟨100⟩ particles show high contrast, which is mainly due to the fact that in the latter case four superstructure reflections contribute to the image.

The visibility depends only weakly on the position in the foil, but there might be an error in size determination.

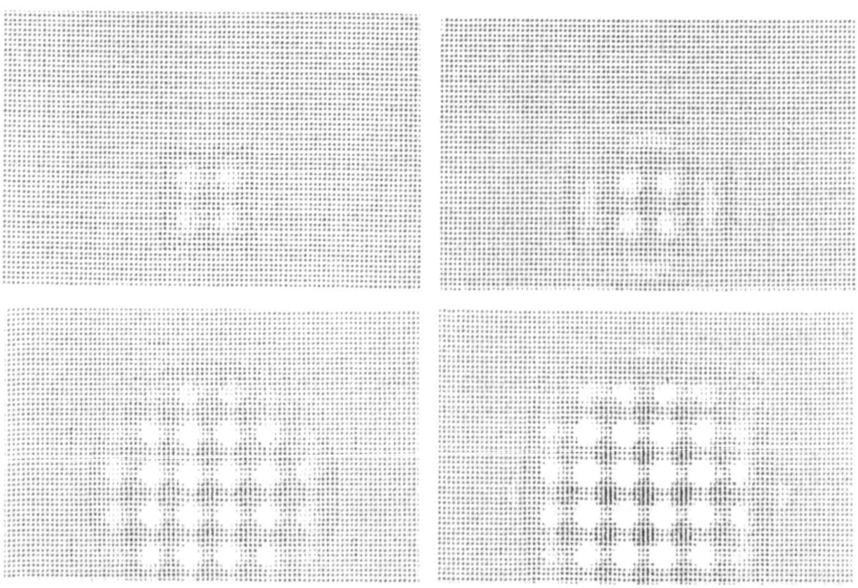

Fig. 2. Ni_3Al particle, foil normal ⟨100⟩ (defocus 105 nm). a) $R = 0.5$ nm, $z_0 = 3$ nm; b) 0.5 nm, 9 nm; c) 1 nm, 3 nm; d) 1 nm, 9 nm

Table 3

The overlap matrix elements V_{ln}(nm^{-1}) for Ni–Ni_3Al

	⟨110⟩				⟨100⟩	
	$l = 1$	8	10		$l = 2/3$	13/14
$n = 2$	0.137	−0.016	0.038	$n = 1$	−0.078	0.029
4	0.0272	−0.003	0.009	9	−0.031	0.011

Table 4

Bloch-wave characteristics for Al

numbers of Bloch waves coupling to the zero beam	1	2
eigenvalue (nm^{-1})	−0.3251	0.0273
extinction length (nm)	21.1	

Table 5

The overlap matrix elements V_{ln} (nm^{-1}) for Al–45 at% Ag

	$l = 1$	2
$n = 1$	−0.3325	−0.1206
$n = 2$	−0.1206	−0.0401

Fig. 3. Ni_3Al particle, foil normal ⟨110⟩ (defocus 105 nm). a) $R = 0.5$ nm, $z_0 = 4$ nm; b) 0.5 nm, 8 nm; c) 1 nm, 4 nm; d) 1 nm, $z_0 = 8$ nm

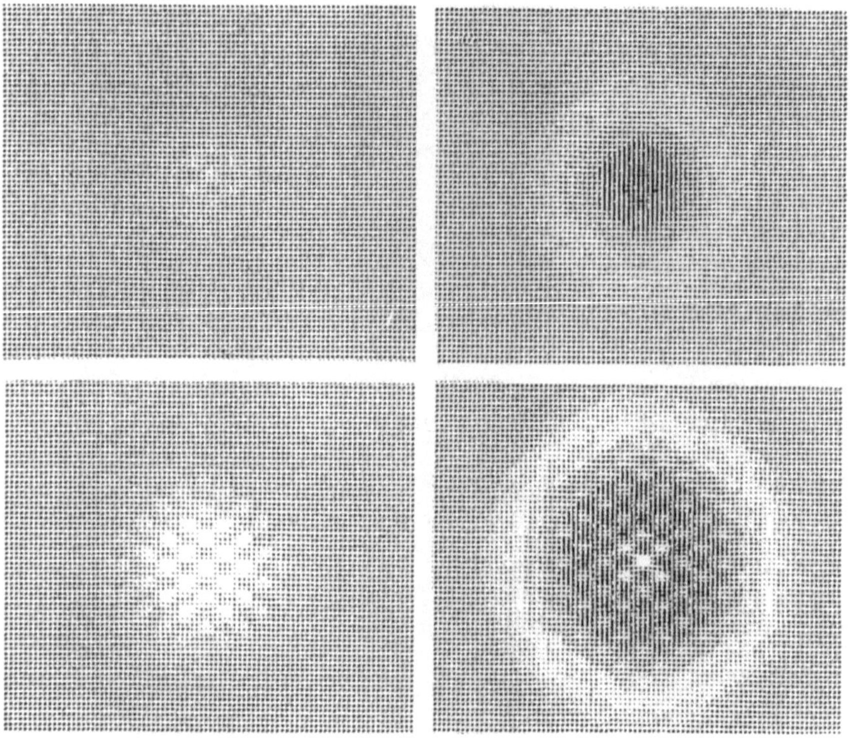

Fig. 4. AlAg particle, foil normal ⟨110⟩ (defocus 248 nm). a) $R = 0.5$ nm, $z_0 = 0.5$ nm; b) 0.5 nm, 10.5 nm; c) 1 nm, 0.5 nm; d) 1 nm, 10.59 nm

4.2 *Results for AlAg*

Here a many-beam image is formed with the four {111}-reflections which have different phase and magnitude for matrix and precipitate. These reflections are excited only for a beam direction along a ⟨110⟩-zone axis. The Bloch-wave characteristics are tabulated in Table 4, the overlap matrix elements in Table 5. Fig. 4 shows precipitates of two sizes in two positions, one near the upper surface, one at half the extinction length. It is rather obvious that, as predicted, the visibility is much better in the latter case, but also that if a particle is visible, it is imaged with its true size, if one takes the size of the inner dark area as particle diameter. So corrections would only be necessary for statistics, e.g. by introduction of an effective foil thickness a bit smaller than the true thickness.

Also in some of the images the particle seems to be facetted, which is a pure artefact by the imaging process.

5. Conclusion

A simple procedure based on a Bloch-wave description of electron scattering was proposed to calculate the image of small coherent precipitates. A perturbational expansion of the additional scattering of the electrons in the precipitate leads to simple expressions which show the dependence of image contrast on the parameters foil thickness, particle position in the foil, and particle size.

Two methods were proposed, one based on z-dependent and one with z-independent potentials. The latter is applicable only for sharp interfaces, whereas the former in principle could be used also for non-trivial z-dependences as e.g. could be possible in spinodal decomposition.

The formalism was applied to two cases, to the precipitation of Ni_3Al in a supersaturated NiAl matrix and to that of Ag-rich Guinier-Preston zones in AlAg.

Acknowledgements

The author thanks P. Haasen, M. Wilkens, M. Rühle and the Göttingen TEM group for discussions and helpful suggestions.

References

[1] F. Ernst, S. Q. Xiao, F. Haider, and P. J. Wilbrandt, Scripta metall. **21**, 1189 (1987).
[2] J. M. Cowley and A. F. Moodie, Acta cryst. **10**, 609 (1961).
[3] D. van Dyck, phys. stat. sol. (b) **72**, 321 (1975).
[4] G. Nihoul and M. H. Pischedda, Phil. Mag. **A57**, 1 (1988).
[5] G. Nihoul and A. Charai, Phil. Mag. **A52**, 407 (1985).
[6] A. Howie and M. J. Whelan, Proc. Roy. Soc. **A263**, 217 (1961).

(Received November 10, 1988)

phys. stat. sol. (a) **111**, 81 (1989)

Subject classification: 61.80 and 71.55; 73.40; S5.11

Institute of Microelectronics Technology and Superpure Materials, Academy of Sciences of the USSR, Chernogolovka[1])

Plasma Stimulated Impurity Redistribution in Silicon

By

S. V. Koveshnikov, E. B. Yakimov, N. A. Yarykin, and V. A. Yunkin

The influence of low temperature (≈100 °C) plasma treatment on the properties of silicon near surface layers is studied. A decrease in concentration of deep level centres as well as the formation of the new vacancy related complexes at a depth of 1 to 10 μm are observed. The changes in the silicon properties are found to be connected closely with the generation of intrinsic point defects. Experimental data confirming the possibility of enhanced gold diffusion are presented and the value of the diffusion coefficient is estimated.

Исследовано влияние низкотемпературной (≈100 °C) плазменной обработки на свойства приповерхностных слоев кремния. Обнаружено, что на глубинах от 1 до 10 μm происходит уменьшение концентрации центров с глубокими уровнями и образование новых вакансионных комплексов. Обсуждается связь между собственными точечными дефектами, генерированными в процессе плазменной обработки, и наблюдаемыми изменениями свойств кремния. Представлены результаты, подтверждающие осуществление ускоренной диффузии золота и оценен его коэффициент диффузии.

1. Introduction

At present the processes of rf plasma etching of semiconductors are widely investigated in connection with the great prospects of their application in VLSI technology [1, 2]. Interest has also been developed in the study of crystal surface damage by the plasma influence. The primary attention was focused on the investigation of very thin (several tens of nm) near surface layers which exhibited an increase in the contents of structural defects [3 to 7] and incorporation of plasma [4, 5] and electrode [5 to 7] particles. However, in spite of numerous works devoted to the problem of plasma induced defects, the changes in the properties of crystals subjected to plasma etching are still not clearly understood.

It should be noted that the energy of ions bombarding the crystal surface is sufficiently high to create intrinsic point defects. An excess concentration of point defects is known to enhance impurity diffusion and/or to stimulate the formation and reconstruction of more complex defects. Taking into account the high mobility of vacancies and interstitials [8, 9] one can expect to observe the resulting changes in the semiconductor properties at considerable distances from the plasma etched surface.

As we have shown recently, plasma treatment at ambient temperatures changes drastically the spectrum of deep levels in silicon at depth up to several μm below the surface [10 to 12]. This paper reports the results of further investigations which confirm the fact that the changes in the properties of silicon under low temperature plasma treatment are closely connected with plasma induced point defects.

[1]) SU-142432 Chernogolovka, Moscow District, USSR.

2. Experimental Procedure

The investigations were performed on 10 to 15 Ω cm n- and p-type silicon single crystals doped with phosphorus and boron, respectively. Some samples were additionally doped with gold, platinum, or sulfur, the others contained defects introduced by high energy ($E = 2.5$ MeV) electron irradiation. The samples were treated by rf (13.56 MHz) plasma in the RIE mode. Argon and a mixture of SF_6–O_2 were used as active gases. The etching time varied from 1 to 30 min. The cathode potential did not exceed 600 V. The sample temperature controlled by the method described elsewhere [13] was lower than 100 °C since the cathode was cooled by water. The deep level spectrum and concentration profiles of electrically active centres were investigated using DLTS on Schottky diodes fabricated by thermal sputtering of gold and aluminium on n- and p-type samples, respectively. The charge carrier concentration was controlled by the C–U technique.

To estimate the contribution of separate constituents of plasma treatment such as ion bombardment, chemical surface reactions, X-ray, and UV radiation different etching modes [5, 14] were used. The crystals subjected to a deliberate treatment in a stream of atomic hydrogen were also investigated. In this case crystals were placed in the region of hydrogen plasma afterglow and their temperature did not exceed 100 °C. The observed effect of neutralization of electrically active centres caused by atomic hydrogen at temperatures lower than used in other works [15, 16] may be ascribed to the lack of a surface damaged layer since the crystals were not bombarded by energetic particles. The plasma etched samples as well as the crystals exposed to hydrogen were thermally annealed in air in the range from 100 to 850 °C.

3. Experimental Results

1. The capacitance–voltage investigations of Schottky diodes formed on plasma etched samples show that the majority charge carrier concentration remains practically the same whilst the concentration of deep level, gold related centres decreases significantly. This is reflected in the lack of a hysteresis of the C^{-2} versus U dependence measured at the temperature of liquid nitrogen on Si⟨Au⟩ samples, while a pronounced hysteresis was obtained on untreated crystals (Fig. 1) The deep level spectra of both the control crystals doped with gold and those subjected to plasma treatment are shown in Fig 2. It is seen that the concentrations of centres with $E_t = E_c$ –

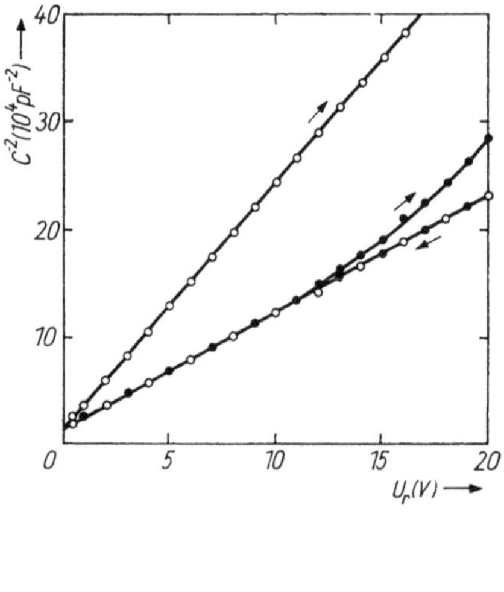

Fig. 1. C–U plot for Schottky diodes formed on starting (○) and plasma treated (●) gold doped n-silicon. Measurements were carried out at 77 K. On cooling the samples were biased at $U = 20$ V

$- 0.54$ eV and $E_t = E_v + 0.35$ eV levels which are usually ascribed to gold atoms [15] are strongly decreased due to plasma etching. The distribution profile of gold related E (0.54) centres is shown in Fig. 3a. As in the case of Si⟨Au⟩ the treatment of Si crystals additionally doped with platinum or sulfur is also responsible for the changes in the deep level spectrum. The concentration of corresponding centres decreases significantly (Fig. 4) in the vicinity (several μm) of the treated surface. Plasma etching of the samples containing defects introduced by high energy electron irradiation leads to a decrease in the concentration of A-centres and divacancies (Fig. 5).

2. Plasma treatment also gives rise to new electrically active defects in the near surface region. The E(0.22) and H(0.26) centres were introduced in the largest concentration in n- and p-type silicon, respectively. Note that such centres were revealed also in the Si samples which did not contain deep level centres before treatment (Fig. 6). As is seen from Fig. 3b the penetration depth of plasma induced defects is nearly equal to the depth where the changes in the concentration of electrically active gold, platinum, and sulfur were observed.

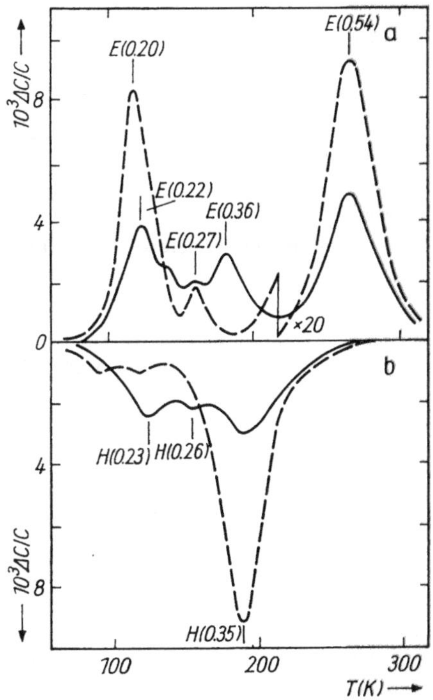

Fig. 2

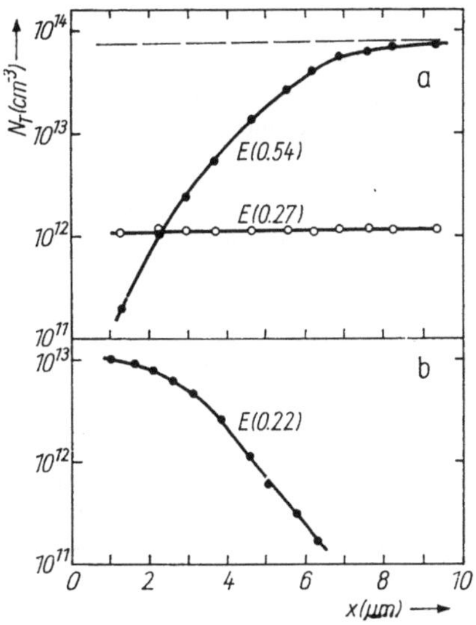

Fig. 3

Fig. 2. DLTS spectra of starting (dashed lines) and plasma treated (solid lines) gold doped silicon: a) n-type; b) p-type

Fig. 3. The distribution profiles of electrically active centres in silicon doped with gold after plasma treatment: a) E(0.27) and gold related E(0.54) centres. The dashed line indicates the initial gold concentration. b) Plasma induced E(0.22) centres which are also observed in gold free silicon

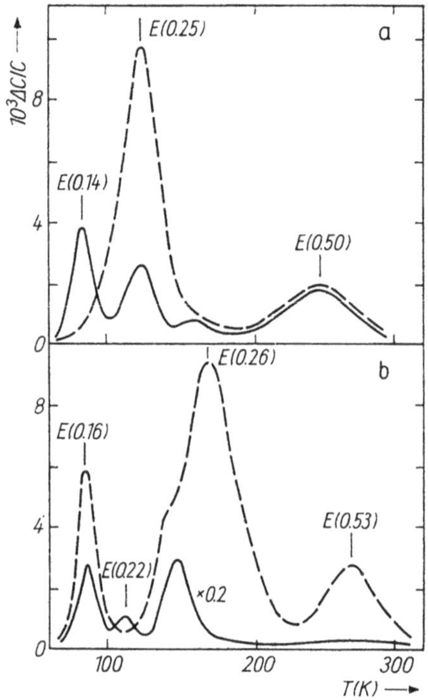

Fig. 4

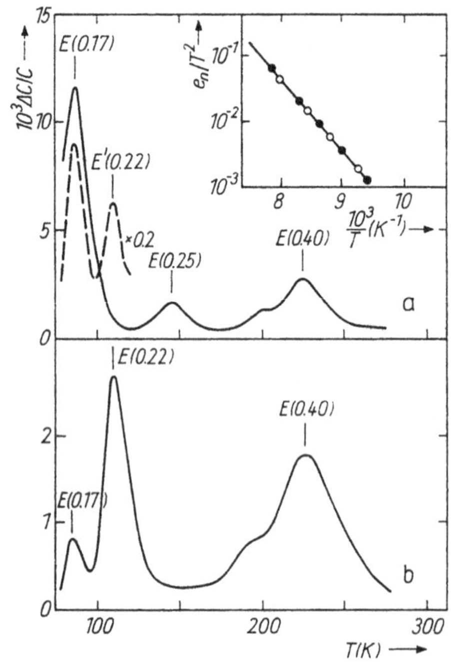

Fig. 5

Fig. 4. DLTS spectra of n-Si crystals doped with a) platinum and b) sulfur, before (dashed lines) and after (solid lines) plasma treatment

Fig. 5. DLTS spectra of high energy electron irradiated n-silicon: a) after irradiation (solid line) and subsequent thermal annealing at 330 °C, 1 h (dashed line); b) after plasma treatment. The inset shows e_n/T^2 (in units of s^{-1} K^{-2}) vs. $1/T$ ($E_c - E_t = 0.22$; $\sigma_n = 2.3 \times 10^{-15}$; ● E′(0.22), ○ E(0.22))

4. Discussion

As seen from the foregoing, low temperature plasma treatment is responsible for the changes of the electrical properties of layers several μm below the etched surface.

Obviously, the changes in the deep level impurity concentration cannot be caused by overheating of the sample during etching. This is confirmed by the presence of their distribution profiles (taking into account that the thermal conductivity of silicon is rather high) and by the data shown in Fig. 5 which indicate that the concentration of the E-centres (with annealing temperature about 150 °C [17]) remains practically unaltered after plasma treatment.

In principle, since only electrically active centres are detected by the DLTS, a decrease in their concentration may be interpreted as a conversion of impurity related centres into an electrically inactive state. Such a neutralization of electrically active defects may be caused, e.g., by atomic hydrogen [18]. Indeed, the measurements carried out on the samples treated by atomic hydrogen show that as in the case of plasma treatment a decrease in the deep level centre concentration occurs in near surface layers of silicon. However, subsequent annealing, e.g., (Fig. 7) of n-Si⟨Au⟩ samples at 600 °C almost completely restores the electrical activity of gold,

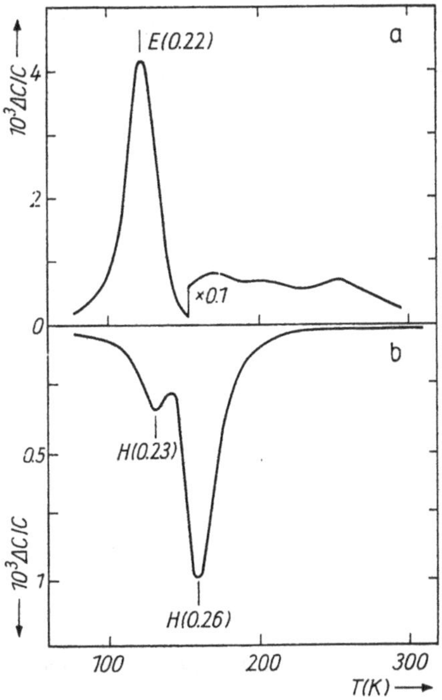

Fig. 6

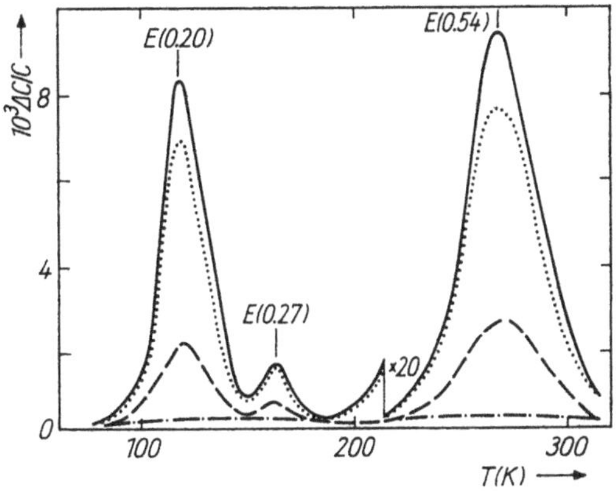

Fig. 7

Fig. 6. DLTS spectra of plasma etched a) n-type and b) p-type silicon

Fig. 7. Changes in the DLTS spectra of n-Si⟨Au⟩ exposed to atomic hydrogen (—·—·—) and subsequent thermal annealing (30 min, — — — 400 °C, ... 600 °C; ——— initial curve)

while similar annealing of plasma etched samples does not affect the distribution profile of electrically active gold. The distribution profile of gold shown in Fig. 3a remains unaltered up to an annealing temperature of 850 °C. Besides, in the samples with radiation defects hydrogen passivates the E-centres, but this does not occur under plasma etching. The above-mentioned facts indicate that all the changes in the deep level centre concentration caused by plasma etching cannot be explained by the influence of atomic hydrogen. Based on the data given in Fig. 2, it is easy to conclude that gold impurity does not form new electrically active complexes. In our opinion, a decrease in the gold related centre concentration is most likely to be explained on the assumption that the impurity diffuses from the bulk to the surface. The reality of enhanced diffusion even at ambient temperature is proved by the following experimental result. The n-type silicon crystal (deep level concentration less than 10^{10} cm^{-3}) with a piece of Au foil on its surface was subjected to plasma etching under the same conditions. The measurements of the DLTS spectrum performed after the formation of the Schottky barriers indicated that the E(0.54) centres with the value of capture cross-section corresponding to gold atoms were introduced by the treatment. The inset in Fig. 8 shows the distribution profile of the E(0.54) centres.

Investigations of the parameters of plasma induced E(0.22) centres show that their energy level, capture cross-section, and annealing temperature (350 °C) are equal to those of the centres formed in electron irradiated silicon after annealing of A-centres (Fig. 5). This enables us to assume that E(0.22) centres are rather simple vacancy-

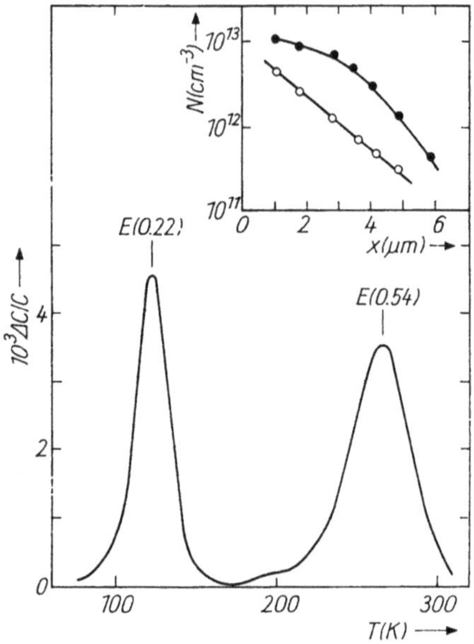

Fig. 8. Introduction of vacancy related E(0.22) complexes and gold related E(0.54) centres during plasma treatment of n-silicon. The inset shows their distribution profiles (○ E(0.54), ● E(0.22))

related complexes formed during generation of point defects in the process of plasma etching.

Since the changes in the properties of near surface layers occur in the mode of reactive ion etching (both in a SF_6–O_2 mixture and in Ar) and are not observed in the plasma etching mode when the samples are placed on the anode, one can conclude that a decisive role in the plasma initiated effects belongs to low energy ion bombardment of the crystal surface. X-ray and UV radiation may also contribute to the observed changes but they alone do not cause a redistribution of impurities. During bombardment of the surface the ions penetrate into a narrow surface layer and create excess vacancy and interstitial concentrations. These point defects displaying high mobility even at room temperature [8, 9] can enhance impurity diffusion. New electrically active complexes E(0.22) are very likely to be formed as a result of the interaction between plasma generated vacancies and other lattice defects.

To evaluate the vacancy diffusion coefficient we must solve the diffusion equation in the region with a variable boundary,

$$\frac{\partial N}{\partial t} = D \frac{\partial^2 N}{\partial x^2} - \frac{N}{\tau}, \tag{1}$$

$$D \left. \frac{\partial N}{\partial x} \right|_{x=Vt} = G. \tag{2}$$

Here N is the defect concentration, D the diffusion coefficient, V the etching rate, τ the defect lifetime, G the defect generation rate near the treated surface. The quasi-stationary regime is achieved some time after the beginning of etching. In this case the solution is found most easily. Turning to a coordinate system connected with a moving crystal surface,

$$z = x - Vt, \tag{3}$$

and assuming

$$\frac{\partial N}{\partial t} = 0 \tag{4}$$

we obtain

$$N(z) = N_0 \, e^{-z/x_0}, \tag{5}$$

where the characteristic length of defect penetration is

$$x_0 = \frac{1}{\frac{V}{2D} + \sqrt{\frac{V^2}{4D^2} + \frac{1}{D\tau}}}. \tag{6}$$

It is easy to show that the distribution profile of the E(0.22) centres formed as a result of vacancy capture is described by the law

$$N_{E(0.22)} \sim \exp(-kz/x_0), \tag{7}$$

where k is the number of vacancies in the complex. From the data given in Fig 8 it follows that in our case the distribution of the E(0.22) centres is satisfactorily described by this law, assuming $x_0/k = 1.5$ μm.

Solving (7) with respect to D, assuming $k = 1$, and taking $V = 4 \times 10^{-7}$ cm s^{-1}, we obtain $D \geqq 5 \times 10^{-11}$ cm^2 s^{-1}. In the case of gold in-diffusion (Fig. 8) with D_{Au} invariable in the region of several μm, the enhanced gold diffusion coefficient is of the same order.

5. Conclusions

The effects of a decrease in deep level centre concentration in Si as well as the formation of point defect related complexes under plasma etching at ambient temperature were observed. The changes in the properties of silicon in near surface layers are closely connected with the generation of intrinsic point defects. The experimental data confirming the possibility of enhanced impurity diffusion are presented. More comprehensive studies are required to clarify the nature of point defects and to develop a microscopic model of their participation in the diffusion processes.

References

[1] J. Dieleman, Thin Solid Films **86**, 147 (1981).
[2] S. J. Fonash, Solid State Technol. **28**, 150 (1985).
[3] J. C. Bean, G. E. Becker, P. M. Petroff, and T. E. Seidel, J. appl. Phys. **48**, 907 (1977).
[4] N. Yabumoto, M. Oshima, O. Michikani, and S. Yoshii, Japan. J. appl. Phys. **20**, 893 (1981).
[5] S. W. Pang, Solid State Technol. **27**, 249 (1984).
[6] H. Matsumoto and T. Sugano, J. Electrochem. Soc. **129**, 2823 (1982).
[7] M. O. Watanabe, M. Taguchi, K. Kanzaki, and Y. Zouta, Japan. J. appl. Phys. **22**, 281 (1983).
[8] T. Y. Tan and U. Gösele, Appl. Phys. A **37**, 1 (1985).
[9] V. V. Emtsev and T. V. Mashovets, Primesi i tochechnie defekty v poluprovodnikakh, Radio i Svyaz, Moskva 1981 (p. 184).
[10] S. V. Koveshnikov, S. V. Redkin, V. V. Starkov, V. A. Yunkin, E. B. Yakimov, and N. A. Yarykin, Mikroelektronika **15**, 462 (1986).

[11] S. V. Koveshnikov, S. V. Redkin, V. V. Starkov, V. A. Yunkin, E. B. Yakimov, and N. A. Yarykin, Poverkh. Fiz. khim Mekh. **11**, 85 (1987).
[12] S. V. Koveshnikov, S. V. Redkin, V. V. Starkov, V. A. Yunkin, E. B. Yakimov, and N. A. Yarykin, in: Proc. of 2. Internat. Autumn Meeting on Gettering and Defect Engineering in Semiconductor Technology, Ed. H. Richter, Inst. for Physics of Semiconductors, Frankfurt (Oder) 1987 (p. 252).
[13] S. S. Lau and R. H. Mills, J. Vacuum Sci. Technol. **9**, 1196 (1972).
[14] S. W. Pang, D. D. Rathman, D. J. Silvesmith, R. W. Mountain, and P. D. De Graff, J. appl. Phys. **54**, 3272 (1983).
[15] S. J. Pearton and A. J. Tavendale, Phys. Rev. B **26**, 7105 (1982).
[16] A. Mogro-Campero, R. P. Love, and R. J. Schubert, Electrochem. Soc. **132**, 2006 (1985).
[17] S. D. Brotherton and P. Bradley J. appl. Phys. **53**, 5720 (1982).
[18] S. J. Pearton, J. W. Corbett, and T. S. Shi, Appl. Phys. A **43**, 153 (1987).

(Received May 31, 1988)

phys. stat. sol. (a) **111**, 89 (1989)

Subject classification: 66.30; 72.15; S1.2

Institut für Physik des Max-Planck-Institutes für Metallforschung, Stuttgart[1])

Diffusion of Hydrogen in Nickel Based Alloys

By

K. YAMAKAWA[2]), B. HOHLER, and H. KRONMÜLLER

The recovery process of hydrogen dissolved in excess in nickel based alloys is studied by means of the hydrogen quenching method and electrical resistance measurement after isochronal and isothermal annealing treatments. A recovery stage of the electrical resistance due to hydrogen in the alloys shows similar characteristics as observed in resistance decay curves due to H in pure nickel, e.g., size dependence of the recovery rate and the shape of the isothermal recovery curve. Diffusion coefficients of hydrogen are obtained for the alloys in the range of 240 to 360 K with a similar method used for pure nickel. The values obtained for high temperatures are nearly the same as those for pure nickel, but at low temperature, especially for 0.10% Cu and 0.10% Fe alloys, are clearly larger than those in pure nickel. Since the diffusivity of hydrogen is depressed by trapping at alloying elements in alloys, this anomalous behaviour suggests that another process, e.g., grain boundary diffusion, in addition to bulk lattice diffusion becomes important in the low temperature range.

Der Erholungsprozeß von gelöstem überschüssigem Wasserstoff in Nickel-Legierungen wird mit der Wasserstoffabschreckungsmethode und mit Messungen des elektrischen Widerstands nach isochronen und isothermen Temperbehandlungen untersucht. Eine Erholungsstufe des elektrischen Widerstands durch den Wasserstoff in den Legierungen zeigt ähnliche Charakteristiken, wie sie in den Widerstandsabklingkurven aufgrund von H in reinem Nickel beobachtet werden, d. h. Größenabhängigkeit der Erholungsrate und der Form der isothermen Erholungskurve. Diffusionskoeffizienten von Wasserstoff werden für die Legierungen im Bereich von 240 bis 360 K mit einer ähnlichen Methode erhalten, wie sie für reines Nickel benutzt wurde. Die Werte für hohe Temperaturen sind nahezu dieselben wie die für reines Nickel, sind jedoch bei niedriger Temperatur, insbesondere für 0,10% Cu- und 0,10% Fe-Legierungen deutlich größer als die in reinem Nickel. Da das Diffusionsvermögen von Wasserstoff durch Anlagerung an Legierungselementen in den Legierungen erniedrigt wird, weist dieses anomale Verhalten darauf hin, daß ein anderer Prozeß, d. h. Korngrenzendiffusion zusätzlich zur Volumengitterdiffusion im Niedertemperaturbereich wesentlich wird.

1. Introduction

In recent years there has been a growing interest in the diffusivity of hydrogen in metals and alloys and a large number of measurements of the diffusivity has been published [1]. Among these investigations, hydrogen trapping by solute impurity atoms is observed causing a lowering of the diffusivity. For f.c.c. alloys in contrast to b.c.c. alloys only few results are available especially in Ni based alloys. By magnetic after-effect and internal friction measurements one local jump to nearest neighbour sites is studied as the elementary process of long-range diffusion. The magnetic relaxation due to the reorientation of diatomic metal–H complexes has been investigated previously for Ni based alloys containing Fe and Cu etc. [2 to 4]. From these results the

---

[1]) Heisenbergstr. 1, D-7000 Stuttgart 80, FRG.

[2]) Permanent address: Faculty of Engineering, Hiroshima University, Saijoh, Higashi-Hiroshima 724, Japan.

diffusion coefficients of hydrogen are determined from the annealing of the relaxation amplitude. The parameters obtained for the reorientation of metal–H complexes describe the local hopping mode of the MeH complex. These parameters in general do not correspond to the long-range diffusion of hydrogen in Ni based alloys.

In pure Ni the recovery process of hydrogen dissolved in excess is studied by measurements of the electrical resistance. The evaporation of hydrogen from the specimen is described by a model calculation and the diffusion coefficient is obtained with high accuracy [5, 6]. For the precise analysis of the electrical resistivity recovery due to solute hydrogen the resistivity of the alloy specimen in the hydrogen free state is desired as low as possible, because both the hydrogen solubility and resistivity per unit concentration of hydrogen are not so high.

The resistivity increase by alloying is reported in previous papers [7, 8]. The amount is about 4 nΩm/at% and 30 nΩm/at% for Ni–Fe, Ni–Cu, and Ni–Ti dilute alloys, respectively. Therefore Fe, Cu, and Ti are chosen as the solute elements in dilute alloys. The solubilities of the elements, Fe, Cu, and Ti in Ni are sufficiently high for these alloys.

In the present paper the diffusion experiments of hydrogen are done for Ni based alloys by using the hydrogen quenching technique similar as applied previously in pure Ni [5, 6]. The resistivity decay curves are analysed by the same method as applied for pure Ni giving the diffusion coefficients for the alloys. The diffusion coefficients for alloys with 0.10 at% impurities are larger than those of pure Ni especially at the low temperatures investigated. This enhancement of diffusivity at low temperature is discussed in relation to grain boundary diffusion.

2. Experimental Procedure

The recovery process of hydrogen introduced in excess in Ni based alloys is studied by measuring the electrical resistivity of the wire specimens at 4.2 K during the annealing.

The alloys, Ni–0.10 at% Cu, Ni–0.10 at% Fe, Ni–1.0 at% Fe, and Ni–0.05 at% Ti are produced by melting in an induction furnace and vacuum casting the melt into molybdenum crucibles. The used raw materials are Ni, Cu, Fe, and Ti in nominal purity, 99.998%, 99.999%, 99.998%, and 99.97%, respectively, supplied by Koch-Light Labo. Co. Ltd. for Ni and Cu, by Johnson Matthey Co. Ltd. for Fe and Material Research Co. for Ti. The chemical composition of prepared alloys are analysed by atomic absorption spectroscopy and the results are tabulated in Table 1.

The pre-annealing of the alloy specimens and of pure Ni is performed at 480 to 500 °C in hydrogen gas atmosphere. The sizes of the wire specimens are between 0.050 and 0.100 mm in radius. The experimental apparatus and the method in the present experiment are similar ones as those used before for the hydrogen quenching and diffusion ex-

Table 1

Chemical composition of Ni based alloys by atomic absorption spectroscopy

	concentration (at%)
Ni–0.1% Cu	0.102 ± 0.02
Ni–0.1% Fe	0.106 ± 0.02
Ni–1.0% Fe	1.038 ± 0.02
Ni–0.05% Ti	0.049 ± 0.02

periment in pure Ni [5, 6]. The specimen in a specimen chamber is heated and quenched forhydrogenation in liquid hydrogen or cooled hydrogen atmosphere. The quenching temperatures are between 1000 and 1200 K.

The electrical resistance of the specimen is measured in liquid helium by standard four-probe method under an applied longitudinal magnetic field of 500 Oe. The isothermal resistance decays are investigated between 240 and 360 K and the isochronal annealings are done in helium gas atmosphere up to 420 K.

3. Experimental Results

3.1 *Isochronal annealing*

The isochronal resistance decay curves for Ni–0.10% Cu, 0.10% Fe, and 1.0% Fe, $\Delta R/\Delta R_i$, where ΔR_i and ΔR are the amount of the experimental total resistance change and the residual resistance due to solute hydrogen, respectively, are shown for different sized specimens in the upper parts of Fig. 1. In the lower part their temperature derivatives $-\mathrm{d}(\Delta R/\Delta R_i)\,\mathrm{d}T$ are shown. In the hydrogen quenched alloy specimens the resistivity decreases also remarkably similar to the case of pure Ni. The peak around 180 K, which appears in pure Ni is not observed in these alloys. In these figures the decay curves for specimens with large radii ($a = 0.078$ mm) shift to higher temperatures than those for small radii ($a = 0.050$ mm). The shape of the peak is similar to that of pure Ni but the peak of the alloys is a little bit broader.

3.2 *Isothermal annealing*

The hydrogen annealing stage is investigated by isothermal annealing, in which the specimen is brought up to the desired annealing temperature after hydrogen quenching.

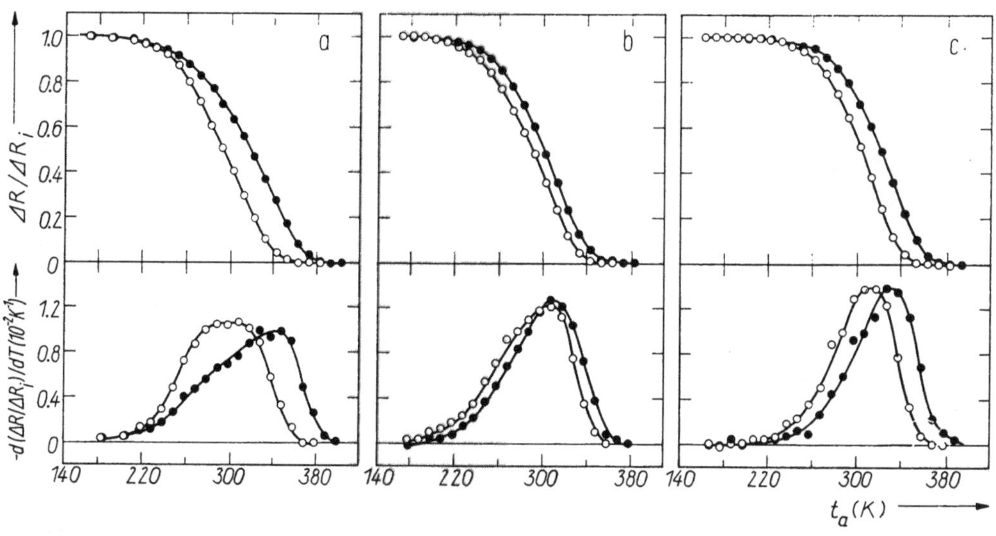

Fig. 1. Isochronal resistance decay curves and their temperature derivatives of hydrogen quenched Ni based alloys for two specimen radii, $a = 0.500$ and 0.078 mm. The heating rate is 10 K/10 min. a) Ni–0.10% Cu, b) Ni–0.10% Fe, and c) Ni–1.04% Fe. In all alloys the recovery rates of 0.050 mm specimens are faster than those for 0.078 mm specimens. ○ $a = 0.050$, ● 0.078 mm

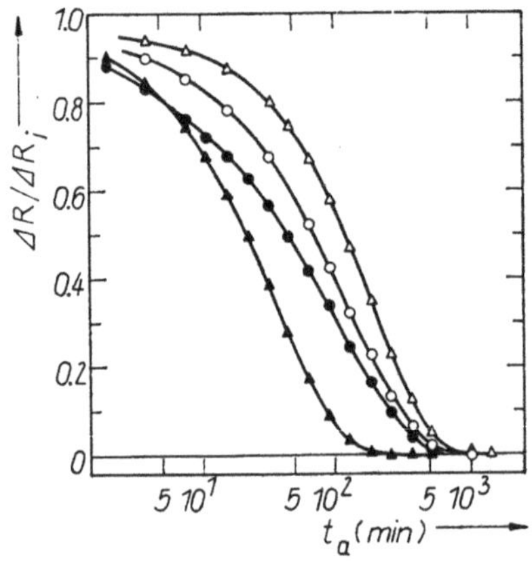

Fig. 2. Isothermal resistance decay curves for various specimen radii of Ni–0.10% Cu wires at the same temperature of 312.2 K. The smaller the size becomes, the faster the recovery rate is. △ a = 0.100, ○ 0.090, ● 0.078, ▲ 0.050 mm

Isothermal resistance decay curves for hydrogen quenched Ni–0.10% Cu are shown in Fig. 2 for four different specimen radii at the same temperature 312.2 K. Although the annealing treatment is the same, the decay rates of the resistance extremely differ from one another corresponding with the results of the size dependence of the above

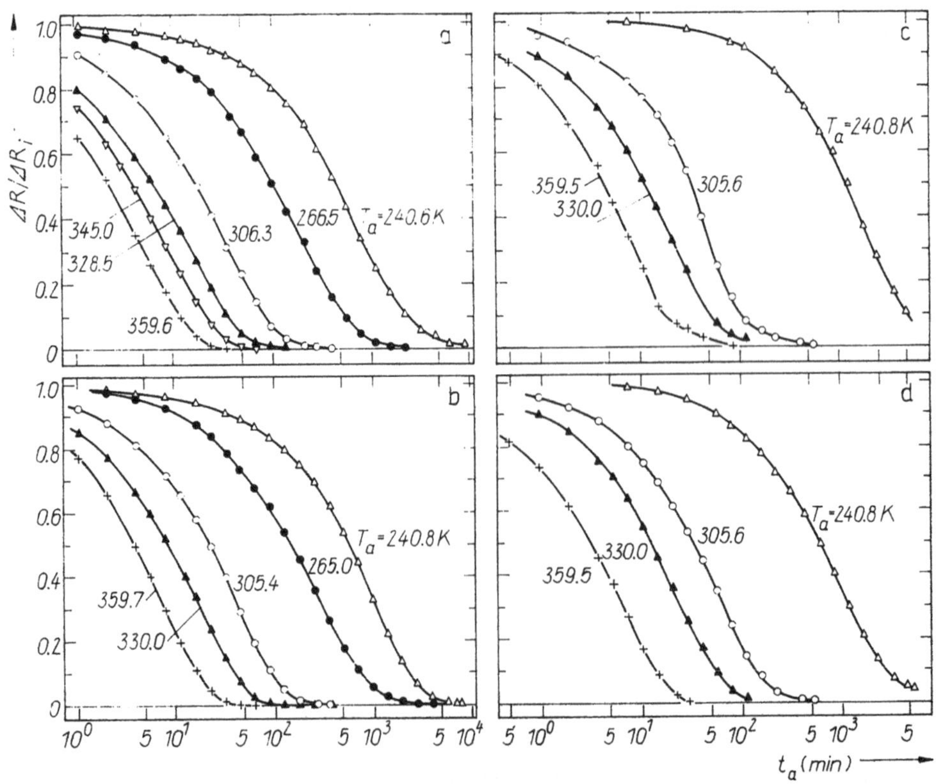

Fig. 3. Isothermal resistance decay curves at several temperatures for Ni based alloys. The specimen radius is a = 0.050 mm. a) Ni–0.10% Cu, b) Ni–0.10% Fe, c) Ni–1.04% Fe, and d) Ni–0.05% Ti

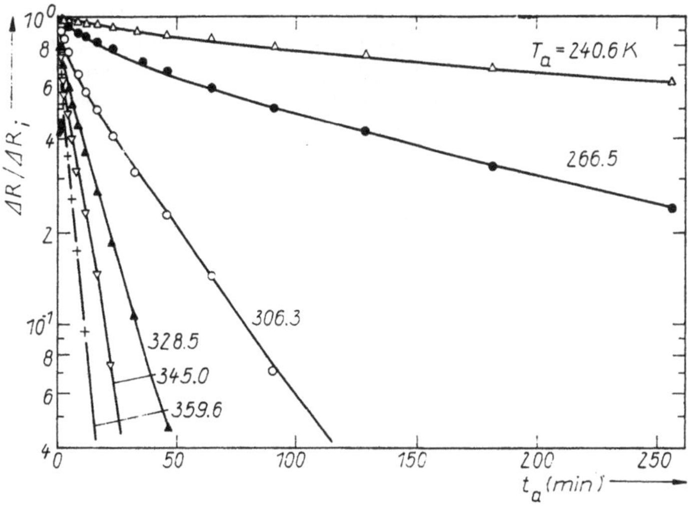

Fig. 4. Isothermal decay curves at several temperatures for Ni–0.10% Cu replotted with linear time scale from Fig. 3a. Solid curves are calculated ones from the method described for pure Ni [5, 6]

isochronal decay curves. The decay rates for hydrogen quenched specimens with different radii decrease with specimen size similar to pure Ni.

Isothermal decay curves for the hydrogen quenched Ni–0.10% Cu, 0.10% Fe, and 0.05% Ti alloy are shown in Fig. 3 for various annealing temperatures. The specimen radius, a, is 0.050 mm for each alloy if not especially noted. The decay curves are replotted with the linear time scales for the analysis to determine the diffusion constant. The figure corresponding to Fig. 3a is shown in Fig. 4. The curves are bent downward which suggests that the process does not obey simple exponential decay similar to the case of pure Ni. It must be noted here that the curves are normalized with respect to ΔR_i which is the amount of the total experimental resistance change due to quenched-in hydrogen.

4. Analysis and Discussion

The recovery stage which appears in the hydrogen quenched Ni based alloys arises from the hydrogen diffusion, because they show a similar behaviour as hydrogen quenched pure Ni [5, 6]. The size dependence of the recovery rate appears also for these alloys in isochronal and isothermal decay curves; i.e., the smaller the specimen radius is, the faster the decay rate becomes. Therefore the main sink for hydrogen atoms is the external surface of the specimen similar to pure Ni. At first the annealing stage is analysed by the method reported already for hydrogen diffusion in pure Ni specimen. The hydrogen atoms are quenched in a specimen with uniform distribution by ideal quenching (infinite quenching rate). By an annealing treatment the hydrogen atoms diffuse to the specimen surface and escape from the specimen forming a concentration gradient in the specimen following the diffusion equation. The finite quenching rate in the experiment is taken into account as part of the annealing. The diffusion coefficient is obtained from a comparison between the experimental isothermal decay curves and calculated isotherms. A precise description of the calculation and analysis already has been given in previous papers [5, 6].

The characteristic feature of these curves of alloy specimens is that the base resistance ratio, $R_B/\Delta R_0$, is larger than those for pure Ni, where R_B and ΔR_0 are the residual resistance of the specimen from which hydrogen is fully annealed out and the amount of ideal total resistance change for infinite quenching rate, respectively. Usually the values of $R_B/\Delta R_0$ are 3 to 6, 1 to 2, 5 to 7, and 12 to 16 for Ni–0.10% Cu,

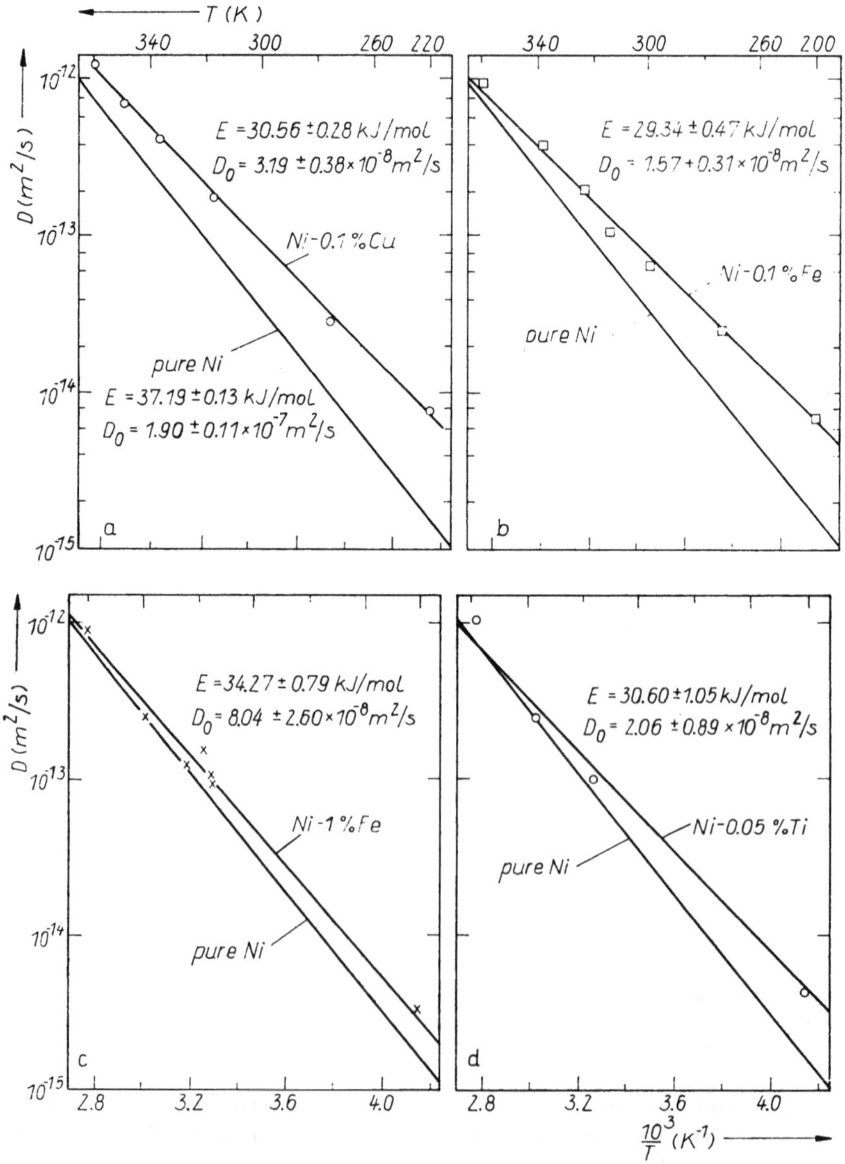

Fig. 5. The obtained diffusion coefficient of hydrogen against reciprocal temperature for Ni based alloys with that for pure Ni for the specimens of 0.050 mm in radius. The diffusion coefficients for the alloys are clearly larger than for pure Ni at low temperatures. a) 0.1% Cu; b) 0.1% Fe; c) 1.0% Fe; d) 0.05% Ti. The full lines correspond to Asvhenius plots using the activation parameters given in Table 3

0.10% Fe, 1.0% Fe, and 0.05% Ti, respectively. As shown in Fig. 9 of a previous paper [5] the calculated isothermal decay curves for large $R_B/\Delta R_0$ are bent downward much lesser. In fact, all decay curves show the is tendency comparable with that of pure Ni. Part of the isotherms are shown in Fig. 4 for Ni–0.10% Cu specimens. The solid curves are calculated ones. The agreement between the experiment and calculation is very good also for other alloys except the Ni–Ti alloy. As the residual resistivity increase by impurity atom addition is large in Ni–0.05% Ti alloy as mentioned before, the hydrogen resistance (ΔR_i or ΔR_0) relative to the residual resistance (R_B) becomes smaller than in other Ni based alloys. Therefore the accuracy of the hydrogen resistance measurement becomes worse. From these data fitting the diffusion coefficients are obtained for various annealing temperatures similar to pure Ni.

The diffusion coefficients of hydrogen are shown in Fig. 5 a to d as Arrhenius plots together with the results for pure Ni. The diffusion coefficients for the alloys at high temperature in the investigated temperature range are nearly the same as the values of pure Ni as shown in the figure. The values for Ni–0.10% Cu and Ni–0.10% Fe at low temperature, however, are clearly larger than that of pure Ni. The values for Ni–1.0% Fe are nearer to the values for pure Ni than for 0.10% Cu and 0.10% Fe. For Ni–0.05% Ti the coefficient shows large scattering due to the smaller ratio of the hydrogen resistance as compared to the total specimen resistance mentioned before. The enhancement of diffusivity by alloying at low temperature is a very strange behaviour in hydrogen diffusion in solids if we consider the trapping model in which the hydrogen atoms are trapped at alloying impurity atoms. In this model the hydrogen diffusion coefficient must become smaller than that in pure material [9, 10]. For the hydrogen trapping model at impurity atoms in dilute alloys over a limited temperature range, Hohler and Kronmüller [9] obtained the diffusion coefficient of the alloy, D_A, as

$$D_A = D_{0A} \exp\left(-\frac{E_A}{RT}\right), \tag{1}$$

$$E_A = E + \frac{C_{MeH}}{C_H} E_B , \tag{2}$$

$$\ln D_{0A} = \ln D_0 + \ln \frac{1 - \dfrac{C_{MeH}}{C_H} + \dfrac{C_{MeH}}{C_H}\dfrac{E_B}{RT}}{1 - 6C_{MeH}}, \tag{3}$$

$$C_{MeH} = C_H \frac{\{6C_{Me}/(1-6C_{Me})\}\, e^{-S_B/R}\, e^{E_B/RT}}{1 + \dfrac{6C_{Me}}{1-6C_{Me}}\, e^{-S_B/R}\, e^{E_B/RT}}. \tag{4}$$

Here E, D_0, and C_H are the activation energy, pre-exponential factor of diffusion, and hydrogen concentration in the specimen, in the case of pure Ni, respectively. C_{Me}, C_{MeH}, E_B, and S_B denote the substitutional impurity concentration, the metallic impurity–hydrogen complex concentration in equilibrium, and the binding energy and binding entropy of the complex [9]. The values D_A are calculated with $S_B = 0$ for Ni–0.10% Cu, 0.10% Fe, 1.0% Fe, and 0.05% Ti alloy, respectively, and are shown in Fig. 6 together with the Arrhenius plot for $D = 1.90 \times 10^{-7} \exp(37.19\ \text{kJ mol}^{-1}/RT)\ \text{m}^2/\text{s}$ of pure Ni. The used parameters are tabulated in Table 2 and 3. On the basis of the trapping model the diffusion coefficient in the alloys must become considerably smaller than in pure Ni.

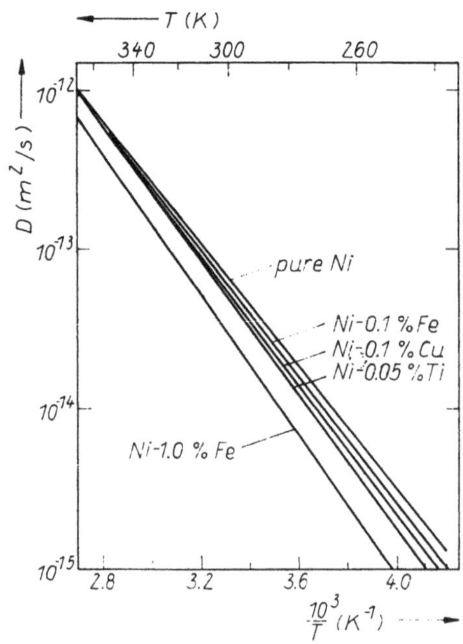

Fig. 6. Calculated diffusion coefficients for the hydrogen trapping model as described from (1) to (4)

From the above-mentioned argument the diffusion of hydrogen atoms in the alloys may not be caused by one process but by two processes, i.e., lattice diffusion and boundary diffusion. Nevertheless the straight lines are obtained by Arrhenius plotting for various Ni based alloys. The activation energy, E, and the pre-exponential factor, D_0,

Table 2

The parameters for the calculation of the diffusion coefficients in Fig. 5. E_B has been obtained by Hohler and Kronmüller [9]

	C_{Me} (at%)	E_B (kJ/mol)
Ni-0.1% Cu	0.1	9.17
Ni-0.1% Fe	0.1	7.72
Ni-1.0% Fe	1.0	
Ni-0.05% Ti	0.05	11.7

Table 3

Diffusivity constants of hydrogen

	D_0 (10^{-8} m²/s)	E (kJ/mol)	temperature range (K)
Ni-0.1% Cu	3.19 ± 0.38	30.56 ± 0.28	240 to 360
Ni-0.1% Fe	1.57 ± 0.31	29.34 ± 0.47	240 to 360
Ni-1.0% Fe	8.04 ± 2.60	34.27 ± 0.79	240 to 360
Ni-0.05% Ti	2.06 ± 0.89	30.60 ± 1.05	240 to 360
Ni(b)	19.0 ± 1.1	37.19 ± 0.13	220 to 330

are evaluated by using the method of least squares for the temperature range between 240 and 360 K and are shown with the values for pure Ni in Table 3. Due to the anomalous high diffusivity at low temperatures the activation energies and pre-exponential factors become smaller for the alloys than for pure Ni.

Robertson investigated hydrogen diffusion in pure Ni with various grain sizes and concluded that grain boundary diffusion of hydrogen did not take place in his investigated temperature range (300 to 770 K) [11]. Recently Tsuru and Latanison detected the grain boundary diffusion of hydrogen in pure Ni [12]. They reported that the diffusion coefficient of hydrogen for boundary diffusion was approximately 60 times larger than for lattice diffusion and boundary diffusion was about one thousandth of the lattice diffusion at probably room temperature (they did not mention about their measuring temperature) following the amount of grain boundary in their specimen. This extremely small contribution from boundary diffusion of hydrogen is responsible to no detectable boundary diffusion above room temperature by Robertson. Because the activation energy in boundary diffusion of hydrogen is smaller than in lattice diffusion the boundary diffusion becomes important at low temperatures. For Ni–1.0% Fe the decrease of lattice diffusion of hydrogen by trapping at the alloying element may be cancelled with the increase of the boundary diffusion. This fact means that the ratio of boundary diffusion increases from pure Ni to Ni–0.10% Cu and Ni–0.10% Fe but the ratio increases less from Ni–0.10% Fe to 1.0% Fe.

The grain size of pure Ni and of the alloys are observed by an optical microscope after pre-annealing at 1000 °C for 5 h. The size is widely spread from part to part of the specimen especially for pure Ni. The mean sizes are about 125, 65, 70, 50, and 50 μm for pure Ni, Ni–0.10% Cu, Ni–0.10% Fe, and Ni–1.0% Fe, respectively. In fact the size varies consistently with the above discussion.

It is interesting to clarify the grain boundary diffusion of hydrogen to carry out similar diffusion experiments for single crystals and polycrystals at low temperatures.

Acknowledgements

The authors would like to express their appreciation to W. Maisch and R. Henes for their technical assistance and to the other technical staff of the Max-Planck-Institut for continuous support during the course of this experiment. The authors furthermore wish to acknowledge Profs. Drs. A. Seeger and H. Schultz, Max-Planck-Institut für Metallforschung, and Prof. Dr. E. E. Fujita, Osaka University, Japan, for continuous encouragement throughout this study. The financial support to a visiting scientist by the Max-Planck-Institut is gratefully acknowledged by K.Y.

References

[1] J. Völkl and G. Alefeld, Hydrogen in Metals, Vol. 1, Ed. G. Alefeld and J. Völkl, Springer-Verlag, Heidelberg and New York 1978 (p. 321).
[2] B. Hohler and H. Kronmüller, Phil. Mag. **45**, 607 (1982).
[3] B. Hohler and H. Kronmüller, Phil. Mag. **43**, 1189 (1981).
[4] B. Hohler and H. Schreyer, J. Phys. F **12**, 857 (1982).
[5] K. Yamakawa, M. Tada, and F. E. Fujita, Japan J. appl. Phys. **15**, 769 (1976).
[6] K. Yamakawa and F. E. Fujita, Japan J. appl. Phys. **16**, 1747 (1977).
[7] T. Farrell and D. Greig, J. Phys. C **1**, 1359 (1968).
[8] A. Fert and I. A. Cambell, J. Phys. F **6**, 849 (1976).
[9] B. Hohler and H. Kronmüller, Z. phys. Chem., N.F. **114**, 93 (1979).
[10] R. A. Oriani, Acta metall. **18**, 147 (1970).
[11] W. M. Robertson, Z. Metallk. **64**, 436 (1973).
[12] T. Tsuru and R. M. Latanision, Scripta metall. **16**, 575 (1982).

(Received October 27, 1988)

phys. stat. sol. (a) **111**, **99** **(1989)**

Subject classification: 61.70; 68.25

Institute of Agriculture, Ryazan[1])

Development of Eshelby Concept in the Solution of Boundary Problems for Elastic Solids with Defects

By

A. F. VLADIMIROV and E. N. MOOS

A theoretical research of the elastic fields of defects lying near the free and interface boundaries is carried out. The Somigliana surface dislocation method is developed as the general view point of using the surface continuum distributions of dislocations solving the boundary problem of the theory of defects. New solutions of the boundary problem are obtained for a halfspace and a plate, and also for sandwich-like systems by means of double surface dislocations. The correlation between the Somigliana vectors and the surface relief is determined. Methods of calculating the self and interaction energy of defects in a halfspace are elaborated.

Eine theoretische Untersuchung der elastischen Felder der an freien Grenzen und an Zwischenphasengrenzen liegenden Defekte wird durchgeführt. Die Methode der Somiglianschen Oberflächenversetzung wird als verallgemeinerter Gesichtspunkt zur Ausnutzung der kontinuierlichen Oberflächenversetzungsverteilungen bei der Lösung des Grenzwertproblems der Defekttheorie entwickelt. Neue Lösungen des Grenzenproblems werden für einen Halbraum und für eine Platte sowie für Sandwichsysteme mit Hilfe von Doppeloberflächenversetzungen gewonnen. Dabei wird ein Zusammenhang zwischen den Somiglianschen Vektoren und dem Oberflächenrelief festgestellt. Eine Methodik zur Berechnung der Eigenenergie und der Wechselwirkungsenergie der Defekte in einem Halbraum wird ausgearbeitet.

1. Introduction

The finding of elastic fields of defects lying near free or interface boundaries is the content of the boundary problem (BP) of the theory of defects [1]. In the last few years the surface continuum distributions of dislocations and disclinations have been often used for the solution of the BP [2 to 12]. In [2, 3] the term "surface dislocation" (SD) for this idea is introduced. Many authors choose the geometry of defects on a surface beforehand and compose the integral equation for their density by means of the boundary condition. For example, the SDs are being modelled by straight dislocations [2 to 6], dislocation and disclination loops [7], fan-shaped dislocations from the end of a dislocation piercing the surface [8 to 10]. Methods of the BP solution in [2 to 10] are adapted to defects of specific geometry.

In [11, 12] the general solution of the BP in a half-space is investigated by means of a general view, namely the Somigliana dislocation with the searched variable vector of the discontinuity of the displacement field over all the boundaries of a solid body. The method of the surface dislocation of Somigliana (SDS) [11, 12] reduces the BP to the problem of finding the elastic fields in an infinite solid which includes this dislocation. In [7] the Somigliana dislocations are also used, but their geometry is chosen beforehand, namely it is a distribution of concentric circular loops with a set variable Burgers vector.

---

[1]) SU-390044 Ryazan, USSR.

In the present paper the SDS method is developed in a form more convenient for many applications, and the BPs for a half-space, plate, bicrystal, film on a semi-infinite underlying solid are solved in a new fashion. Also methods of calculating the energy of defects in a half-space are presented.

2. Theory and Results

2.1 The BP solution for the case of a free boundary by means of the SDS method

The general elastic field of magnitude $\mathcal{F}_G$ (displacement, elastic distortion, or stress) is found as the superposition

$$\mathcal{F}_G = \mathcal{F}_D + \mathcal{F} , \tag{1}$$

where $\mathcal{F}_D$ is the field of a defect in an infinite medium, $\mathcal{F}$ the relaxation field which is caused by the free surface S with the normal $\boldsymbol{n}$. In all boundary points $\boldsymbol{x}_S$ the relaxation stress field satisfies the boundary condition

$$\boldsymbol{\sigma}(\boldsymbol{x}_S) \cdot \boldsymbol{n}(\boldsymbol{x}_S) + \boldsymbol{t}(\boldsymbol{x}_S) = \boldsymbol{0} , \tag{2}$$

where $\boldsymbol{t}$ is the surface tension field in an infinite medium.

In the previous research [11, 12] the Somigliana dislocation with unknown discontinuity vector ξ was described by the plastic distortion tensor. The fields $\mathcal{F}_G$ and the vector ξ were found from the simultaneous equations for an elastic medium with defects [13] which were supplemented by the condition (2). The relaxation fields $\mathcal{F}$ were not zero all over the space and only part of the vector ξ was connected with the displacement of the free surface. The SDS method may be improved so that the fields $\mathcal{F}$ were equal to zero outside the restricted body and the Somigliana vector was equal to the relaxation displacement of the free surface. Let us rest on the Gebbia theorem given by Eshelby [1] in a modern way. Let the relaxation stress $\boldsymbol{\sigma}$ balance the surface tension $[-\boldsymbol{t}]$ according to condition (2), and the relaxation displacement of the surface be equal to $\boldsymbol{\beta}$. The Gebbia-Eshelby theorem means (Fig. 1): the elastic state in a body with a boundary S on which the surface tension $[-\boldsymbol{t}]$ and the surface displacement $\boldsymbol{\beta}$ are given, is the same as it would be in the same part of an infinite medium in which a layer of body forces of the surface density $[-\boldsymbol{t}]$ and a Somigliana dislocation with the discontinuity vector $\boldsymbol{\beta}$ of displacements are placed; moreover elastic fields $\mathcal{F}$ from (1) are equal to zero outside the restricted part of the medium.

Eshelby connects the Gebbia theorem with the boundary integral equation method of the theory of elasticity, noticing that it is not an easy mathematical method. Therefore, in [1] some particular methods of the BP solutions are reviewed. The connection of the surface dislocation method with the boundary integral equation is also discussed in [2, 14].

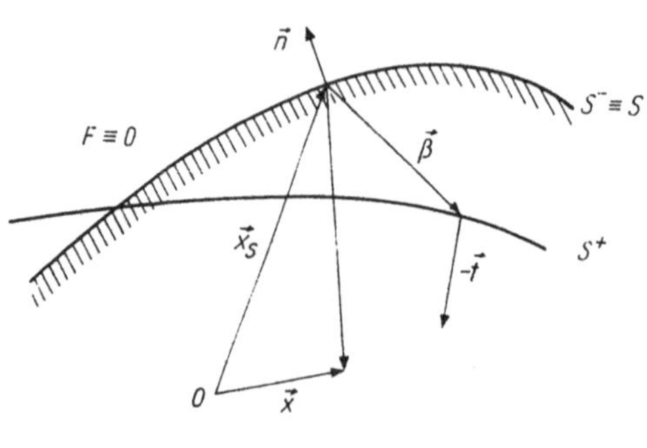

Fig. 1. The solution of the BP for the free surface S by means of the SDS method. The boundary S^- of the outer part rests in the state S. After applying the surface tensions $[-\boldsymbol{t}]$ the boundary of the body secures the state S^+ when displacing along the Somigliana vector $\boldsymbol{\beta}$. 0 is the coordinate system zero, $\boldsymbol{x}$ are the points in the space

However, as for an initial mathematical organization of the BP by means of the SDS method and for its solution as well the use of the theory of elasticity with defects in an infinite medium [13] is more simple and suitable. Let us write the Somigliana dislocation as the tensor of plastic distortion by means of Dirac's δ-function,

$$\boldsymbol{\xi}(\boldsymbol{x}) = -\int \delta(\boldsymbol{x} - \boldsymbol{x}_S)\, \mathrm{d}\boldsymbol{S} \otimes \boldsymbol{\beta}(\boldsymbol{x}_S)\,. \tag{3}$$

We regard the surface tensions $[-\boldsymbol{t}]$ as body forces,

$$\boldsymbol{f}(\boldsymbol{x}) = -\int \delta(\boldsymbol{x} - \boldsymbol{x}_S)\, \boldsymbol{t}(\boldsymbol{x}_S)\, \mathrm{d}S\,, \tag{4}$$

and take into account the existence of a jump of the displacement field $\boldsymbol{u}$,

$$\boldsymbol{\beta}(\boldsymbol{x}_S) = \lim_{(\boldsymbol{x}-\boldsymbol{x}_S)\cdot\boldsymbol{n}\to -0} \boldsymbol{u}(\boldsymbol{x})\,. \tag{5}$$

The complete system of equations for the determination of the fields $\mathcal{F}$ and $\boldsymbol{\beta}$ will consist of the equilibrium equation for the stress tensor $\boldsymbol{\sigma}$ with the right-hand part of (4) with negative sign, Hooke's law, the structure of elastic distortion tensor $\boldsymbol{\beta}$; one more peculiarity of the SDS method will be the inclusion of (5) in the above-said system,

$$\boldsymbol{\sigma}(\boldsymbol{x}) \cdot \nabla = -\boldsymbol{f}(\boldsymbol{x})\,, \qquad \boldsymbol{\sigma}(\boldsymbol{x}) = \mathbf{c} \cdot\cdot\, \boldsymbol{\beta}(\boldsymbol{x})\,, \qquad \boldsymbol{\beta}(\boldsymbol{x}) = -\boldsymbol{\xi}(\boldsymbol{x}) + \nabla \otimes \boldsymbol{u}(\boldsymbol{x})\,,$$
$$\boldsymbol{\beta}(\boldsymbol{x}_S) = \lim_{(\boldsymbol{x}-\boldsymbol{x}_S)\cdot\boldsymbol{n}\to -0} \boldsymbol{u}(\boldsymbol{x})\,. \tag{6}$$

Here $\mathbf{c}$ is the tensor of elastic moduli; points mean the convolution of tensor indexes.

For plane, free surfaces the system (6) is solved in an algebraic form by means of the Fourier $\boldsymbol{k}$-representation. Let us solve it for the lower half-space of a Cartesian co-ordinate system $0x_1x_2x_3$ with the basis $\boldsymbol{e}_1$, $\boldsymbol{e}_2$, $\boldsymbol{e}_3$, considering the plane $0x_1x_2$ with the normal vector $\boldsymbol{e}_3$ as the boundary S. Performing in (6) the Fourier transformation and neglecting the argument $\boldsymbol{k} = k_1\boldsymbol{e}_1 + k_2\boldsymbol{e}_2 + k_3\boldsymbol{e}_3$ for brevity, one get $\boldsymbol{\sigma} \cdot i\boldsymbol{k} = \boldsymbol{t}(\boldsymbol{k}_0)$, $\boldsymbol{\sigma} = \mathbf{c} \cdot\cdot\, \boldsymbol{\beta}$,

$$\boldsymbol{\beta} = \boldsymbol{e}_3 \otimes \boldsymbol{\beta}(\boldsymbol{k}_0) + i\boldsymbol{k} \otimes \boldsymbol{u}\,, \qquad \boldsymbol{\beta}(\boldsymbol{k}_0) = \lim_{x_3\to -0} \frac{1}{2\pi} {\textstyle\int\!\!\!\!-} \boldsymbol{u}(\boldsymbol{k}) \exp{(ik_3x_3)}\, \mathrm{d}k_3\,, \tag{7}$$

where $\boldsymbol{k}_0 = k_1\boldsymbol{e}_1 + k_2\boldsymbol{e}_2$, and the line on the integral means the principial value; the method of the $\boldsymbol{t}(\boldsymbol{k}_0)$ calculation is given in Appendix A. When solving system (7) some second-rank tensors are used which are defined in the Appendix, (B1) to (B6), moreover $\boldsymbol{\Gamma}$ is the Green tensor for an infinite medium. The result of solving for the lower half-space is

$$\boldsymbol{\beta}(\boldsymbol{k}_0) = \mathbf{A}^{-1} \cdot \mathbf{Y} \cdot [-\boldsymbol{t}(\boldsymbol{k}_0)]\,, \qquad \boldsymbol{u} = i\boldsymbol{\Gamma} \cdot \boldsymbol{\Pi}(\boldsymbol{k}, \boldsymbol{e}_3) \cdot \boldsymbol{\beta}(\boldsymbol{k}_0) + \boldsymbol{\Gamma} \cdot [-\boldsymbol{t}(\boldsymbol{k}_0)]\,. \tag{8}$$

The BP for the upper half-space is solved by analogy to (8),

$$\boldsymbol{\beta}'(\boldsymbol{k}_0) = \mathbf{A}^{*-1} \cdot \mathbf{Y} \cdot \boldsymbol{t}(\boldsymbol{k}_0)\,, \qquad \boldsymbol{u}' = -i\boldsymbol{\Gamma} \cdot \boldsymbol{\Pi}(\boldsymbol{k}, \boldsymbol{e}_3) \cdot \boldsymbol{\beta}'(\boldsymbol{k}_0) + \boldsymbol{\Gamma} \cdot \boldsymbol{t}(\boldsymbol{k}_0)\,, \tag{9}$$

where the star means complex conjugation.

The simultaneous equations (6) may be also analytically solved for a plate with the composite boundary $S = S_1US_2$ (Fig. 2). The solution in $\boldsymbol{k}$-representation is the following:

$$\begin{aligned}
\boldsymbol{\beta}_1(\boldsymbol{k}_0) &= \mathbf{N}^{-1} \cdot \{\mathbf{H} \cdot [-\boldsymbol{t}_1(\boldsymbol{k}_0)] + \mathbf{P} \cdot [-\boldsymbol{t}_2(\boldsymbol{k}_0)]\}\,, \\
\boldsymbol{\beta}_2(\boldsymbol{k}_0) &= \mathbf{N}^{*-1} \cdot \{\mathbf{P}^* \cdot [-\boldsymbol{t}_1(\boldsymbol{k}_0)] + \mathbf{H}^* \cdot [-\boldsymbol{t}_2(\boldsymbol{k}_0)]\}\,, \\
\boldsymbol{u} &= i\boldsymbol{\Gamma} \cdot \boldsymbol{\Pi}(\boldsymbol{k}, \boldsymbol{e}_3) \cdot \boldsymbol{\beta}_1(\boldsymbol{k}_0) - i\boldsymbol{\Gamma} \cdot \boldsymbol{\Pi}(\boldsymbol{k}, \boldsymbol{e}_3) \cdot \boldsymbol{\beta}_2(\boldsymbol{k}_0) \exp{(ik_3h)} + \\
&\quad + \boldsymbol{\Gamma} \cdot [-\boldsymbol{t}_1(\boldsymbol{k}_0)] + \boldsymbol{\Gamma} \cdot [-\boldsymbol{t}_2(\boldsymbol{k}_0)] \exp{(ik_3h)}\,.
\end{aligned} \tag{10}$$

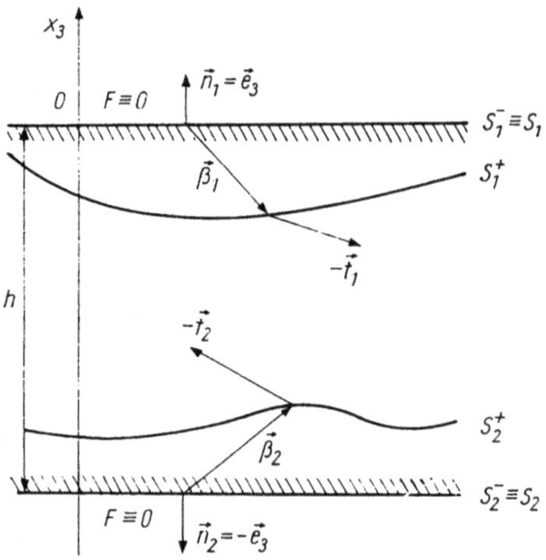

Fig. 2. The solution of the BP for the plate of thickness h. The boundary $S_1^- U S_2^-$ of the outer part rests in the state $S_1 U S_2$. After applying the surface tensions $[-\boldsymbol{t}_1]$ and $[-\boldsymbol{t}_2]$ the boundaries of the body displace along the Somigliana vectors $\boldsymbol{\beta}_1$ and $\boldsymbol{\beta}_2$ in the states S_1^+ and S_2^+. The elastic fields outside the plate are zero

Here second-rank tensors are used,

$$\mathbf{N} = \mathbf{A} - \mathbf{B} \cdot \mathbf{A}^{*-1} \cdot \mathbf{B}^* , \qquad \mathbf{H} = \mathbf{Y} - \mathbf{B} \cdot \mathbf{A}^{*-1} \cdot \mathbf{Z}^* ,$$

$$\mathbf{P} = \mathbf{Z} - \mathbf{B} \cdot \mathbf{A}^{*-1} \cdot \mathbf{Y} ,$$

the tensors $\mathbf{A}$, $\mathbf{Y}$, $\mathbf{B}$, $\mathbf{Z}$ are defined in (B4) to (B8).

A remarkable peculiarity of the solutions (8) to (10) is that they automatically take into account the equality to zero of fields outside the considered bounded body in the $\boldsymbol{x}$-representation. One more significant result is the analytical expressions for the relaxation displacements $\boldsymbol{\beta}$ of free surfaces, the knowledge of which may have an independent interest for the experimental research work of the relief of the surface of the defect crystal.

2.2 The BP solution for a bicrystal

Let us imagine two crystals with the tensors of elastic moduli $\mathbf{c}_1$ and $\mathbf{c}_2$, which have the common interface boundary $S \equiv 0x_1x_2$ with the normal $\boldsymbol{n} = \boldsymbol{e}_3$ without any defects (Fig. 3). After introducing defects at least in one part of the bicrystal their common surface will take the position S'. It is possible to go into this state differently, in two stages.

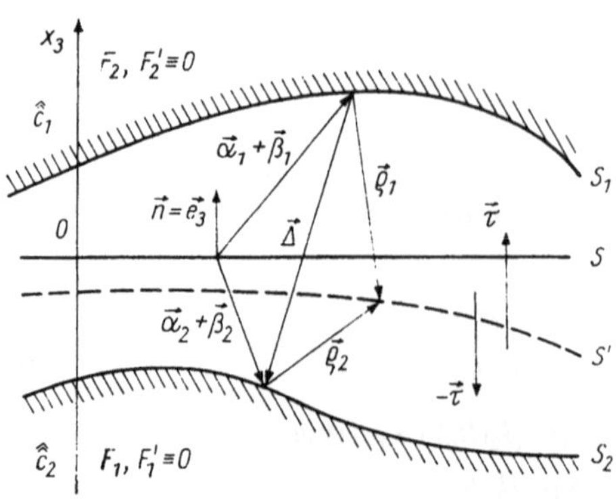

Fig. 3. The solution of the BP for the bicrystal. After cutting along the plane S and introducing defects the free boundaries of the half-spaces will secure the states S_1 and S_2 (the first pair of BPs) when displacing along the vectors $(\boldsymbol{\alpha}_J + \boldsymbol{\beta}_J)$ $(J = 1, 2)$. The welded surfaces S_1 and S_2 will displace along the Somigliana vectors $\boldsymbol{\varrho}_1$ and $\boldsymbol{\varrho}_2$ forming the common surface S' and creating the surface tensions $\boldsymbol{\tau}$ on it (the second pair of BPs)

First, we cut the bicrystal along S and introduce in each part the required defects. The finding of elastic fields at this stage may be reduced to the solution of two BPs for the upper and lower half-spaces with free boundaries, that is finding the fields $\mathcal{F}_{GJ}$ ($J = 1, 2$) by (1) using the solutions (8) and (9). The complete displacements of the surfaces S_1 and S_2 will consist of two addends $(\boldsymbol{\alpha}_J + \boldsymbol{\beta}_J)$, where $\boldsymbol{\alpha}_J$ is the surface displacement in the J-th crystal which is infinite. $\boldsymbol{\beta}_J$ are the corresponding Somigliana vectors. As a result, the magnitude of the splitting between S_1 and S_2 may be found,

$$\boldsymbol{\Delta} = (\boldsymbol{\alpha}_2 + \boldsymbol{\beta}_2) - (\boldsymbol{\alpha}_1 + \boldsymbol{\beta}_1) . \tag{11}$$

Second, let us connect the surfaces S_1 and S_2 over the corresponding points and put them in the state S' together, that will lead to the emergence of some additional elastic fields $\mathcal{F}'_J$. So the complete elastic fields $\mathcal{F}_{BJ}$ of every part of the crystal will consist unlike (1) of the three terms,

$$\mathcal{F}_{BJ} = \mathcal{F}_{DJ} + \mathcal{F}_J + \mathcal{F}'_J . \tag{12}$$

Now the BP is reduced to finding the fields $\mathcal{F}'_J$. Over the surface S there will be a still unknown field of surface tensions $\boldsymbol{\tau}$ securing the equilibrium condition

$$\boldsymbol{\sigma}'_1 \cdot \boldsymbol{n} = \boldsymbol{\sigma}'_2 \cdot \boldsymbol{n} = \boldsymbol{\tau} . \tag{13}$$

Besides the condition of the continuity of the displacement field must be satisfied, it transforms into the following condition for the second relaxation displacements of the surfaces:

$$\boldsymbol{\varrho}_1 - \boldsymbol{\varrho}_2 = \boldsymbol{\Delta} . \tag{14}$$

The magnitude $\boldsymbol{\tau}$ being known the finding of the $\mathcal{F}'_J$ fields is represented by two problems of the previous subsection for the upper and lower half-spaces. Therefore, it is sufficient to find the magnitude $\boldsymbol{\tau}$ over the magnitude $\boldsymbol{\Delta}$ known from (11) in order to reduce the BP for the bicrystal to the already known problems. Let us use the first expressions from (8) and (9) which connect surface tensions and surface displacements,

$$\boldsymbol{\varrho}_1(\boldsymbol{k}_0) = \hat{\mathbf{A}}_1^{*-1} \cdot \hat{\mathbf{Y}}_1 \cdot [-\boldsymbol{\tau}(\boldsymbol{k}_0)] , \tag{15}$$

$$\boldsymbol{\varrho}_2(\boldsymbol{k}_0) = \hat{\mathbf{A}}_2^{-1} \cdot \hat{\mathbf{Y}}_2 \cdot \boldsymbol{\tau}(\boldsymbol{k}_0) . \tag{16}$$

Subtracting (15) from (16) by means of (14) one can get the field of the surface tensions

$$\boldsymbol{\tau}(\boldsymbol{k}_0) = - [\hat{\mathbf{A}}_2^{-1} \cdot \hat{\mathbf{Y}}_2 + \hat{\mathbf{A}}_1^{*-1} \cdot \hat{\mathbf{Y}}_1]^{-1} \cdot \boldsymbol{\Delta}(\boldsymbol{k}_0) . \tag{17}$$

Thus, the result of this subsection is the new BP solution for the bicrystal by means of reducing it to the solution of two pairs of BPs for the half-spaces. Moreover $\mathcal{F}'_1 \equiv 0$ for $x_3 < 0$ and $\mathcal{F}'_2 \equiv 0$ for $x_3 > 0$.

2.3 The BP solution for a film on a semi-infinite underlying solid

This solution may be done in two stages, too (Fig. 4), and the elastic fields consist of terms like (12). The first relaxation surface displacements $\boldsymbol{\beta}$, $\boldsymbol{\beta}_1$, $\boldsymbol{\beta}_2$ are determined by (9) and (10). According to these formulae the second relaxation displacements are determined by using the surface tensions $\boldsymbol{\tau}$ applied to the interface boundary and the zero boundary tensions on the free boundary S_2,

$$\begin{aligned} &\boldsymbol{\varrho}(\boldsymbol{k}_0) = \hat{\mathbf{A}}_1^{*-1} \cdot \hat{\mathbf{Y}} \cdot [-\boldsymbol{\tau}(\boldsymbol{k}_0)] , \\ &\boldsymbol{\varrho}_1(\boldsymbol{k}_0) = \hat{\mathbf{N}}_2^{-1} \cdot \hat{\mathbf{H}}_2 \cdot \boldsymbol{\tau}(\boldsymbol{k}_0) , \qquad \boldsymbol{\varrho}_2(\boldsymbol{k}_0) = \hat{\mathbf{N}}_2^{*-1} \cdot \hat{\mathbf{P}}_2 \cdot \boldsymbol{\tau}(\boldsymbol{k}_0) . \end{aligned} \tag{18}$$

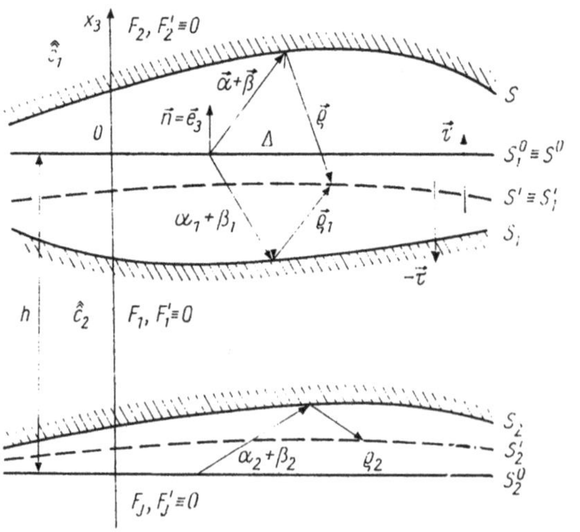

Fig. 4. The solution of the BP for a film c_2 on the underlying solid c_1. S^0, S_1^0, S_2^0 are the initial points of the departure of the planes before the introduction of defects. S, S_1, S_2 are the states of the free surfaces after introducing the defects. S', S_1', S_2' are the final states of the surfaces after welding (all α, β, Δ, ϱ have to be read as vectors)

The surface tension field $\boldsymbol{\tau}$ is determined using the magnitude of the splitting Δ by

$$\boldsymbol{\tau}(\boldsymbol{k}_0) = -[\hat{\mathbf{A}}_1^{*-1} \cdot \hat{\mathbf{Y}}_1 + \hat{\mathbf{N}}_2^{-1} \cdot \hat{\mathbf{H}}_2]^{-1} \cdot \Delta(\boldsymbol{k}_0) . \tag{19}$$

Notice that the second relaxation addends $\mathcal{F}'_J$ satisfy conditions

$$\mathcal{F}'_1 \equiv 0 \quad \text{for} \quad x_3 < 0 , \qquad \mathcal{F}'_2 \equiv 0 \quad \text{for} \quad x_3 > 0 \quad \text{and} \quad x_3 < -h .$$

2.4 The method of calculating the self-energy and the interaction energy of defects in a half-space

Let us assume the elastic self-energy of a defect of finite geometrical form to be finite. It may be attained by using the model of a normally not concentrated defect [15]. In accordance with (1) the general elastic distortion of a defect consists of the two addends

$$\boldsymbol{\beta}_G = \boldsymbol{\beta}_D + \boldsymbol{\beta} . \tag{20}$$

Then the energy as a quadratic form of the distortion will consist of the three terms

$$W = W_D + W_S + W_{DS} , \tag{21}$$

where W_D is the self-energy of a defect in the bounded part of the solid, W_S the energy of the relaxation field, W_{DS} the interaction energy of self-field and relaxation field.

Let the initial defect in an infinite medium create the fields of surface tensions $\boldsymbol{t}$ and surface displacements $\boldsymbol{\alpha}$ on the surface $S \equiv 0x_1x_2$, and the surface tensions $[-\boldsymbol{t}]$ create the relaxation displacements $\boldsymbol{\beta}$ of the surface of the bounded body. We calculate the two last energy terms in (21) by means of the surface integrals and use the two-dimensional Parceval equality

$$W_S = \frac{1}{2} \int [-\boldsymbol{t}^*(\boldsymbol{x}_0)] \cdot \boldsymbol{\beta}(\boldsymbol{x}_0) \, d^2\boldsymbol{x}_0 = \frac{1}{8\pi^2} \int [-\boldsymbol{t}^*(\boldsymbol{k}_0)] \cdot \boldsymbol{\beta}(\boldsymbol{k}_0) \, d^2\boldsymbol{k}_0 , \tag{22}$$

$$W_{DS} = \int \boldsymbol{t}^*(\boldsymbol{x}_0) \cdot \boldsymbol{\beta}(\boldsymbol{x}_0) \, d^2\boldsymbol{x}_0 = \frac{1}{4\pi^2} \int \boldsymbol{t}^*(\boldsymbol{k}_0) \cdot \boldsymbol{\beta}(\boldsymbol{k}_0) \, d^2\boldsymbol{k}_0 , \tag{23}$$

where $\boldsymbol{x}_0 = x_1\boldsymbol{e}_1 + x_2\boldsymbol{e}_2$. Summarising (22) and (23) one transforms (21) to

$$W = W_D - W_S . \tag{24}$$

The magnitude W_D may be found as the difference of the elastic energy of a defect all over the space and the elastic energy of the opposite half-space. As the elastic fields in the opposite half-space are regular the corresponding energy may be calculated by means of the surface integral over $0x_1x_2$. When using the three- and two-dimensional Parceval equalities one gets

$$W_D = \frac{1}{2} \int \boldsymbol{\beta}_D^*(\boldsymbol{x}) \,..\, \mathrm{c} \,..\, \boldsymbol{\beta}_D(\boldsymbol{x}) \, d^3\boldsymbol{x} - \frac{1}{2} \int [-\boldsymbol{t}^*(\boldsymbol{x}_0)] \cdot \boldsymbol{\alpha}(\boldsymbol{x}_0) \, d^2\boldsymbol{x}_0 = \\ = \frac{1}{16\pi^3} \int \boldsymbol{\beta}_D^*(\boldsymbol{k}) \,..\, \mathrm{c} \,..\, \boldsymbol{\beta}_D(\boldsymbol{k}) \, d^3\boldsymbol{k} + \frac{1}{8\pi^2} \int \boldsymbol{t}^*(\boldsymbol{k}_0) \cdot \boldsymbol{\alpha}(\boldsymbol{k}_0) \, d^2\boldsymbol{k}_0 \,. \tag{25}$$

The second term in (25) will always be negative. For example, the defect from the lower half-space which creates the positive displacement $\boldsymbol{\alpha}$, creates also the state of compression, i.e. the negative $\boldsymbol{t}$ field. Taking (22) and (25) into account, we copy (24) in detailed form in the $\boldsymbol{k}$-representation,

$$W = \frac{1}{16\pi^3} \int \boldsymbol{\beta}_D^*(\boldsymbol{k}) \,..\, \mathrm{c} \,..\, \boldsymbol{\beta}_D(\boldsymbol{k}) \, d^3\boldsymbol{k} + \frac{1}{8\pi^2} \int \boldsymbol{t}^*(\boldsymbol{k}_0) \cdot [\boldsymbol{\alpha}(\boldsymbol{k}_0) + \boldsymbol{\beta}(\boldsymbol{k}_0)] \, d^2\boldsymbol{k}_0 \,. \tag{26}$$

Let us consider the interaction of two defects D1 and D2 in the same half-space. Their interaction energy will be finite if at least one of the defects has finite geometrical form. The other defect may be an infinite dislocation, in particular, crossing the surface. Let the first defect with the elastic distortion $\boldsymbol{\beta}_{D1}$ create the field of the surface tension $\boldsymbol{t}_1$ on the surface $0x_1x_2$, and the second one with the elastic distortion $\boldsymbol{\beta}_{D2}$ create the displacement $(\boldsymbol{\alpha}_2 + \boldsymbol{\beta}_2)$ of this surface. Then from (26), taking into account Betti's theorem, one can get the expression for the interaction energy of the defects in a half-space,

$$W_{12} = \frac{1}{8\pi^3} \int \boldsymbol{\beta}_{D1}^*(\boldsymbol{k}) \,..\, \mathrm{c} \,..\, \boldsymbol{\beta}_{D2}(\boldsymbol{k}) \, d^3\boldsymbol{k} + \frac{1}{4\pi^2} \int \boldsymbol{t}_1^*(\boldsymbol{k}_0) \cdot [\boldsymbol{\alpha}_2(\boldsymbol{k}_0) + \boldsymbol{\beta}_2(\boldsymbol{k}_0)] \, d^2\boldsymbol{k}_0. \tag{27}$$

Proceeding from the methods described in this subsection and from (26) and (27) one can get analogous expressions for the self-energy and the interaction energy of any defects in a plate, a bicrystal, and a sandwich-like system.

3. Discussion of Results and Conclusion

The key result of this paper is the elaboration of a new SDS method for the determination of the relaxation elastic fields of defects for a solid with a free boundary by means of composing and solving the simultaneous equations (6). The peculiarities of the method are the following: 1. the problem of a bounded body is reduced to that of an infinite body using the Gebbia-Eshelby's theorem; 2. the surface tensions $\boldsymbol{t}$ are presented as body forces (4) which are introduced in the right-hand side of the equilibrium equation; 3. the Somigliana dislocation is described by the tensor of a plastic distortion; 4. the condition (5) which connects the Somigliana vector $\boldsymbol{\beta}(\boldsymbol{x}_S)$ with the displacement field $\boldsymbol{u}(\boldsymbol{x})$ is inserted in the simultaneous equations; 5. the Fourier transformation makes the system (6) to be solvable algebraically for a half-space and a plate ((8) to (10)).

The SDS method is a general one as compared with other methods in which the surface distributions of defects are used [2 to 10] as the above method is applicable to defects with any plastic distortion. Therefore, let us consider the connection of the SDS method with the other general methods of the theory of elasticity.

During the last twenty years the method of the boundary integral equation has been developed for the solution of the BP [2, 14, 16]. We write the Maiti and Makan equation [16] using the symbols of the present paper,

$$\boldsymbol{u}(\boldsymbol{x}) = \int \{[\boldsymbol{t}(\boldsymbol{x}_S) - \boldsymbol{t}'(\boldsymbol{x}_S)] \cdot \boldsymbol{\Gamma}(\boldsymbol{x} - \boldsymbol{x}_S) - [\boldsymbol{\beta}(\boldsymbol{x}_S) - \boldsymbol{\beta}'(\boldsymbol{x}_S)] \cdot \boldsymbol{\Pi}(\boldsymbol{n}, \nabla_S) \cdot \cdot \boldsymbol{\Gamma}(\boldsymbol{x} - \boldsymbol{x}_S)\} \, \mathrm{d}S \,, \tag{28}$$

where $\boldsymbol{t}$ and $\boldsymbol{\beta}$ are the surface tensions and displacements on the inner side of the surface, $\boldsymbol{t}'$ and $\boldsymbol{\beta}'$ are the corresponding magnitudes on the external side of the surface: the gradient ∇_S acts on $\boldsymbol{x}_S$. The surface dislocation methods [2 to 12] may be obtained from (28) when $\boldsymbol{t} = \boldsymbol{t}'$. In this case the elastic fields may differ from zero all over the space. Let us introduce the magnitude $2\boldsymbol{b}(\boldsymbol{x}_S) = \boldsymbol{\beta}(\boldsymbol{x}_S) - \boldsymbol{\beta}'(\boldsymbol{x}_S)$. In [2] a wrong conclusion is made: $\boldsymbol{\beta} = \boldsymbol{b}$, $\boldsymbol{\beta}' = -\boldsymbol{b}$. From the formula of the kind (5) one can obtain

$$\boldsymbol{\beta}(\boldsymbol{k}_0) = \boldsymbol{b}(\boldsymbol{k}_0) + \frac{i}{\pi} \int \boldsymbol{\Gamma} \cdot \boldsymbol{\Pi}(\boldsymbol{k}, \boldsymbol{e}_3) \, \mathrm{d}k_3 \cdot \boldsymbol{b}(\boldsymbol{k}_0) \,,$$

$$\boldsymbol{\beta}'(\boldsymbol{k}_0) = -\boldsymbol{b}(\boldsymbol{k}_0) + \frac{i}{\pi} \int \boldsymbol{\Gamma} \cdot \boldsymbol{\Pi}(\boldsymbol{k}, \boldsymbol{e}_3) \, \mathrm{d}k_3 \cdot \boldsymbol{b}(\boldsymbol{k}_0) \,. \tag{29}$$

If in (28) $\boldsymbol{t}' = \boldsymbol{0}$, $\boldsymbol{\beta}' = \boldsymbol{0}$, one can get Somigliana's identity or Betti's second identity. In this case the elastic fields outside the bounded body are clearly zero. Making the Fourier transformation of Somigliana's identity one can get (8) to (10). But the simultaneous equations (6) being equivalent to this identity are more convenient for this purpose.

The SDS method allows to get the Fourier images of the Green tensors for the plane BP. For example, substituting the first expression from (8) in the second one and transposing the result, one can get

$$\boldsymbol{u}(\boldsymbol{k}) = [-\boldsymbol{t}(\boldsymbol{k}_0)] \cdot \{[\mathbf{I} + \mathbf{Y} \cdot \mathbf{A}^{*-1} \cdot \boldsymbol{\Pi}(\boldsymbol{e}_3, i\boldsymbol{k})] \cdot \boldsymbol{\Gamma}\} \,. \tag{30}$$

The tensor in the curly brackets (30) is the Green tensor of the lower half-space for the solution of the BP with a set surface tensions. From (8) it also follows that

$$\boldsymbol{u}(\boldsymbol{k}) = \boldsymbol{\beta}(\boldsymbol{k}_0) \cdot \{[\mathbf{A}^* \cdot \mathbf{Y}^{-1} + \boldsymbol{\Pi}(\boldsymbol{e}_3, i\boldsymbol{k})] \cdot \boldsymbol{\Gamma}\} \,. \tag{31}$$

The tensor in the curly brackets (31) is the Green tensor of the lower half-surface for the solution of the BP with a set surface displacements. Analogically one can get the Green tensor for some other plane BPs.

One of the general methods for the solution of the BP in an isotropic half-space and a plate is Sneddon's method [17] in which the two-dimensional Fourier transformation is used. In the SDS method the three-dimensional one is used. Therefore, the latter method is more simple algebraically and it allows to get the BP solution in the general anisotropic case, too.

The calculation of the elastic fields in multi-layer, heterogeneous systems is an important application of the SDS method. In this paper the method has been demonstrated for a bicrystal and a film on a semi-infinite underlying solid. Now it is easy to generalize this method for heterogeneous systems with arbitrary number of layers. First, the boundaries of each layer are assumed free, and the BPs for the corresponding plates and (or) half-spaces are solved by means of expressions of the kind (8) to (10). Second, the boundaries are connected again, and the surface tensions and displacements are found by means of formulae of the kind (15) to (19). Then the elastic fields $\mathcal{F}'_J$ from (12) are calculated by the corresponding formulae (7) to (12). The method offered here is more general than the one given in [18]. The method of calculation of the elastic fields of any continuously distributed and isolated defects in a bicrystal is also proposed in [19]. Although the elastic fields in [19] consist of three addends analogous to (12) in the present paper, the latter are quite different. For instance, even if a defect is only in the lower half-space, nevertheless one needs to calculate six addends for each elastic magnitude in [19] (three in every half-space). By means of the SDS method one finds only four addends: $\mathcal{F}_1'$, $\mathcal{F}_{D2}$, $\mathcal{F}_2$, $\mathcal{F}_2'$. So the SDS method for heterogenous systems is shorter than the method of [19]. It also differs from a scheme of the BP solution in [1] in which it is suggested to place the same defect in media with different elastic constants.

Formulae (26) and (27) for the calculation of self-energy and interaction energy of the defects in a half-space contain two addends, the first one is the corresponding energy in an infinite solid. In the case of a plate instead of the second term there will be two analogous ones corresponding to the two boundaries. If the defect is only in the lower half-space of a bicrystal, then a third term having the integrand $[-\boldsymbol{\tau}^*(\boldsymbol{k}_0) \cdot \Delta(\boldsymbol{k}_0)]$ should be added to (26) and (27) by analogy with the second one.

4. Conclusion

A new general method, namely the SDS method, is proposed for the investigation of the elastic fields of defects and for the calculation of their self-energy and interaction energy in homo- and heterosystems with the plane boundaries. The close connection of this method with other general methods of the theory of elasticity is ascertained.

Appendix A

The Fourier image of the surface tension field $\boldsymbol{t}(\boldsymbol{k}_0)$ is calculated as follows: The elastic fields of a defect described by the plastic distortion tensor $\boldsymbol{\zeta}$ are defined in $\boldsymbol{k}$-representation by means of the simultaneous equations

$$\boldsymbol{\sigma}_D \cdot i\boldsymbol{k} = \boldsymbol{0}\,, \qquad \boldsymbol{\sigma}_D = \mathbf{c} \cdot\cdot\, \boldsymbol{\beta}_D\,, \qquad \boldsymbol{\beta}_D = -\boldsymbol{\zeta} + i\boldsymbol{k} \otimes \boldsymbol{u}_D\,. \tag{A1}$$

We find the expression for $\boldsymbol{t}(\boldsymbol{k}_0)$ in the plane $x_3 = h$ making the reverse Fourier transformation over k_3,

$$\boldsymbol{t}(\boldsymbol{k}_0) = \frac{1}{2\pi} \int \boldsymbol{\sigma}_D(\boldsymbol{k}) \cdot \boldsymbol{e}_3 \exp(ik_3 h)\, dk_3\,. \tag{A2}$$

Expressing $\boldsymbol{u}_D$, $\boldsymbol{\beta}_D$, $\boldsymbol{\sigma}_D$ from (A1) and substituting $\boldsymbol{\sigma}_D$ in (A2) one gets

$$\boldsymbol{t}(\boldsymbol{k}_0) = -\frac{1}{2\pi} \int [\boldsymbol{\Psi} \cdot \boldsymbol{e}_3 - \boldsymbol{\Pi}(\boldsymbol{e}_3, \boldsymbol{k}) \cdot \boldsymbol{\Gamma} \cdot \boldsymbol{\Psi} \cdot \boldsymbol{k}] \exp(ik_3 h)\, dk_3\,, \tag{A3}$$

where $\boldsymbol{\Psi} = \mathbf{c} \cdot\cdot\, \boldsymbol{\zeta}$.

Appendix B

The meaning of some second-rank tensors and their appearance for an isotropic medium with shear modulus μ and Poisson's ratio ν ($\mathbf{I}$ is the unit tensor, $\mathbf{I}_3^{\pm} = \boldsymbol{e}_3 \otimes \boldsymbol{k}_0 \pm \boldsymbol{k}_0 \otimes \boldsymbol{e}_3$) are as follows:

$$\boldsymbol{\Pi}(\boldsymbol{p}, \boldsymbol{g}) \equiv \boldsymbol{p} \cdot \mathbf{c} \cdot \boldsymbol{g} = \mu \left[\frac{2\nu}{1 - 2\nu}\, \boldsymbol{p} \otimes \boldsymbol{g} + \boldsymbol{g} \otimes \boldsymbol{p} + \boldsymbol{p} \cdot \boldsymbol{g}\mathbf{I}\right]. \tag{B1}$$

$$\boldsymbol{\Gamma} \equiv \boldsymbol{\Pi}^{-1}(\boldsymbol{k}, \boldsymbol{k}) = \frac{1}{\mu k^2}\left[\mathbf{I} - \frac{\boldsymbol{k} \otimes \boldsymbol{k}}{2(1 - \nu)\, k^2}\right]. \tag{B2}$$

$$\mathbf{A} \equiv \mathbf{I} - \frac{i}{\pi} \int \boldsymbol{\Gamma} \cdot \boldsymbol{\Pi}(\boldsymbol{k}, \boldsymbol{e}_3)\, dk_3 = \mathbf{I} - \frac{i(1 - 2\nu)}{2(1 - \nu)\, k_0}\, \mathbf{I}_3^-\,, \tag{B3}$$

$$\mathbf{A}^{-1} = \mathbf{I} + \frac{(1-2\nu)^2}{(3-4\nu)}\left(\mathbf{e}_3 \otimes \mathbf{e}_3 + \frac{\mathbf{k}_0 \otimes \mathbf{k}_0}{k_0^2}\right) + \frac{2i(1-\nu)(1-2\nu)}{(3-4\nu)\,k_0}\,\mathbf{I}_3^- . \tag{B4}$$

$$\mathbf{Y} \equiv \frac{1}{\pi}\int \boldsymbol{\Gamma}\,\mathrm{d}k_3 = \frac{1}{\mu k_0}\left[\mathbf{I} - \frac{\mathbf{e}_3 \otimes \mathbf{e}_3}{4(1-\nu)} - \frac{\mathbf{k}_0 \otimes \mathbf{k}_0}{4(1-\nu)\,k_0^2}\right]. \tag{B5}$$

$$\mathbf{A}^{-1}\cdot\mathbf{Y} = \frac{1}{\mu k_0}\left[\mathbf{I} - \nu\mathbf{e}_3 \otimes \mathbf{e}_3 - \nu\,\frac{\mathbf{k}_0 \otimes \mathbf{k}_0}{k_0^2} + \frac{i(1-2\nu)}{2k_0}\,\mathbf{I}_3^-\right]. \tag{B6}$$

$$\mathbf{B} \equiv \frac{i}{\pi}\int \boldsymbol{\Gamma}\cdot\boldsymbol{\Pi}(\mathbf{k}, \mathbf{e}_3)\exp(ik_3h)\,\mathrm{d}k_3 =$$

$$= \exp(-k_0h)\left\{-\left[\mathbf{I} - \frac{k_0h}{2(1-\nu)}\,\mathbf{e}_3 \otimes \mathbf{e}_3 - \frac{h\mathbf{k}_0 \otimes \mathbf{k}_0}{2(1-\nu)\,k_0}\right] + \right.$$

$$\left. + \frac{i}{2(1-\nu)}\left(\frac{1-2\nu}{k_0}\,\mathbf{I}_3^- + h\mathbf{I}_3^+\right)\right\}. \tag{B7}$$

$$\mathbf{Z} \equiv \frac{1}{\pi}\int \boldsymbol{\Gamma}\exp(ik_3h)\,\mathrm{d}k_3 =$$

$$= \frac{\exp(-k_0h)}{\mu k_0}\left[\mathbf{I} - \frac{1-k_0h}{4(1-\nu)}\,\mathbf{e}_3 \otimes \mathbf{e}_3 - \frac{1+k_0h}{4(1-\nu)}\,\frac{\mathbf{k}_0 \otimes \mathbf{k}_0}{k_0^2} - \frac{ih}{4(1-\nu)}\,\mathbf{I}_3^+\right]. \tag{B8}$$

References

[1] J. D. ESHELBY, in: Dislocations in Solids, Vol. 1, Ed. F. R. N. NABARRO, North-Holland Publ. Co., Amsterdam 1979 (p. 167).
[2] M. J. MARCINKOWSKI, phys. stat. sol. (a) 78, 401 (1983).
[3] K. JAGANNADHAM and M. J. MARCINKOWSKI, J. Mater. Sci. **15**, 709 (1980).
[4] A. E. ROMANOV and V. I. VLADIMIROV, phys. stat. sol. (a) **63**, 109 (1981).
[5] V. I. VLADIMIROV, A. L. KOLESNIKOVA, and A. E. ROMANOV, Fiz. Metallov i Metallovedenie **60**, 1106 (1985) (in Russian).
[6] A. L. KOLESNIKOVA, N. D. PRIEMSKI, and A. E. ROMANOV, Preprint FTI No. 869, Izd. FTI, Leningrad 1984 (in Russian).
[7] A. L. KOLESNIKOVA and A. E. ROMANOV, Preprint FTI No. 1019, Izd. FTI, Leningrad 1986 (in Russian).
[8] J. LOTHE, V. L. INDENBOM, and V. A. CHAMROV, phys. stat. sol. (b) **111**, 671 (1982).
[9] A. Y. BELOV, V. A. CHAMROV, V. L. INDENBOM, and J. LOTHE, phys. stat. sol. (b) **119**, 565 (1983).
[10] A. Y. BELOV and V. A. CHAMROV, Metallofiz. **9**, 68 (1987) (in Russian).
[11] A. F. VLADIMIROV and E. N. MOOS, in: Exoelectron Emission and Its Application, Izd. Tbilisi State University, Tbilisi 1985 (p. 85) (in Russian).
[12] A. F. VLADIMIROV and E. N. MOOS, in: 7th Conf. Processes of Growth and Syntheses of Semiconductor Crystals and Films, Izd. Acad. Sci. USSR, Novosibirsk 1986 (p. 28) (in Russian).
[13] R. DE WIT, The Continuum Theory of Disclinations, Izd. Mir, Moscow 1977 (p. 38) (in Russian).
[14] N. J. ALTIERO and S. D. GAVAZZA, J. Elasticity **10**, 1 (1980).
[15] A. F. VLADIMIROV and E. N. MOOS, phys. stat. sol. (a) **94**, 581 (1986).
[16] M. MAITI and G. R. MAKAN, J. Elasticity **3**, 45 (1973).
[17] W. NOWACKI, The Theory of Elasticity, Izd. Mir, Moscow 1975 (p. 248, 264) (in Russian).
[18] R. BONNET, Phil. Mag. **A 51**, 51 (1985).
[19] C. F. HSIEH and J. DUNDURS, Internat. J. Eng. Sci. **11**, 933 (1973).

(Received June 3, 1988; in revised form October 12, 1988)

phys. stat. sol. (a) **111**, 109 (1989)

Subject classification: 61.70 and 62.20; S10.15

Zentralinstitut für Festkörperphysik und Werkstofforschung der Akademie der Wissenschaften der DDR, Dresden[1])

Grain Boundary Microdamage Due to Visco-Elastic Relaxation of Residual Stresses in Alumina Ceramics

By

A. KRELL and W. POMPE

Grain boundary strengths are estimated experimentally for various Al_2O_3 structures. It turns out that grain boundary strength at room temperature is strongly affected by viscose relaxation processes proceeding during cooling the sintered structures. The results are used to discuss associated macroscopic mechanical properties of Al_2O_3 and $Al_2O_3 + ZrO_2$ as the dependence of fracture strength on the sintering atmosphere, but also on testing atmosphere.

Für unterschiedliche Al_2O_3-Gefüge wird experimentell die Korngrenzenfestigkeit bestimmt. Dabei zeigt sich, daß die bei Raumtemperatur gemessene Korngrenzenfestigkeit stark von viskosen Relaxationsprozessen beeinflußt wird, wie sie beim Abkühlen der Sinterwerkstoffe auftreten. Diese Ergebnisse werden genutzt, um das makroskopische mechanische Verhalten von Al_2O_3 und $Al_2O_3 + ZrO_2$ in seiner Abhängigkeit von der Sinteratmosphäre und den Prüfbedingungen zu diskutieren.

1. Introduction

The macroscopic mechanical behaviour of ceramic materials is strongly influenced by structure, fracture toughness K_{Ic}^{gb}, and strength σ_f^{gb} of the grain boundaries. The properties of the grain boundaries depend especially on the sintering regime. For hot pressed materials with a carbon contamination of 100 ppm it was shown that oxidation can significantly change the defect structure of grain boundaries by pore swelling as a consequence of internal gas pressure [1]. Detailed investigations of the grain boundary fracture toughness of alumina [2, 3] showed remarkable variations of the grain boundary defect structure after conventional sintering in reducing or oxidizing atmospheres. As established for hot pressed ceramics, in sintered alumina we also assume internal stresses to be the driving forces for such structural changes occurring during sintering and subsequent annealing treatments. First of all we consider stresses induced by thermal mismatch or (in single phase non-cubic materials) thermal expansion anisotropy (TEA) to be the dominating parameter. In order to estimate the extent of those stresses in the high temperature range, the role of diffusionally assisted visco-elastic relaxation has to be considered [4].

We expect a contrary influence of sintering parameters on residual stresses and induced defect structure: Decreasing grain boundary viscosity (high diffusion coefficients) should be associated with decreasing residual stresses and, consequently, with an increasing characteristic time of formation for larger flaws. Thus, we assume that two different kinds of defect formation on grain boundary facets can operate:

(i) in the range of high viscosity, generation of single larger microcracks,

(ii) at lower viscosity, cumulative weakening of many grain boundaries by formation of very small micropores due to shear stress components arising from residual stress relaxation.

---

[1]) Helmholtzstr. 20, DDR-8027 Dresden, GDR.

Microcracking in ZrO_2 toughened ceramics predominantly occurs as intergranular grain boundary fracture. Initiation of the tetragonal/monoclinic ZrO_2 transformation is nucleation-controlled and stress-assisted [5]. Since in zirconia toughened alumina (ZTA) these stresses arise from thermal expansion mismatch [6] under the influence of diffisionally assisted visco-elastic relaxation processes [4], grain boundary structures of the alumina matrix are of outstanding importance for the transformation behaviour.

To test these assumptions, the present study was intended to investigate changed relaxation capabilities of alumina grain boundaries after different sintering regimes as well as consequences for the mechanical behaviour at room temperature. To this purpose we varied the amount of MgO additives and the reducing/oxidizing character of the sintering atmosphere. An enhancement of possible effects was to be studied by additional high-temperature annealing. Observed correlation between the defect structure of the grain boundary, its toughness K_{Ic}^{gb} and strength σ_f^{gb}, and microscopic mechanical properties are used for a quantitative estimation of the effects.

2. Methods

Vickers indentations provide straight radial cracks which were used to determine the percentage of intergranular fracture PIF. Grain boundary fracture toughness K_{Ic}^{gb} can be calculated from PIF as reported previously [2]. Transmission electron microscopy gives typical flaw sizes for grain boundary microfracture, allowing an estimation of σ_f^{gb} in the surface region of the specimens (Table 1). Information on typical flaw sizes and geometries and on grain boundary microstructures as obtained by high resolution TEM of the discussed here materials is given in [2] and [7]. The transformation behaviour of ZrO_2 particles, however, will be dominated by presumably higher bulk values of K_{Ic}^{gb}. Those K_{Ic}^{gb} should be determined from the percentage of intergranular fracture (PIF) measured on macroscopic fracture surfaces generated at low crack velocity. Kirchner and Gruver found PIF $\approx$ 12 to 27% for hot-pressed [8] and PIF $\approx$ $\approx$ 8% for air-sintered Al_2O_3 [9]. With the relation given in [2] this also — as in Table 1 — results in a higher K_{Ic}^{gb} for the material sintered in air, but the absolute values of $K_{Ic}^{gb} \approx 2.4$ MPa $m^{1/2}$ are considerably higher than those measured on as-sintered surfaces. The associated grain boundary strength for the bulk of relatively impure Al_2O_3 can be evaluated as $\sigma_f^{gb} \approx$ 2600 to 3000 MPa.

Young's modulus was determined from the resonance frequency of bars with dimension $4 \times 6 \times 50$ mm³. Macroscopic strength σ_f and toughness K_{Ic} were measured in three-point bending with span length 25 to 30 mm and cross-head speed 0.5 (σ_f) and

Table 1
Influence of MgO concentration and sintering atmosphere on grain boundary fracture toughness K_{Ic}^{gb} and grain boundary strength σ_f^{gb} of alumina structures with average grain sizes 2.5 to 4.5 µm and relative densities 98 to 99%

	MgO (wt%)	sintering atmosphere	K_{Ic}^{gb} (MPa m$^{1/2}$)	σ_f^{gb} (MPa)
A	0.15	air	1.02	1270
B	0.05	H_2	0.83 to 0.98	1050
C	0.15	H_2	0.75	840
D	0.30	H_2	0.3 to 0.4	470 to 550

1.0 (K_{Ic}) mm/min as averages of six to ten as-sintered samples. Notch tip radius of K_I-specimens was about 0.1 mm; diamond-wheel notched bars of $Al_2O_3 + ZrO_2$ were annealed at 1350 °C (1 h in air) before testing.

3. Results

3.1 Grain boundary strength and relaxation effects

Table 1 gives previous results [2, 3] for the grain boundary toughness and strength of alumina ceramics. Both parameters diminish with rising MgO concentration and if sintering is performed in reducing atmosphere, i.e. under conditions which promote high-temperature diffusion. To explain this behaviour, replica micrographs (Fig. 1) show that diminishing K_{Ic}^{gb} are associated with a growing concentration of grain boundary microdefects (10 to 50 nm in diameter), most of them being pores. Auger spectroscopy confirmed this characterization of the microdefects as being no second-phase particles. Analysis of the spatial distribution of Auger emission originating from added Mg or from impurities (Si, Ca) was performed on grain boundary facets in fracture surfaces with electron probe focus varying between 0.1 and 100 μm: whereas more

a b c

Fig. 1. Density of micropores on Al_2O_3 grain boundaries (two-stage C-Pt replica taken from fracture surfaces generated at 20 °C): a) low dopant concentration/H_2 sintering (structure B in Table 1), b) raised MgO concentration/H_2 sintering (structure C), c) MgO concentration as b) but sintered in air (structure A)

extended, 1 to 20 μm large areas with locally enhanced Mg concentrations (0.5%) were often observed, there were no indications of smaller, spot-like occurrences of Mg, Si, or Ca enrichments. Further support for the idea of grain boundary micropores originates from small angle neutron scattering results published recently by Page et al., who demonstrated that such small pores grow on alumina boundaries with [10] and even without [11] amorphous phases. Whereas from our observations it is impossible to decide how far the development of micropores on grain boundaries is affected by locally occurring amorphous phases, the important implication of those latter publications [10, 11] is the general independence of such micropore damage of the presence of amorphous phases. On the other hand, it is clear, of course, that the quantitative nucleation and growth kinetics of micropores will be promoted by amorphous phases in grain boundaires.

Often we observed rather different concentrations of micropores on neighbouring grain facets (Fig. 2). The reducing influence of such structures on K_{Ic}^{gb} is readily understood. The heterogeneous appearance points to visco-elastic relaxation of stresses [4] and possibly heterogeneously distributed solid solutions as reasons for the growing microporosity under conditions of intensified high-temperature diffusion.

Of course, instead of being relaxation induced all those micropores observed in cooled structures might have been developed already during isothermal sintering, and different pore densities might result from the influence of grain boundary chemistry on sintering. To test our idea of pore growth during the cooling cycle, specimens of type C (Table 1, Fig. 1b) have been comparatively annealed for 10 min in air and then cooled slowly inside the furnace or taken out of it at 1560 °C to be fan-cooled, respectively. Indeed, Fig. 3 reveals extensive pore growth for the slowly cooled specimens only.

Quite plausibly, avoidance of intensive pore growth by fast cooling increased K_{Ic}^{gb} from 0.75 MPa $m^{1/2}$ (Table 1) to 1.29 MPa $m^{1/2}$. The macroscopic fracture toughness was increased by 10% and fracture strength by 30%. Comparing ground and unground specimens, special experiments demonstrated that macroscopic, quenched-in residual stresses have no influence on the strength [12].

Further evidence for viscose relaxation processes in alumina ceramics with residual stresses resulting from thermal expansion anisotropy (TEA) originates from annealing experiments (Table 2), which reduced Young's modulus by about 6%.

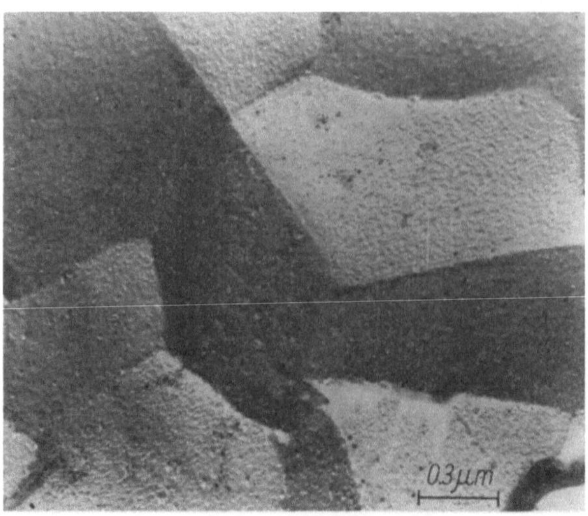

Fig. 2. Heterogeneous distribution of micropores on neighbouring facets of alumina grains. C-Pt replica

Fig. 3. C-Pt replicas taken from fracture surfaces of C-type alumina (Table 1). a) Slowly cooled within the furnace, b) quenched from 1560 °C in air with a fan

3.2 Grain boundary strength and crack propagation

Macroscopic fracture strengths were measured in vacuum and in usual humid air. Table 3 compares the results for three of the structures listed in Table 1. High-temperature diffusion on sintering is assumed to increase in the direction A → C, B → C, which is the direction of rising inert strength. On the other hand, these are the directions of diminishing grain boundary toughness, corresponding to a stronger strength reduction in air (Table 3, [13]).

Table 2

Effect of annealing at 1350 °C/1 h/air on Young's modulus E of Al_2O_3 + 0.16 wt% MgO. (0) as-sintered, (1) annealed specimen state. Density is 3.91 after sintering in air, and 3.93 g/cm³ when sintered in H_2

sintering atmosphere	E (GPa)		E_1/E_0
	(0)	(1)	
air	376	351	0.93
H_2	392	369	0.94

Table 3

Fracture strength of aluminas measured in vacuum (10^{-2} Pa) and in humid air. All specimens have similar K_{Ic} of 5.2 ± 0.2 MPa $m^{1/2}$. Denotations of structural types are the same as in Table 1

	σ_f^{gb} (MPa)	σ_f^{vac} (MPa)	σ_f^{air} (MPa)	$\frac{\sigma_f^{air} - \sigma_f^{vac}}{\sigma_f^{vac}}$
A	1270	458 ± 65	430 ± 45	−0.06
B	1050	515 ± 65	410 ± 40	−0.20
C	840	620 ± 110	403 ± 45	−0.35

Comparing ZTA ceramics sintered in hydrogen or in air, X-ray analysis gave higher monoclinic ZrO_2 content after air sintering. For Al_2O_3 + 11.5 Vol% ZrO_2 we found $x_m = 0.75$ (air) and $x_m = 0.67$ (H_2), if ZrO_2 had been added as a powder with average agglomerate size 1.1 μm. Nevertheless, Fig. 4 [3] reveals higher Young's moduli for the air-sintered specimens in spite of their higher x_m.

Obviously, in the case of a relatively coarse ZrO_2 dispersion, the air-sintered alumina matrix can better bear transformation induced residual stresses without relaxation by microcracking. The explanation which appeals most to the authors is the high grain boundary fracture toughness of air-sintered alumina.

4. Discussion

Diminishing grain boundary strengths at increasing dopant concentration (Table 1) is associated with increasing microporosity (Fig. 1, 2). Thus, the generation and existence of such small pores at high temperatures have to be explained.

Specific surface energy $\gamma = 0.2$ to 1.0 J/m² (from K_{Ic}^{gb} in Table 1) and a pore radius of $R = 25$ nm result in a Laplace stress of about 8 to 40 MPa, which is lower than the average TEA residual stresses in Al_2O_3 (X-ray measurements gave 60 to 100 MPa). Following the argumentation of Howlett and Brook [1] we assume that micropores

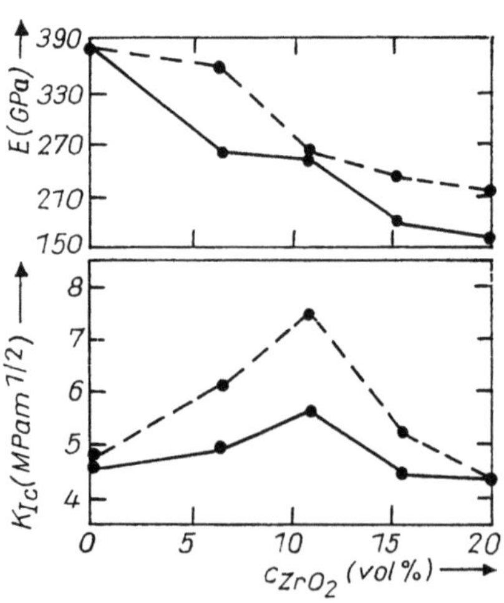

Fig. 4. Young's modulus and fracture toughness of $Al_2O_3 + ZrO_2$, fabricated from coarse ZrO_2 powder (1.1 μm) and sintered in air or in hydrogen, respectively [3]. – – – 1600 °C/25 h/air; ——— 1625 °C/1 h/H_2

grow by a volume diffusion process. We describe the change of the pore radius r by

$$\frac{\mathrm{d}r}{\mathrm{d}t} = \frac{D_l \Omega}{r k T} \left(p_{\mathrm{int}} + \sigma_{\mathrm{res}} - \frac{2\gamma}{r} \right), \tag{1}$$

where D_l is the lattice diffusion coefficient, p_{int} the internal gas pressure in pores, σ_{res} a residual tensile stress acting hydrostatically, and γ the gas/solid interfacial energy. Atomic volume, Boltzmann constant, and absolute temperature are termed Ω, k, T. Unlike the treatment in [1] for hot-pressed materials, σ_{res} and the Laplace stress $2\gamma/r$ cannot be neglected in comparison to p_{int}. On the contrary, keeping in mind that $p_{\mathrm{int}} \sim (T/T_0)(r_0/r)^3$ we assume that p_{int} can be excluded as a dominating driving force for pore growth under conditions of conventional sintering — although it might be essential for pore nucleation. If, for instance, we try to explain our experiments by the action of p_{int}, a decreasing microdamage of the grain boundaries in samples sintered under H_2 should be observed. The temperature dependence of thermal stress is for $T < T_A$ in structures with Young's modulus E and thermal expansion mismatch $\Delta\alpha$,

$$\sigma_{\mathrm{res}} \approx E\, \Delta\alpha (T_A - T)\,, \tag{2}$$

T_A is the temperature where the structure is yet free of stress. If the relaxation behaviour at high temperatures as calculated in [4] is approximated by T_A as an effective temperature, this T_A depends on cooling rate $\dot{T}$, diffusion coefficients of lattice and grain boundaries, and, of course, on grain size. Below T_A on cooling the specimen the driving force σ_{res} for pore growth increases, but, on the other hand, D_l/kT decreases. Depending on the cooling rate, the superposition of these two mechanisms yields some growth of pre-existing pore nuclei (with size r_A at T_A). A rough estimate using (1) yields for the pore size r_T at temperature T

$$r_T^2 - r_A^2 = \frac{2\Omega}{k\dot{T}} \int_T^{T_A} \frac{\mathrm{d}T'}{T'} D_l(T')\, \sigma_{\mathrm{res}}(T')\,. \tag{3}$$

Thus the relative change of the pore size $\Delta r/r_0$ increases as

$$\frac{\Delta r}{r_A} \sim \frac{\bar{D}_l (T_A - T)}{r_A^2 \dot{T}}, \tag{4}$$

where $\bar{D}_l$ is an effective diffusion coefficient averaged according to (3) [14]. Therefore, we conclude that micropores might be generated in the course of viscous relaxation processes when cooling the sintered structure.

Such a mechanism should be the more effective, the higher the diffusion coefficients are. In this way the dependence of room temperature σ_f^{gb} and K_{Ic}^{gb} on high-temperature diffusion (promoted by MgO additives or reducing atmosphere, cf. Table 1) is well understood. The associated cumulative growth of micropores on tension stressed grain facets is demonstrated in Fig. 1. Additional annealing at 1350 °C (Table 2) enhances this process. The observed drop of Young's modulus is related with a high density of micropores at those grain boundaries. Due to the interconnecting topology of the grain boundary network the decrease of grain boundary stiffness markedly influences the macroscopic Young's modulus of the sample even without measurable change of its density.

Since in inert environment a direct correlation exists between σ_f and K_{Ic} [13], the increasing inert strengths documented in Table 3 for structures with rising high-temperature diffusion, but constant K_{Ic}, must be associated with decreasing flaw sizes responsible for instability. The size of these flaws is in the order of 50 μm. Thus, while high diffusion coefficients reduce the size of such single flaws, they give rise to extensive relaxation phenomena generating many additional (very small) grain boundary microdefects. The resulting low K_{Ic}^{gb} causes a strong reduction of strength in humid air due to subscritical crack growth (Table 3).

Whereas in our experiments with $Al_2O_3 + ZrO_2$ Mg dopants did never enhance x_t, the observed higher stability of tetragonal ZrO_2 in H_2-sintered samples compared with structures sintered in air might be caused by stoichiometric changes. Additionally, we suppose relaxation of TEA residual stresses to be an additional factor. Higher diffusion coefficients in H_2-sintered ZTA reduce the tensile stresses generated by thermal mismatch and complicate the nucleation process. If, however, some tetragonal ZrO_2 transforms in H_2-sintered specimens, the lower grain boundary strength is less able to bear the resulting stresses than after sintering in air, and a lower Young's modulus is measured as a consequence of spontaneous microcracking (Fig. 4). The lower macroscopic toughness of such H_2-sintered structures is thus readily explained by lower microstrength (σ_f^{gb} critical stress for microcracking), lower residual stresses around monoclinic ZrO_2 particles (due to stress relaxation by spontaneous microcracking), and within the alumina matrix (due to viscous relaxation).

Quite naturally, this effect is lost if the ZrO_2 particles themselves acquire high stability, e.g. due to their smaller grain sizes resulting in shorter diffusion path of H species within ZrO_2 grains and in an increased influence of diffusional and stoichiometric processes on the stability of tetragonal ZrO_2. Consequently, the importance of the matrix grain boundary strength for spontaneous transformation induced microcracking and Young's modulus is diminished under such conditions. It is, therefore, no surprise that investigating zirconia toughened aluminas with ZrO_2 grain sizes below 1 μm in the sintered structure (0.04 μm in the raw material) we always measured higher strengths and toughnesses after sintering in hydrogen than in air. Although the explanation of additional atmospheric influences on ZrO_2 stability at very small grain size is clearly beyond the scope of the present paper, it should be pointed out that, because of the changed importance of the alumina grain boundary strength, there is no contradiction between such findings and our suggestions concerning the behaviour of larger, less stable ZrO_2 grains (Fig. 4). The basic idea to be emphasized is that visco-elastic relaxations in the grain boundaries of a ceramic matrix like Al_2O_3 do not only result in micropore generation and in the formation of weaker boundaries more susceptible to brittle microfracture at room temperature — additionally, the relaxation of residual stresses affects the transformation behaviour of metastable phases as unstabilized or partially stabilized ZrO_2.

These suggestions are further substantiated by experimental results recently published by Witek and Butler [15]. They examined structures made from Al_2O_3 + + 15 to 20% ZrO_2 (5% Y_2O_3) without (A2Z) and with 3% glass constituents (A3Z), where the less pure composition A3Z showed the more extensive fracture induced phase transformation of ZrO_2 — due to the larger ZrO_2 grain size in that structure. However, on cooling just that less pure material A3Z exhibited a higher stability against spontaneous transformation, which behaviour, therefore, cannot be dominated by specific properties of the ZrO_2 particles themselves. The explanation which appeals most to the present writers is the preferred visco-elastic relaxation of thermal misfit stresses in A3Z, causing a reduction of these transformation-initiating residual stresses. On macroscopic fracture, however, this mechanism is of minor importance, since residual misfit stresses are always low compared to the localized stresses generated by the macroscopic load.

5. Conclusions

Visco-elastic, diffusion-assisted grain boundary deformation in alumina had been discussed by Blendell and Coble [4] as a relaxation mechanism of TEA residual stress. We observe grain boundary microdamage (pores 10 to 50 nm), which becomes increasingly severe under conditions of rising high-temperature diffusion. Since residual stress-induced deformation seems to be the most probable reason for that microdamage, we take viscous relaxation for a mechanism of fundamental importance. It dominates not only TEA stress but also the characteristic time of defect formation. Therefore, the mechanical properties at room temperature as grain boundary strength and thus subcritical crack growth and microcracking-induced stress relaxation around dispersed transforming ZrO_2 particles, markedly depend on the sintering regime. Influencing the stress and strain fields around tetragonal ZrO_2 in an alumina matrix, viscous grain boundary relaxation is an important factor for transformation nucleation.

Acknowledgement

The authors wish to thank Dipl-Phys. T. Weiß, who provided Fig. 3 and examined grain boundary toughness of fan-cooled specimens.

References

[1] S. Howlett and R. J. Brook, in: Proc. 2nd Irish Durability and Fracture Conf., Nihe, Limerick, 1984, Ed. J. D. Bolton and S. Hampshire (p. 320).
[2] A. Krell, J. Woltersdorf, E. Pippel, and D. Schulze, Phil. Mag. A51, 765 (1985).
[3] A. Krell, E. Pippel, and J. Woltersdorf, Phil. Mag. A53, L11 (1986).
[4] J. E. Blendell and R. L. Coble, J. Amer. Ceram. Soc. 65, 174 (1982).
[5] M. Rühle and A. H. Heuer, in: Advances in Ceramics, Vol. 12, Ed. N. Claussen, M. Rühle, and A. H. Heuer, Amer. Ceram. Soc., Columbus (Ohio) 1984 (p. 14).
[6] W. M. Kriven, ibid (p. 64).
[7] E. Pippel and J. Woltersdorf, Phil. Mag. A56, 595 (1987).
[8] H. P. Kirchner and R. M. Gruver, J. Mater. Sci. 14, 2110 (1979).
[9] H. P. Kirchner and R. M. Gruver, J. Amer. Ceram. Soc. 63, 169 (1980).
[10] R. A. Page, J. Lankford, K. S. Chan, K. Hardman-Rhyne, and S. Spooner, J. Amer. Ceram. Soc. 70, 137 (1987).
[11] R. A. Page and J. Lankford, J. Mater. Sci. 22, 2521 (1987).
[12] K. Lippmann and A. Krell, Sprechsaal 121, 268 (1988).
[13] A. Krell and W. Pompe, Mater. Sci. Engng. 89, 161 (1987).
[14] W. Pompe and H.-A. Bahr, Silikattechnik 38, 97 (1987).
[15] S. R. Witek and E. P. Butler, J. Amer. Ceram. Soc. 69, 523 (1986).

(Received November 6, 1987; in revised form September 8, 1988)

phys. stat. sol. (a) **111**, 119 (1989)

Subject classification: 61.70 and 62.50; S1.2

U.S. Army Materials Technology Laboratory, Watertown[1])

Fracture in Titanium Diboride

By

D. M. VANDERWALKER

Transmission electron microscopy is used to examine the structure of shock-loaded titanium diboride. At a shock pressure of 1.7 GPa, fracture occurs by the nucleation and splitting of $\boldsymbol{b}$ = = [001] dislocation loops in the basal plane. The tensile stress of the shock pulse opens the dislocation loops into microcracks. The dislocation loops form when point defects spontaneously condense at impurities. Dislocation loops are also left by moving dislocations in a deformation band and nucleate at pressure boundaries.

Transmissionselektronenmikroskopie wird benutzt, um die Struktur von Schock-belastetem Titandiborid zu untersuchen. Bei einem Schock-Druck von 1,7 GPa tritt Bruch auf durch Bildung und Aufspaltung von $\boldsymbol{b}$ = [001]-Versetzungsschleifen in der Basisebene. Die Dehnungsspannungen der Schockimpulse öffnen die Versetzungsschleifen in Mikrorisse. Die Versetzungsschleifen bilden sich, wenn die Punktdefekte spontan an Störstellen kondensieren. Versetzungsschleifen werden auch von wandernden Versetzungen in einem Deformationsband zurückgelassen und bilden sich an Druckgrenzen.

1. Introduction

Failure by ballistic impact is a subject of importance [1, 2]. The shock waves introduce deformation and by the interaction of incident and reflected waves, spall. Fracture can be brittle or ductile depending on whether the fracture stress exceeds the yield stress [3]. In some materials, there is a brittle–ductile transition temperature. The application of hydrostatic pressure can lower the flow stress [4]. At high strain rates, dynamic stress–strain relations apply [1] with the elastic regime defined as below the Hugoniot elastic limit (HEL).

The brittle fracture of solids has been modeled by Griffith [5] to explain the difference between the theoretical cohesive strength and the experimental value. He assumed that glass contains flaws which create stress concentrations. For crack propagation to occur, the total elastic energy must exceed the surface energy of the crack. The stress at the microcrack contributes to the total elastic energy, reducing the applied stress needed for failure. The dislocation models of brittle fracture describe the process as microcrack formation by dislocation coalescence at low-angle grain boundaries, twins, and slip bands [6 to 8]. In hexagonal materials, brittle fracture has been proposed to occur as a result of crystalline anisotropy [9] or by cleavage in (001) plane [10] .In some cases, Sohncke's critical normal stress law applies as a criterion for cleavage fracture [11, 12].

Gust et al. [13] measured the shock properties of several light-weight intermetallics. The Hugoniot elastic limits were found to be in the 5.3 to 8.7 GPa range. The spall strength of brittle materials varies since its value is 3.0 GPa in glass [14] while 0.3 and 0.2 GPa in alumina [15] and titanium diboride [16], respectively. Transmission

[1]) Watertown, Massachusetts 02172, USA.

electron microscopy was used to investigate deformation produced by shock waves in TiB_2 [17]. Annealing experiments show by the observation of large stacking fault configurations that a high concentration of point defects is produced by the stress pulse. This paper is on fracture of titanium diboride far below the HEL.

2. Experimental Procedure

Plate-slap experiments were performed on the 50 mm gas/powder gun at the University of Dayton Research Institute [16]. A $2.1 \times 2.1 \times 1.27$ cm TiB_2 plate, mounted in a Ni frame, was impacted by Cu star-shaped flyer at a velocity of 80 m/s (shot 1). At a peak stress of 1.7 GPa, spall occurred. As evidenced there is a spall ring on the Ni frame. The recovered sample was sectioned, cut into 3 mm discs, and ion beam thinned for transmission electron microscopy (TEM). Samples from TiB_2 plates impacted with a W bullet were also included in this work (shot 2). The TEM was performed on the JEM 200CX transmission electron microscope. X-ray STEM microanalysis and electron energy loss spectroscopy (EELS) was used for element identification. The dislocation were characterized by Burgers vector analysis and trace analysis.

3. Results

The shock-loaded titanium diboride samples were found to contain dislocation loops and split loops or voids. Fig. 1 is a bright-field image taken from a region with a high density of voids. The voids have a hexagonal projection in the basal plane (Fig. 2). Fig. 3 shows the difference in contrast between a dislocation loop, on the left side of the micrograph, and split loops on the right. The voids have white contrast in the center. There is a trail of dislocation loops behind the dislocation intersection in Fig. 4a. The dislocation loops were found to nucleate at impurity particles (Fig. 4b). At the edge of more than one dislocation loop there is strong contrast from the particle interface. The STEM X-ray results show that solid particles contain Fe and/or Mo. X-ray spectra from dislocation loops or voids contain Fe peaks (Fig. 5a) and EELS spectra from loops the nitrogen peak (Fig. 5b).

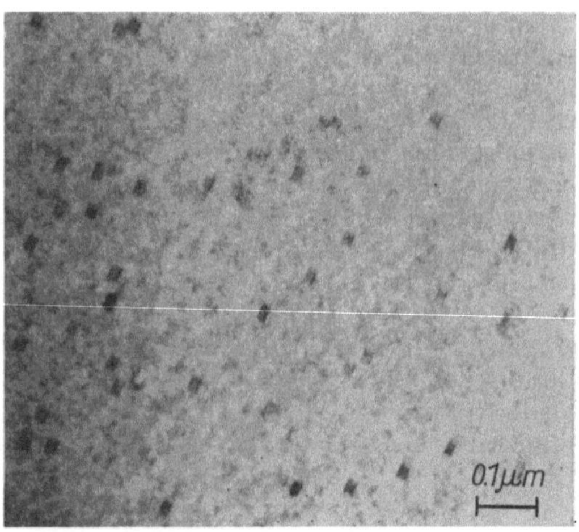

Fig. 1. A high density of voids in a TiB_2 plate (shot 2)

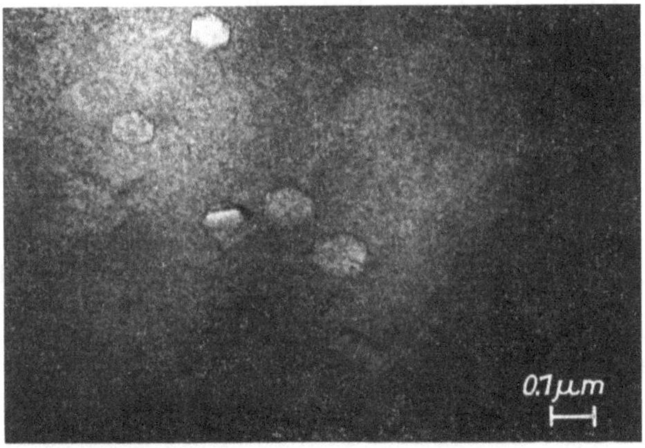

Fig. 2. Hexagonal split loops in a TiB_2 sample (shot 1)

Dislocation loops form a trail in the basal plane in the deformation band in Fig. 6. There are straight segments along the $\xi = [00.1]$ direction attached to loops of Burgers vector $\boldsymbol{b} = [00.1]$. Fig. 7 is a micrograph of an extended dislocation configuration formed by the dissociation of $\boldsymbol{b} = [00.1]$ into two 1/2[00.1] partials. Dislocation loops nucleate on an internal surface at the edge of a deformation zone in Fig. 8. The interface is not a grain boundary since the selected area diffraction patterns are identical on each side.

4. Discussion

A model for brittle fracture of ceramics can be deduced from the experimental evidence. Fracture occurs by the nucleation and splitting of dislocation loops of Burgers vector $\boldsymbol{b} = [00.1]$. The dislocation loops form spontaneously by condensation of point defects emitted by dislocation jogs [18, 19]. The tensile stress from the shock pulse opens the dislocation loops into microcracks.

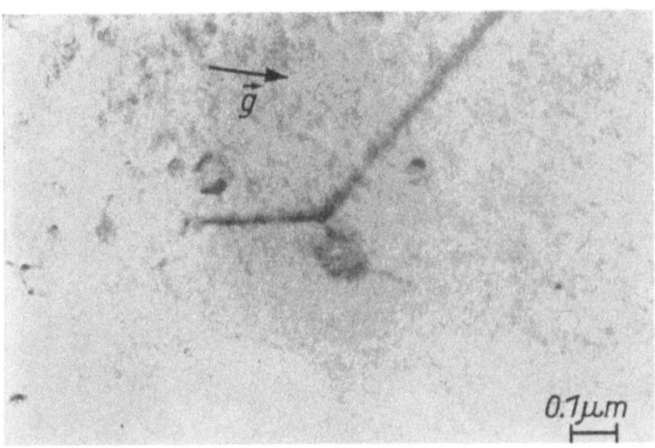

Fig. 3. Dislocation loops, hexagons, and voids from annealed TiB_2 sample (shot 2). The voids are distinguished by the white contrast at the center

Examine the thermodynamics of dislocation loop nucleation. The free energy change [18] for homogeneous loop nucleation has two contributions, the first term is the elastic energy per unit length needed to create the loop, the second is the chemical potential associated with vacancy condensation. The second term depends on vacancy formation enthalpy [18],

$$G = 2\pi R \frac{W}{L} - \frac{\pi R^2}{v_a} \bar{G} ,$$

where W/L is the elastic energy per unit length, $\bar{G} = W_v/2$ the chemical potential, W_v the vacancy formation energy, R the loop radius, d the interplanar spacing, and $v_a = b^3$ the volume associated with one vacancy. The elastic energy for a $\boldsymbol{b} = [001]$ dislocation loop can be evaluated from the following expression [18]:

$$W = \frac{2\pi R \mu b^2}{4\pi(1-\nu)} \left(\ln \frac{4R}{\varphi} - 1\right),$$

where μ is the shear modulus, $\boldsymbol{b}$ the Burgers vector, ν Poisson's ratio, R the loop radius, φ the dislocation core radius. The vacancy formation enthalpy can be approximated by the Lindemann equation [20],

$$\theta = C' \left(\frac{\Delta W_v}{M V^{2/3}}\right)^{1/2}$$

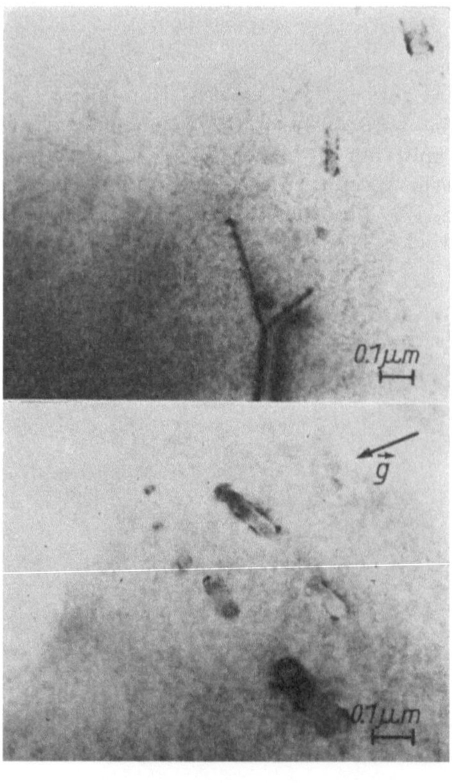

a

b

Fig. 4. a) Dislocation loops near dislocation intersection. b) Loop nucleation at impurity particles (shot 1)

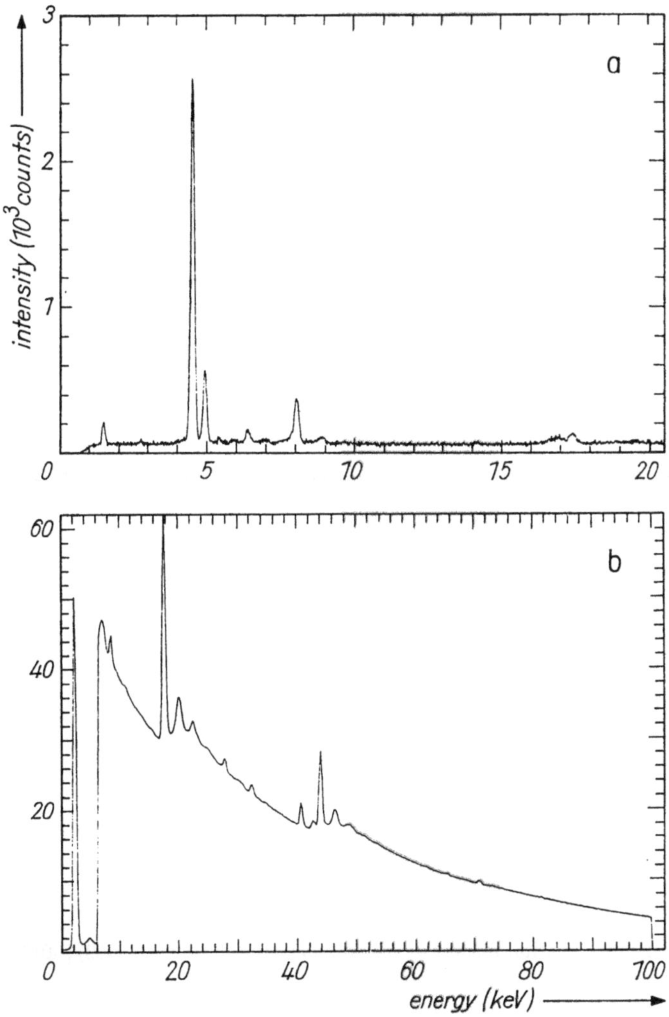

Fig. 5. a) X-Ray spectra taken from voids or loops which contain the Fe peak. b) EELS spectra from voids or dislocation loops which contain the nitrogen edge

with M the molecular weight, V the molar volume, θ the Debye temperature, and $C' = 34.3$ [20]. When the Debye temperature $\theta = 1024$ K [21], the vacancy formation energy of the TiB_2 molecule is 6.6 eV. The second term in the free energy expression is a large negative term. The equation predicts spontaneous loop nucleation. About 10^{-15} J of energy is released when 40 vacancies condense.

There is an additional negative term associated with heterogeneous nucleation of a dislocation loop at an impurity cluster if binding energy is released by point defect–impurity bonding. The tendency for vacancy–impurity bonding is thought [22] to depend on impurity valence in quenched gold because loops nucleate at Mg^{+2}, tetrahedra at Al^{+3}, and neither at Cu^{+1}. In TiB_2, loops nucleate at Fe and N impurity clusters. Fracture is more probable in grains containing impurities and those at orientations with stress wave components parallel to c-axis.

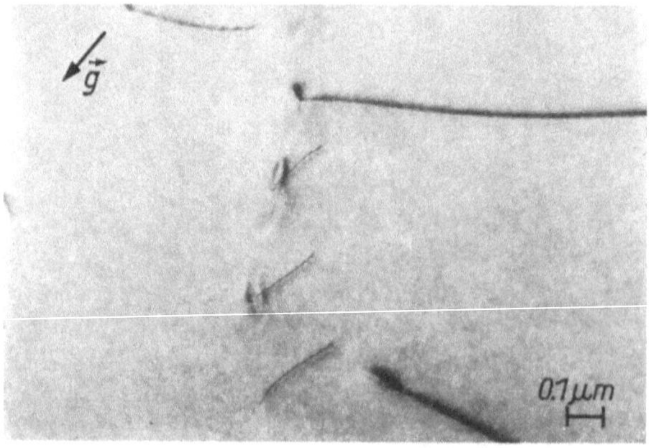

Fig. 6. Dislocation loops of Burgers vector $\boldsymbol{b} = [00.1]$ are connected to dislocations in this deformation band (shot 1)

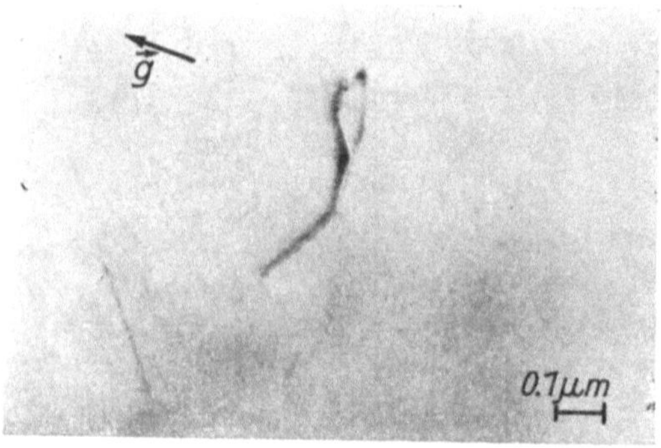

Fig. 7. An extended dislocation configuration formed by the dissociation of [001] → (1/2) [001] + + (1/2) [001] (shot 1)

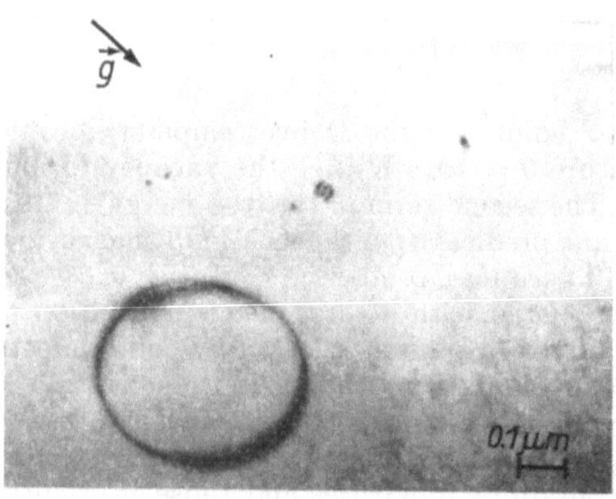

Fig. 8. Dislocation loops nucleate at pressure boundary (shot 1)

The point defects produced in deformation will control the size and distribution of dislocation loops and thus the fracture process. The vacancy and interstial concentrations have been found to depend on both strain rate and total strain [23, 24]. The applied forces acting on dislocation jogs is a controlling factor since there is a large driving force for loop nucleation.

Fracture can occur along a deformation band by a similar mechanism. In some cases, dislocations which propagate from a source have segments on two slip planes. In Fig. 6, dislocation loops of Burgers vector [00.1] in the basal plane are connected to segments in the prismatic plane oriented along the $\xi = [00.1]$ direction. A large microcrack can form along the band if a tensile stress splits the loops open.

Extended jogs which form in TiB_2 may be fracture initiation sites much like twins [2] at higher impact velocities. A variety of dislocation structures have been observed in TiB_2. In hexagonal materials, there are two types of stacking faults, that produced by collapse and shear with displacement vector $\boldsymbol{R} = (1/6)$ [20.3] and pure shear $\boldsymbol{R} = [100]$ [18, 25]. It is possible for a stress component along $\langle 100 \rangle$ to induce the (1/2) [00.1] loop to transform to a configuration with a (1/6) $\langle 203 \rangle$ stacking fault. $(1/2)\,[00.1] + (1/3)\,\langle 100 \rangle \rightarrow (1/6)\,\langle 20.3 \rangle$. The large stacking fault in Fig. 7, bound by (1/2) [00.1] partials, is an example of one observed configuration. In this case, large split loops or sheets may form from extended jogs.

At the edge of the deformation zones, there are often internal surfaces which separate the deformed region from the perfect crystal. There is no misorientation between crystals on each of the surfaces. It is thought to be a pressure boundary where the incident compression wave ends or interferes with a wave reflected from a grain boundary. Fig. 8 shows the internal surface decorated with dislocation loops. Loop nucleation occurs at the pressure gradient because at this position, there is an abrupt change in the equilibrium vacancy concentration, which depends on pressure. The boundary is likely to be the positive/negative pressure boundary.

5. Summary

1. The brittle fracture of titanium diboride can occur by the nucleation and splitting of dislocation loops. A tensile stress from the shock pulse opens the loops into microcracks.

2. The high vacancy formation energy of TiB_2, 6eV, causes there to be a large driving force for loop nucleation. Dislocation loops and voids nucleate at impurity clusters such as Fe and N and are formed by moving dislocation in a deformation band. There exists pressure boundaries at the edge of deformation zones on which loops nucleate.

References

[1] M. H. Rice, R. G. McQueen, and J. M. Walsh, Solid State Phys. 6, 1 (1958).
[2] M. A. Meyers and L. E. Murr, Shock-Waves and High Strain Rate Phen in Metals Concepts and Applications, Plenum Press, New York 1981 (p. 487).
[3] A. H. Cottrell, Proc. Roy. Soc. **A276**, 1 (1963).
[4] J. Castaing, P. Veyssiere, L. P. Kubin, and J. Rabier, Phil. Mag. **44**, 1407 (1981).
[5] A. A. Griffith, Phil. Trans. Roy. Soc. **A221**, 163 (1921).
[6] A. H. Cottrell, Trans. AIME **212**, 192.
[7] R. L. Bell and R. W. Cahn, Proc. Roy. Soc. **A239**, 494 (1957).
[8] A. N. Stroh, Phil. Mag. **3**, 597 (1958).
[9] A. P. Green and J. Sawkill, J. nuclear Mater. **3**, 101 (1961).
[10] D. J. Burr and N. Thompson, Phil. Mag. **7**, 1773 (1962).
[11] N. J. Petch, Progr. Metal Phys. **5**, 1 (1954).

[12] E. Schmid and W. Boas, The Plasticity of Crystals, F. A. Hughes & Co. Ltd., London 1950 (p. 169).
[13] W. H. Gust, A. C. Holt, and E. B. Royce, J. appl. Phys. **44**, 550 (1973).
[14] Z. Rosenberg, D. Yaziv, and S. Bless, J. appl. Phys. 58, 3249 (1985).
[15] Z. Rosenberg, G. Luttwak, Y. Yeshurun, and Y. Parton, J. appl. Phys. **54**, 2147 (1983).
[16] N. S. Brar, S. J. Bless, and D. Yaziv, UDR-TR-87-149 (1988).
[17] D. M. Vanderwalker and W. J. Croft, J. Mater. Res. **3**, 761 (1988).
[18] J. P. Hirth and J. Lothe, Theory of Dislocations, John Wiley & Sons, New York 1982 (p. 612).
[19] R. W. Balluffi and A. V. Granato, Dislocation in Solids, Ed. F. R. N. Nabarro, North-Holland Publ. Co., 1979 (p. 82).
[20] R. A. Swalin, Thermodynamics of Solids, John Wiley & Sons, New York 1972 (p. 276).
[21] D. Singh and Y. P. Varshni, Acta. cryst. **A38**, 854 (1982).
[22] I. A. Johnston, P. S. Dobson, and R. E. Smallman, Proc. Roy. Soc. **A315**, 231 (1970).
[23] H. Kressel and M. Brown, J. appl. Phys. **38**, 1618 (1967).
[24] J. J. Dick and D. L. Styris, J. appl. Phys. **46**, 1602 (1975).
[25] M. H. Loretto and R. E. Smallman, Defect Analysis in Electron Microscopy, Chapman and Hall, London 1975 (p. 70).

(Received June 17, 1988; in revised form September 12, 1988

phys. stat. sol. (a) **111**, 127 (1989)

Subject classification: 61.40 and 64.70; 62.40; S1.1; S1.2

Institute of Physics and Chemistry of Metals, Silesian University, Katowice[1]) *(a) and Institute of Physical Metallurgy and Welding, Silesian Polytechnical University, Gliwice*[2]) *(b)*

Activation Parameters of Linear Heating Crystallisation in $Ni_{70.3}Fe_{4.7}Si_{15}B_{10}$ Glassy Alloy Determined Using the Internal Friction Method

By

M. Kłosek (a), J. W. Moroń (a), and T. Połoczek (b)

Using the internal friction method the effective activation energy E of crystallisation during linear heating is determined for the glassy alloy $Ni_{70.3}Fe_{4.7}Si_{15}B_{10}$. E values are obtained: from the Kissinger formula 2.49 eV, from the Augis-Bennett formula 2.51 eV. It is shown that this agreement is not accidental. The pre-exponential factor in the Arrhenius law K_0 is calculated for four alloys investigated in the literature by the IF method. Values obtained lay in a very wide interval 10^{14} to 10^{27} s^{-1}. It is shown that ln K_0 increases approximately linearly with E.

Die effektive Aktivierungsenergie E der Kristallisation während linearer Erwärmung wird in amorphen $Ni_{70,3}Fe_{4,7}Si_{15}B_{10}$-Legierungen mit Hilfe der inneren Reibung untersucht. Für E werden folgende Werte ermittelt: aus der Kissinger-Formel 2,49 eV, aus der Formel von Augis und Bennett 2,51 eV. Es wird gezeigt, daß diese Übereinstimmung nicht zufällig ist. Der präexponentielle Koeffizient K_0 des Arrheniusgesetzes wird für vier Legierungen, die in der Literatur mit der Methode der inneren Reibung untersucht wurden, berechnet. Die erhaltenen Werte liegen in einem sehr breiten Bereich, 10^{14} bis 10^{27} s^{-1}. Es wird gezeigt, daß ln K_0 mit F fast linear ansteigt.

1. Introduction

In recent years the phenomenon of crystallisation in glassy alloys has attracted considerable attention. This phenomenon is of major significance both from the aspect of basic research and of the practical utilisation of these alloys. Apart from the standard methods normally used in these studies (e.g. DTA, DSC) researchers have lately begun to make use of the internal friction method (IF) which is very sensitive to all structural changes in the specimen. When applying this method with continuous heating of the specimen, at the temperature at which crystallisation takes place most rapidly, on the curves of the reciprocal quality factor of the pendulum as a function of temperature $Q^{-1}(T)$ there appears a high IF peak, usually lying above the temperature T_g of the glassy transition: this peak is associated with a minimum of the shear modulus (G) or Young's modulus, which are proportional to the square of the frequency f of the pendulum vibrations (e.g. [1, 2]). With increase in heating rate v_g the temperature T_p of this peak shifts towards higher values of T, but this temperature T_p is independent of the pendulum vibrations frequency.

The crystallisation peak should be distinguished from the normally very high IF peaks situated immediately above T_g $\left(Q^{-1}(T_p) \approx 1\right)$, which shift with change in

[1]) Bankowa 12, PL-40-007 Katowice, Poland.

[2]) Konarskiego 18a, PL-44-100 Gliwice, Poland.

frequency. It would appear that the most important criterion making it possible to distinguish crystallisation peaks from peaks of the second type, which in the literature are associated with the so-called viscoelastic [3 to 5] or viscoplastic [6] relaxation, is whether or not there occurs at T_p, for the same heating rate, an exothermal DTA or DSC peak; for crystallisation a peak of this kind occurs (e.g. [7]), while for viscoelastic relaxation the DTA or DSC peak does not occur [8]. Sometimes it is also possible from direct observation to determine if the observed peak comes from a phase transition or is related to viscoelastic relaxation [9].

A general study has recently been made of IF peaks appearing near T_g in glassy alloys, as reported in [10]. These determinations have principally been based on curves of T_p as a function of v_g or of f, i.e. parameters important from the point of view of a method for estimating the effective activation energy of crystallisation or viscoelastic relaxation.

Up to now the IF method has only been used for the determination of the effective activation energy of crystallisation E for two alloys, i.e. $Ni_{68.7}Cr_{6.6}Fe_{2.65}Si_{7.8}B_{14}C_{0.25}$ (alloy 1) [1, 2] and $Ni_{74}P_{16}B_6Al_4$ (alloy 2) [11]; in the present paper E is also estimated for a third alloy, $Fe_{40}Ni_{40}P_{10}B_{10}$ (alloy 3), based on data given in [7]. In all three cases additional properties of these peaks indicate that the first IF maximum above T_g is associated with crystallisation [10]. In these investigations, measurements were repeated for a fourth alloy, i.e. $Ni_{70.3}Fe_{4.7}Si_{15}B_{10}$ (alloy 4).

2. Samples and Procedure

Tests were conducted on ribbon samples of nonferromagnetic $Ni_{70.3}Fe_{4.7}Si_{15}B_{10}$ of cross-section 5×0.03 mm^2 and length about 30 mm, produced at the Institute of Materials Science and Engineering of the Warsaw Technical University using a standard melt-spinning method. Measurements were made for two series of samples: 1. as-quenched samples, 2. samples additionally aged for 2.2×10^7 s at room temperature. For the first series, X-ray examination showed that this alloy is amorphous. Measurements conducted for the second series confirmed that the samples do not undergo crystallisation during ageing: for series 1 samples the height of the first peak in the vicinity of T_g was 0.026, and for series 2 samples —0.027. It is noteworthy that for series 1 samples the temperature of the peak was 710 K, and for series 2 samples it was 711 K ($v_g = 0.033$ K/s) indicating clearly that the test material exhibited satisfactory chemical homogeneity.

Internal friction was determined using an inverted Kê-type pendulum. Measurements were made in a vacuum of 0.013 Pa in the temperature range from 300 to 1000 K. The rate of linear heating applied was from 0.010 to 0.050 K/s and the measuring frequency of the pendulum was about 0.5 Hz.

3. Results

For samples from series 1 curves of $Q^{-1}(T)$ and $G(T) \sim [f(T)]^2$ were determined for five different heating rates. Fig. 1 shows the results obtained for $v_g = 0.033$ K/s. Two peaks may be seen on the IF curve, i.e. A at 710 K and C at 838 K. Three inflections may also be seen, i.e. A_0 at 650 K, B at 725 K, and D at 880 to 900 K. The curve $f^2(T)$ has a horizontal part in the interval 300 to 400 K, a wide maximum in the vicinity of 500 K, two minima A′ and C′ and a distinct inflection at D′; less distinct inflection points may be seen at A_0 and B′. Minima and inflection points on the modulus curve occur at temperatures very close to the corresponding temperatures of peaks

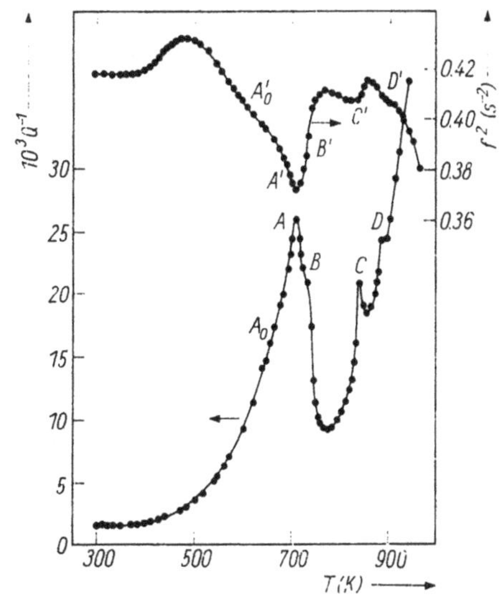

Fig. 1. Curves $Q^{-1}(T)$ and $f^2(T)$ determined for the tested amorphous alloy in as-quenched state; $v_g = 0.033$ K/s, $f = 0.54$ Hz at peak A

and inflection points on $Q^{-1}(T)$. During the second measurement, carried out for the same sample beginning at room temperature, the IF curve increases and the $G \sim f^2$ curve decreases monotonically with T, as is found for crystals, for which neither migrational relaxation nor a phase transition occurs in the tested temperature range. For the remaining heating rates curves of a similar form were found.

In Table 1 the peak temperatures (T_p) for maxima A and C are given.

X-ray examination showed that after heating the sample to 504 K the radiograph made at room temperature does not differ from that for the as-quenched sample. A broad maximum is observed, typical for an amorphous alloy. However, after heating to 650 K, on the background of this maximum there appears a small, sharp reflection coming from the (111) planes of the crystalline nickel. After heating to higher temperatures the amorphous maximum reduces and vanishes, while the Ni(111) reflection grows; further reflections from nickel and intermetallic compounds appear.

Results obtained using the IF method and X-ray examination, in particular the sharp minimum of the modulus A' seen in Fig. 1 (the theory of this effect is given, e.g., in [12]) and also the appearance and increase of the Ni(111) reflection after heating to 650 K and higher temperatures, indicates that the first high IF peak in the alloy $Ni_{70.3}Fe_{4.7}Si_{15}B_{10}$ is associated with the crystallisation of nickel in the amorphous phase. On the curve of $Q^{-1}(T)$ the peaks and inflection points occurring above 710 K (Fig. 1) are associated with the precipitation of crystalline intermetallic compounds or alternatively with phase transitions in the crystalline phase.

From measurements conducted for samples of series 1 and 2 it was possible to evaluate the influence of the difference between the highest T_f temperature from the previous measurement ($T_f > T_p$) and T_p, on the form of the curve $Q^{-1}(T)$ plotted

Table 1

Peak temperatures (in K) obtained for various heating rates

v_g (K/s)	0.010	0.018	0.027	0.033	0.050
A	692	694	699	710	712
C	823	829	831	838	840

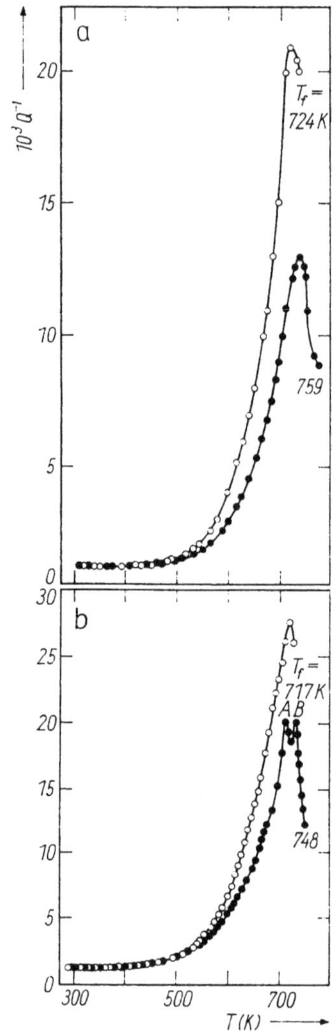

Fig. 2. Part of curves $Q^{-1}(T)$ obtained for the $Ni_{70.3}Fe_{4.7}$·$Si_{15}B_{10}$ alloy by measurement to ever higher temperatures; a) series 1, b) series 2; T_f the highest measurement temperature

during the next measurement. In the literature the opinion has been voiced that internal friction associated with crystallisation vanishes during heating to a temperature several degrees higher than T_p, while IF coming from viscoelastic relaxation exhibits good repeatability even for $T_f - T_p = 10$ K or more (e.g. [10], and also Fig. 2 in [8]).

Fig. 2 a and b show the results obtained for two values of T_f. It may be seen that on Fig. 2b, alongside the exposed peak B there appears peak A of considerably smaller height (sample from series 2). On Fig. 2a, however, only peak B is to be seen (sample from series 1). In the case of Fig. 2b the first measurement was conducted up to $T_f = 717$ K ($T_f - T_p = 6$ K), and in the case of Fig. 2a up to $T_f = 724$ K ($T_f - T_p = 14$ K). Thus heating to 6 K above T_p significantly reduces the peak height, while heating to 14 K above T_p causes this peak to vanish.

This result provides an additional confirmation that in the tested alloy the high peak in the vicinity of T_g is associated with crystallisation.

4. Discussion

4.1 Notes on the formulae used for the calculation of activation energy of crystallisation in glassy alloys

In the literature dealing with the use of the IF method for the determination of the effective activation energy of crystallisation in glassy alloys during linear heating for which

$$T = T_0 + v_g t \tag{1}$$

(T_0 initial temperature, t time of heating) use is made of one of the two formulae which can be derived from the Johnson-Mehl-Avrami equation of transition kinetics, i.e.

$$x = 1 - \exp[-(Kt)^n] \tag{2}$$

in which

$$K = K_0 \exp\left(-\frac{E}{k_B T}\right). \tag{3}$$

In (2) and (3), n is an integer or half-integer depending on the growth mechanism and the relative dimensions of the crystals, K_0 is the pre-exponential factor, k_B the Boltzmann constant, and x the volume fraction crystallised. The first of these formulae is the commonly used Kissinger formula [13],

$$\ln\left(\frac{T_p^2}{v_g}\right) = \frac{E}{k_B}\frac{1}{T_p} + \text{const}, \tag{4}$$

while the second formula, obtained first by Augis and Bennett [14], and next derived using slightly different simplifying assumptions by Yinnon and Uhlmann in a paper dealing with the calculation of E for crystallisation in glassy alloys [15], has the form

$$\ln\left(\frac{T_p - T_0}{v_g}\right) = \frac{E'}{k_B}\frac{1}{T_p} - \ln K_0, \tag{5}$$

where E' is the effective activation energy calculated from (5).

According to Yinnon and Uhlmann [15] the Kissinger formula is not, in principle, suitable for describing the given effect, since when deriving it from (1) to (3) the factor

$$at = \frac{E(T_p - T_0)}{k_B T_p^2}$$

occurring in the derivative with respect to time in (2) is neglected as having a value small in comparison with 1 (formula (5) in [2]), which is not strictly true; e.g. for alloy 1, this factor is greater than 25 [2]. This simplification is not introduced in the case of (5) given in [14]. It is worth adding that another method of deriving (5) is given in [2]. This formula was found from a more general equation for the activation energy, as a very reasonable approximation, and is introduced in a more lucid manner, also without making the assumption $T_0 \ll T_p$.

In [2] it was observed that both (4) and (5), within the limits of error, give the same value of E for crystallisation for alloy 1 (Table 2). Since a similar result was found in this present paper for alloy 3 (Table 2), (4) and (5) were compared. After rearranging, we obtain

$$\ln \frac{T_p^2}{T_p - T_0} = \frac{-\Delta E}{k_B}\frac{1}{T_p} + \text{const}'; \tag{6}$$

in this formula $\Delta E = E' - E$ (E from the Kissinger formula). After differentiating with respect to T_p the following expression is found for ΔE:

$$\Delta E = k_B T_p \frac{T_p - 2T_0}{T_p - T_0} \tag{7}$$

and

$$\frac{\Delta E}{E} = \frac{k_B T_p}{E}\frac{T_p - 2T_0}{T_p - T_0}. \tag{8}$$

Activation energies for glassy alloys are of the order of 2 to 4 eV (see Section 4.2), while values of T_p observed by the IF method lie in the limits 600 to 800 K. Hence the first fraction on the right-hand side of (8) is of the order of 0.02, while the second

Table 2

Crystallisation activation parameters obtained for four alloys investigated in tho literature by the IF method

No.	alloy	ref.	Augis-Bennett formula			Kissinger formula		remarks
			E' (eV)	ln K_0	K_0 (s^{-1})	E (eV)	const	
1	$Ni_{68.7}Cr_{6.6}Fe_{2.65}Si_{7.8}B_{14}C_{0.25}$	[1, 2]	3.74 ± 0.12	49 ± 2	1.7×10^{21}	3.72 ± 0.12	−(41 ± 2)	calculations repeated in this paper
2	$Ni_{74}Al_4P_{16}B_6$	[11]	4.05 ± 0.20*)	64 ± 2*)	5.4×10^{27}*)	4.2	−(57 ± 2)*)	*) values calculated in this paper from Fig. 3 in [11]
3	$Fe_{40}Ni_{40}P_{10}B_{10}$	this paper	2.66 ± 0.20	37 ± 4	1.2×10^{16}	2.64 ± 0.20	−(30 ± 4)	calculated in this paper from Fig. 1 in [7]
4	$Ni_{70.3}Fe_{4.7}Si_{15}B_{10}$	this paper	2.51 ± 0.65	32 ± 11	6.4×10^{13}	2.49 ± 0.65	−(24 ± 11)	—

is less than one and further lowers the value of $\Delta E/E$. Thus it may be seen that in the given measurement conditions (i.e. values of v_g used), (4) and (5) give the same values for the activation energy of crystallisation for metallic glasses.

Considering the magnitude of the constant in the Kissinger formula, the following facts should be noted. Kissinger gives his equation in differential form, i.e.

$$\frac{\mathrm{d}\ln\left(\frac{T_p^2}{v_g}\right)}{\mathrm{d}\left(\frac{1}{T_p}\right)} = \frac{E}{k_B}. \tag{9}$$

This expression, after integration, gives the Kissinger formula including the unexplained constant (formula (4)). According to Yinnon and Uhlmann [15] this constant has a value

$$\text{const} = -\ln \frac{k_B A K_0}{E}, \tag{10}$$

where

$$A = [-\ln(1 - x)]^{(n-1)/n} \tag{11}$$

and x corresponds to the temperature of the peak.

4.2 Calculation of the effective activation energy of crystallisation

On Fig. 3 and 4 the relations $\ln[(T_p - T_0)/v_g]$ and $\ln(T_p^2/v_g)$ are shown as functions of $1/T_p$ for peaks A and C, respectively, from Fig. 1. As may be seen, in agreement with (4) and (5) straight lines are obtained. Making use of the positions of the straight lines obtained for peak A the values of E', $\ln K_0$, K_0, E, and the constant were determined. The values obtained are collected in Table 2 together with the values for the same parameters found in the present paper for alloys 1, 2, 3 estimated from the following data: for 1 from Table 1 in [2], for 2 from Fig. 3 in [11], for 3 from Fig. 1 in [7].

For peak C a value $E = (5.0 \pm 0.7)$ eV was obtained from the Kissinger formula (4), generally used in the case of chemical reactions and phase transformations in crystalline solids.

Values for the effective activation energy of crystallisation for glassy alloys, calculated on the basis of IF tests (2 to 4 eV) do not differ, in principle, from the values

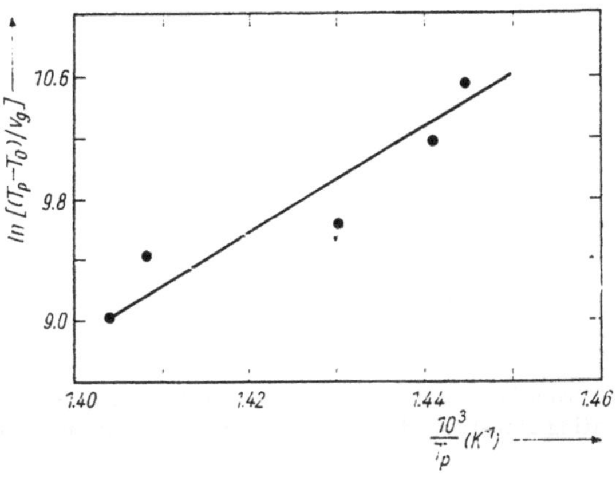

Fig. 3. Curve of $\ln((T_p - T_0)/v_g)$ vs. $(T_p)^{-1}$ for peak A ($T_0 = 300$ K)

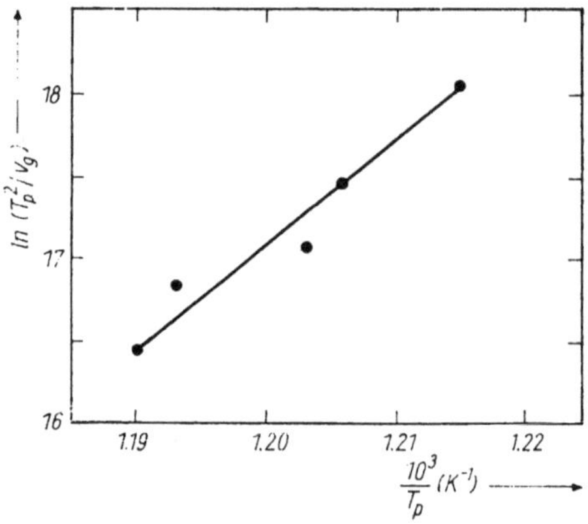

Fig. 4. Curve of $\ln (T_p^2/v_g)$ vs. $(T_p)^{-1}$ for peak C

obtained by other methods, mainly DTA and DSC (e.g. [16 to 18]). It is also noteworthy that in [19] using the DTA method for alloy 1 a value of $E = 3.8$ eV (v_g up to 0.083 K/s) was obtained, that is the same value as found in [1, 2] using the IF method ($v_g = 0.017$ to 0.050 K/s).

Considering the constant term in (4) and (5), independent of T_p, it is clear that in the equation of Augis and Bennett it contains only the pre-exponential constant K_0 from (3), while the constant in (4) has a very complex form (cf. (10) and (11)); apart from K_0, other physical magnitudes are found here. Moreover, a direct comparison of $\ln K_0$ and const in Table 2 indicates that they differ.

Values of K_0 obtained from (5) given in Table 2 vary quite considerably, lying in a wide interval from 10^{14} to 10^{27} s^{-1}. The significance of this is difficult to elucidate now from the literature in which K_0 is not generally mentioned. It is worth adding that recently, as reported in [20], using the DSC method two crystallisation peaks were observed in the alloy $Fe_{78}B_9Si_{13}$, while for K_0 a value of 4×10^{28} or 7×10^{25} s^{-1} was obtained, i.e. values lying in the upper interval in Table 2.

On Fig. 5 $\ln K_0$ is shown as a function of E obtained for four glassy alloys using the IF method; points obtained for the alloy $Fe_{78}B_9Si_{13}$ as reported in [20] and for alloys

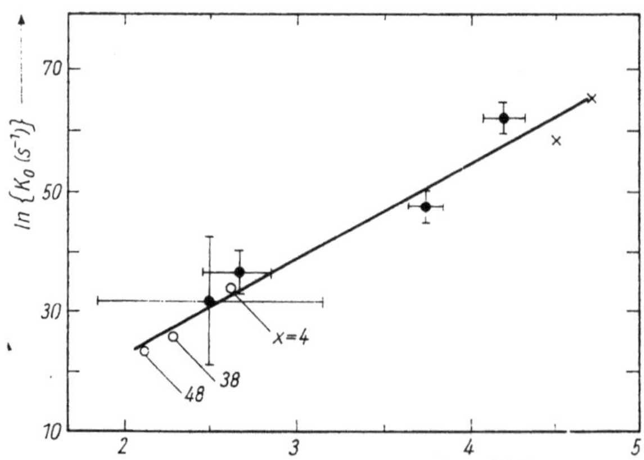

Fig. 5. $\ln K_0$ as a function of E; ● obtained from IF measurements, ○ obtained recently in [21] by IF (one heating rate) and DTA (two heating rates) for alloys $Co_{78-y}Fe_ySi_9B_{13}$, × obtained by DSC in [20]; correlation coefficient 0.984

$Co_{78-y}Fe_ySi_9B_{13}$ [21] are also plotted. As may be seen, ln K_0 is approximately proportional to the effective energy of activation.

The constants in (4) and (5) appear to suggest, when also taking into consideration (10) and (11), that the constant n can be calculated directly. On this assumption, making use of the data in Table 2 and also of (10), first of all A was calculated for the four alloys. Next n was determined from (11), making the assumption, similarly as in [15], that $x = 0.63$ (see p. 268 in [15]). The obtained values differ very little from each other and amount to $n \approx 0.002$, i.e. are certainly too small.

In the literature it is assumed that the constant n lies in the interval [3/2, 4] (e.g. [15]). It was recently reported in [20] that for the alloy $Fe_{78}B_9Si_{13}$ for the first phase transition $n = 2.0$ and for the second $n = 4.0$, i.e. the values obtained in the present paper are three orders of magnitude smaller.

The reasoning and calculations presented in Sections 4.1 and 4.2 thus indicate clearly that although both (4) and (5) give, in principle, the same values for the effective activation energy of crystallisation of amorphous metals, the Kissinger formula does not give correct values of n in the case when making use of (10) and (11) and of the value of K_0 determined from (5).

4.3 Further remarks

As has already been stressed, on Fig. 1 the curve of the modulus passes through a fairly wide maximum near to 500 K, which does not correspond to any anomaly on the curve $Q^{-1}(T)$. A similar effect has been observed for other alloys (e.g. in [22, 23]) and also for a series of alloys $Co_{78-y}Fe_ySi_9B_{13}$ where $0 \leq y \leq 74$ [21]. The correct interpretation of this maximum could be given by the comments formulated in [22]: the less or more flat maximum on the curve $G(T)$ is, in principle, a pseudo-peak, and its value is associated with an increase in G due to structural relaxation at lower temperatures and with a drop in G arising due to phase transition phenomena (e.g. [12]). The whole course of this effect takes place against the background of a monotonic drop in the modulus value with temperature.

5. Conclusions

1. Basing on measurement results obtained during linear heating, using the internal friction (IF) method, the effective activation energy of crystallisation E was determined for the glassy alloy $Ni_{70.3}Fe_{4.7}Si_{15}B_{10}$. From the Kissinger and Augis-Bennet formulae approximately the same values were obtained, i.e. 2.49 and 2.51 eV, respectively.
2. It was shown that this result is not accidental, and is associated with the mathematical form of the formulae, despite the fact that, in principle, the Kissinger formula should not be used in the case of crystallisation in glassy metals.
3. For the pre-exponential constant K_0 in the Arrhenius law for diffusion processes associated with crystallisation, from four alloys investigated in the literature using the IF method, values were obtained which differ widely, i.e. from 10^{14} to 10^{27} s^{-1}. These values increase markedly with E, and at the same time ln K_0 is approximately proportional to the effective activation energy.
4. It was ascertained that values of the constant n in the Johnson-Mehl-Avrami law calculated from the Kissinger and Augis-Bennett relations and from formulae given by Yinnon and Uhlmann, were smaller by more than three orders of magnitude than the theoretical and experimental values quoted in the literature.
5. An explanation is given of the origin of the flat maximum on the curves of elastic modulus as a function of temperature which occurs for many glassy alloys. This maximum does not correspond to any anomaly on the curves of IF as a function of T.

Acknowledgement

The authors wish to express their thanks to J. Janik, M.Sc. for carrying out part of the measurements.

References

[1] T. Poloczek, Ł. Cieślak, and J. W. Moroń, in: Internal Friction and Magnetic After-Effects in Solids, Silesian University Press, Katowice 1985 (p. 24) (in Polish).
[2] T. Poloczek, J. W. Moroń, and Ł. Cieślak, Archiwum Nauki o Materiałach 7, 229 (1986) (in English).
[3] H. S. Chen and N. Morito, J. non-crystall. Solids 72, 287 (1985).
[4] H. S. Chen and N. Morito, Proc. 5th Internat. Conf. Rapidly Quenched Metals, Würzburg 1984, Vol. 1, North-Holland Publ. Co., Amsterdam 1985 (p. 715).
[5] H. R. Sinning and F. Haessner, Scripta metall. 20, 1541 (1986).
[6] J. Perez, J. Physique 46, C10-427 (1985).
[7] Lin Deming, Zhang Jinxiu, and Luo Laizhong, J. Physique 46, C10-477 (1985).
[8] He Yizhen and Li Xiao-Guang, phys. stat. sol. (a) 99, 115 (1987).
[9] E. Bonetti, E. Lanzoni, and G. Paparo, J. Physique 46, C10-481 (1985).
[10] J. W. Moroń and M. Kłosek, phys. stat. sol. (a) 106, K13 (1988).
[11] Li Xiao-Guang and He Yizhen, phys. stat. sol. (a) 95, 467 (1986).
[12] R. Kuśka, phys. stat. sol. (a) 88, 151 (1985).
[13] H. Kissinger, Ann. Chem. 29, 1702 (1957).
[14] J. Augis and J. E. Bennett, J. thermal Anal. 13, 283 (1978).
[15] H. Yinnon and D. R. Uhlmann, J. non-crystall. Solids 54, 253 (1983).
[16] H. S. Chen, Rep. Progr. Phys. 43, 355 (1980).
[17] U. Köster, Z. Metallk. 75, 691 (1984).
[18] Yi Qun Gao and S. H. Whang, J. non-crystall. Solids 70, 85 (1985).
[19] T. Poloczek, J. Ryś, and Ł. Cieślak, Proc. 3-rd Polish Seminar Magnetic Amorphous Materials, Kazimierz nad Wisła 1985, Lublin Technical University Press, Lublin 1985 (p. 35) (in Polish).
[20] M. D. Baró, S. Surinach, M. T. Clavaguara-Mora, and N. Clavaguara, J. non-crystall. Solids 69, 105 (1984).
[21] G. Haneczok, M. Kłosek, J. W. Moroń, and K. Nowakowska, preliminary results.
[22] T. Poloczek, G. Haneczok, and J. W. Moroń, J. Physique 44, C9-115 (1983).
[23] T. Poloczek, Ł. Cieślak, and J. W. Moroń, J. Physique 44, C9-121 (1983).

(Received April 19, 1988; in revised form August 10, 1988)

phys. stat. sol. (a) **111**, 137 (1989)

Subject classification: 62.20; S4

Metal Physics Unit, Solid State Physics Department, National Research Centre, Cairo[1])

Plastic Instability during Combined Torsion–Tension Deformation in Al–0.72% Si Alloy

By

T. H. YOUSSEF, F. A. SAADALAH, and I. K. BISHAY

Pure tensile and combined torsion–tension deformation experiments are made using Al–0.72% Si samples of different grain diameters. Plastic instability behaviour is observed in the case of combined torsion–tension deformation. The onset and disappearance of this instability is found to depend on some parameters, as mode of deformation, applied axial tensile stress, working temperature, sample grain diameter, and quenched-in vacancies.

Reine Dehnungs- und kombinierte Torsions–Dehnungs-Deformationsexperimente werden an Al–0,72% Si-Proben verschiedenen Korndurchmessers durchgeführt. Es wird plastisches Instabilitätsverhalten im Fall der kombinierten Torsions–Dehnungsdeformation beobachtet. Es wird gefunden, daß das Einsetzen und Verschwinden dieser Instabilität von einigen Parametern, wie Deformationsmode, angelegter axialer Tensionsspannung, Arbeitstemperatur, Probenkorndurchmesser und den eingeschreckten Leerstellen abhängt.

1. Introduction

The comparison of plastic deformation mechanisms in different deformation modes, such as tension, torsion, compression, etc. is a problem of considerable scientific importance [1]. If it were understood, quantitative predictions could be made, on the basis of simple tests, of the behaviour of material under complex multiaxial boundary conditions. The dependence of the rate of strain hardening on deformation mode has been a serious gap in any useful theory of large strain plasticity.

Heavily deformed metals and alloys are susceptible to plastic instability, i.e. strain localization, mainly when the mode of deformation changes [2]. In previous work the authors [3] showed the existence of instability in Al–0.72% Si alloy when it was pre-annealed at temperature below 300 °C before twisting it. To throw more light on the characteristics of this instability, this work was done to study the hardening mechanisms controlling the tensile strain ($\Delta L/L_0$) associated with pure tension as well as combined torsion-tension deformation in polycrystalline Al–0.72% Si. The study is of technological interest as the Al–Si alloy is used in the industry of aluminium cables for electric power transfer.

2. Experimental Results

This study was carried out using Al–Si of commercial purity. It was supplied by Helwan Company for non-ferrous metals (Egypt) in the form of wires of 2.15 mm diameter.

---

[1]) Dokki, Cairo, Egypt.

Precise chemical analysis gave

Table 1

element	Al	Si	Fe	Mn
content (wt%)	99.24	0.72	0.03	0.01

The samples were swaged into wires of 0.5 mm diameter by successive wire drawing. Wires of grain diameters covering the range 0.013 up to 0.048 mm were prepared by annealing for 1 h at different temperatures. In order to reveal the grains of the studied samples they were electrolytically polished and etched by a method described elsewhere [4]. The variation of the attained average grain diameter with annealing temperature is shown in Fig. 1. From this figure it is clear that the average grain diameter increased with annealing through two rates: a relatively small rate in the temperature range from room temperature (RT) up to 300 °C and a higher rate in the temperature range above 300 °C.

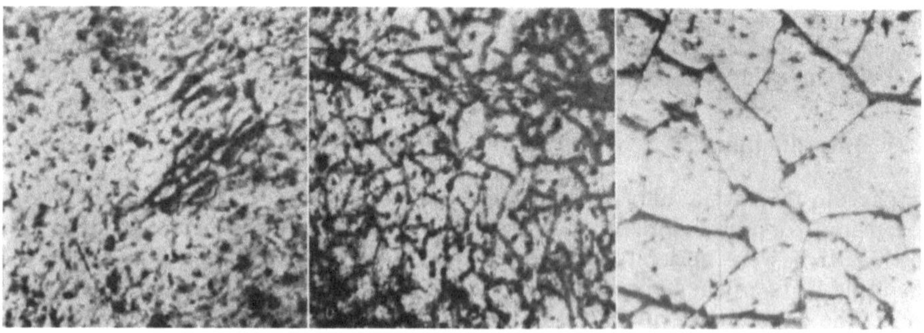

RT 300 °C 500 °C

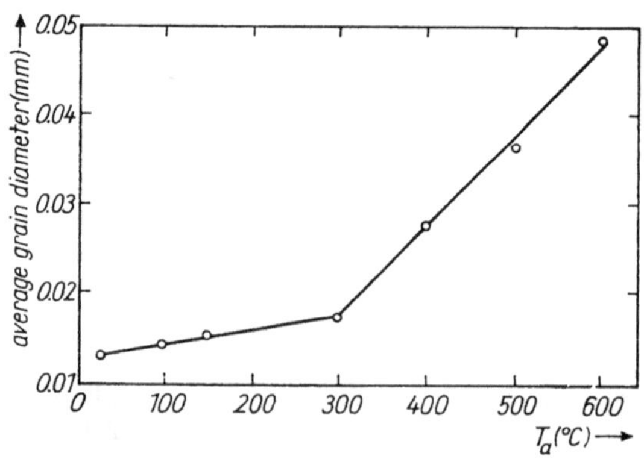

Fig. 1. Dependence of average grain diameter on the annealing temperature T_a (annealing time 60 min)

2.1 Pure tensile deformation

Experimental sets of tensile stress–strain curves were taken at the different working temperatures: RT, 120, 225, 350, and 430 °C, using the torsion machine described elsewhere [5]. The tested wire was heated in situ. All the observations were carried out at a constant strain rate ≈ 0.7 mm/min. The results showed that the stress-strain curves were extremely sensitive to working temperature (T_0). At RT the wire sample was brittle and changed to ductile with increasing both working temperature and/or grain diameter. Also sets of stress–strain curves for samples of different grain diameters were measured at RT, after being quenched from 400 °C. In all cases the stress–strain curves were found parabolic around the yield stress. Accordingly the coefficient of work hardening (X_p) for these samples was calculated as $X_p = \text{stress}^2/\text{strain}$ [6] in the parabolic stress range. The dependence of calculated X_p on grain diameters for quenched and non-quenched samples tested at RT is shown in Fig. 2b from which it could be deduced that quenching decreased X_p and that the rate of this decrease varies with grain diameter. It could also be observed that X_p for non-quenched samples depend on working temperature (T_w). The dependence of (X_p) on working temperature for samples of different grain diameters, non-quenched, is shown in Fig. 2a. From this figure it could be easily seen that increasing working temperature decreased X_p and that the rate of this decrease depends on grain di-

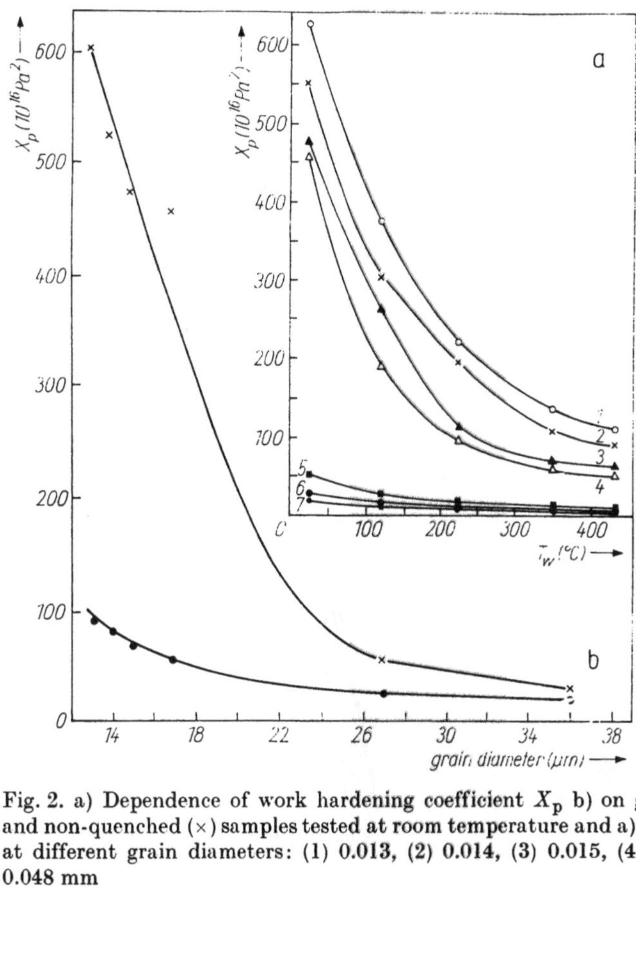

Fig. 2. a) Dependence of work hardening coefficient X_p b) on grain diameters for quenched (●) and non-quenched (×) samples tested at room temperature and a) on working temperature for wires at different grain diameters: (1) 0.013, (2) 0.014, (3) 0.015, (4) 0.017, (5) 0.027, (6) 0.036, (7) 0.048 mm

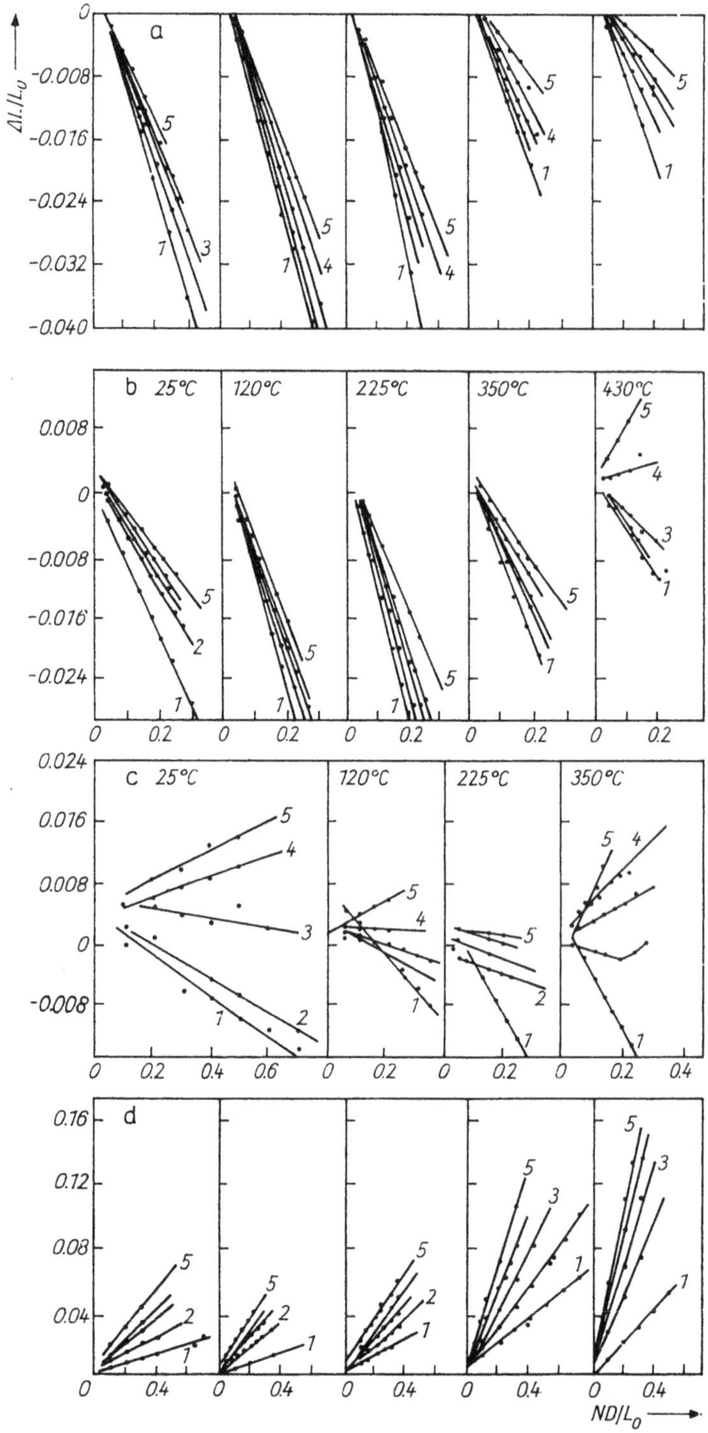
a
b
c
d
ΔL/L₀
ND/L₀
25°C
120°C
225°C
350°C
430°C
0
-0.008
-0.016
-0.024
-0.032
-0.040
0.008
0.024
0.016
0.16
0.12
0.08
0.04
0.2
0.4
0.6

Fig. 3a to d

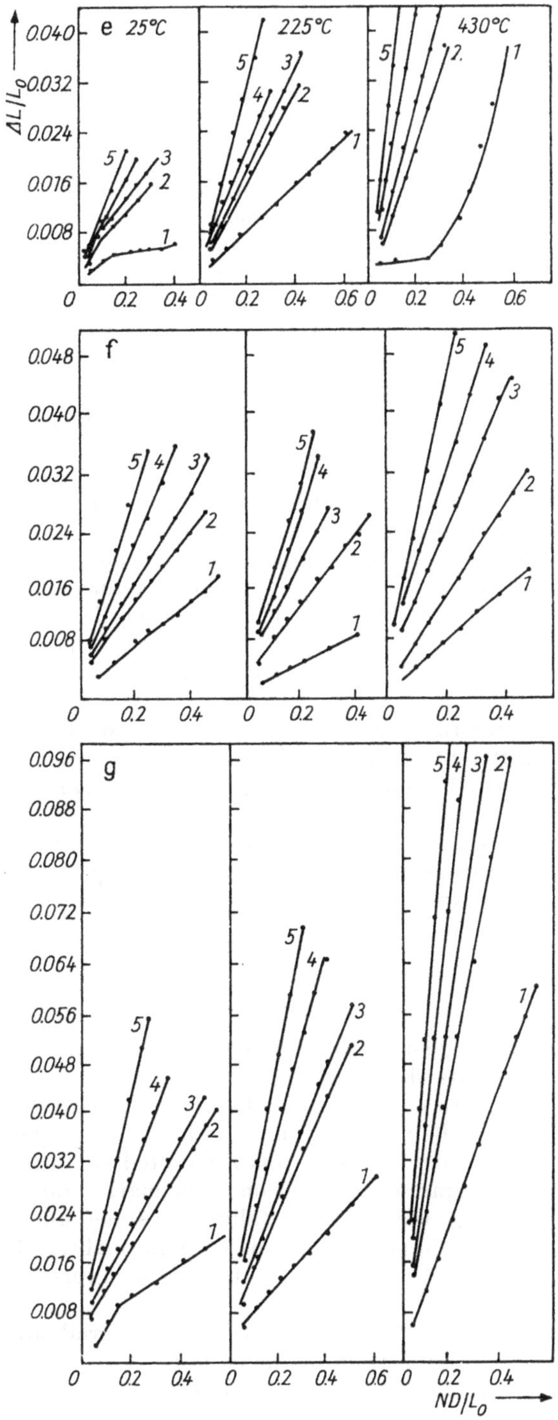

Fig. 3. Variation of tensile strain ($\Delta L/L_0$) with torsional deformation (ND/L_0) for samples of different grain diameters (G.D.) tested at different working temperatures and different tensile stresses: (1) 6.11; (2) 10.2, (3) 10.2, (4) 14.5, (5) 17.1 MPa. Left-hand side: non-quenched samples; right-hand side: quenched samples. a, e) grain diameter 13, b) 15, c, f) 17, d, g) 36 μm

ameter; it is relatively high for samples of grain diameters 0.013, 0.015, and 0.017 mm, while it is of smaller rate for samples of grain diameters 0.027, 0.036, and 0.048 mm, i.e. the rate of decrease is high in samples pre-annealed in the temperature range RT to 300 °C, and smaller for samples pre-annealed in the higher temperature range.

2.2 *Combined torsion–tension deformation*

Wires of different grain diameters were subjected to uniform twisting using the torsion machine [5]. The degree of torsional deformation given was measured [5, 6] by the dimensionless quantity (ND/L_0), N being the number of turn twists, D the wire diameter (0.5 mm), and L_0 the original wire length (50 mm). During twisting the wires were axially loaded by stresses 6.11, 10.19, 12.0, 14.5, and 17.1 MPa. These stresses were within the elastic range of the tested wires. The length variation (ΔL) associated with combined torsion–tension deformation was successfully measured optically till fracture with accuracy of $\pm 1/10^{-2}$ mm. The variation of tensile strain ($\Delta L/L_0$) with torsional deformation (ND/L_0) for samples of grain diameters 0.013, 0.015, 0.017, and 0.036 mm, each tested at a different working temperature is shown in Fig. 3a. Fig. 3b shows the variation of tensile strain with torsional deformation for samples of grain diameters 0.013, 0.017, and 0.036 mm, pre-quenched from 400 °C before twisting.

From these two figures (Fig. 3 a and b), it is clear that

a) in wires of grain diameters 0.013, 0.015, or 0.017 mm, a negative increase in tensile strain ($\Delta L/L_0 < 0$), i.e. contraction, was observed at torsional deformation. This negativity ($\Delta L/L_0 < 0$) decreased by increasing the applied axial tensile stress, the working temperature, and/or the average grain diameter. It could be seen that the negative results ($\Delta L/L_0 < 0$) are of relatively small values compared with the positive results in ($\Delta L/L_0 > 0$);

b) the relation between the torsional deformation (ND/L_0) and the accompanying strain ($\Delta L/L_0$) is nearly linear for the different applied axial tensile stresses studied;

c) by increasing the axial tensile stress, the tensile strain ($\Delta L/L_0$) accompanying the torsional deformation increased;

d) increasing grain diameter caused an increase in the resulting tensile strain;

e) no negative tensile strain ($\Delta L/L_0 < 0$) was observed for quenched samples, and the values of the attained positive tensile strain ($\Delta L/L_0 > 0$) are relatively larger than those corresponding to non-quenched samples.

3. Discussion

It was found previously [3] that the annealing spectrum of Al–0.72% Si is composed of two stages, in accordance with the results shown in Fig. 1. The recovery stage (RT up to 300 °C), where the grains grow with a small rate and are characterized by negative tensile strain ($\Delta L/L_0 < 0$) — instability behaviour — when the samples were subjected to combined torsion–tension deformation, was attributed to silicon clustering on the grain boundaries as well as grain misorientation in the matrix during torsion–tension deformation. The second stage was the recrystallization stage (above 300 °C), accompanied by increasing rate of grain growth. In this stage normal tensile strain ($\Delta L/L_0 > 0$) took place when the samples were subjected to combined torsion–tension deformation.

In this work the results showed that the instability behaviour decreases gradually and transforms homogeneously to positive tensile strain by increasing

(i) the tensile stress applied during combined torsion–tension deformation;
(ii) sample grain diameter;
(iii) the working temperature.

The increase in tensile strain $(\Delta L/L_0)$ by increasing the applied axial tensile stress implies that the responsible reaction is stress dependent. This reaction allows dislocations to move over larger distances leading to an increase in the tensile strain. Since Al is a metal of high stacking fault energy [7, 8], cross slipping of dislocations would take place by twisting [8, 9]. Torsional deformation in Al wires produces a large number of dislocations which interact with each other and form jogs [10]. Jogs also have a tendency to cluster under the action of applied tensile stress [11], this process leads to a decrease in jog density, and as a result the dislocation motion by cross slipping would be easier. It was also found that the energy activating cross slip motion decreased by increasing the applied tensile stress [12], i.e. the probability of cross slipping of dislocations would increase with increasing applied tensile stress leading to an increased rate of jog clustering. Consequently the rate of annihilation and coalescence of these jogs would also be increased.

Another effect is that the axial tensile stress applied during torsional deformation for Al helps in orienting the deformed grains [8] along the tensile axis. Consequently the increased value of applied axial tensile stress might facilitate the grain orientation and thus lead to the observed increased value of tensile strain. The role of the orientation mechanism on $(\Delta L/L_0)$ is quite evident during torsion–tension deformation. For large-grained samples the number of grains per unit volume is relatively smaller, and thus the work done for grain orientation decreases, consequently the tensile elongation of grains increases.

The observed increase in tensile strain by increasing the applied working temperature would be interpreted in terms of thermally activated cross slipping [12]. Generally, thermal agitation might cause the formed jogs to be more easily clustered, and as a result the rate at which they were annihilated or coalesced would increase, leading to an increase in $(\Delta L/L_0)$.

The increase of $(\Delta L/L_0)$ in quenched samples, and absence on instability behaviour, even for samples annealed in the recovery stage, might be due to the fact that, on quenching Al–0.72% Si alloy, vacancy clusters are formed which act as preferential nucleation sites for Si precipitates [13, 14] distributed in the matrix and not clustered on the grain boundaries. The gathering of Si atoms in clusters seems to decrease the back stress in the matrix leaving it relaxed and so the grain boundaries being free allow the process of grain orientation during combined torsion–tension deformation. Consequently a relatively larger elongation was observed for quenched samples. The observed difference between the values of tensile strain $(\Delta L/L_0)$ for samples annealed, and for those annealed and quenched, Fig. 3 a, b clarify the observed variation in work hardening coefficient X_p Fig. 2, which points towards the effect of quenched-in vacancies in dragging Si atoms, decreasing the Si effect as locking stress, and leading to an enhancement of the grain orientation process.

References

[1] C. Tome, G. R. Canova, U. F. Kocks, N. Christodoulou, and J. J. Jonas, Acta metall. **32**, 1637 (1984).
[2] A. Szökefalvi-Nagy, Scripta metall. **16**, 1009 (1982).
[3] T. H. Yussef, F. A. Saadalah, and I. K. Bishay, phys. stat. sol. (a) **102**, 273 (1987).
[4] T. H. Youssef and F. A. Saadalah, Aluminium **60**, E 188/E 189 (1984).
[5] T. H. Youssef and G. Grais, Scripta metall. **7**, 983 (1973).
[6] T. H. Youssef and M. S. Sakr, phys. stat. sol. (a) **63**, 85 (1981).
[7] R. E. Smallman, Modern Physical Metallurgy, 3rd ed., Butterworths, Washington/London 1976.

[8] D. B. HOLT, Acta metall. 7, 446 (1959).
[9] T. H. YOUSSEF, phys. stat. sol. (a) **3**, 801 (1970).
[10] A. S. KEH, Direct Observations of Imperfections in Crystals, AIME/Interscience Publishers, New York 1962 (p. 213).
[11] J. FRIEDEL, Dislocations, Pergamon Press 1964.
[12] G. SCHOECK and A. SEEGER, Defects in Crystalline Solids, The Physical Society, London 1955 (p. 340).
[13] E. OZAWA and H. KIMURA, Mater. Sci. Eng. 8, 327 (1971).
[14] G. DLUBEK, O. BRÜMMER, and P. HAUTOJÄRVI, Acta metall. **34**, 661 (1986).

(Received November 12, 1987; in revised form August 2, 1988)

phys. stat. sol. (a) **111**, 145 (1989)

Subject classification: 64.80; 72.20; 75.50; S11.2

Department of Semiconductors, Higher Institute of Chemical Technology, Sofia[1])

Influence of the Microstructure on Some Microwave Properties of Substituted Nickel Ferrites

By

D. VLADIKOVA, L. ILKOV, and S. KARBANOV

The influence of the substituents on the microstructure of nickel ferrites from the system $Ni_{1-x_1-x_2-x_3-x_4}Ca_{x_1}Cu_{x_2}Co_{x_3}Zn_{x_4}Mn_{x_5}Al_{x_6}Fe_{x_7}O_4 + y\,SiO_2$ where $0 \leqq x_1 \leqq 0.02$; $0 \leqq x_2 \leqq 0.3$; $0 \leqq x_3 \leqq 0.06$; $0 \leqq x_4 \leqq 0.5$; $0 \leqq x_5 \leqq 0.06$; $0 \leqq x_6 \leqq 0.2$; $1.8 \leqq x_7 \leqq 2$; $0 \leqq y \leqq 0.3$ wt%; $x_6 + x_7 = 2$; is investigated. Electrical resistance measurements are performed by a complex impedance method which permits the evaluation of the contributions due to the resistance of the bulk material and the grain boundaries. It is found that the dielectric and resonance magnetic losses can be improved to a certain degree by the average grain size.

Исследовано влияние добавок на микроструктуру никелевых ферритов системы $Ni_{1-x_1-x_2-x_3-x_4}Ca_{x_1}Cu_{x_2}Co_{x_3}Zn_{x_4}Mn_{x_5}Al_{x_6}Fe_{x_7}O_4$, где $0 \leqq x_1 \leqq 0{,}02$; $0 \leqq x_2 \leqq 0{,}3$; $0 \leqq x_3 \leqq 0{,}06$; $0 \leqq x_4 \leqq 0{,}5$; $0 \leqq x_5 \leqq 0{,}06$; $0 \leqq x_6 \leqq 0{,}2$; $1{,}8 \leqq x_7 \leqq 2$; $0 \leqq y \leqq 0{,}3$ wt%; $x_6 + x_7 = 2$. Применяя комплексный импедансный метод оценен вклад объемного материала и границ зерен в общее сопротивление образцов. Показано, что контролированием размеров зерен можно уменьшить диэлектрические и магнитные потери ферритов.

1. Introduction

The microstructure of the substituted nickel ferrites depends on the sintering conditions [1 to 5] and the deviations from stoichiometry [5 to 8] as well as on the presence of flux components (CuO, Bi_2O_3, V_2O_5) [2, 3, 9] or additives suppressing the mobility of the grain boundaries (BaO, CaO) [10 to 12]. The data about the relation between microstructure and microwave properties in substituted nickel ferrites are insufficient. They concern mainly the dependence of their high power characteristics on the grain size [7, 8, 13, 14]. The influence of the sintering conditions on the specific electrical resistance and thus on the dielectric loss angle tangent tg δ_ε [15] has been explained mostly by the changes of the conductivity [2, 16, 17]. Naik and Polar [1] have connected the reduction of the resistance of nickel–zinc ferrites at higher sintering temperatures with some differences in the microstructure. They have used the two-probe method, which gives the total sample resistance ϱ, determined by the resistance of the grains (bulk material) ϱ_g, of the grain boundaries ϱ_{gb}, and of the electrodes ϱ_{el}: $\varrho = \varrho_g + \varrho_{gb} + \varrho_{el}$ [18, 19]. For this reason the influence of the grain boundaries on ϱ has not been estimated.

The effect of the pores on ΔH can be quantitatively evaluated [20] in contrast to the effect of other inhomogeneities such as additional phases and grain boundaries. Some data about the improvement of ΔH in nickel ferrites with the increase of the grain size from 0.17 up to 30 μm have been obtained in [21]. It has to be taken into account that the fine-grained samples have been prepared by the hot pressing technique.

---

[1]) Bld. Kl. Ohridski 8, BG-1156 Sofia, Bulgaria.

The present work aims at investigating some possibilities for the formation of a desired microstructure which could improve the dielectric losses and the resonance linewidth of substituted nickel ferrites prepared by the standard ceramic technology. The flux components accelerate the densification of the sintered material [2, 3] and thus improve ΔH [20]. As Bi_2O_3 and V_2O_3 form a separate phase between the grains [3, 9], we have chosen copper, which enters the spinel lattice. The possibility of obtaining a fine-grained microstructure was checked by the addition of a grain growth inhibiter. The combination of $CaO + SiO_2$ was examined.

2. Experimental Procedure

Substituted nickel ferrites of the composition: $Ni_{1-x_1-x_2-x_3-x_4}Ca_{x_1}Cu_{x_2}Co_{x_3}Zn_{x_4}Mn_{x_5}\cdot$ $\cdot Al_{x_6}Fe_{x_7}O_4 + ySiO_2$ where $0 \leqq x_1 \leqq 0.02$; $0 \leqq x_2 \leqq 0.3$; $0 \leqq x_3 \leqq 0.06$; $0 \leqq x_4 \leqq$ $\leqq 0.5$; $0 \leqq x_5 \leqq 0.06$; $0 \leqq x_6 \leqq 0.2$; $1.8 \leqq x_7 \leqq 2$; $0 \leqq y \leqq 0.3$ wt%; $x_6 + x_7 = 2$; were prepared by the standard ceramic technology. Compositions with $x_2 < 0.10$ were prefired in air for 6 h at 1173 K and those with $x_2 \geqq 0.10$ — at 1073 K. The samples were sintered in oxygen at different conditions given in Table 1. The heating and cooling rates were 200 and 100 K/h, respectively.

For the characterization of the samples, techniques of X-ray diffraction (XRD) performed on a diffractometer HZ-4C using CoK_α radiation and electron probe microanalysis (EPMA) on a SEM JEOL Superprobe 733 with an energy dispersive spectrometer EG & G Ortec System 5000 were used. The microstructure was observed on a SEM JEM 100. The average grain size $\bar{d}$ was determined by counting the number of grain boundaries intercepted by measured lengths of random straight lines drawn on the microphotographs of polished and etched samples. The densities d_{exp} were measured by the hydrostatic method. The X-ray densities $d_{X\text{-ray}}$ were calculated from the data of the lattice parameters obtained by the XRD analysis. The porosity p defined as [5]

$$p = \left(\frac{d_{X\text{-ray}} - d_{exp}}{d_{X\text{-ray}}}\right) \times 100\% \qquad (1)$$

was calculated.

The electrical resistance of the ferrite samples was determined by the complex impedance method [18, 19]. The impedance Z of a ferrite sample measured over a range of frequencies ω describes a characteristic locus of three semi-circles in the complex plane (Fig. 1). It has an equivalent electrical circuit. The intercepts of the three semi-circles with the real axis are related directly to the resistance values of ϱ_g, ϱ_{gb}, and ϱ_{el}. The measurement system used [22] includes Solatron 1174 a frequency response analyser (FRA) which provides a voltage measurement and a circuit which determines the impedance. FRA is interfaced with a microcomputer which enables

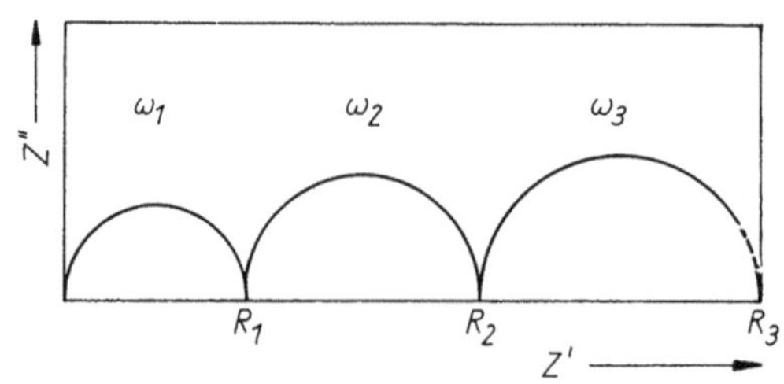

Fig. 1. Idealized impedance spectrum of a polycrystalline ferrite in the complex plane ($\omega_1 > \omega_2 > \omega_3$)

Table 1

Composition, sintering conditions (S.C.), densities, porosity, average grain size, resonance linewidth, and dielectric losses of substituted nickel ferrites

sample No.	index	composition	S.C. (K/h)	d_{exp} (10^{-3} kg/m³)	$d_{X\text{-ray}}$ (10^{-3} kg/m³)	p (%)	$\bar{d}$ (μm)	ΔH (kA/m)	10^4 tg δ_ε
1	M1	$NiFe_2O_4$	1403/3	5.19	5.39	3.7	1.9	59	—
2			1403/10	5.18	5.39	3.9	2.1	—	—
3	M2	$NiFe_{1.8}O_4$	1403/3	—	—	—	1.9	—	—
4			1403/10	5.20	5.13	—	2.0	69	—
5	M3	$NiMn_{0.06}Fe_{1.8}O_4$	1403/10	5.21	5.19	—	2.0	—	—
6	M4	$NiAl_{0.2}Fe_{1.8}O_4$	1373/3	5.14	5.31	3.2	2.0	—	—
7			1453/10	5.14	5.31	3.2	2.2	—	—
8	M5	$Ni_{0.97}Co_{0.03}Mn_{0.04}A_{0.2}Fe_{1.8}O_4$	1403/10	5.05	5.43	5.4	2.0	—	—
9			1443/10	5.03	5.34	5.8	2.2	48	22
10	M6	$Ni_{0.95}Ca_{0.02}Co_{0.03}Mn_{0.04}Al_{0.2}Fe_{1.8}O_4$ +	1403/3	5.05	5.34	5.4	1.4	—	—
11		+ 0.3 wt% SiO_2	1403/10	5.01	5.34	6.2	1.5	55	25
12	M7	$Ni_{0.5}Zn_{0.5}Fe_{1.8}O_4$	1403/3	5.12	5.10	—	2.6	—	—
13			1443/10	—	—	—	3.9	—	—
14	M8	$Ni_{0.7}Cu_{0.3}Fe_{1.8}O_4$	1243/10	5.31	5.14	—	6.6	26	8
15			1293/6	5.31	5.14	—	39.0	17	22
16			1323/3	—	—	—	71.6	—	—
17	M9	$Ni_{0.875}Cu_{0.10}Co_{0.025}Mn_{0.035}Fe_{1.9}O_4$	1393/10	5.15	5.34	3.3	2.9	28	6
18			1393/20	5.18	5.34	3.0	9.1	23	10
19	M10	$Ni_{0.88}Cu_{0.10}Co_{0.02}Mn_{0.02}Fe_{1.95}O_4$	1423/10	5.19	5.37	3.3	11.2	—	100
20			1423/3	5.17	5.39	3.7	4.5	—	60
21	M11	$Ni_{0.7}Cu_{0.4}Fe_{1.8}O_4$	1173/6	5.42	5.29	—	2.8	46	77
22			1173/6 air quenched	5.48	5.29	—	1.9	72	35

measurements to be performed automatically with five points per decade in the frequency range 10^6 to 10^{-2} Hz in the temperature interval 298 to 583 K in heating and cooling cycles.

ΔH and tg δ_ε were measured by a resonance method described in [23].

3. Results and Discussion

3.1 *Influence of substitutions on the microstructure*

The experimental results showed that the pure stoichiometric nickel ferrite $NiFe_2O_4$ has a fine-grained microstructure with an average grain size of about 2 µm (Table 1, composition M1). It is independent of the applied sintering conditions (Table 1, samples 1, 2) as well as on the iron deficiency (Table 1, composition M2), which is in agreement with [2, 6].

The substitution of Fe^{3+} for Al^{3+} did not change the microstructure (Table 1, composition M4), which remained independent of the applied sintering conditions (Table 1, samples 8, 9).

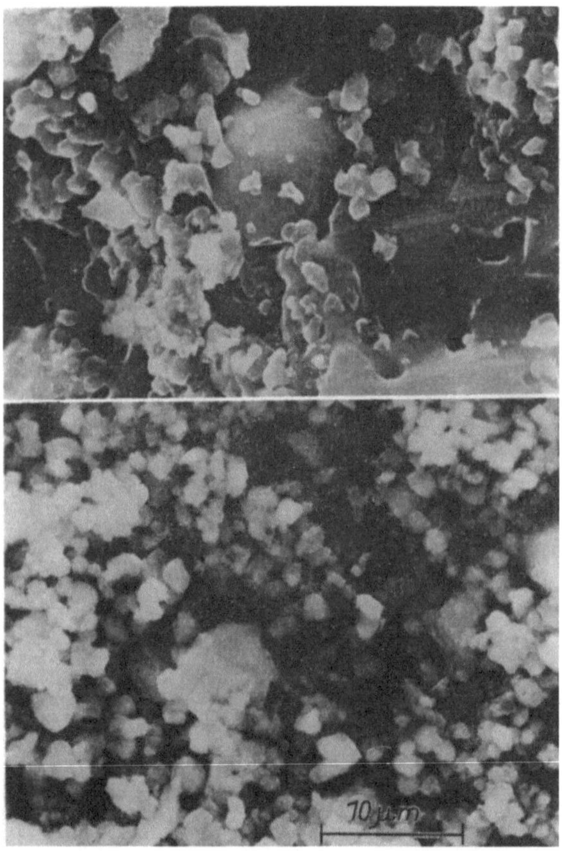

a

b

Fig. 2. Microstructure of a) $Ni_{0.97}Co_{0.03}Mn_{0.01}Al_{0.2}Fe_{1.8}O_4$ (sample 8) and b) $Ni_{0.95}Ca_{0.02}Co_{0.03}Mn_{0.04}$ · · $Al_{0.02}Fe_{1.8}O_4$ + 0.3 wt% SiO_2 (sample 11)

Additions of small amounts of manganese and cobalt did not influence the average grain size, but increased the porosity (Table 1, composition M5). We supposed that the higher density of M1 and M2 (Table 1) was due to the mechanism of exaggerated pore growth [6, 14] which was impeded by the presence of cobalt and manganese. The pores additionally retard the grain growth [6]. This can even lead to a discontinuous grain growth [6], which was observed for M5 (Fig. 2a). A fine-grained microstructure can be obtained by the addition of grain growth inhibiters [7, 8]. In this case discontinuous grain growth remains possible by the dissolution of the addition in the crystal lattice [7]. In order to produce material with $\bar{d} < 2\ \mu m$, we have investigated the influence of $CaO + SiO_2$. As calcium partially enters the spinel lattice, we have tried to obtain uniform grain size by simultaneous addition of CaO and SiO_2. The modified composition (Table 1, composition M6) was X-rayed and found to be single-phase. The EPMA showed the presence of a small quantity of a second phase rich in Ca and Si with the following composition (expressed as oxides): CaO 24 wt%; SiO_2 64 wt%; Fe_2O_3 8 wt%; NiO 4 wt%. The analysis of the main phase confirmed a uniform distribution of the components with a lack of calcium. It was suggested that the presence of SiO_2 facilitated the formation of the calcium silicate phase between the grains, which brought a uniformly sized fine-grained microstructure with $\bar{d} < 2\ \mu m$ (Table 1), (Fig. 2b).

a

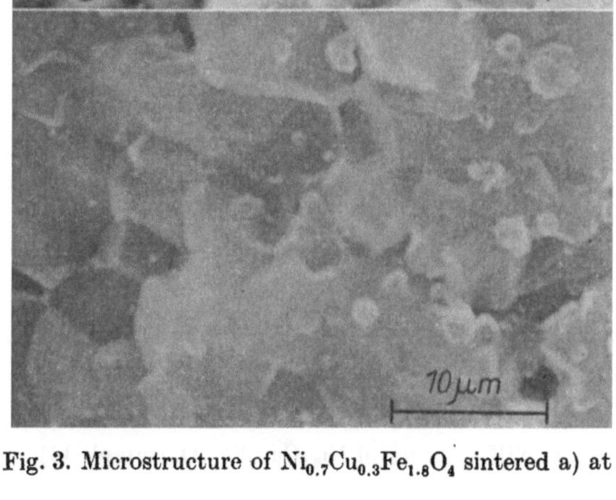

b

Fig. 3. Microstructure of $Ni_{0.7}Cu_{0.3}Fe_{1.8}O_4$ sintered a) at 1323 K for 3 h, b) at 1243 K for 10 h

In accordance with [1, 4, 5] the microstructure of the nickel-zinc ferrites was found to be dependent on the sintering conditions (Table 1, composition M7, samples 12, 13).

The substitution with copper strongly affected the microstructure. Materials with $\bar{d}$ within the range of 2 to 80 μm (Fig. 3) and high densities obtained at lower firing temperatures were prepared (Table 1, compositions M8 to M10).

As can be seen from Table 1, d_{exp} of compositions with iron deficiency (compositions M2, M3, M7, M8) is higher than $d_{X\text{-ray}}$. The XRD analysis confirmed the presence of a single spinel phase. Compositions M3, M7, and M8 were investigated by EPMA. No additional phase was detected in M3 and M7, but the quantity of nickel in the ferrite calculated as NiO was found to be decreased by 4 and 8 wt%, respectively. The presence of the NiO phase with density ($d_{NiO} = 6.8 \times 10^3$ kg/m³ [24]) higher than that of the ferrite (Table 1) and small grains which could not be registrated by the analysis, was proposed.

The EPMA of M8 (sample 14) indicated a second-phase rich in copper (Fig. 4) with the following composition (calculated as oxides): CuO 91 wt%; NiO 2 wt%; Fe_2O_3 7 wt%. The copper content of the ferrite phase calculated as CuO was found to be decreased by 10 wt%. As the density of CuO ($d_{CuO} = 6.4 \times 10^3$ kg/m³ [24]) is higher than that of the ferrite (Table 1), the values of d_{exp} could be explained by the presence of the additional phase.

a b c

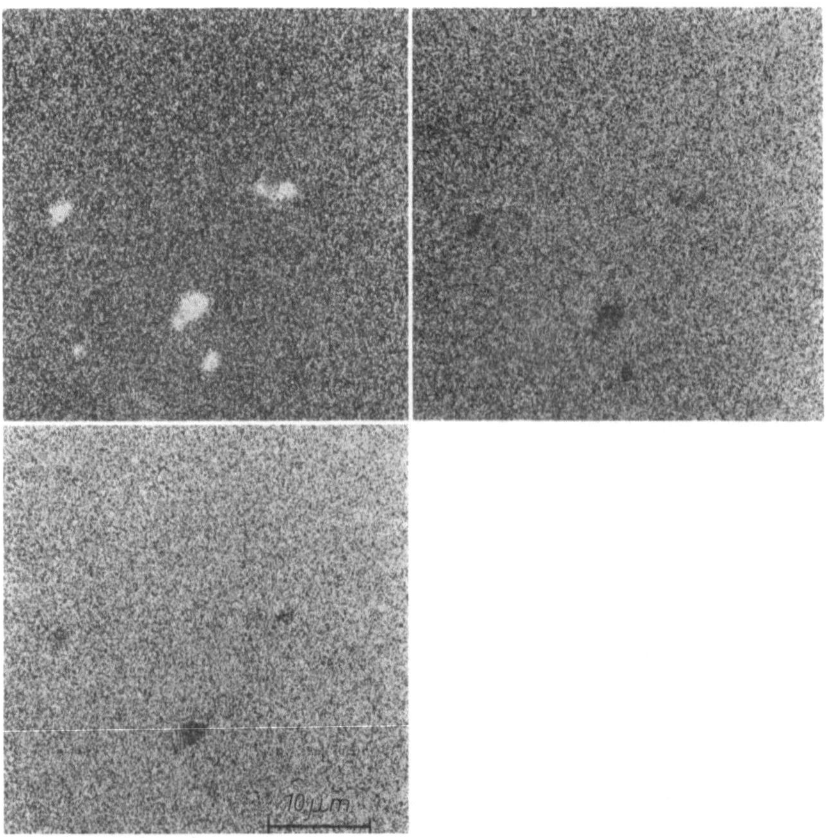

Fig. 4. Characteristic X-ray a) CuK_α, b) FeK_α, c) NiK_α, images of $Ni_{0.7}Cu_{0.3}Fe_{1.8}O_4$ (sample 14)

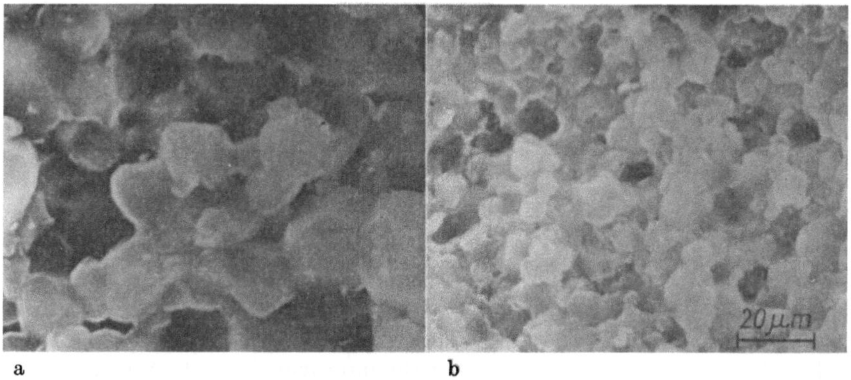

Fig. 5. Microstructure of $Ni_{0.88}Cu_{0.10}Co_{0.02}Mn_{0.02}Fe_{1.95}O_4$ sintered at 1423 K a) for 10 h, b) for 3 h

3.2 Influence of the microstructure on the dielectric losses and the resonance linewidth

The influence of the grain boundaries on the dielectric losses was examined by the impedance measurements of composition M10 (Table 1). As the sintering temperature influences the conduction mechanism [16] and thus the bulk resistivity, we have fired the samples at one and the same temperature for different sintering times (Table 1, samples 19, 20). In this way we have controlled the average grain size (Fig. 5). As was expected [22, 25], the experimentally obtained spectra (Fig. 6) differed from the ideal one (Fig. 2). Because of the small overlap between the semi-circles the resistance values were easily determined. The distinction of ϱ_g from ϱ_{gb} permitted the calculation of the activation energy for the conductivity of the bulk material. It was found that the temperature dependence of ϱ_g obeys the relation

$$\varrho_g = \varrho_0 \exp\left(\frac{\Delta E}{kT}\right), \tag{2}$$

where ΔE is the activation energy and k the Boltzmann constant. The plot lg ϱ_g versus $10^3/T$ (Fig. 7) is a straight line from the slope of which an activation energy

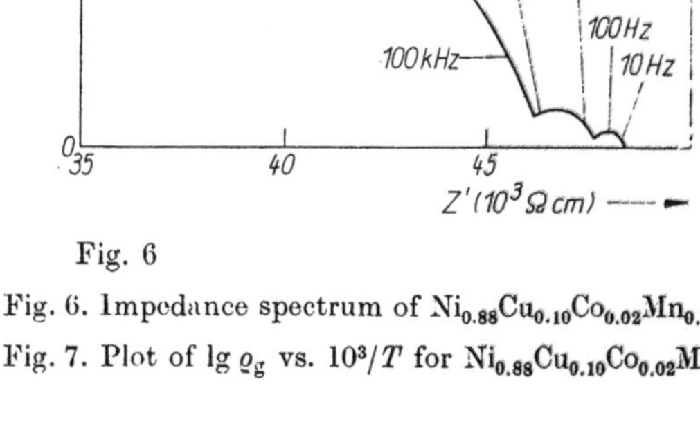

Fig. 6

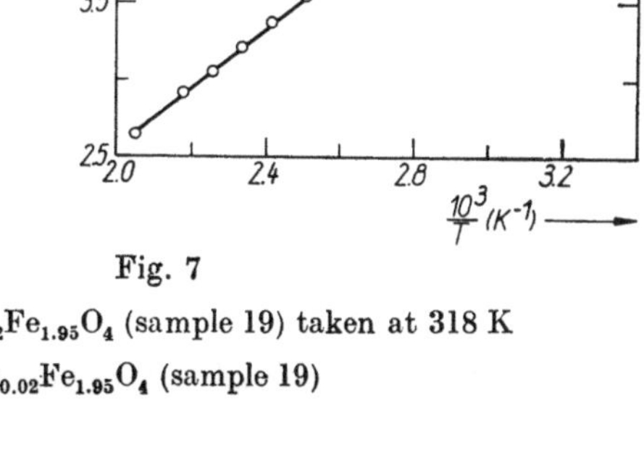

Fig. 7

Fig. 6. Impedance spectrum of $Ni_{0.88}Cu_{0.10}Co_{0.02}Mn_{0.02}Fe_{1.95}O_4$ (sample 19) taken at 318 K

Fig. 7. Plot of lg ϱ_g vs. $10^3/T$ for $Ni_{0.88}Cu_{0.10}Co_{0.02}Mn_{0.02}Fe_{1.95}O_4$ (sample 19)

Table 2
Resistance, average grain size, and dielectric losses of $Ni_{0.88}Cu_{0.10}Co_{0.02}Mn_{0.02}Fe_{1.8}O_4$

sample No. from Table 1	$\bar{d}$ (μm)	ϱ (10^5 Ω cm)	ϱ_g (10^5 Ω cm)	ϱ_{gb} (10^{-5} Ω cm)	10^4 tg δ_ε
19	11.2	1.24	1.07	0.17	100
20	4.5	3.87	1.12	2.75	60

of 0.37 eV was calculated. The room temperature resistance of sample 19 with $\bar{d} = 11.2$ μm was rather low and consequently higher dielectric losses were measured in comparison with those obtained for sample 20 with $\bar{d} = 4.5$ μm (Table 2). The values of ϱ, ϱ_g, and ϱ_{gb} for the two samples show that the improvement of tg δ_ε is due to the reduction of the grain boundary resistance, which is correlated with the microstructural changes. An increase in grain size is accompaned by a decrease in boundary area. An analogous dependence between tg δ_ε and $\bar{d}$ wias observed for composition M9 (Table 1). Taking into account the low sintering temperature and the oxidizing atmosphere, we have attributed the differences in the dielectric losses for samples 14 and 15 of M8 (Table 1) mainly to the change of the average grain size. The results obtained show (Table 1) that tg δ_ε of materials with $\bar{d}$ of the order of 10 μm can be reduced about two times by decreasing the av erage grain size.

The investigations of the influence of the m icrostructure on the resonance linewidth concerned inhomogeneities due to the presence of a second phase and grain boundaries. As can be seen on Table 1, despite its higher measured density, composition M2 has a broader linewidth than M1. This result confirms the presence of a second phase in M2 and its negative influence on ΔH. The observed second phase in composition M6 also deteriorates the linewidth in comparison with M5. The experimental results for M8 and M9 show (Table 1) that the resonance linewidth of copper-containing materials can be additionally decreased by 20 to 30% with increasing average grain size.

The low values of ΔH and tg δ_ε for composition M8 (Table 1) seemed somehow contradictory because of the presence of an additional phase as well as of the higher copper concentration, which should provide an enhanced conductivity [4]. For the explanation of this experimental result, additional samples with composition

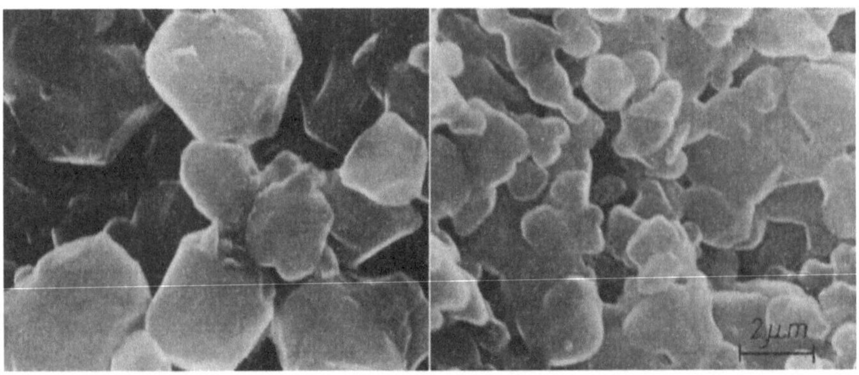

a b

Fig. 8. Microstructure of $Ni_{0.7}Cu_{0.4}Fe_{1.8}O_4$, rate of cooling a) 100, K/h, b) 100 K/min

$Ni_{0.7}Cu_{0.4}Fe_{1.8}O_4$ were prepared (Table 1, composition M11). One of them was sintered at normal conditions (Table 1, sample 21), while the other one was air quenched (cooling rate 100 K/min) (Table 1, sample 22) and annealed for removing internal stresses. The measured density of M11 was higher than the X-ray one. The value obtained for sample 22 was bigger than that for sample 21 (Table 1). EPMA performed for the two samples determined an additional phase rich in copper. The Cu content in the ferrite phase calculated as CuO was reduced by 23 and 29 wt%, respectively, for samples 21 and 22. This result as well as the different values for d_{exp} could be explained by the bigger amount of the additional phase in sample 22. This sample has a smaller average grain size (Table 1, Fig. 8). The differences in the microstructure are reflected in the resonance and dielectric losses: the material with smaller grains and bigger second phase demonstrated a broader linewidth and lower tg δ_ε (Table 1). The results obtained explain the complex influence of copper on the properties of composition M7. The positive effect of higher density and larger average grain size on ΔH dominates over the negative influence of the second phase, which brings about a decrease of the copper concentration in the octahedral sublattice and thus reduces the possibility [4] for enhanced conduction. The improved density suppresses the overoxidation of Ni^{2+} during cooling. The low dielectric losses show that the negative influence of the bigger grains has to be weaker than the above-mentioned positive contributions.

4. Conclusions

The results obtained show that substituted nickel ferrites with an average grain size less than 2 μm can be achieved by the ceramic technology for compositions which do not contain copper and are modified by a grain growth inhibiter such as $CaO + SiO_2$. Materials with high density and large grain size easily controllable by the sintering conditions are developed by copper addition. Their dielectric and magnetic losses can be regulated by the average grain size. As the influence of the grain boundaries on ΔH is opposite to that on tg δ_ε, the microstructure has to be considered with the desired properties of the ferrite material.

Acknowledgements

One of the authors (L.I.) would like to thank Prof. B. Steele, Dr. N. Bonanos, and Dr. J. Kilner for their kind hospitality and helpful discussions during his stay at the Imperial College London, Department of Materials Science, where the electrical conductivity measurements were carried out.

References

[1] A. B. Naik and J. I. Powar, Indian J. pure appl. Phys. **23**, 436 (1985).
[2] L. G. Van Uitert, J. appl. Phys. **27**, 723 (1956).
[3] S. Gasiorek and J. Kulikowski, J. Magnetism magnetic Mater. **26**, 295 (1982).
[4] D. R. Secrist and H. L. Turk, J. Amer. Ceram. Soc. **53**, 12, 683 (1970).
[5] S. K. Kurtz and F. M. A. Capray, J. appl. Phys. **51**, 11, 5745 (1980).
[6] P. J. L. Reijnen, Sci. Ceram. **4**, 169 (1968).
[7] A. Broese Van Groenou, P. F. Bongers, and A. L. Stuijts, Mater. Sci. Engng. **3**, 317 (1968/69).
[8] G. H. Jonker and A. L. Stuijts, Philips tech. Rev. **32**, 79 (1971).
[9] R. Narayan, R. B. Tripathi, B. K. Das, and G. C. Jain, J. Mater. Sci. **18**, 1583 (1983).

[10] M. DROFENIK, S. BESENICAR, and D. KOLAR, Amer. Ceram. Soc. Bull. **63**, 699 (1984).
[11] S. BESENICAR and M. DROFENIK, Brit. Ceram. Trans. J. **84**, 199 (1985).
[12] C. PASNICU, D. CONDURACHE, and E. LUCA, phys. stat. sol. (a) **76**, 145 (1983).
[13] Q. H. F. VREHEN, J. appl. Phys. **40**, 1849 (1969).
[14] C. BORGHESE and R. ROVEDA, J. appl. Phys. **40**, 4791 (1969).
[15] C. HECK, Magnetic Materials and Their Applications, Butterworths, London 1974.
[16] L. G. VAN UITERT, J. chem. Phys. **23**, 1289 (1955).
[17] C. G. KOOPS, Phys. Rev. **83**, 121 (1951).
[18] E. SCHULER, G. GIRAUD, and M. KLEITZ, J. chem. Phys. **70**, 1309 (1973).
[19] J. E. BAUERLE, J. Phys. Chem. Solids **36**, 57 (1969).
[20] P. E. SEIDEN, J. appl. Phys. **34**, 1606 (1963).
[21] J. J. GREEN, J. S. WAUGH, and B. J. HEALY, J. appl. Phys. **35**, 1006 (1964).
[22] J. D. FACTOR, Ph. D. Thesis, Imperial College, London 1984.
[23] S. IVANOV and K. BACHEV, Bulg. J. Phys. **11**, 513 (1984).
[24] I. T. GORONOVSKII, U. P. NAZARENKO, and E. F. NEKRIACH, Kratkii spravochnik po khimii, Naukova Dumka, Kiev 1965.
[25] M. KLEITZ, H. BERNARD, E. FERNANDEZ, and E. SCHOULER, Adv. Ceramics **3**, 310 (1981).

(Received June 3, 1988)

phys. stat. sol. (a) **111**, 155 (1989)

Subject classification: 64.60; S10.1

Laboratoire de Spectrométrie Physique, Saint-Martin-d'Hères[1]) *(a) and*
Laboratoire de Minéralogie et Cristallographie associé au C.N.R.S.,
Université Pierre et Marie Curie, Paris[2]) *(b)*

Stress-Induced Effects in the Incommensurate Phase of Quartz

By

M. VALLADE (a) and MONIQUE PETIT (b)

The stress-induced changes of the amplitudes and of the modulation wave-vectors of the "triple-q" incommensurate phase of quartz-like crystals are calculated using a phenomenological theory, in the small stress limit. The temperature dependence of these changes and in particular their behaviour at the "triple-q"–"single-q" transition are also discussed.

Les changements des amplitudes et des vecteurs d'onde de modulation induits par une contrainte dans la phase incommensurable „triple-q" de cristaux du type du quartz sont calculés en utilisant une théorie phénoménologique, dans la limite des faibles contraintes. La dépendance en température de ces changements, et en particulier leur comportement à la transition „triple-q"–„mono-q" sont également discutés.

1. Introduction

In recent years incommensurate (inc) phases have been found in many different systems [1]. They correspond to a spontaneous modulation of a physical property (magnetization, charge density, atomic displacements, etc.) along one or several directions. When several symmetry related, non-colinear modulation wave vectors are simultaneously present, the inc structure is called a "multi-q" structure. Several examples of multi-q phases have been observed either in magnetic systems (Nd [2]), charge density wave crystals (2H–$TaSe_2$) [3], or dielectrics ($BaNaNb_2O_{15}$ [4], proustite [5], Berlinite [6, 7], and quartz [8]). In these two latter crystals a triple-q structure has been observed both in real space (by electron microscopy [9, 10], and in the reciprocal space (by neutron [11], γ [12], and X-ray diffraction [13]). The occurrence of this triple-q structure has been successfully explained by a Landau-type phenomenological theory which includes a gradient coupling of the order parameter η of the α–β structural phase transition with elastic strains [14, 15]. In particular the model correctly predicts a state with three wave vectors $\boldsymbol{q}_i$ of equal length at 120° from each other and tilted from the y-type symmetry directions by a temperature-dependent angle. This rotation angle of the triangle structure which can take with an equal probability two opposite signs $\pm\delta\varphi_0$, gives rise to a "macro-domain structure" which has actually been observed [9, 10, 13]. The rotation angle $\delta\varphi_0$, which is expected to vary as the amplitude ϱ of the modulation, has been recently accurately measured as a function of temperature [16].

[1]) Boite postale no. 87, F-38402 Saint-Martin-d'Hères Cedex, France.
[2]) Tour 16, 4 place Jussieu, F-75252 Paris, France.

In a recent letter [17] we have shown that a uniaxial stress in the basal plane of the hexagonal structure is able to change the inc structure from 3-q to 1-q. This new phase transition which has indeed been observed in quartz by different techniques (neutron, X-rays [18], optics [19]), can be easily understood in the frame of the phenomenological theory when the coupling of the inc order parameter with the symmetry breaking components of the stress is taken into account. The theoretical phase diagram predicts two straight transition lines (β-1q and 1q-3q) which cross each other at the zero stress inc transition temperature T_i, in fair agreement with that observed experimentally. (A detailed study [16, 18], however, indicates that the 1-q state exists even at zero stress over a very small range of temperature (0.05 K); this point is not yet understood.) For stresses applied in the 3-q state and not strong enough to induce the 3-q–1-q transition, some modifications of the triangular structure are expected to be induced. This point was only briefly reported in [17] and the purpose of the present paper is to give a more detailed theoretical analysis of these induced effects, and in particular to calculate the changes in the amplitude of the three modulations and in the orientations and lengths of the corresponding wave vectors. The discontinuities of these parameters at the first-order 3-q–1-q phase transition are also calculated in the frame of our phenomenological theory.

2. Phenomenological Theory of the inc Phase of Quartz under Uniaxial Stress

In order to explain the β-inc phase transition at zero stress the Landau free energy is expanded as a functional of the order parameter $\eta(r)$ of the α–β phase transition and of the elastic strain field $U_{ij}(r)$. This functional contains a gradient coupling term

$$a\left(\frac{\partial\eta}{\partial x}\,(U_{xx} - U_{yy}) - 2\,\frac{\partial\eta}{\partial y}\,U_{xy}\right) \tag{1}$$

which was shown by Aslanyan and Levanyuk [14] to be at the origin of the inc instability. After eliminating the elastic degrees of freedom through the usual minimization procedure, one arrives at an expression of the free energy of the following form [15]:

$$F = \frac{1}{2}\sum_{\boldsymbol{q}_j} A(\boldsymbol{q}_j)\,\eta_{\boldsymbol{q}_j}\eta_{-\boldsymbol{q}_j} + \frac{1}{3}\sum_{\boldsymbol{q}_j,\boldsymbol{q}_l,\boldsymbol{q}_m} G(\boldsymbol{q}_j,\boldsymbol{q}_l,\boldsymbol{q}_m)\,\eta_{\boldsymbol{q}_j}\eta_{\boldsymbol{q}_l}\eta_{\boldsymbol{q}_m}\,\delta(\boldsymbol{q}_j+\boldsymbol{q}_l+\boldsymbol{q}_m) +$$
$$+ \frac{1}{4}\sum_{\boldsymbol{q}_j,\boldsymbol{q}_l,\boldsymbol{q}_m,\boldsymbol{q}_n} B\eta_{\boldsymbol{q}_j}\eta_{\boldsymbol{q}_l}\eta_{\boldsymbol{q}_m}\eta_{\boldsymbol{q}_n}\,\delta(\boldsymbol{q}_j+\boldsymbol{q}_l+\boldsymbol{q}_m+\boldsymbol{q}_n) + \dots \tag{2}$$

with

$$A(\boldsymbol{q}_j) = A(T - T_i) + h(q_j^2 - q_0^2)^2 + \Delta q_j^2 \cos^2(3\varphi_j)\,, \tag{3a}$$

$$G(\boldsymbol{q}_j,\boldsymbol{q}_l,\boldsymbol{q}_m) = i\sum_{i=j,l,m} (Gq_i^3 + G'q_i)\cos(3\varphi_i)\,, \tag{3b}$$

where $q_j = |\boldsymbol{q}_j|$ and φ_j is the angle of $\boldsymbol{q}_j$ with the two fold crystallographic axis $0x$. The q dependence of the quartic coefficient B is ignored for the sake of simplicity and B is assumed to be positive.

As shown in [17] the introduction of an applied stress with components in the X–Y plane leads to the following additional terms in the quadratic part of F:

$$-\sum_{\boldsymbol{q}_j}\left(D(\sigma_{xx}+\sigma_{yy}) + Kq_j^2(\cos(2\varphi_j)\,(\sigma_{xx}-\sigma_{yy}) + \sin(2\varphi_j)\,(\sigma_{xy}+\sigma_{yx}))\right)\eta_{\boldsymbol{q}_j}\eta_{-\boldsymbol{q}_j}\,. \tag{4}$$

This term is the most important one for small stresses, which is the only case we consider in the following. At this stage it is most convenient to introduce dimensionless reduced parameters (as already done in [20])

$$\frac{q}{q_0} \to q; \quad \sqrt{\frac{B}{hq_0^4}}\,\eta_q \to \eta_q; \quad \frac{B}{hq_0^4}\,F \to F; \quad \frac{A(T_i - T)}{hq_0^4} \to t; \quad \frac{\Delta}{hq_0^2} \to \Delta; \quad (5)$$

$$\frac{q_0}{\sqrt{Bh}}\,G \to G; \quad \frac{1}{q_0\sqrt{Bh}}\,G' \to G'; \quad \frac{D(\sigma_{xx} + \sigma_{yy})}{hq_0^4} \to \bar{\sigma};$$

$$\frac{K(\sigma_{xx} - \sigma_{yy})}{hq_0^2} \to \sigma'; \quad \frac{K(\sigma_{xy} + \sigma_{yx})}{hq_0^2} \to \sigma''.$$

The thermodynamical Gibbs potential corresponding to the sum of (2) and (4), has to be minimized at constant temperature and stress with respect to the complex amplitudes η_q corresponding to all the possible wave vectors $\boldsymbol{q}$. Two kinds of solution can be found:

2.1 *Single-q state*

In this case the order parameter $\eta(\boldsymbol{r})$ can be written as

$$\eta(\boldsymbol{r}) = 2\varrho \cos(\boldsymbol{qr} + \psi) \quad (6)$$

and the corresponding thermodynamic potential is

$$G_1 = ((-t + (q^2 - 1)^2 + \Delta q^2 \cos^2(3\varphi)) - 2(\bar{\sigma} + q^2(\sigma' \cos(2\varphi) + \sigma'' \sin(2\varphi))))\,\varrho^2 + \tfrac{3}{2}\,\varrho^4 + \dots \quad (7)$$

It is clear that the non-symmetry breaking component of the stress, $\bar{\sigma}$, has the only role of changing the transition temperature T_i and can be incorporated into the term t. In the following we shall use the notation $\bar{t} = t + 2\bar{\sigma}$.

The minimization equations

$$\frac{\partial G_1}{\partial \varrho} = \frac{\partial G_1}{\partial \varphi} = \frac{\partial G_1}{\partial q} = 0 \quad (8)$$

bave the following solutions, up to first order in σ 'or σ'':

a) $\sigma' \neq 0$; $\sigma'' = 0$ (pure compressive stress along the x or y direction)

$$\varphi_1 = \frac{\pi}{2} + k\pi; \qquad q_1 = 1 - \sigma'; \quad \varrho_1 = \left(\frac{\bar{t} - 2\sigma'}{3}\right)^{1/2}, \quad (9a)$$

$$\varphi_2 = \frac{\pi}{2} + \left(\frac{2\pi}{3} - \varepsilon\right) + k\pi; \quad q_2 = 1 + \frac{\sigma'}{2}; \quad \varrho_2 = \left(\frac{\bar{t} + \sigma'}{3}\right)^{1/2}, \quad (9b)$$

$$\varphi_3 = \frac{\pi}{2} - \left(\frac{2\pi}{3} - \varepsilon\right) + k\pi; \quad q_3 = 1 + \frac{\sigma'}{2}; \quad \varrho_3 = \left(\frac{\bar{t} + \sigma'}{3}\right)^{1/2} \quad (9c)$$

with $\varepsilon = \sigma'/3\sqrt{3}\,\Delta$.

Equation (9a) corresponds to an equilibrium energy $G_1 = -(\bar{t} - 2\sigma')^2/6$ and is then the minimum 1-q solution for $\sigma' < 0$, whereas (9b) and (9c) correspond to $G_1 = -(\bar{t} + \sigma')^2/6$ and are degenerate solutions for $\sigma' > 0$.

b) $\sigma' = 0$; $\sigma \neq 0$ (pure shear stress)

$$\varphi_2 = \frac{\pi}{2} + \frac{2\pi}{3} + \varepsilon + k\pi; \quad q_2 = 1 + \frac{\sqrt{3}}{2}\sigma''; \quad \varrho_2 = \left(\bar{t} + \frac{\sqrt{3}}{2}\sigma''\right)^{1/2}, \tag{10a}$$

$$\varphi_3 = \frac{\pi}{2} - \frac{2\pi}{3} + \varepsilon + k\pi; \quad q_3 = 1 - \frac{\sqrt{3}}{2}\sigma''; \quad \varrho_3 = \left(\bar{t} - \frac{\sqrt{3}}{2}\sigma''\right)^{1/2} \tag{10b}$$

with

$$\varepsilon = \sigma''/9\Delta\,.$$

Equation (10a) and (10b) correspond to $G_1 = -(\bar{t} \pm (\sqrt{3}/2)\,\sigma'')^2/6$ and are solutions for $\sigma'' > 0$ and $\sigma'' < 0$, respectively.

2.2 *Triple-q state*

The order parameter is then of the form

$$\eta(\boldsymbol{r}) = \sum_{j=1}^{3} 2\varrho_j \cos(\boldsymbol{q}_j\boldsymbol{r} + \psi_j) \tag{11}$$

and the Gibbs potential is of the form

$$\begin{aligned} G_3 = \sum_{j=1}^{3} &((-t + (q_j^2 - 1)^2 + \Delta q_j^2 \cos^2(3\varphi_j)) - 2q_j^2(\sigma' \cos(2\varphi_j) + \\ &+ \sigma'' \sin(2\varphi_j)))\,\varrho_j^2 + \tfrac{4}{3}\sum_{j=1}^{3}(Gq_j^3 + G'q_j)\cos(3\varphi_j)\sin\psi\varrho_1\varrho_2\varrho_3 \times \\ &\times \delta(\boldsymbol{q}_1 + \boldsymbol{q}_2 + \boldsymbol{q}_3) + \tfrac{3}{2}((\varrho_1^2 + \varrho_2^2 + \varrho_3^2)^2 + 2(\varrho_1^2\varrho_2^2 + \varrho_2^2\varrho_3^2 + \varrho_3^2\varrho_1^2)) + \dots \end{aligned} \tag{12}$$

with $\psi = \psi_1 + \psi_2 + \psi_3$. It is important to note that the cubic term vanishes, unless $\sum_j \boldsymbol{q}_j = 0$. It was shown [15] that this term is necessary to stabilize the 3-q state relative to the 1-q state. Therefore, the minimization of (12) must be performed under the constraints

$$\sum_{j=1}^{3} q_j \cos\varphi_j = \sum_{j=1}^{3} q_j \sin\varphi_j = 0\,. \tag{13}$$

We are then lead to introduce two Lagrange multipliers λ and μ and the minimization equations are

$$\begin{aligned} &\frac{\partial G_3}{\partial\varphi_j} + q_j(\lambda\sin\varphi_j - \mu\cos\varphi_j) = 0; \qquad \frac{\partial G_3}{\partial\varrho_j} = 0\,, \\ &\frac{\partial G_3}{\partial q_j} - (\lambda\cos\varphi_j + \mu\sin\varphi_j) = 0; \qquad \frac{\partial G_3}{\partial\psi_j} = 0\,. \end{aligned} \tag{14}$$

To solve (13) and (14) it is convenient to use the symmetry adapted variables,

$$\begin{aligned} &\bar{x} = x_1 + x_2 + x_3\,, \\ &x' = x_1 - \tfrac{1}{2}(x_2 + x_3)\,, \\ &x'' = \frac{\sqrt{3}}{2}(x_2 - x_3)\,; \qquad x_j = \varphi_j, \varrho_j, q_j\,. \end{aligned} \tag{15}$$

For small stresses, all expressions can be linearized with respect to σ' (or σ'') and to the deviations δx_j from the zero-stress equilibrium state. In particular since $(\partial G_3/\partial x_j')_0 = 0$, one can write in (14)

$$\left(\frac{\partial G_3}{\partial x_j}\right) \approx \sum_i \left(\frac{\partial^2 G_3}{\partial x_j\, \partial x_i}\right)_0 \delta x_i + \left(\frac{\partial^2 G_3}{\partial x_j\, \partial \sigma_i}\right)_0 \sigma_i\,, \tag{16}$$

where the subscript 0 refers to the zero-stress state. Equation (13) then leads to

$$\begin{aligned} \delta q' &= \delta\varphi''\,, \\ \delta q'' &= -\delta\varphi'\,. \end{aligned} \tag{17}$$

The main advantage of the transformation (15) is to allow a decoupling between the variables $\bar{x}$ and the other ones. After a long, but straightforward calculation one gets (to the lowest order in ϱ_0)

a) For $\sigma' \neq 0$; $\sigma'' = 0$,

$$\delta\bar{\varrho} = \delta\bar{\varphi} = \delta\bar{q} = 0\,, \tag{18a}$$

$$\delta\varrho' = -\frac{9\Delta + 4}{\Gamma}\,\frac{\sigma'}{\varrho_0}\,, \qquad \delta\varrho''_{\pm} = \pm\,\frac{4(9\Delta - 4)\,|G + G'|}{9\Delta\Gamma}\,\sigma'\,, \tag{18b}$$

$$\delta\varphi'_{\pm} = \pm\,\frac{4\,|G + G'|}{\Gamma}\,\frac{\sigma'}{\varrho_0}\,, \qquad \delta\varphi'' = \frac{3}{\Gamma}\left(4 - p\left(1 - \frac{2G}{3(G + G')}\right)\right)\sigma'\,. \tag{18c}$$

with $\Gamma = 3(p - 6)\,\Delta + 4(p - 2)$ and $p = 8(G + G')^2/(9\Delta)$.

The signs which appear in (18) correspond to the two possible values of the phase ψ, and thus to the two kinds of "macrodomain" of rotation in the zero stress 3-q state.

From (17) and (18) the initial variable δx_j can be easily deduced

$$\begin{aligned} \delta x_1 &= \tfrac{2}{3}\,\delta x'\,, \\ \delta x_2 &= -\frac{1}{3}\,\delta x' + \frac{\delta x''}{\sqrt{3}}\,, \\ \delta x_3 &= -\frac{1}{3}\,\delta x' - \frac{\delta x''}{\sqrt{3}}\,. \end{aligned} \tag{19}$$

One can note that

$$\begin{aligned} \delta\varrho_1^+ &= \delta\varrho_1^-\,, & \delta\varrho_2^{\pm} &= \delta\varrho_3^{\mp}\,, \\ \delta q_1^+ &= \delta q_1^-\,, & \delta q_2^{\pm} &= \delta q_3^{\mp}\,, \\ \delta\varphi_1^+ &= -\delta\varphi_1^-\,, & \delta\varphi_2^{\pm} &= -\delta\varphi_3^{\mp}\,. \end{aligned} \tag{20}$$

Therefore the σ' component of the stress does not lift completely the degeneracy of the various amplitudes and is not expected to influence the macrodomain texture.

b) For $\sigma' = 0$; $\sigma'' \neq 0$,

$$\delta\bar{\varrho} = \delta\bar{\varphi} = \delta\bar{q} = 0\,, \tag{21a}$$

$$\delta\varrho'_{\pm} = \pm\,\frac{4(9\Delta - 4)\,|G + G'|}{9\Delta\Gamma}\,\sigma''\,, \qquad \delta\varrho'' = \frac{9\Delta + 4}{\Gamma}\,\frac{\sigma''}{\varrho_0}\,, \tag{21b}$$

$$\delta\varphi' = \frac{3}{\Gamma}\left(4 - p\left(1 - \frac{2G}{3(G + G')}\right)\right)\sigma''\,, \qquad \delta\varphi''_{\pm} = \mp\,\frac{4\,|G + G'|}{\Gamma}\,\frac{\sigma''}{\varrho_0}\,. \tag{21c}$$

Instead of (20) we have then

$$\begin{aligned} \delta\varrho_1^+ &= -\delta\varrho_1^- , & \delta\varrho_2^\pm &= -\delta\varrho_3^\mp , \\ \delta q_1^+ &= -\delta q_1^- , & \delta q_2^\pm &= -\delta q_3^\mp , \\ \delta\varphi_1^+ &= \delta\varphi_1^- , & \delta\varphi_2^\pm &= \delta\varphi_3^\mp . \end{aligned} \tag{22}$$

In this case the stress σ'' lifts completely the degeneracy between the various $\delta\varrho_j^\pm$ and can thus influence the macrodomain texture.

At constant stress and temperature, the stress-induced effects are controlled by the dimensionless parameters G, G', and Δ. In particular, they involve the parameter p which characterizes the degree of stability of the 3-q state at $\sigma = 0$: For $p < p_0 = 6(\Delta + 4/9)/(\Delta + 4/3)$ the determinant of the Hessian matrix $(\partial^2 G_3/(\partial x_i\, \partial x_j))_0$ which is proportional to Γ, is negative and the 3-q state is absolutely unstable. For $6 > p > p_0$, the 3-q state is locally stable but less stable than the 1-q state and for $p > 6$ the 3-q state is the most stable one. (For $p > 15$ however the zero stress β–3-q transition becomes first order and our expansion (12) is no longer sufficient.) The increase of Γ with p, which tends to decrease the stress-induced effects, reflects this increase of "rigidity" of the 3-q state.

From a comparison of the temperature dependence of the zero-stress quantities with experimental results [16], one deduces that $\Delta < 1$ and $G(G + G') < 0$. With, $p > 6$, this allows to determine completely the signs of the induced effects in (18) and (21). The results are schematically summarized in Fig. 1.

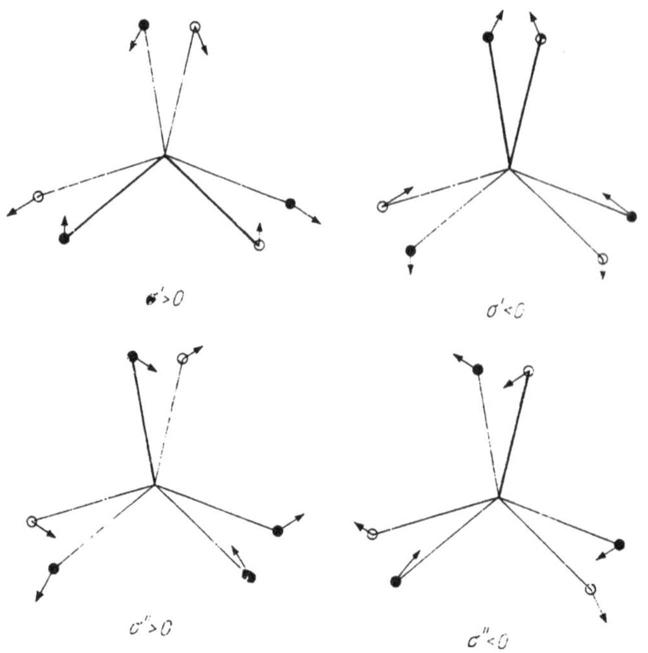

Fig. 1. Schematic representation of the stress-induced changes of the modulation wave-vectors in the "triple-q" state of quartz. The arrows indicate the changes of the length and of the orientation of the wave vectors, the extremities of which are represented by small circles: ● for macrodomains (+) and ○ for macrodomains (−). The heavy lines correspond to the waves with the largest amplitudes

2.3 Approach to the 3-q-1-q transition

Let us now discuss the temperature dependence of the stress-induced effects. As discussed in [17], the results of (18) and (21) are valid, for a given stress up to a temperature $T_{13}(\sigma)$ which corresponds to the onset of the 1-q state and is determined by the equality of the Gibbs potentials G_3 and G_1. Since G_1 and G_3 are of the general form

$$G_1 = -\tfrac{1}{3}\,(\bar{t} + \alpha\sigma)^2\,, \tag{23a}$$

$$G_3 = -\tfrac{3}{2}\,(\bar{t}^2)/(15 - p) + 0(\sigma^2)\,. \tag{23b}$$

The reduced temperature $\bar{t}_{13}$ corresponding to $T = T_{13}(\sigma)$ is given by

$$\bar{t}_{13} \approx \alpha c \sigma \tag{24}$$

with

$$c = \sqrt{15 - p}/(3 - \sqrt{15 - p})\,. \tag{25}$$

From (18) and (21), one can see that the stress-induced changes either vary with temperature as ϱ_0^{-1} (i.e. as $\bar{t}^{-1/2}$) or are constant. The former are maximum at $\bar{t} = \bar{t}_{13}$ and, by taking (24) into account, are proportional to $\sigma^{1/2}$ on the transition line. The latter are proportional to σ along this line. It is interesting to calculate the discontinuities of the various parameters at $t = t_{13}(\sigma)$. Let us consider, for example, the case $\sigma' \neq 0$, $\sigma'' = 0$. For $\sigma' < 0$, the most stable 1-q state corresponds to $\varphi_1 = \pi/2$. As a precursor of this state, the amplitude ϱ_1 grows with $|\sigma'|$ at the expense of ϱ_2 and ϱ_3, in the 3-q state.

At $\bar{t}_{13}$, the ϱ_1 amplitudes in the 1-q state and in the 3-q state are, respectively,

$$(\varrho_1)_{1\text{-}q} = \left(\frac{\bar{t}_{13} + 2\,|\sigma'|}{3}\right)^{1/2} = \left(\frac{2(c+1)}{3}\right)^{1/2} |\sigma'|^{1/2}\,, \tag{26a}$$

$$(\varrho_1)_{3\text{-}q} = \left(\frac{\bar{t}_{13}}{15 - p}\right)^{1/2} \left(1 + \frac{2(9\Delta + 4)}{3\Gamma}\,\frac{|\sigma'|}{\bar{t}_{13}}\,(15 - p)\right) |\sigma'|^{1/2}\,. \tag{26b}$$

So that the ratio of these quantities is independent of the stress,

$$\frac{(\varrho_1)_{3\text{-}q}}{(\varrho_1)_{1\text{-}q}} = \frac{1}{(15 - p)^{1/4}} \left(1 + \frac{(9\Delta + 4)\,(15 - p)}{3\Gamma c}\right) = R\,. \tag{27}$$

One can note that this ratio can exceed unity when p is close to 15. This, however, means that the zero stress β–3-q transition is nearly tricritical so that the temperature range over which ϱ_0 is proportional to $\sqrt{\bar{t}}$ becomes vanishingly small. (When $\bar{t}_{13}$ lies outside this range our calculations are no longer valid.) In particular, just at the tricritical point ($p = 15$). $\varrho_0 \sim \bar{t}^{1/4}$; $G_3 \sim \bar{t}^{3/2}$ and one can easily show that $\bar{t}_{13} \sim |\sigma'|^{4/3}$ so that the ratio $(\varrho_1)_{3\text{-}q}/(\varrho_1)_{1\text{-}q}$ increases as $|\sigma'|^{-1/6}$ when $|\sigma'| \to 0$.

Concerning the changes in orientation and in length of the wave vectors at $t_{13}(\sigma')$ one gets

$$(\varphi_1^{\pm})_{3\text{-}q} - (\varphi_1)_{1\text{-}q} = \pm(\delta\varphi_0 - \tfrac{2}{3}\,|\delta\varphi'|) = \pm\Phi\sqrt{|\sigma'|} \tag{28}$$

and

$$(q_1)_{3\text{-}q} - (q_1)_{1\text{-}q} = (\delta q_0 - \tfrac{2}{3}\,\delta q') = Q\,|\sigma'|\,, \tag{29}$$

where Φ and Q can be evaluated using (18) and (24).

φ_1 is thus more sensitive to the stress than q_1 for small stresses.

For $\sigma' > 0$ the most stable 1-q states correspond to the wave vectors q_2 or q_3.

The largest modulation amplitudes are then $\varrho_2^+ = \varrho_3^-$ for $\Delta > (4/9)$ so that in a given macrodomain (say +) the component ϱ_2^+ becomes preponderant before the other (ϱ_3^+) and the converse is true in the opposite domain; therefore the macrodomains of rotation tend to become also domains of 1-q states (with q_2 and q_3 wave vectors, respectively). At the 3-q–1-q transition temperature $\bar{t}_{13}$ one has

$$\frac{(\varrho_2^\pm)_{3\text{-}q}}{(\varrho_2)_{1\text{-}q}} \equiv \frac{(\varrho_3^\mp)_{3\text{-}q}}{(\varrho_3)_{1\text{-}q}} = R \pm R' \sqrt{\sigma'}\,, \tag{30}$$

$$(\delta\varphi_2^\pm)_{3\text{-}q} - (\delta\varphi_2)_{1\text{-}q} = -(\delta\varphi_3^\pm)_{3\text{-}q} + (\delta\varphi_3)_{1\text{-}q} = \pm \frac{\Phi\sqrt{\sigma'}}{\sqrt{2}} + \Phi'\sigma'\,, \tag{31}$$

$$(\delta q_2^\pm)_{3\text{-}q} - (\delta q_2)_{1\text{-}q} = (\delta q_3^\pm)_{3\text{-}q} - (\delta q_3)_{1\text{-}q} = \frac{Q}{2}\,\sigma' \mp Q'\sqrt{\sigma'}\,, \tag{32}$$

where R', Φ', Q' can be easily calculated using (9), (18), (19), and (24).

In this case both the orientation and the length of the modulation wave vectors varies as $\sigma^{1/2}$ at small σ'.

3. Concluding Remarks

The calculations presented in this paper could in principle be quantitatively tested experimentally since the modulation wave vectors and the corresponding amplitudes are related to the position and to the intensity of the satellite peaks in X-ray or neutron diffraction. Experiments performed with neutrons at I.L.L. [17] and with the X-ray synchrotron facility at L.U.R.E. [18] indeed indicate that some stress-induced distortions of the 3-q state take place before the 3-q–1-q transition is reached. Neutron diffraction experiments were performed with a limited resolution and an accurate determination of the stress-induced changes of the satellites reflections was not possible. A better resolution was achieved using the X-ray synchrotron source but, in preliminary experiments, the temperature and stress control was not sufficient to allow a quantitative determination of the distortions and a definite comparison with our calculations.

On the other hand, the Landau theory used in this paper is only valid in the so-called "classical regime", i.e. outside the critical region around T_i where critical fluctuations are important. It is possible, in particular, that the unability of the Landau theory to explain the observation of a 1-q state at zero stress very close to T_i is related to an instability of the 3-q state against the 1-q state, driven by critical fluctuations. Recent renormalization group calculations [21], however, do not point out such an instability but rather conclude to a first-order phase transition at T_i. An alternative explanation for the occurrence of a 1-q state is the influence of random strains due to impurities. Further accurate measurements of the satellite diffraction peaks under stress and the use of quartz samples of different purity are necessary to fully elucidate this question.

References

[1] R. BLINC and A. P. LEVANYUK (Ed.), Incommensurate Phase in Dielectrics, North-Holland Publ. Co., 1986.
[2] B. LEBECH, J. appl. Phys. **52**, 2019 (1981).
[3] D. E. MONCTON, J. D. AXE, and F. J. DI SALVO, Phys. Rev. Letters **45**, 576 (1980).
[4] J. SCHNECK, J. C. TOLEDANO, R. WHATMORE, and F. N. AINGER, Ferroelectrics **36**, 327 (1981).
[5] S. S. KHASANOV and V. S. H. SHEKHTMAN, Ferroelectrics **67**, 55 (1986).

[6] J. P. Bachheimer, B. Berge, G. Dolino, P. Saint Gregoire, and C. M. E. Zeyen, Solid State Commun. **51**, 55 (1984).
[7] A. Zarka, B. Capelle, and J. C. Jumas, C.R. Acad. Sci. 302, Serie II, 523 (1986).
[8] G. Dolino, in [1], Vol. 2, Chap. **16** (p. 205).
[9] G. van Tandeloo, J. van Landuyt, and S. Amelinckx, phys. stat. sol. (a) **33**, 723 (1976).
[10] E. Snoeck, C. H. Roucau, and P. Saint-Gregoire, J. Physique **47**, 2041 (1986).
[11] G. Dolino, J. P. Bachheimer, B. Berge, and C. M. E. Zeyen, J. Physique **45**, 361 (1984).
[12] G. Dolino and P. Bastie, Phys. Rev. B **31**, 2857 (1985).
[13] K. Gouhara and N. Kato, J. Phys. Soc. Japan **53**, 2177 (1984); J. Phys. Soc. Japan **54**, 1868 (1985).
[14] T. A. Aslanyan and A. P. Levanyuk, Solid State Commun. **31**, 547 (1974).
[15] T. A. Aslanyan, A. P. Levanyuk, M. Vallade, and J. Lajzerowicz, J. Phys. C **16**, 6705 (1983).
[16] P. Bastie, F. Mogeon, and C. M. E. Zeyen, to be published in Phys. Rev. (1988) and private communication.
[17] G. Dolino, P. Bastie, B. Berge, M. Vallade, J. Bethke, L. P. Regnault, and C. M. E. Zeyen, Europhysics Letters **3**, 601 (1987).
[18] A. Zarka, B. Capelle, M. Petit, G. Dolino, and B. Berge, J. appl. Cryst. **21**, 72 (1988).
[19] J. P. Bachheimer, J. Phys. C **19**, 5509 (1986).
[20] M. Vallade, V. Dvorak, and J. Laizerowicz, J. Physique **48**, 1171 (1987).
[21] O. Biham, D. Mukamel, J. Toner, and X. Zhu, Phys. Rev. Letters **59**, 2439 (1987).

(Received August 24, 1988)

phys. stat. sol. (a) **111**, 165 (1989)

Subject classification: 62.20; 62.40; S9.11

Faculty of Technology, Kanazawa University[1])

Relation between Plastic Deformation and the Blaha Effect for Alkali Halide Crystals

By

T. OHGAKU and N. TAKEUCHI

The Blaha effect includes both, the effect of oscillatory dislocation on the plastic deformation and the effect of plastic deformation on the internal friction. It is found from the Blaha effect measurement for alkali halide crystals that the dislocation moves during plastic deformation by overcoming two kinds of obstacles such as tetragonal defects and forest dislocations. The Blaha effect measurement is very useful for studying the plastic deformation of materials.

Der Blaha-Effekt beinhaltet zwei Effekte, den Einfluß der Schwingungen der Versetzungen auf die plastische Verformung und den Einfluß der plastischen Verformung auf die innere Reibung. Bei Messungen des Blaha-Effekts in Alkalihalogenideinkristallen wird gefunden, daß die Versetzungen sich während der plastischen Verformung durch Überwinden zweier Arten von Hindernissen, tetragonalen Defekten und Waldversetzungen, bewegen. Die Messungen des Blaha-Effekts sind für die Erforschung der plastischen Verformung fester Körper von großem Nutzen.

1. Introduction

The Blaha effect, which was found by Blaha and Langenecker when an oscillation of 800 kHz was superimposed during the deformation of Zn single crystals [1, 2], is the phenomenon that the static flow stress decreases under superposition of ultrasonic oscillatory stress during plastic deformation. In addition to the decrease in stress, its frequency independence [2, 3], the temperature rise [4, 5], and the decrease in work hardening rate [4, 6] were observed under the superposition of oscillations. This effect has been widely applied for industrial purpose, e.g. wire drawing or rolling [7 to 9].

The Blaha effect has two aspects: one is the plastic deformation and the other is the internal friction measurement. Therefore, by the Blaha effect measurement much information which is given by the above two techniques, can be obtained at once. Particularly, it is expected more useful for the investigation of plastic deformation. So far this effect has been explained by the oscillatory stress superposition mechanism [3, 10 to 13].

Recently, we have reported that the Blaha effect of alkali halide single crystals is caused by the increase in average length of a dislocation segment under the application of oscillatory stress with small amplitude [5, 14]. For the strain rate cycling test between strain rates of $\dot{\varepsilon}_1$ and $\dot{\varepsilon}_2$ with superimposed oscillation, this mechanism gives the relation

$$\Delta\sigma' = \left[\left(\frac{\dot{\varepsilon}_2}{\dot{\varepsilon}_1}\right)^{1/m^*} - 1\right](\sigma_e - \Delta\sigma)\,, \tag{1}$$

[1]) Kodatsuno 2-40-20, Kanazawa 920, Japan.

where $\Delta\sigma'$ is the stress change due to the strain rate cycling test, m^* the stress–velocity exponent, σ_e the effective stress, $\Delta\sigma$ the stress decrement due to the superposition of oscillatory stress [14]. The effective stresses and the stress-velocity exponents of some alkali halide crystals were obtained by this method [14, 15]. However, when the amplitude of oscillatory stress increases and the stress change $\Delta\sigma'$ due to the strain rate cycling test under oscillation approach zero, this mechanism gives rise to the unreasonable result that the average length of a dislocation segment becomes infinitely large because of its inverse proportionality to $\Delta\sigma'$. This unreasonableness may have been introduced by the assumption that the ratio of the separation between two strong obstacles to that between two weak ones becomes infinite [5]. Actually, when the stress amplitude becomes large, deviation from a linear relation was found between $\Delta\sigma$ and the internal friction Q^{-1} as well as between $\Delta\sigma'$ and $\Delta\sigma$ [5]. This reveals a limit of our mechanism: To explain the Blaha effect at higher amplitude, another mechanism is required. In this paper, we attempt to explain the relation between the plastic deformation and the Blaha effect on the basis of the experimental results.

2. Experimental

The specimens were obtained by cleaving the ingots which were grown from the melt in air and annealing for 24 h at 740, 700, 680, and 690 °C for NaCl, KCl, NaBr, and KBr, respectively, followed by cooling to room temperature at 40 K/h. Their dimensions were either 4 × 7 × 14 or 5 × 5 × 15 mm³. No appreciable size effect was observed in the present experiment. The details of the internal friction measurement and the strain rate cycling test during the plastic deformation were described in previous papers [5, 14]. The strain rate cycling tests were performed between cross head speeds of 5 and 20 and between 20 and 100 μm/min. The former is denoted by case I and the latter is case II. All experiments were carried out at room temperature.

3. Results

Fig. 1 shows the results of the present investigation for KCl, in which the relation between $\Delta\sigma'$ and $\Delta\sigma$ clearly deviates from linearity for $\Delta\sigma$ of about 1.1 MPa both in case I and II. This investigation is similar to the results in [14], where the break of linearity between $\Delta\sigma'$ and $\Delta\sigma$ appeared at $\Delta\sigma$ of 0.8 MPa. Other linearities with slopes which are smaller than those at small $\Delta\sigma$ appear when $\Delta\sigma$ is larger than 1.1 MPa. Values of $\Delta\sigma'$ on some curves with small strain are negative when $\Delta\sigma$ is large. Any two straight lines at the same strain and small amplitude do not necessarily intersect

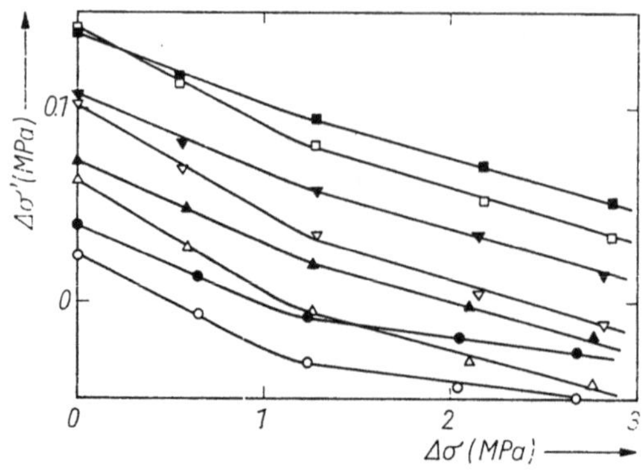

Fig. 1. Relation between $\Delta\sigma'$ and $\Delta\sigma$ for KCl. ○, ● ε = 8, △, ▲ 12, ▽, ▼ 16, and □, ■ 20%; open symbols correspond to case II and full symbols to case I

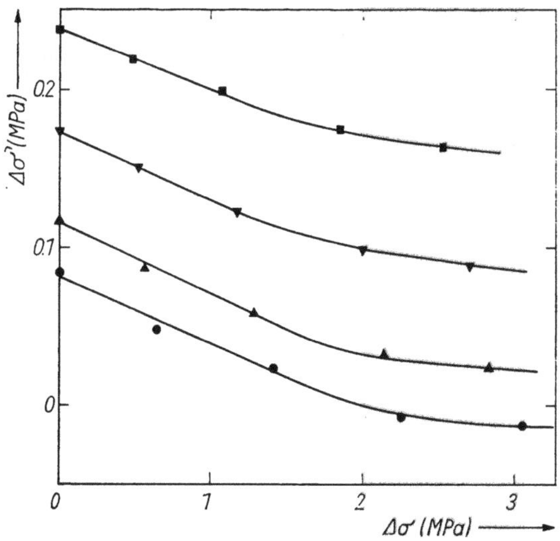

Fig. 2. Relation between $\Delta\sigma'$ and $\Delta\sigma$ for NaCl in case I. ● $\varepsilon = 4$, ▲ 8, ▼ 12, and ● 16%

on the abscissa. These results do not correspond to the previous ones [14]. The reason is probably that the choice of the two strain rates in the strain rate cycling test for the present work is different from that for [14]. The break of proportionality between Q^{-1} and $\Delta\sigma$ in KCl was also observed for $\Delta\sigma$ of about 1 MPa, as reported in [5].

Fig. 2 shows that the linearity between $\Delta\sigma'$ and $\Delta\sigma$ for NaCl in case I breaks at $\Delta\sigma$ of about 2 MPa similarly to that for KCl. The proportionality between $\Delta\sigma$ and Q^{-1} for NaCl as well as for KCl breaks at this $\Delta\sigma$ (see Fig. 6 in [5]). Values of $\Delta\sigma'$ at 4% strain are again negative when $\Delta\sigma$ is larger than 2 MPa.

Fig. 3 shows the relation between $\lg \Delta\sigma$ and $\lg Q^{-1}$ for NaBr at different strain values. The amplitude-independent part of internal friction was negligible for NaBr compared with its amplitude-dependent part. It should be remarked that two curves at strains of 2.5 and 9.4% have unit slope when $\Delta\sigma$ is smaller than 0.5 MPa. The two curves deviate from the proportionality in each manner at $\Delta\sigma$ larger than 0.5 MPa. The straight line at the strain of 16% has a slope of about 2. It is also shown in Fig. 4

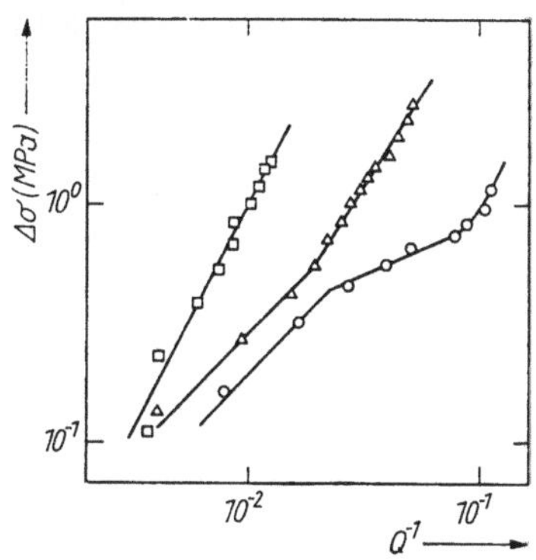

Fig. 3

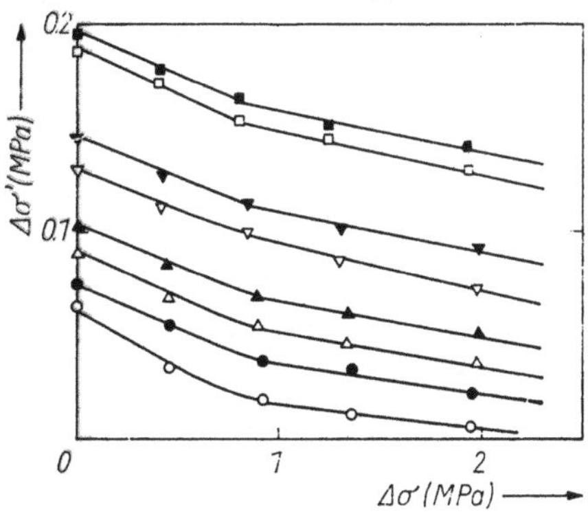

Fig. 4

Fig. 3. lg–lg plots of $\Delta\sigma$ and Q^{-1} for NaBr. ○ $\varepsilon = 2.5$, △ 9.4, and □ 16%

Fig. 4. Relation between $\Delta\sigma'$ and $\Delta\sigma$ for NaBr, ○, ● $\varepsilon = 8$, △, ▲ 12, ▽, ▼ 16, and □, ■ 20%; open symbols correspond to case II and full symbols to case I

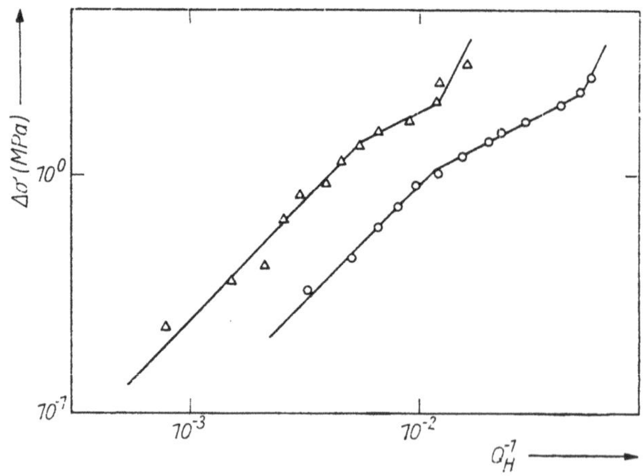

Fig. 5. lg–lg plots of $\Delta\sigma$ and Q_H^{-1} for KBr. ○ $\varepsilon = 5.7$ and △ 12%

that the lines representing the relation between $\Delta\sigma'$ and $\Delta\sigma$ bend at $\Delta\sigma \approx 0.8$ MPa both in cases I and II.

Fig. 5 and 6 show the results for KBr in which the proportionality between lg Q_H^{-1} and lg $\Delta\sigma$ and the linearity between $\Delta\sigma'$ and $\Delta\sigma$ break at $\Delta\sigma \approx 1.2$ MPa, where Q_H^{-1} is the amplitude-dependent part of internal friction. Negative $\Delta\sigma'$ values are observed at small strains and large $\Delta\sigma$ for the present case of KBr as well as for the cases of both KCl and NaCl.

The values of m^* obtained by (1) for KCl, NaCl, and NaBr are around 30 both in cases I and II. On the contrary, m^* for KBr is on the average around 26 in case I and 50 in case II. This means that m^* depends strongly on the strain rate for KBr.

4. Discussion

The relation between lg $\Delta\sigma$ and lg Q_H^{-1} is shown in Fig. 7, which has been drawn from the collected experimental results for KCl, NaCl, NaBr, and KBr crystals. The strain ε increases in the order of increasing subscripts of ε. The relation is divided into three regions denoted by I, II, and III. The experimental results correspond to ε_1, ε_2, and ε_3 for KCl, ε_3 and ε_4 for NaCl, ε_3, ε_4, and ε_5 for NaBr, and ε_2 and ε_3 for KBr.

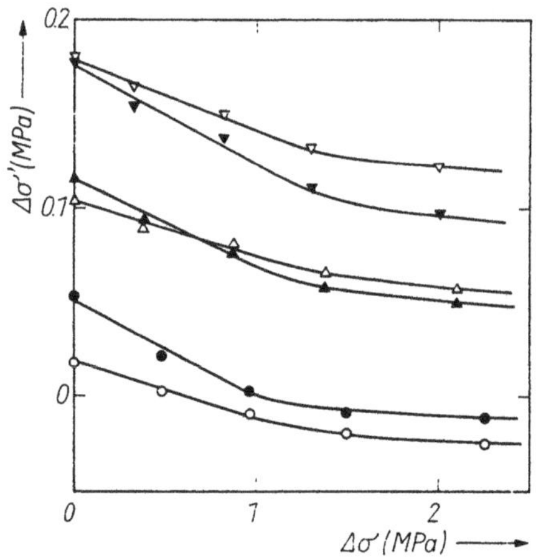

Fig. 6. Relation between $\Delta\sigma'$ and $\Delta\sigma$ for KBr. ○, ● $\varepsilon = 8$, △, ▲, 15, and ▽, ▼ 20%; open symbols correspond to case II and full symbols to case I

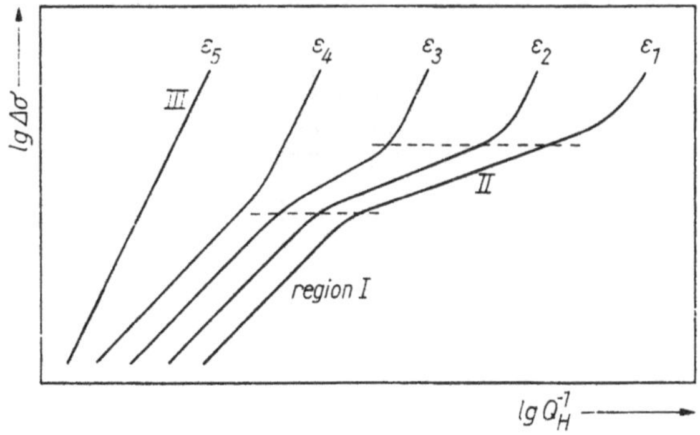

Fig. 7. Schematic relation between ln $\Delta\sigma$ and ln Q_H^{-1}. The strain ε increases with its suffix. The relation is divided into three regions I, II, and III

We assume that there is a relation

$$Q^{-1} = A\sigma_a^n$$

at constant $\Delta\sigma$ in regions I and III, where A is a constant, σ_a the applied stress, and n the exponent. As can be seen from Fig. 8, $n = -2$ in region III for all crystals without NaCl for which region III does not clearly appear in our test because the stress amplitude is not enough to cause region III. This independence of n on the kind of crystal suggests that Q_H^{-1} in region III depends only on the dislocation density, because it is known that the flow stress is proportional to the square root of the dislocation density [16, 17]. Therefore, Q_H^{-1} in region III is considered to be inversely proportional to the dislocation density. Region II should vary according to region III. On the other hand, in region I $n \approx 0$, -0.42, -0.72, and -1.7 for NaCl, NaBr, KCl, and KBr, respectively. This indicates that n in region I decreases with increasing lattice constant.

The relation between $\Delta\sigma'$ and $\Delta\sigma$ obtained from the experiments (Fig. 1, 2, 4, and 6) is shown in Fig. 9. The curves shift upward with the increase of strain and bend at a critical $\Delta\sigma$. This $\Delta\sigma$ corresponds to the point which separates regions I and II in Fig. 7. Therefore, we identify the small $\Delta\sigma$ region in Fig. 9 with region I in Fig. 7 and the larger $\Delta\sigma$ regions with regions II and III.

Anyway, it is found that both the relations between $\Delta\sigma$ and Q_H^{-1} as well as $\Delta\sigma$ and $\Delta\sigma'$ are divided into essentially two regions, respectively, where different mechanisms are considered to dominate. We will discuss the situations as follows.

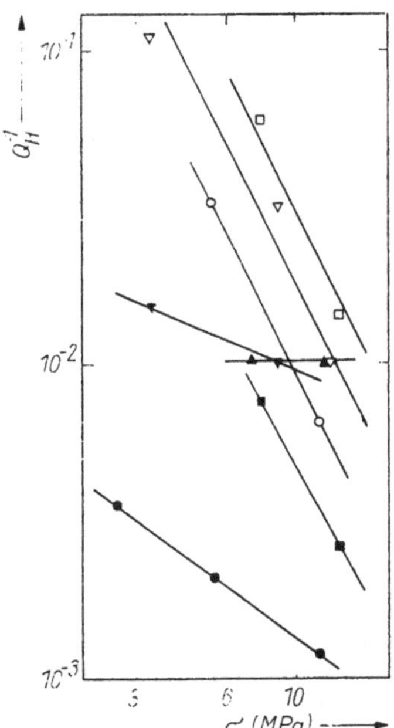

Fig. 8. lg–lg plots of Q_H^{-1} and σ_a at constant $\Delta\sigma$. ○ ● for KCl, ▲ NaCl, ▽ ▼ NaBr, and □ ■ KBr; open symbols correspond to region III and full symbols to region I

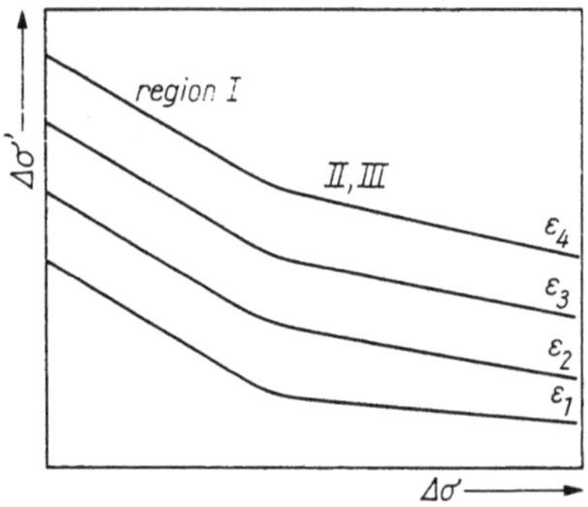

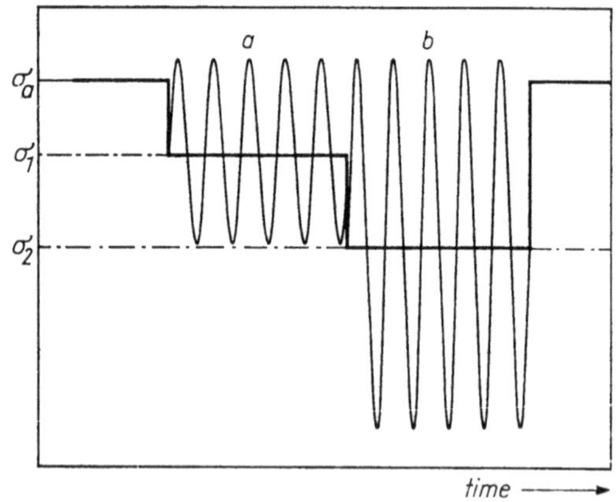

Fig. 9

Fig. 10

Fig. 9. Schematic relation between $\Delta\sigma'$ and $\Delta\sigma$. The strain ε increases with its suffix. The curves shift upward with the strain

Fig. 10. Variation of the applied stress when an oscillatory stress is superimposed

When the oscillatory stress is applied, the applied stress decreases, as shown in Fig. 10, in which the region denoted a corresponds to region I where the stress amplitude is small. In this case, a proportionality between Q_H^{-1} and $\Delta\sigma$ and a linear decrease of $\Delta\sigma'$ with $\Delta\sigma$ are observed. Moreover, the average length of dislocation segments is considered to increase by the application of oscillatory stress during the plastic deformation, as reported in [5, 14]. When the dislocation moves upward in Fig. 11a, it oscillates between two solid curves, where the dots are weak pinning points and the open circles strong ones. The dash-dotted line corresponds to the stress, σ_1, in region a of Fig. 10. The dislocation breaks away from the weak pinning points and its length increases with the stress amplitude. When the dislocation breaks away from all weak pinning points between the two strong ones, the average lengths of dislocation segments approach a constant. Then, it is considered that $\Delta\sigma'$ approaches a constant.

Fig. 11. Schematic illustration of obstacles to dislocation motion and dislocation configuration. Dots are weak obstacles and open circles strong ones

However, Q_H^{-1} rapidly increases, followed by a very slow increase, instead of approaching a constant. This is explained as follows: When the oscillatory stress amplitude increases and the applied stress decreases further in the b-region of Fig. 10, the strong pinning point at the centre of the dislocation loop happens to be unpinned backward as shown in Fig. 11b. We do not consider that the dislocation is unpinned forward, because the unpinning forward can carry the dislocation forward, but this does not contribute to Q_H^{-1}. The reason is that the dislocation unpinned forward does not return to the original site. As a result, Q_H^{-1} rapidly increases in region II. Q_H^{-1} saturates gradually as the saturation in unpinning backward proceeds. Then, the increase of Q_H^{-1} with the stress amplitude slows down in region III. When the average separation between weak pinning points is denoted by l_c and that between strong ones L_n as shown in Fig. 11, the appearances of regions I, II, and III should depend on the ratio between l_c and L_n and the strength of pins. The ratio between l_c and L_n varies depending on the strain and on the kind of crystal. This results in the variation of Q^{-1} depending on the strain and on the specimen.

Now, we assume, following Appel [18] and Guiu [19],

$$\sigma_a = \sigma_i + \sigma_p + \sigma_s ,$$

and for the strain rate sensitivity,

$$\left(\frac{\partial\sigma}{\partial \lg \dot{\varepsilon}}\right)_a = \left(\frac{\partial\sigma}{\partial \lg \dot{\varepsilon}}\right)_i + \left(\frac{\partial\sigma}{\partial \lg \dot{\varepsilon}}\right)_p + \left(\frac{\partial\sigma}{\partial \lg \dot{\varepsilon}}\right)_s ,$$

where σ_i is the internal stress, σ_p the part of the effective stress due to impurities which act as weak pinning points, σ_s the strain-dependent part of the effective stress, and the suffixes of the strain rate sensitivity have the same meanings. Then, $\Delta\sigma'$ can be expressed by a similar superposition rule,

$$\Delta\sigma' = \Delta\sigma_i' + \Delta\sigma_p' + \Delta\sigma_s' ,$$

when the tests in cases I and II are carried out. We consider that $\Delta\sigma_p'$ corresponds to the linearly decreasing part of $\Delta\sigma'$ at small stress amplitude in Fig. 9. $\Delta\sigma_p'$ decreases linearly with $\Delta\sigma$ in region I and is zero at the critical $\Delta\sigma$, where the dislocation breaks away from all weak pinning points between strong ones. Therefore, $\Delta\sigma_p'$ without oscillation is the difference between $\Delta\sigma'$ at zero stress amplitude and $\Delta\sigma'$ at the critical point which distinguishes regions I and II. This $\Delta\sigma'$ is approximately independent of the strain, as can be seen from Fig. 1, 2, 4, and 6. $\Delta\sigma_s'$ dominates in regions II and III, where $\Delta\sigma_p'$ is zero. $\Delta\sigma_s'$ increases with the increasing strain and dislocation density. This consideration agrees with the relation in [18],

$$\frac{1}{v} = \frac{1}{v_p} + \frac{1}{v_s},$$

where v is the activation volume, v_p that due to impurities, and v_s that due to the dislocation cutting process. Although $\Delta\sigma_i'$ is assumed to be zero in the previous paper [14], the presence of negative $\Delta\sigma'$ at the small strain and large stress amplitude in Fig. 1, 2, and 6 suggests that $\Delta\sigma_i'$ is not necessarily zero, but it may take a negative value. Li reported that the effective internal stress depends on the flow stress [20]. According to Li, the average internal stress may decrease with increasing strain rate. When m^* is nearly 30 and the maximum internal stress is close to the flow stress, the calculation based on [20] indicates that the influence of the strain rate cycling in our test to the internal stress is negligible. Dotsenko [21] reported that the proceeding of the inhomogeneous internal stress field during stress relaxation is responsible for the increase in internal stress. This inhomogeneous internal stress field is considered to decrease with increasing strain rate.

We conclude that the Blaha effect measurement can distinguish the effect of weak obstacles such as tetragonal defects on the dislocation motion from that of strong ones such as forest dislocations, when the dislocation moves with the aid of thermal fluctuations and the internal stress depends on strain rate for alkali halide crystals.

References

[1] F. BLAHA and B. LANGENECKER, Naturwissenschaften **42**, 556 (1955).
[2] F. BLAHA and B. LANGENECKER, Acta metall. **7**, 93 (1959).
[3] T. OHGAKU and N. TAKEUCHI, J. Mater. Sci. **20**, 2245 (1985).
[4] H. MAEDA, A. NAKAGIRI, and M. NAKAI, J. Japan. Inst. Metals **33**, 1213 (1969).
[5] T. OHGAKU and N. TAKEUCHI, phys. stat. sol. (a) **105**, 153 (1988).
[6] O. IZUMI, K. OYAMA, and Y. SUZUKI, Trans. Japan. Inst. Metals **7**, 158 (1966).
[7] E. MORI and M. INOUE, Bull. Japan. Inst. Metals **7**, 27 (1968).
[8] C. E. WINSPER and D. H. SANSOME, J. Inst. Metals **97**, 274 (1969).
[9] A. E. EAVES, A. M. SMITH, W. J. WATERHOUSE, and D. H. SANSOME, Ultrasonics **13**, 162 (1975).
[10] G. S. BAKER and S. H. CARPENTER, J. appl. Phys. **38**, 1586 (1967).
[11] G. KAISER and W. PECHHOLD, Acta metall. **17**, 527 (1969).
[12] T. ENDO, K. SUZUKI, and M. ISHIKAWA, Trans. Japan. Inst. Metals **20**, 706 (1979).
[13] A. V. KOZULOV, N. S. MOLDYUK, and S. I. SELITSER, Soviet Phys. – Solid State **28**, 1008 (1986).
[14] T. OHGAKU and N. TAKEUCHI, phys. stat. sol. (a) **102**, 293 (1987).
[15] T. OHGAKU and N. TAKEUCHI, phys. stat. sol. (a) **106**, K19 (1988).
[16] J. HESSE and L. W. HOBBS, phys. stat. sol. (a) **14**, 599 (1972).
[17] W. FRANK, Mater. Sci. Engng. **6**, 121, 132 (1970).
[18] F. APPEL, phys. stat. sol. (a) **25**, 607 (1974); **31**, 615 (1975).
[19] F. GUIU, Phil. Mag. **21**, 365 (1970).
[20] J. C. M. LI, Dislocation Dynamics, Ed. A. R. ROSENFIELD, G. T. HAHN, A. L. BEMONT, and R. I. JAFFEE, McGraw-Hill Publ. Co., New York 1968 (p. 87).
[21] V. I. DOTSENKO, phys. stat. sol. (b) **93**, 11 (1979).

(Received July 27, 1988)

phys. stat. sol. (a) **111**, 173 (1989)
Subject classification: 61.80 and 68.20; S1.1, S4

Departamento de Materiales, Comisión Nacional de Energía Atómica, Buenos Aires[1]) *(a)*,
Instituto de Física, Universidade Federal do Rio Grande do Sul, Porto Alegre[2]) *(b)*,
and Department of Materials Engineering,
Virginia Polytechnic Institute and State University, Blacksburg[3]) *(c)*

Argon Irradiation of Sn Thin Layers Deposited on Fe Substrates

By

R. PASIANOT (a), G. GALAMBOS (a), E. J. SAVINO (a), L. AMARAL (b), M. BEHAR (b), D. FARKAS (c), and S. RAO (c)

Iron polycrystals deposited with thin films of Sn (100 nm approximately) are irradiated with 300 and 400 keV argon ions at the temperatures of liquid hydrogen, room, and 180 °C. After irradiation, the specimens are examined using optical and scanning microscopy, X-ray microprobe, X-ray diffraction, and He-Rutherford back scattering. The surface structures obtained by irradiation at different temperatures are compared with those resulting from isothermal treatments. It is concluded that at liquid hydrogen temperature the major effect of the irradiation is a large and inhomogeneous, localized sputtering of Sn. Also sputtering dominates at room temperature, although it appears to be homogeneous. Finally at 180 °C, Sn islands and FeSn, $FeSn_2$ crystals develop at the Fe surface. $FeSn_x$ precipitates form between 300 to 400 °C when the Sn–Fe bilayer is annealed. Light particle irradiation seems to be effective in reducing thermodynamic barriers for nucleation of the above mentioned precipitates at the Fe–Sn interface.

Auf Eisenpolykristalle aufgebrachte dünne Sn-Schichten (etwa 100 nm) werden mit 300 und 400 keV-Argon-Ionen bei Temperaturen des flüssigen Wasserstoff, Raumtemperatur und 180 °C bestrahlt. Nach der Bestrahlung werden die Proben mittels optischer und Rastermikroskopie, Röntgenmikrosonde, Röntgenbeugung und He-Rutherfordrückstreuung untersucht. Die bei der Bestrahlung unter unterschiedlicher Temperatur erhaltenen Oberflächenstrukturen werden mit den nach isothermischer Temperung erhaltenen verglichen. Es wird angenommen, daß bei der Temperatur des flüssigen Wasserstoff der Haupteffekt der Bestrahlung eine große und inhomogene lokalisierte Abstäubung von Sn ist. Zerstäubung dominiert auch bei Zimmertemperatur, obgleich es homogen zu sein scheint. Schließlich entwickeln sich bei 180 °C, Sn-Inseln und FeSn- und $FeSn_2$-Kristalle. $FeSn_x$-Präzipitate bilden sich zwischen 300 und 400 °C, wenn die Sn–Fe-Doppelschicht ausgeheilt wird. Bestrahlung mit leichten Teilchen scheint bei der Reduzierung thermodynamischer Barrieren für die Keimbildung der oben erwähnten Präzipitate an der Fe–Sn-Grenzfläche effektiv zu sein.

1. Introduction

Baumvol [1, 2] and Giacomozzi et al. [3] report that deposition of thin surface layers of high Sn concentration onto an otherwise pure Fe substrate notably improve the wear and corrosion properties of the metal surface. Fe substrates can be covered with appropriate thin Sn layers by either direct implantation [1] or vapour deposition followed by light particle irradiation [2, 3], the second process being inherently cheaper than the first. In both cases the energy necessary for mixing the pure species is provided by irradiation. However, the ion beam mixing of inhomogeneous irradiation targets is not yet fully understood and in order to developed technological applica-

[1]) Avda. Libertador 8250, 1429 Buenos Aires, Argentina.
[2]) Campus do Vale, 90049 Porto Alegre, RS, Brazil.
[3]) Blacksburg, VA 24061, USA.

tions of the above processes, further research is needed. The observed improvement of Fe surface properties by the addition of Sn has generated a series of studies of this system. Gratton et al. [4], Dionisio et al. [5], and Farkas et al. [6] studied the effects of direct Sn implantation and/or vapour deposition followed by light particle irradiation on the surface structure. With respect to microstructural effects of irradiation it was found that [4, 5]:

1. In addition to the pure metals, $FeSn_2$, FeSn, and SnO_2 precipitates or agglomerates seem to appear at the surface. The last precipitate, SnO_2, results from O absorption.

2. The enhancement of oxidation and wear resistance with respect to pure Fe, is due to the creation of $FeSn_x$ and SnO_2 precipitates at the surface. The $FeSn_x$ phases also increase the surface hardness.

3. In the case of thin Sn layers deposited on the Fe surface and subsequently irradiated with light particles, the thin layer topology seems to be modified both by irradiation and by temperature effects.

4. Results from CEMS (conversion electron Mössbauer spectroscopy) [5] suggest that the $FeSn_x$ phase that appears at the surface differs from the phase predicted by the equilibrium diagram. Dionisio et al. [5] suggest the possibility that this is an amorphous phase.

Several questions remain open about the evolution of the Fe–Sn microstructure under irradiation. We shall address in this work some of those points, mainly in relation to light particle irradiation of thin Sn layers deposited on Fe. From the more fundamental point of view as well as for practical applications, it is important to know the mixing efficiency of irradiation and understand the physical processes behind mixing. Results from light particle irradiation experiments at different temperatures, energies and doses are used to answer some of these questions. An important unanswered question is the one raised by Dionisio et al. [5] about the possibility of nucleating new phases under irradiation, i.e. phases that are not reported in the equilibrium phase diagram. Both, irradiation and temperature effects, can give rise to new phases and change the microstructural morphology of the system. It is important therefore, to be able to study the two types of phenomena separately and understand the effect of each of them in the microstructural morphology. Irradiation may also enhance the kinetics of phase growth at the surface. The resulting structure will be influenced by thermo-chemical effects plus any contribution due to surface tension and irradiation-deposited energy. In this respect, the Fe–Sn system is ideal for studying these effects on phase transformation and microstructure evolution. The low melting temperature of Sn makes it possible to perform experiments at relatively large temperatures. Also the Fe–Sn alloys can be easily identified using standard analytical techniques.

2. Specimen Preparation and Analysis

Fe–Sn specimens for this work were prepared at the Departmento de Materiales, Comisión Nacional de Energía Atómica laboratories (CNEA from now on) and irradiated with 300 to 400 keV Ar at the Porto Alegre High Voltage Engineering Europe 400 keV accelerator (HVEE-400). Rutherford backscattering (RBS) measurements were also performed at the HVEE-400; while metallographic analysis, scanning electron microscopy (SEM), X-ray diffraction (XRD) measurements, etc. were performed at CNEA and Virginia Polytechnic Institute and State University (VPI). The cylindrical, 99.99% pure Fe, specimens were annealed at 850 °C in an Ar atmosphere. The surface was diamond polished and cleaned with alcohol and ultrasonic methods. After polishing, a 100 nm Sn layer thick was vapour deposited on the Fe surface. The chamber

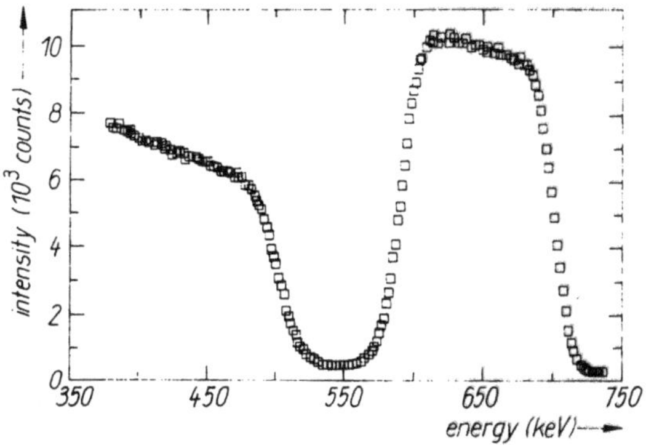

Fig. 1. Rutherford backscattering spectrum from the as-deposited specimens

sustains a vacuum of 5×10^{-5} Torr. The surface of some specimens was cleaned by sputtering with Ar atoms. This cleaning before depositing Sn enhances the adhesion of the Sn layer, although this enhancement does not seem to affect the results below.

As-deposited and irradiated specimens were studied using several techniques. SEM metallographic observation show that the surface of the as deposited specimen is homgeneous. The thickness of the layer was determined using resonance measurements with a quartz crystal oscillator. Extra confirmation of the thickness measurements were also provided by RBS and a characteristic RBS spectrum is shown in Fig. 1. From the sharp appearance of the back step in the spectrum corresponding to Sn, it can be conculded that the Fe–Sn surface is well defined, within the roughness of the polished Fe surface, and the Sn layer width appears as constant. XRD measurements show the Sn film to be strongly textured, with the [200] direction perpendicular to the crystal surface.

3. Thermal Annealing Results

Isothermal annealing treatments of the specimens were performed at 200, 250, 300, 400, and 500 °C. The surface structure evolution after each heat treatment was fol-

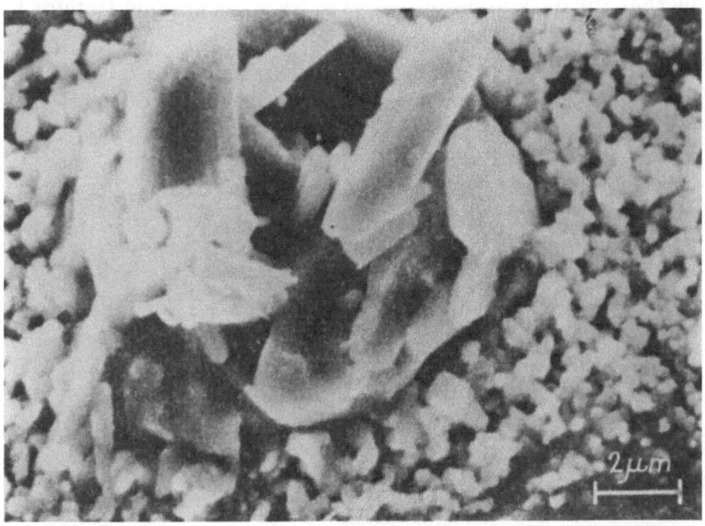

Fig. 2. SEM micrograph of a specimen heated to 400 °C. This micrograph shows FeSn crystals which have grown from Sn islands on the Fe surface

Table 1
Irradiation conditions

specimen set	energy (keV)	temperature	doses (10^{16} cm^{-2})
I	300	20 K	≤ 8
II	300	RT	5 and 7
III	300	180 °C	2 to 8
IV	400	180 °C	6 and 8

lowed by metallographic and SEM observations. SEM observations show that within experimental error, the surface morphology remains invariant for temperatures up to 200 °C. In addition, even after several hours of treatment at 200 °C, the RBS spectrum remains identical to that of the as-deposited sample. However, for an annealing treatment at 250 °C, i.e. 20 °C above the melting temperature of Sn, islands of Sn nucleate at the Fe surface. A clear growth of microcrystals on the Fe surface is observed after a 300 to 400 °C anneal, SEM images of those crystals are shown in Fig. 2. XRD measurements show that they are FeSn alloy precipitates.

4. Irradiation Effects

Several as-deposited Fe–Sn specimens were bombarded with energetic Ar particles in the HVEE-400. The current density was kept constant at 1 μA/cm² and the irradiation energy and specimen temperature was varied over a wide range of values. Table 1 shows the irradiation conditions used for the different set of experiments (I to IV).

The evolution of chemical composition in the surface region of the irradiated specimens was followed using RBS. SEM and XRD measurements of the surface were also performed in order to supplement the RBS results. For the 20 K irradiation, we observed a rapid deterioration of the surface, as indicated by the SEM images shown in Fig. 3. XRD results show that the dark areas observed in SEM correspond to pure Fe

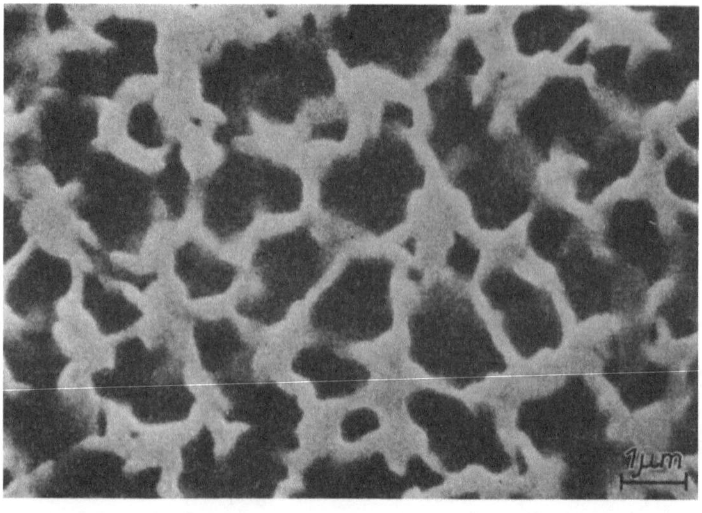

Fig. 3. SEM micrograph of a specimen irradiated at 20 K with a dose of 8×10^{16} Ar/cm²

and the light areas to Sn. They also show that the pure Sn phase retains its originally as deposited [200] texture and no extra phases besides metallic Fe and Sn, are formed. This results indicate that low-temperature Ar irradiation modifies the morphology of the surface primarily due to strong inhomogeneous sputtering of Sn atoms. Chemical analysis using RBS to determine concentration versus depth profile was not possible due to the irregular surface. SEM micrographs of room temperature irradiated specimens (set II) show no roughening of the surface. The corresponding RBS spectra for a specimen before and after 5×10^{16} Ar irradiation are shown in Fig. 4. The relative distribution of Sn and Fe as a function of depth in the specimen is also shown in Fig. 4c). A deconvolution routine, developed by Scherer et al. [7], was used to determine the concentration profiles.

In the 180 °C irradiated specimens (sets III and IV), Sn based islands appear with crystalline and metallographic structures similar to the structures found in specimens annealed at temperatures above 250 °C. The islands seem to be better developed in the 300 keV irradiated specimens (set III) than in the 400 keV specimens (set IV);

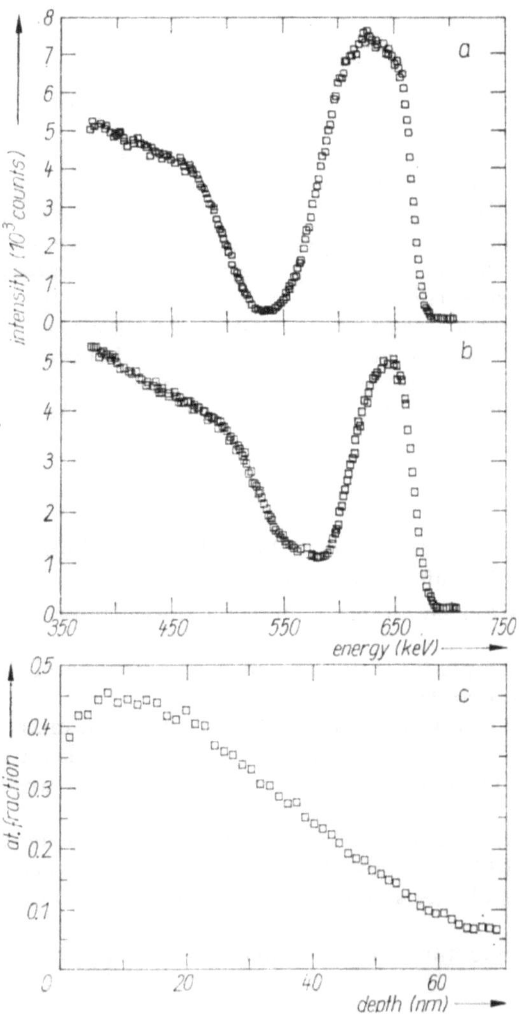

Fig. 4. a) RBS of the specimen before irradiation. b) RBS after an irradiation of 5×10^{16} Ar/cm^2 at room temperature. c) Sn concentration vs. depth for the case of b)

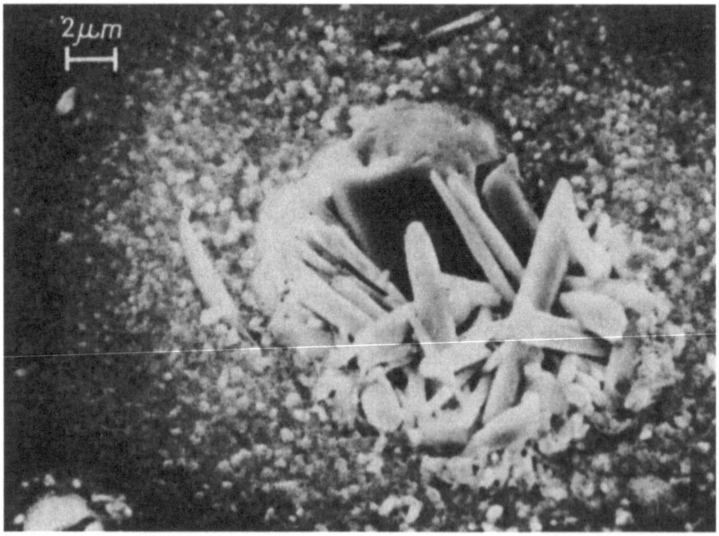

Fig. 5. SEM of a specimen irradiated at 180 °C to a dose of 6×10^{16} Ar/cm^2

the SEM micrograph of a typical specimen of set III irradiated to a dose of $6 \times \times 10^{16}$ Ar atoms/cm^2 is shown in Fig. 5. XRD analysis shows the cylindrical crystals and islands observed in the figure to be FeSn crystals with traces of $FeSn_2$ compound. It can be seen in Fig. 5 that the crystalline structure observed at the Fe surface matches exactly with the one shown in Fig. 2, observed during thermal annealing experiments. However the corresponding thermal treatments were performed at temperatures of at least 100 °C above the irradiation temperature. Although one can argue that some heating due to irradiation is expected at the surface, we found this effect to be smaller than 20 °C. One must conclude, therefore, that the nucleation and growth of $FeSn_x$ phases in these specimens result mainly from irradiation with light Ar particles.

5. Summary

We studied the behaviour of thin films of Sn deposited on Fe under thermal and Ar irradiation treatments. Our experimental results indicate the following:

1. Structural changes due to irradiation are observed only for doses larger than 4×10^{16} Ar atoms/cm^2.

2. Low temperature (20 K) irradiation produces inhomogeneous sputtering of Sn from the surface, leaving large pits in the Sn layer.

3. Room temperature irradiation at 300 keV produces a relatively uniform mixing of Fe and Sn in the first 25 nm from the surface. In addition, the original Sn layer is largely depleted by atomic sputtering. This effect can be clearly seen by comparing Fig. 4a with 4b.

4. 180 °C irradiation produces Sn islands on the surface. As irradiation dose accumulates, those islands evolve into rod-like crystals of FeSn with traces of $FeSn_2$ precipitate. 300 keV irradiation produces a larger amount of $FeSn_x$ phases than 400 keV irradiation.

5. $FeSn_x$ structures are obtained during annealing treatments only when the annealing temperature is above 250 °C.

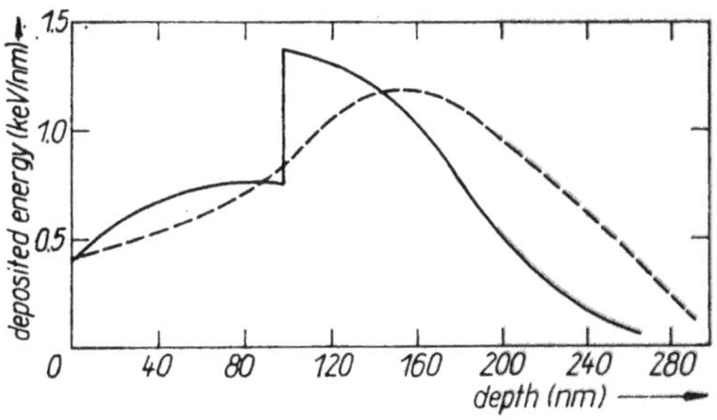

Fig. 6. Predicted damage for 300 (solid line) and 400 keV (dashed line) Ar irradiation of a 100 nm Sn layer on Fe

6. Fe must diffuse under irradiation at a fast rate into Sn for $FeSn_x$ precipitates to nucleate and growth at relatively low temperature. Therefore irradiation seems to reduce the migration barrier at the interface between pure Fe and pure Sn and/or increase the diffusivity of Fe into Sn.

7. The diffusion of Sn into Fe is very small both under thermal treatment or irradiation. However, in regions where Fe–Sn islands are formed, some diffusion takes place compensating for the depletion of Fe in Sn.

8. No evidence was found for the formation of amorphous or new phase during irradiation.

The decision to perform the Ar irradiations at 300 and 400 keV was based on numerical results of damage distribution simulation. The distributions were simulated using a computer program based on TRIM [8]. In Fig. 6, the damage distribution profiles calculated for 300 and 400 keV Ar irradiation of a 100 nm Sn layer on Fe are shown. It can be seen that, for the 300 keV irradiation, the maximum damage occurs at the Sn–Fe interface; while for the 400 keV irradiation the maximum occurs a little beyond the Fe–Sn interface, in the Fe substrate. The difference in the microstructure evolution rate observed at the two irradiation energies can be explained in terms of the differences in the location of the damage energy maxima.

6. Conclusions

A coherent picture of the effects of irradiation with light particles consistent with the above findings seems to emerge. At low temperatures (20 K), a large sputtering of Sn atoms is observed. The sputtering effect seems to be localized in some regions of the crystal surface. Pits are observed in the regions where Sn was sputtered and the Fe surface is exposed. The inhomogeneity in the damage must be influenced by large strains at the Sn interface. The strain depends on local differences in adhesion between Sn and Fe and might be the cause of the Sn sputtering being preferential in some regions. Once a small discontinuity is nucleated by this preferential sputtering the growth of a pit is self-sustained by the now relatively larger area for atomic sputtering. At room temperature, although the sputtering of Sn atoms still remains large, the atoms appear to have enough mobility to reaccommodate the local sputtering effects and the surface remains smooth. However, although the Sn atoms are very mobile at this temperature, the adherence of the Sn layer to the Fe surface is relatively strong, i.e. the atomic mobility of Sn remains below the mobility needed for wetting the surface of Fe. At high temperatures (180 °C), the sputtering of Sn is still important but the atomic mobility is much larger and, possibly, radiation enhanced. At these temperature ranges Sn seems to flow like a liquid on the surface of Fe, nucleating into drop shape islands. Before large Sn islands are created by the flow of Sn, it is possible that, in the first stages of irradiation, some FeSn and $FeSn_2$ embryos are nucleated at the Fe–Sn interface, especially for the 300 keV irradiation. These embryos are expected to have a pinning effect on the Sn layer, and the Sn islands are expected to nucleate precisely at such sites. In the Fe–Sn interface region, the embryos grow into stable nuclei that later evolve into the observed rod-like crystals. This last effect must be mainly due to thermal diffusion because the irradiation dose at the base of the relatively large Sn islands is negligible. Therefore, we conclude that, while only the intermetallic phases reported in the binary Fe–Sn equilibrium diagram are observed at 300 and 400 keV Ar irradiation, the radiation deposited energy at the Fe–Sn interface strongly enhances the kinetics of nucleation of chemically stable compound phases.

Acknowledgements

The authors would like to thank I. J. R. Baumvol for suggesting this work. We want also to thank C. Scherer and J. P. Biersack for allowing us to use their computer programs. This work was partially financed by Proyecto Multinacional de Investigación en Materiales, Organization of American States, Comisión Nacional de Energiá Atómica (CNEA); Consejo Nacional de Investigaciones Científicas y Técnicas (CONICET), Argentina; Conselho Nacional de Pesquisas (CNPq), Brazil; National Science Foundation (NSF), U.S.A. and the Office of Naval Research, Division of Materials Science, U.S.A.

References

[1] I. J. R. Baumvol, Hyperfine Interactions **10**, 1123 (1981).
[2] I. J. R. Baumvol, phys. stat. sol. (a) **67**, 287 (1981).
[3] F. Giacomozzi, L. Guzman, A. Molinari, A. Tomasi, E. Voltolini, and L. M. Gratton, Mater. Sci. Engng. **69**, 341 (1985).
[4] L. M. Gratton, L. Guzman, and A. Molinari, J. Microscopie Spectroscopie electron. **8**, 293 (1983).
[5] P. H. Dionisio, C. Scherer, S. R. Texeira, and I. J. R. Baumvol, Nuclear Instrum. and Methods **B16**, 345 (1986).
[6] D. Farkas, R. Pasianot, M. Rangaswamy, and E. J. Savino, Nuclear Instrum. and Methods **B16**, 183 (1986).
[7] C. Scherer, S. R. Teixera, and I. J. R. Baumvol, private communication.
[8] J. P. Biersack and L. G. Häggmark, Nuclear Instrum. and Methods **174**, 257 (1980).

(Received July 13, 1988)

phys. stat. sol. (a) **111**, 181 (1989)

Subject classification: 73.60; S8

Laboratoire de Physique des Matériaux et Composants de l'Electronique, Faculté des Sciences et Techniques, Université de Nantes[1])

Etude de la résistivité de couches minces polycristallines de $MoSe_2$

Par

J. C. BERNÈDE, A. MALLOUKY et J. POUZET

Des couches minces de $MoSe_2$ déposées par pulvérisation diode continue sont traitées par recuits, en présence de vapeur de sélénium puis sous vide, de façon à obtenir des couches polycristallines stoechiométriques. La composition et la pureté des couches sont contrôlées par spectroscopie d'électrons. Leur homogénéité est étudiée à la microsonde électronique. La conductivité est mesurée entre 80 et 700 K. Les résultats expérimentaux montrent que les couches sont homogènes et que la quantité d'oxygène présente dans les couches est très faible. Les résultats des mesures électriques mettent en évidence l'existence de deux domaines de températures. A haute température ($T > 250$ K) la conductivité est limitée par les joints de grains. Aux basses températures la variation de la conductivité avec la température vérifie une loi en $T^{-1/4}$ ce qui correspond à une conduction par sauts à distance variable entre états localisés.

$MoSe_2$ thin films deposited by dc diode sputtering are annealed under selenium atmospheric pressure to obtain stoichiometric polycrystalline films. The composition, purity, and homogeneity of the films are checked by electron spectroscopy and electron microprobe analysis. The temperature dependence of the electrical conductivity is measured between 80 and 700 K. The experimental results show that the films are homogeneous and that the oxygen content in the layers is very low. The electrical data reveal the existence of two temperature regions. At high temperature ($T > 250$ K) the conductivity is limited by grain boundary scattering. At low temperatures the conductivity varies with temperature according to a $T^{-1/4}$ law corresponding to hopping conduction via localized states.

1. Introduction

Le disélénuire de molybdène ($MoSe_2$) est un membre de la famille des dichalcogénures des métaux de transition du groupe VIA ($MoSe_2$, WS_2, WSe_2 etc.). Il cristallise dans une structure molybdénite hexagonale de type lamellaire. Il a été largement étudié à l'état de monocristaux [1]. Cependant le composé $MoSe_2$ ne cesse de susciter un intérêt croissant en raison de ces possibilités d'applications dans le domaine des cellules solaires photoélectrochimiques [2 à 4]. Pour des raisons économiques il serait intéressant d'obtenir du $MoSe_2$ en couches minces. A ce jour seuls Bichsel et Levy [5] ont réalisé, par pulvérisation, des couches de ce matériau. Ils ont montré qu'il était possible d'obtenir des couches de stoechiométrie correcte en déposant par pulvérisation radio fréquence de type magnétron. Mais les cristallites sont petits et leur grand axe *c* est parallèle au substrat. Si la résistivité de leur couche dépend peu de leur composition elle évolue en fonction des traitements thermiques réalisés après le dépôt.

Une technique de dépôt de couches minces de $MoSe_2$ stoechiométrique, mise au point au laboratoire, a été développée antérieurement [6, 7]. Ici, après contrôle de la

---

[1]) 2, rue de la Houssinière, F-44072 Nantes Cédex, France.

pureté de la couche réalisée, par recherche des traces d'oxygène, et de leur homogénéité par étude de la répartition des constituants, une étude de la conductivité a été réalisée. Les résultats sont interprétés en termes de conductivité par sauts et d'effets de joints de grains suivant le domaine de température.

2. Techniques expérimentales

Les couches minces sont déposées par pulvérisation diode continue, puis enrichies en sélénium par recuit sous vapeur de sélénium. Ensuite on obtient la stoechiométrie, par recuit sous vide dynamique des couches précédemment enrichies en sélénium. La totalité du processus d'obtention de ces couches a été exposée par ailleurs (7), aussi n'y reviendrons nous que très succinctement. Les substrats utilisés lors du dépôt des couches — à savoir, mica, silice, verre, molybdène — sont nettoyés chimiquement avant la mise sous vide, puis chauffés à 400 K pendant 1 h avant le dépôt. Une prépulvérisation de plusieurs minutes précède systématiquement le dépôt lui même. Les couches pulvérisées étant mal cristallisées et déficitaires en sélénium, elles sont rechargées en sélénium par recuit 24 h à 770 K sous vapeur de sélénium. Un recuit ultérieur sous vide dynamique à une température comprise entre 650 et 700 K pendant 10 à 24 h permet, pour les plus hautes températures et pour une durée suffisante de recuit, d'obtenir des couches cristallisées très proches de la stoechiométrie.

La structure cristalline des couches minces a été contrôlée par diffraction de rayons X à l'aide d'un diffractomètre à compteur. La taille des grains des couches a été estimée statistiquement, à partir des photos obtenues au microscope à balayage, ou suivant la méthode dite "à élargissement des pics de rayons X à mi hauteur" [7]. Les images X sont obtenues à partir d'un microscope à balayage équipé d'une microsonde électronique. L'étude de composition des couches a été menée par microsonde électronique et spectroscopie d'électrons (XPS) [7]. Les analyses XPS ont été réalisées à l'aide d'un spectromètre Leybold-LHS 10 (Université de Nantes, C.N.R.S.). L'étude par spectroscopie d'électrons (XPS) a été effectuée à l'aide d'une source de rayons X opérant dans les conditions suivantes: anticathode magnésium (1253.6 eV), 10 kV à 10 mA. La résolution en énergie est meilleure que 1 eV. De même la pureté des couches a été contrôlée par analyse XPS. L'épaisseur des couches est mesurée par palpeur mécanique. Pour les mesures électriques les contacts ont été évaporés thermiquement sur les couches de $MoSe_2$. Plusieurs métaux ont été testés (In, Au, Mn, Al). C'est l'or qui a donné les résultats les plus satisfaisants, quant à l'ohmicité des contacts [8]. La conductivité des couches a été mesurée, à l'aide d'un électromètre, entre 80 et 700 K.

3. Résultats expérimentaux

3.1 Conductivité électrique

Après qu'une couche ait subi le processus complet d'élaboration, elle se caractérise par une remarquable stabilité des ses propriétés électriques (Fig. 1). Par ailleurs on peut constater que l'allure des courbes est identique, quelles que soient la nature du substrat et la composition de la couche (Fig. 2 à 4). Le comportement identique des couches quelque soit le substrat (verre, mica ou silice) corrobore l'étude de la structure qui montrait que les résultats (taille des grains, orientation etc.) ne différaient pas pour ce type de substrats [8]. Un écart, même important, à la stoechiométrie ne conduit pas à des changements plus significatifs que ceux rencontrés pour deux couches stoechiométriques différentes (Tableau 1). Tous ces résultats ont été obtenus pour des couches d'épaisseur comprise entre 0,5 et 7,5 μm, c'est à dire un domaine où la conductivité, comme la taille des grains, n'est plus sensible à l'épaisseur (Tableau 1).

Tableau 1

Conductivité des couches en fonction de leur épaisseur et de leur composition (voir [9] et [10] pour les monocristaux)

échantillon	épaisseur (μm)	composition (% at)		nature du substrat	σ (Ω^{-1} cm^{-1})
		Se	Mo		
61	1,1	70	30	verre	10^{-3}
55	1,1	68,75	21,25	verre	$1,14 \times 10^{-3}$
56	0,9	72	28	verre	9×10^{-4}
17	2	76	24	verre	9×10^{-3}
13	2,5	70	30	verre	8×10^{-5}
59	1,6	71	29	verre	$2,6 \times 10^{-4}$
86	0,7	66	34	verre	10^{-2}
9	0,5	67,7	32,3	verre	$0,3 \times 10^{-3}$
20	6	66,4	33,6	verre	10^{-3}
22	4,3	66	34	mica	$0,7 \times 10^{-3}$
23	4,3	65,7	34,3	silice	$6,3 \times 10^{-3}$
27	7	66	34	verre	$3,8 \times 10^{-3}$

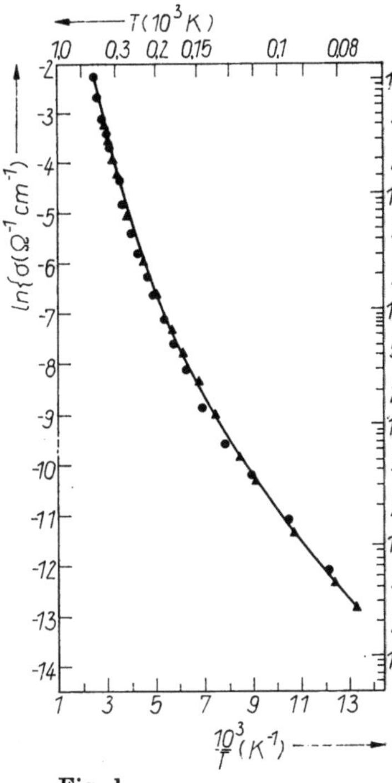

Fig. 1

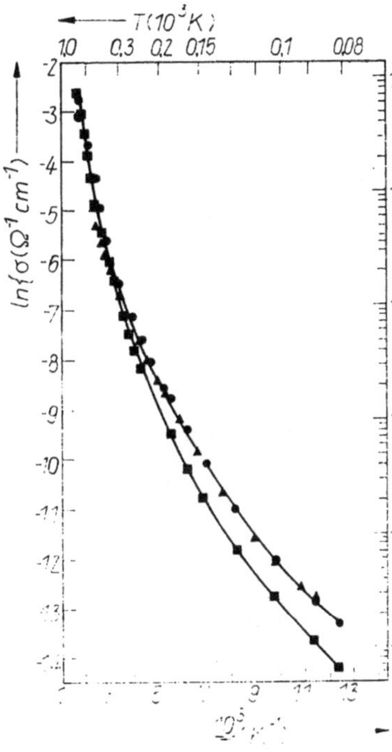

Fig. 2

Fig. 1. Evolution de la conductivité électrique d'une couche de $MoSe_2$ avec la température. ● mesures effectuées à la date t_1, ▲ $t_1 + 1$ année (les couches ont été conservées à l'air ambiant à l'abri des poussières)

Fig. 2. Evolution de la conductivité électrique avec la température pour trois couches stoechiométriques différentes: ● 66, ■ 67,7, ▲ 66,4% Se

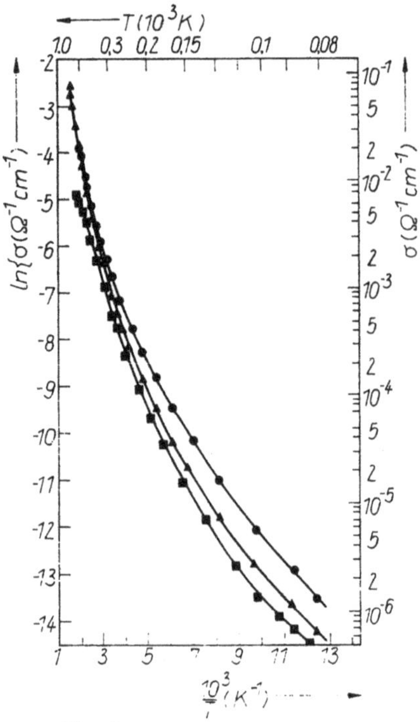

Fig. 3

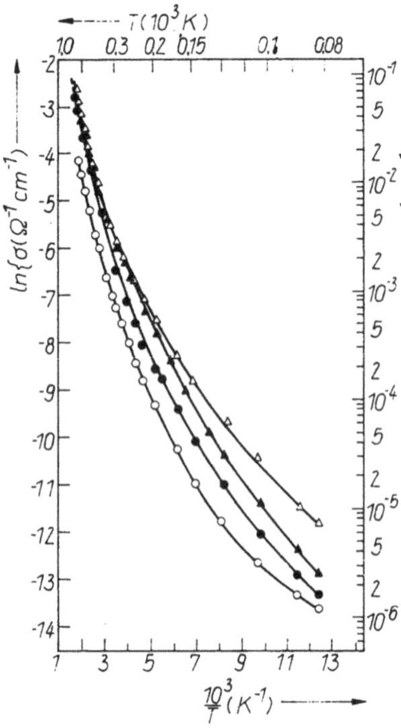

Fig. 4

Fig. 3. Evolution de la conductivité électrique avec la température. Effet de la nature du substrat: ▲ verre, ■ mica, ● silice

Fig. 4. Evolution de la conductivité électrique avec la température. Effet d'un écart à la stoechiométrie: ● 66,4, ▲ 67,7, ○ 71, △ 77,5% Se

Par ailleurs on peut voir que la dépendance en température de la conductivité ne suit pas une loi d'Arrhenius puisque $\partial \ln \sigma / \partial T$ augmente avec la température.

La conductivité est inférieure d'un facteur cent au moins à celle des monocristaux puisqu'aussi bien, celle-ci, quoique très variable d'un échantillon à l'autre, est comprise entre 0,04 et 4 Ω^{-1} cm^{-1} [1, 9, 10]. Ces résultats typiques des matériaux polycristallins nous ont conduits à nous interroger sur la pureté et l'homogénéité des couches obtenues.

3.2 Contrôle de la pureté et de l'homogénéité des couches

L'analyse ESCA, avant la recharge en sélénium des couches, montre que la raie Mo 3d présente une énergie de liaison égale à celle du MoO_3 (Fig. 5a) [7]. Après le complet déroulement du traitement thermique, l'énergie de liaison de la raie Mo 3d est bien celle du molybdène lié au sélénium dans la structure $MoSe_2$ (Fig. 5e) [7]. Il était cependant nécessaire de pousser plus loin nos investigations quant à une éventuelle pollution des couches par l'oxygène, pur ou sous forme H_2O, celle-ci pouvant fortement influencer leurs propriétés [11]. Pour ce faire, nous avons recherché l'oxygène présent dans la couche en fonction de la profondeur, par enregistrement des spectres XPS successifs obtenus après chaque décapage ionique. Le décapage a été réalisé à l'aide d'un canon à ions travaillant dans les conditions suivantes: une pression

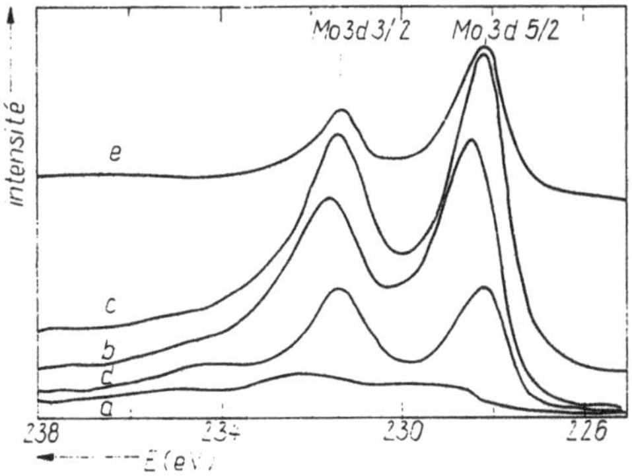

Fig. 5

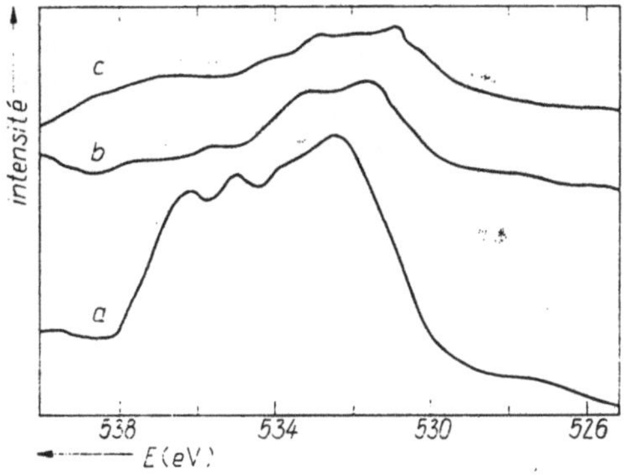

Fig. 6

Fig. 5. Evolution en énergie de la raie Mo 3d du molybdène en fonction du temps de décapage d'une couche mince de $MoSe_2$ non rechargée en sélénium. Durée du décapage: (a) 0, (b) 1, (c) 11, (d) 21 min, (e) raie Mo 3d après recharge en sélénium et décapage de 5 min

Fig. 6. Evolution de l'intensité de la raie O 1s en fonction du temps de décapage d'une couche mince de $MoSe_2$ traitée thermiquement (a) avant décapage, (b) après décapage de 1 min, (c) après décapage de 5 min

de 5×10^{-4} Pa, un courant d'émission de 10 mA, une tension d'accélération de 5 kV. La totalité de la surface des échantillons est alors soumise au faisceau d'ions Ar^+.

La Fig. 6 montre l'évolution de la raie O 1s en fonction du temps de décapage. On peut voir qu'après 1 min de décapage, l'intensité de la raie a fortement diminué, ce qui montre que l'essentiel de l'oxygène détecté en surface des couches correspond à la pollution systématique d'une couche mince lors de sa mise à l'air, quelle que soit sa nature.

La trace d'oxygène qui persiste après décapage, quoique faible et non quantifiable, reste constante, quelle que soit la durée du décapage. Elle correspond probablement à de l'oxygène qui a diffusé le long des joints de grains, lors de la mise à l'air des échantillons cristallisés et rechargés en sélénium. Le décalage en énergie du pic à 532 eV au fur et à mesure du décapage montre que l'on passe d'un oxygène physiadsorbé de pollution en surface de l'échantillon à un oxygène chimiadsorbé piégé aux joints de grains.

La Fig. 5 montre l'évolution de l'énergie de liaison du pic Mo 3d avec le temps de décapage dans le cas d'une couche non rechargée (donc excédentaire en molybdène). Il est clair que très rapidement la raie quitte la position correspondant à MoO_3 [12]. Il n'y a donc oxydation du molybdène qu'en surface de la couche, c'est à dire lors de la mise à l'atmosphère ambiante et non lors du dépôt par pulvérisation.

De même le contrôle de l'homogénéité des couches est fondamental pour l'interprétation des variations de leur conductivité avec la température. Ce contrôle de la répartition du sélénium et du molybdène a été effectué par image X (Fig. 7).

Les images X (Fig. 7b et c) montrent que les éléments constituants de la couche sont répartis de façon homogène (les taches noires détectées correspondent à des trous présents dans la couche, comme le montre la Fig. 7a). Ceci est confirmé par l'analyse de composition effectuée par microsonde, les résultats obtenus variant au maximum de $\pm 1\%$, d'un point à l'autre, pour une même couche.

Il faut noter que les analyses de composition, effectuées sur les mêmes échantillons, donnent des résultats identiques à 1% près, quelle que soit la technique utilisée,

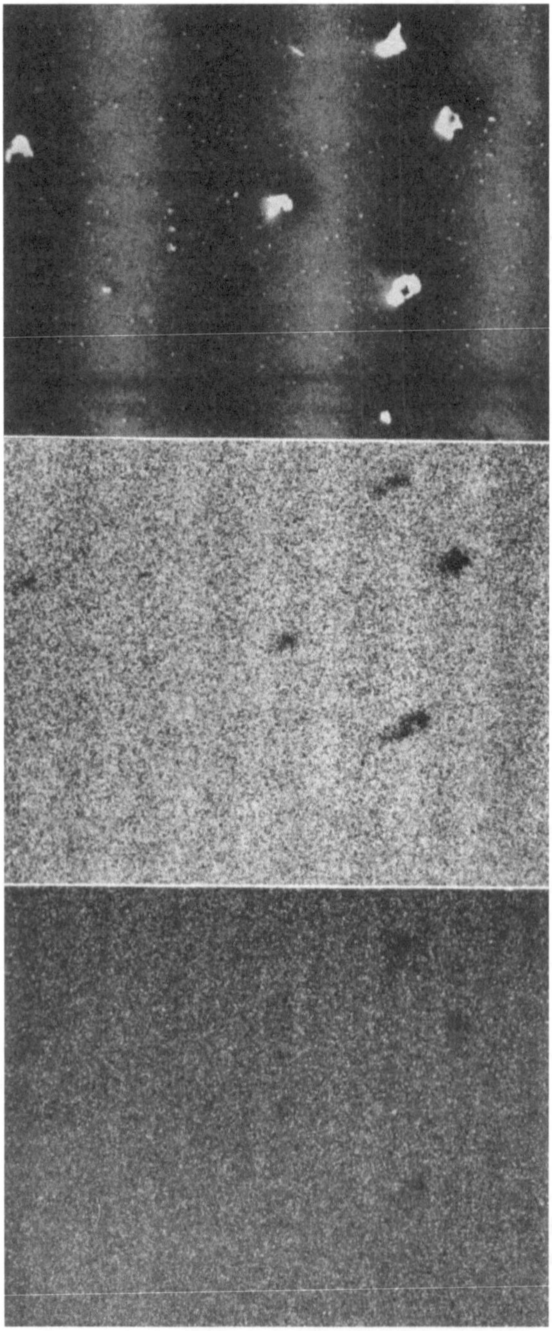

a

b

c

Fig. 7. Images d'une couche mince de $MoSe_2$ (800×). a) Image au microscope électronique à balayage, b) image X de la répartition du sélénium, c) image X de la répartition du molybdène

XPS ou microsonde électronique. La première analyse étant une technique concernant la surface, la seconde le volume (1 μm^3), ceci confirme que la recharge stoechiométrique de la couche n'est pas simplement un effet de surface, mais bien de volume.

4. Discussion

L'étude de la conductivité électrique a été réalisée sur des couches préalablement caractérisées par mesure d'absorption optique, diffraction de rayons X et analyse ESCA [7].

Toutes ces techniques de caractérisation convergeaient vers la conclusion suivante: les couches obtenues sont des couches polycristallines de $MoSe_2$, de largeur de bande interdite de l'ordre de celle du monocristal (1,15 à 1,26 eV), de composition quasi stoechiométrique et de structure hexagonale [7]. Quoique la composition de certaines couches du Tableau 1 soit plus proche de $MoSe_3$ que de $MoSe_2$ les diagrammes de diffractions X n'ont permi de mettre en évidence que des raies issues du $MoSe_2$. Bien que la résolution des techniques d'analyses utilisées soit insuffisante pour trancher, on peut penser que l'excès de sélénium s'accumule aux joints de grains. Il faut de plus noter que les résultats concernant ces couches non stoechiométriques ne sont donnés (Tableau 1) que pour rendre compte de la faible dépendance de la conductivité des couches avec la composition, la majorité des couches étant proches de la stoechiométrie. La technique de détermination de la taille des cristallites, à partir de la largeur des pics de diffraction de rayons X au goniomètre à compteur, montre que ceux-ci se développent préférentiellement suivant l'axe *c* [7]. Une mesure statistique de la taille des grains parallèles au substrat (Fig. 8) montre que l'on peut estimer leur dimension, de l'ordre de 500 nm suivant l'axe *c* tandis qu'elle est très inférieure suivant les plans parallèles à l'axe *c* (10 à 30 nm [7]).

Ceci fait que l'on est très loin des grains cubiques supposés dans les divers modèles de conductivité pour les couches polycristallines [13 à 16]. D'où la nécessité de se limiter à des interprétations plus qualitatives que quantitatives des résultats.

La faible quantité d'oxygène détectée dans les couches après élaboration complète et l'absence de MoO_3, malgré sa présence en surface lors des phases intermédiaires du processus, peut s'expliquer par une expulsion de l'oxygène lors du recuit.

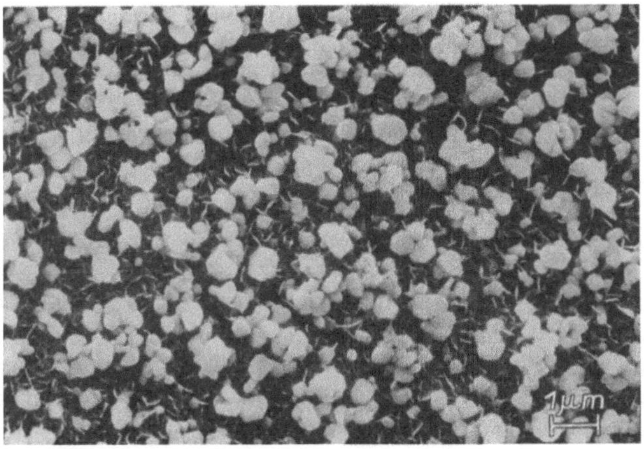

Fig. 8. Visualisation d'une couche en fin d'élaboration (substrat en silice)

Ce phénomène a déjà été rencontré par d'autres équipes dans le cas de siliciures de métaux réfractaires [17 à 19]. Dans le cas présent des réactions d'oxydo réduction peuvent expliquer le phénomène.

En présence de soufre à 200 °C, le MoO_3 donne du MoO_2 et à 400 °C du MoS_2 [20]. L'électronégativité (échelle de Pauling) du soufre étant très proche de celle du sélénium 2,5 et 2,4 respectivement, on peut penser qu' à 500 °C, le MoO_3 est de même façon réduit par le sélénium pour donner du $MoSe_2$.

Les Fig. 1 à 4 montrent que la conductivité augmente avec la température. Cependant l'accroissement est moindre dans le domaine 77 à 250 K.

De nombreux auteurs [21 à 24] ont montré qu'aux alentours de 200 K il se produit un changement de type de conductivité pour les couches polycristallines.

A basse température la variation de la conductivité avec la température suit une loi en $T^{-1/4}$, correspondant à une conduction par sauts à distance variable thermiquement activés, suivant la loi de Mott [25, 21, 22, 24]. Aux température supérieures, c'est la résistivité des joints de grains qui est prédominante [21 à 24]. Ici la Fig. 9 montre qu'aux températures égales ou inférieures à 200 K la loi [25]

$$\sigma = \sigma_0 \exp\left\{-\left[\frac{16\alpha^3}{kT\,N(E)}\right]^{1/4}\right\} \tag{1}$$

est respectée, avec $\sigma_0 \sim A/T^{1/2}$, $\alpha = 10^7\ \mathrm{cm}^{-1}$, k constante de Boltzmann ($8{,}625 \times$ $\times\ 10^{-5}$ eV) et $N(E)$ densité d'états localisés.

L'existence d'états localisés, mise en évidence par le processus de conduction par sauts décrit ci-dessus, est dûe aux imperfections associées à la nature polycristalline des couches.

En prenant $\alpha = 10^7\ \mathrm{cm}^{-1}$ ce qui est la valeur admise communément [25, 22], les valeurs calculées pour $N(E)$, à partir de la pente de la courbe $\sigma T^{1/2} = f(T^{-1/4})$, sont données dans le Tableau 2, de même pour σ_0.

Les valeurs obtenues pour $N(E)$ sont relativement faibles, quoique situées dans la fourchette prévue pour ce type de conductivité, 10^{18} à $10^{21}\ \mathrm{cm}^{-3}\,\mathrm{eV}^{-1}$ [25], comparées à celles rencontrées dans des couches minces polycristallines de $CuInTe_2$ ($8{,}8 \times 10^{20}$ à $6 \times 10^{23}\ \mathrm{cm}^{-3}\,\mathrm{eV}^{-1}$) [22]. Ceci montre la bonne qualité des couches de $MoSe_2$.

Tableau 2

Propriétés de transport électrique des couches minces polycristallines de $MoSe_2$ (recuit sous vide à 650 K pendant 24 h)

échantillon	température du substrat dépot (K)	température et durée du recuit lors de la charge en Se (K)	(h)	σ à température ambiante ($\Omega^{-1}\,\mathrm{cm}^{-1}$)	épaisseur de la couche (μm)	σ_0 ($\Omega^{-1}\,\mathrm{cm}^{-1}$)	$N(E)$ ($\mathrm{cm}^{-3}\,\mathrm{eV}^{-1}$)	$q\varphi_B$ (eV)
CM 13	580	780	24	10^{-4}	2,5	88,6	$1{,}35 \times 10^{19}$	0,29
CM 14	560	780	24	$2{,}53 \times 10^{-3}$	5,5	$2{,}83 \times 10^{6}$	$3{,}36 \times 10^{18}$	0,15
CM 17	540	780	24	$7{,}24 \times 10^{-3}$	2	$36{,}5 \times 10^{3}$	$1{,}06 \times 10^{19}$	0,17
CM 55	400	730	24	$1{,}14 \times 10^{-3}$	1,1	$28{,}8 \times 10^{3}$	$7{,}16 \times 10^{18}$	0,15
CM 56	425	730	24	9×10^{-4}	0,9	$93{,}5 \times 10^{3}$	$5{,}21 \times 10^{18}$	0,25
CM 69	425	770	24	$1{,}78 \times 10^{-3}$	–	194×10^{3}	$5{,}18 \times 10^{18}$	0,18

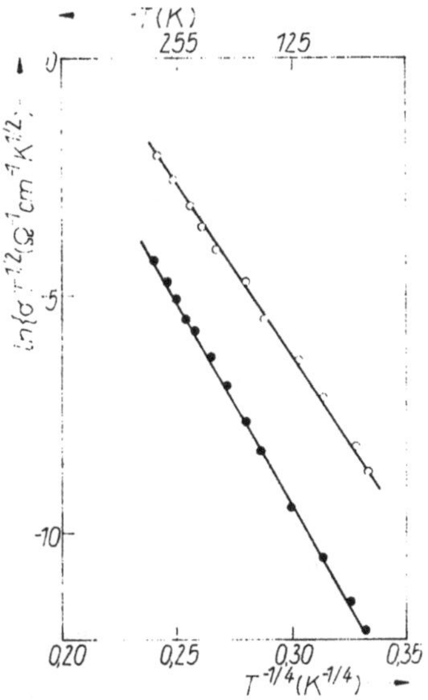

Fig. 9

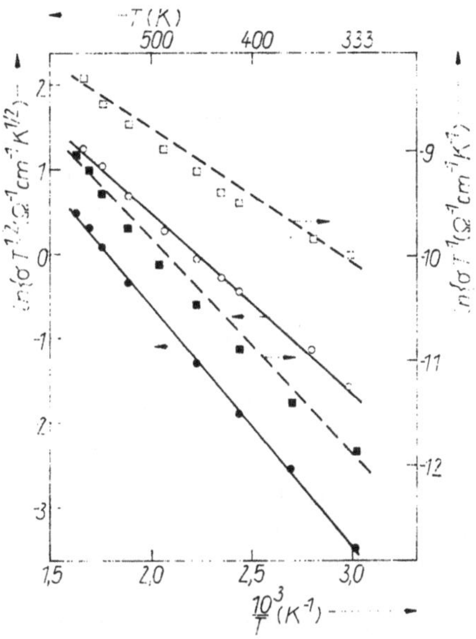

Fig. 10

Fig. 9. Variation de $\sigma T^{1/2}$ en fonction de $T^{-1/4}$. ● échantillon CM 56, ○ CM 17

Fig. 10. Variation de $\sigma T^{1/2}$ en fonction de $10^3/T$: ● CM 56, ○ CM 17; σT^{-1} en fonction de $10^3/T$: ■ CM 56, □ CM 17

Au delà de 250 à 300 K, la loi de la conduction par sauts n'est plus respectée. Les couches étant polycristallines, nous avons vu que leur conductivité est peu sensible à leur composition. Mais, à température ambiante, elle est inférieure de deux à cinq ordres de grandeur à celle attendue pour les monocristaux (Tableau 1). Tout ceci est typique d'une conductivité dominée par les joints de grains. Les théories classiques, concernant les effets de piègeages aux joints de grains, supposent la création d'une région dépeuplée de porteurs, dûe à la capture des porteurs mobiles par les pièges abondants aux joints de grains. Il en découle la formation de barrières de potentiel sur les bords des grains. Il existe alors une concentration critique en accepteurs N^+ (pour un semiconducteur de type p) au-dessous de laquelle les grains sont complètement dépeuplés [15]. Lu et al. [15] ont montré que la conductivité des joints de grains suit les lois suivantes:

$$\sigma \sim T \exp\left(-\frac{E_g/2 + e_T}{kT}\right) \tag{2}$$

quand les grains sont complètement dépeuplés et

$$\sigma \sim T^{-1/2} \exp\left(-\frac{q\varphi_B}{kT}\right) \tag{3}$$

quand les grains ne sont pas complètement dépeuplés avec E_g bande interdite des cristallites, e_T énergie des états pièges aux joints de grains par rapport à E_{i0}, niveau de Fermi intrinsèque au centre du grain, k constante de Boltzmann et φ_B hauteur de la barrière de potentiel au joint de grain.

La Fig. 10 montre que la loi correspondant aux grains non dépeuplés donne une bonne corrélation avec les résultats expérimentaux dans le domaine des hautes températures. On peut donc déduire la hauteur de la barrière de potentiel au joint de grain à partir de la pente de la courbe de la Fig. 10 (●, ○), les valeurs obtenues sont reportées dans le Tableau 2.

On n'observe pas de corrélation notable entre hauteur de barrière de potentiel, température du substrat et température de recuit.

D'autre part, on peut remarquer que les hauteurs de barrière obtenues sont proches de celles rencontrées dans le cas de couches minces polycristallines de sélénium [26].

5. Conclusion

Les résultats des mesures de conductivité électrique, l'étude de l'homogénéité de la répartition du molybdène et du sélénium, ainsi que la recherche d'oxygène, montrent que les couches obtenues sont de bonne qualité. La densité d'états localisés est relativement faible. Cependant elle permet l'apparition, à basse température, d'une conduction par sauts à distance variable activée thermiquement. Au delà de 250 K la conductivité est déterminée par les effets de joints de grains. La disparition de l'oxygène présent initialement sur les couches, lors des recuits, est dûe à la réduction du MoO_3 présent en surface par les vapeurs de sélénium. La croissance en colonne perpendiculaire au plan du substrat des grains entraine, probablement, une forte anisotropie de la conductivité. Ici, du fait des recuits nécessaires à la technique d'obtention des couches mise en oeuvre, il était impossible de mesurer la conductivité transversale. Faute d'électrode inférieure disponible, seule la conductivité dans le plan, qui est déterminée par les barrières intercristallines présentes aux joints de grains, a été étudiée. Une nouvelle technique de dépôt est actuellement à l'étude au laboratoire afin de résoudre ce problème.

Remerciement

Nous remercions M. Bohn (COB) pour sa réalisation de l'étude à la microsonde électronique.

Bibliographie

[1] L. N. Brixner, J. inorg. nuclear Chem. **24**, 257 (1962).
L. N. Brixner et G. Teuffeur, Inorg. Chem. **2**, 992 (1963).
E. Revolinsky et D. J. Beerntsen, J. appl. Phys. **35**, 2086 (1964), J. Phys. Chem. Solids **27**, 523 (1966).
J. A. Wilson et A. D. Yoffe, Adv. Phys. **18**, 193 (1969).
A. M. Goldberg, A. R. Beal, F. A. Levy et E. A. Davis, Phil. Mag. **32**, 367 (1975).
S. H. El-Mahalawy et B. L. Evans, phys. stat. sol. (b) **79**, 713 (1977).
[2] H. Tributsh, Solar Energy Mater. **1**, 257 (1979).
H. Tributsh, T. Sakat et T. Hawar, Electrochim. Acta (London) **26**, 21 (1981).
[3] G. Kline, K. K. Kam, R. Ziegler et B. A. Parkinson, Solar Energy Mater. **6**, 337 (1982).
K. K. Kam et B. A. Parkinson, J. phys. Chem. **86**, 463 (1982).
[4] S. Chandra et R. K. Pandey, phys. stat. sol. (a) **72**, 415 (1982).
[5] R. Bichsel et F. Levy, Thin Solid Films **116**, 67 (1984); **124**, 75 (1985); **131**, 87 (1985).
[6] J. C. Bernède, M. Kettaf, A. Mallouky et R. Messoussi, Internat. Symp. Trends and New Applications in Thin Films, Strasbourg 1987, Soc. Franc. du Vide, Paris 1987 (p. 21).
[7] J. C. Bernède, A. Mallouky et J. Pouzet, Mater. Chem. Phys. **20**, 201 (1988).
[8] M. Mallouky, Thèse 3ème cycle, Nantes 1987.
[9] V. A. Obolonchik, Naukova Dumka, Kiev, **2**, 130 (1970).
[10] M. K. Agarwal, P. D. Patel et O. Vijayan, phys. stat. sol. (a) **78**, 133 (1983).
[11] V. Buck, Thin Solid Films **139**, 157 (1986).

[12] C. D. WAGNER, W. M. RIGGS, L. M. DAVIS, J. F. MOULDER et G. E. MUILENBERG, Handbook of X-Ray Photoelectron Spectroscopy, Perkin Elmer Corp., Phys. Electron. Div., Enden Prairie (Minn.) 1979.
[13] S. Y. SETO, J. appl. Phys. 46, 5247 (1975).
[14] G. BACCARINI, B. RICCO et G. SPADINE, J. appl. Phys. 49, 5565 (1978).
[15] C. C. LU, C. Y. LUAN et J. D. MEINDL, IEEE Trans. Electron Devices 28, 818 (1981).
[16] M. M. MANDURAH, K. C. SARASWAT et T. I. KAMINS, IEEE Trans. Electron Devices 28, 1163 (1981).
[17] S. P. MURARKA et D. B. FRASER, J. appl. Phys. 51, 1593 (1980).
[18] R. J. SCHUTZ et L. R. TESTARDI, J. appl. Phys. 50, 5773 (1979).
[19] S. P. MURARKA et D. B. FRASER, J. appl. Phys. 51, 342, 350 (1980).
[20] P. PASCAL, Compléments au nouveau traité de chimie minérale, Molybdène n° 5, Ed. A. PACAULT et G. PANNETIER, Masson, Paris 1976 (p. 95).
[21] A. M. PHAHLE, Thin Solid Films 41, 235 (1977).
[22] A. L. DAWAR, ANIL KUMAR, R. P. MALL et P. C. MATHUR, Thin Solid Films 112, 107 (1984).
[23] C. JULIEN, M. EDDRIEF, K. KAMBAS et M. BALKANSKI, Thin Solid Films 137, 27 (1986).
[24] D. SRIDEVI et K. V. REDDY, Thin Solid Films 141, 157 (1986).
[25] N. F. MOTT, Phil. Mag. 19, 835 (1969).
N. F. MOTT et E. A. DAVIS, Electronic Processes in Non-Crystalline Materials, 2nd ed. Clarendon Press, Oxfort 1979.
[26] P. BURGAUD, J. C. BERNÈDE, G. SAFOULA et A. AMEZIANE, phys. stat. sol. (a) 95, 721 (1986).

(Received April 15, 1988; in revised form September 6, 1988)

phys. stat. sol. (a) **111**, 193 (1989)

Subject classification: 68.45 and 72.40; S10.1

Institute of Physico-Chemical Problems, V. I. Lenin Belorussian State University, Minsk[1]

Photocurrent Generation and Optical Transitions on Degenerate Cadmium Oxide Photoanodes

By

I. D. MAKUTA, S. K. POZNYAK, A. I. KULAK, and E. A. STRELTSOV

Quantum efficiencies of photocurrent generation (i.e. quantum yields, Y) depending on the electrode potential (E) and irradiation wavelength (λ) are studied for polycrystalline degenerate cadmium oxide (CdO) with variable electron concentrations (n_e). A depletion region (i.e. space charge layer) may be formed and considerable photocurrent (i^{ph}) may be generated ($\lambda < 1200$ nm) in the CdO/electrolyte solution system with appropriate anodic bias. The highest quantum yield (0.8 at $\lambda = 450$ nm) is attained in 0.2 M NaSCN at $n_e = (1$ to $4) \times 10^{19}$ cm^{-3}. With electron concentration increasing to (0.9 to 1.6) $\times 10^{20}$ cm^{-3} it reduces to 0.1 under the same conditions. Here, the CdO energy gaps determined from Y spectra increase due to the Burstein shift from 0.90 eV to 1.10 eV for indirect and from 2.06 to 2.29 eV for direct transitions.

Die Quantenwirkungsgrade der Photostromgeneration (d. h. Quantenausbeuten, Y) in Abhängigkeit vom Elektrodenpotential (E) und Bestrahlungswellenlänge (λ) werden für polykristallines entartetes Kadmiumoxid (CdO) mit unterschiedlichen Elektronenkonzentrationen (n_e) untersucht. Ein Verarmungsgebiet (d. h. Raumladungsschicht) kann sich bilden und ein beträchtlicher Photostrom (i^{ph}) ($\lambda < 1200$ nm) im CdO/Elektrolyt-System bei geeigneter Anodenvorspannung generiert werden. Die höchste Quantenausbeute (0,8 bei $\lambda = 450$ nm) wird in 0,2 M NaSCN bei $n_e = (1$ bis $4) \times 10^{19}$ cm^{-3} erhalten. Mit auf (0,9 bis 1,6) $\times 10^{20}$ cm^{-3} steigender Elektronenkonzentration reduziert sie sich auf 0,1 unter denselben Bedingungen. Hierbei vergrößern sich die CdO-Energielücken, die aus den Y-Spektren bestimmt werden, infolge der Bursteinverschiebung von 0,90 auf 1,10 eV für indirekte und von 2,06 auf 2,29 eV für direkte Übergänge.

1. Introduction

CdO is known to be an n-type semiconductor with nearly metallic conductivity caused by superstoichiometric interstitial cadmium atoms and/or oxygen vacancies [1 to 7]. When the carrier concentration exceeds 10^{18} to 10^{19} cm^{-3}, all donors are fully (i.e. doubly) ionized at room temperature and the Fermi level lies in the conduction band, so degeneracy of CdO occurs [1].

Theoretical calculations of the CdO energy band structure point to one direct optical energy gap ($E_g = 2.3$ eV) and two indirect energy gaps (from 0.8 to 1.2 eV) [8 to 10]. The energy gaps determined experimentally are 2.18 to 2.50 eV for direct and 0.98 to 1.2 eV for indirect transitions [11 to 15], one direct and one indirect transition being usually observed. Intensive plasma absorption and Burstein shift [15] are the factors impeding the investigation of indirect transitions by optical methods, their impact enhancing with increasing degeneracy. To remove these obstacles Koffyberg and Benko [11] have used quantum efficiencies of the photocurrent for the investigation of optical transitions and determined indirect ($E_g = 0.98$ eV) and direct ($E_g = 2.18$ eV) energy gaps in degenerate CdO at electron concentration 9.1×10^{18} cm^{-3} (the Fermi level lies 0.07 eV above the conduction band bottom).

[1]) Minsk 220080, USSR.

We report here a study of photocurrent generation in polycrystalline CdO with variable electron concentrations. Our investigation was intended to find out the dependence of the band gap width and the quantum efficiency of photocurrent generation in CdO-based barrier systems on the degeneracy degree of CdO. This study has required to additionally specify CdO samples by electron concentration, Fermi level location, and other parameters as well as to ascertain the impact of the front contact on the photocurrent generation efficiency and on the formation of a space charge layer in CdO.

2. Experimental Section

Polycrystalline CdO was used in the form of disks (1 cm diameter and 0.1 cm thick) fabricated by compressing CdO powder (of 99.999% purity) at 100 MPa and then annealing for 8 h at 700 to 1000 °C. Microcrystal sizes determined by SEM increase from 3 to 4 μm to 10 to 15 μm when the annealing temperature (T_a) changes from 700 to 1000 °C. Samples were polished with diamond paste, etched in 10 M HNO_3, and washed in doubly distilled water. Ohmic contact with the disk rear side was made by Ga–In alloy with subsequent epoxy resin insulation. The front contacts were realized via aqueous solutions (0.1 M NaOH, 0.2 M KCl, 0.2 M KBr, 0.2 M KI, 0.2 M NaSCN, etc.) in electrochemical measurements or via clamped copper or silver mesh electrodes. External bias was provided by potentiostatic polarization in the CdO/electrolyte solution system and by galvanostatic polarization at the CdO/metal contact.

To determine electron concentration, Hall ($n_{e,H}$) and electrochemical ($n_{e,c}$) measurements were carried out. The Hall constant was dc measured by the Van der Pauw method in a constant magnetic field; $n_{e,c}$ was determined using the bridge technique from the slope of the linear range on C^{-2}–E plots in terms of the Mott-Schottky relationship for degenerate semiconductors [17],

$$C^{-2} = 2\left(E - E_{fb} - \frac{2F}{5\,e}\right)(\varepsilon\varepsilon_0 e n_{e,c})^{-1} + C_H^{-2}\,, \tag{1}$$

where C is the measured capacity, C_H the Helmholtz layer capacity, E the electrode potential (versus SCE), E_{fb} the flat-band potential, F the Fermi energy, $n_{e,c}$ the carrier concentration, e the electronic charge, ε_0 the permittivity of vacuum, ε the dielectric constant ($\varepsilon_{CdO} = 22$ [11]).

The Fermi level location was calculated assuming that all donor states lie on the conduction band bottom and the dependence of their number $g'(E_c)$ on n_e is given by [1]

$$g'(E_c) = n_e\left(\frac{n_e}{c_0}\right)^{-\alpha_0}$$

with c_0 and α_0 as adjustable parameters determined experimentally in [1] for CdO with electron concentrations 1.88×10^{18} to 31.6×10^{18} cm^{-3}. The dependence of the Fermi energy on n_e is given then by

$$\begin{aligned} n_e &= \int \{g(E) + g'(E_c)\}\,[1 + \exp{(E - F)/kT}]^{-1}\,dE \\ &= N_c F_{1/2}(\eta) + \frac{g'(E_c)}{1 + \exp{(-\eta)}}\,, \end{aligned} \tag{2}$$

where $N_e = 2(2\pi m^* kT/h^2)^{3/2}$ is the effective density of states in the conduction band, $g(E) = 4\pi(2m^*/h^2)^{3/2}\,(E - E_c)^{1/2}$, $\eta = (F - E_c)/kT$, $F_{1/2}(\eta)$ is the Fermi-Dirac integral of order 1/2 tabulated e.g. in [18]. We have used electron concentration determined from Hall measurements and effective electron mass (m^*) taken from [14] as the n_e function within the n_e range 7.4×10^{18} to 1.3×10^{20} cm^{-3}.

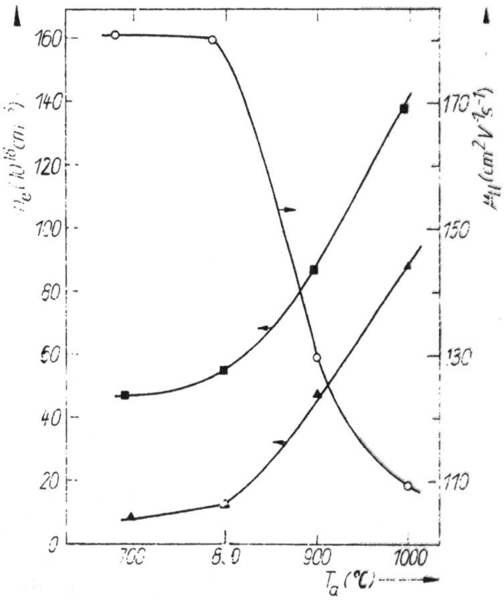

Fig. 1. CdO electron concentration determined from Hall (▲) and capacity (■) data and electron mobility (○) vs. annealing temperature (T_a)

3. Results

Fig. 1. shows an increase of both $n_{e,H}$ and $n_{e,c}$ by an order as the annealing temperature changes from 700 to 1000 °C. The Hall mobility of electrons (μ_H) reduces in the opposite way. It should be noted that the values of $n_{e,c}$ are 1.5 to 2 times larger as compared to the corresponding $n_{e,H}$ values. The reasons for this difference are still unclear. One can merely suppose that higher $n_{e,c}$ values are due to ionization of some additional donors in the field of the depletion region at the CdO/electrolyte solution interface.

It is significant that in polycrystalline CdO μ_H depends on n_e in the same way as in CdO single crystals [1, 16]. Moreover, the difference between the mobility values in CdO single crystals and polycrystalline CdO is not considerable, these values ranging between 150 to 250 and 110 to 190 cm^2/Vs at the same $n_{e,H}$.

Calculations of the Fermi energy by (2) show that all tested CdO samples are fully degenerate and in the annealing temperature range from 700 to 1000 °C the degeneracy degrees are from $\eta = 2.0$ to 11.7 (Table 1).

The degenerate CdO/metal contact is shown to be ohmic and nonphotosensitive. There is no open-circuit photovoltage, and no photocurrent is observed when the polarization voltage amounts to 20 mV (no higher voltages are used because of considerable dark currents).

However, in contact with aqueous solutions of electrolytes (NaOH, NaSCN, etc.) a photocurrent (i^{ph}) may be generated on the CdO electrodes when an external bias is applied (Fig. 2). The dependence of i^{ph} on E is shown to consist of two parts: at

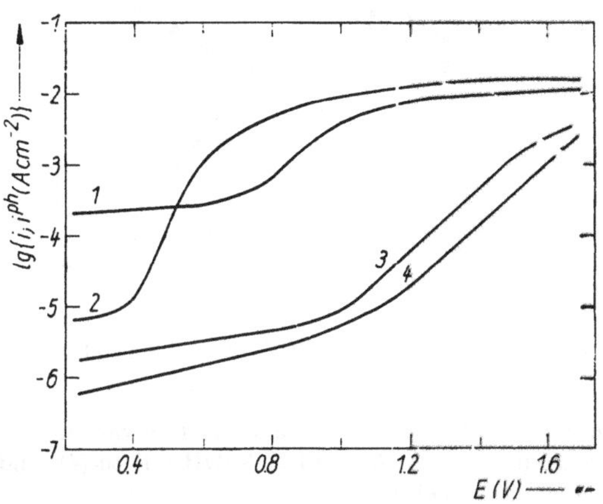

Fig. 2. Photocurrent i^{ph} (1, 2) and dark current i (3, 4) vs. CdO electrode potential E in 0.1 M NaOH (1, 3) and 0.2 M NaSCN (2, 4), $\lambda = 250$ to 700 nm, $J = 70$ mW cm^{-2}. Potential scanning rate is 10 mV/s. $T_a = 700$ °C

$E < E_{on}$ (E_{on} is the electrode potential of CdO photocurrent onset and equals to about 0.4 to 0.5 V in 0.1 M NaOH and 0.2 M NaSCN, 0.6 to 0.7 V in 0.2 M KBr, 0.7 to 0.8 V in 0.2 M KCl, 0.2 V in 0.2 M KI) the i^{ph} value is nearly independent of E, while at $E > E_{on}$ a sharp increase of photocurrent is observed. Dark anodic currents in the CdO/electrolyte solution systems are much less than the ones at the CdO/metal contact.

The photocurrent spectrum depends on the electrode potential: at $E < E_{on}$ the spectrum lies in the UV range, whereas it occurs in a wide spectral range up to about 1200 nm at $E > E_{on}$ (Fig. 3). Photocurrent in the initial part of the i^{ph}–E curves is found to be caused by carrier generation in a thin surface layer of cadmium hydroxide (in NaOH solution) or cadmium hydroxyhalogenides of $Cd(OH)_x\,Hal_y$-type (in solutions containing Hal^- ions) with wide band gaps (3.3 to 4.4 eV). These compounds are spontaneously formed when the CdO electrode is in contact with aqueous solutions of the corresponding electrolytes. Photocurrent generation connected with CdO band gap absorption is observed at $E > E_{on}$. In solutions containing some electron donor species (SCN^-, Br^-, I^-) quantum yields are shown to increase, whereas in SCN^--containing solution the long-term stability of the photocurrent increases as well, the character of photocurrent spectra and i^{ph}–E curves being the same. It should be noted that considerable quantum yields can be achieved in SCN^--containing solution (Table 1). It indicates an effective separation of photogenerated electron–hole pairs with subsequent transport of minority carriers to the CdO/electrolyte solution interface and their rapid engagement in electrochemical reactions.

It is significant that the onset potential E_{on} coincides with the potential of linear growth onset on the C^{-2}–E plot (Fig. 4) indicating the formation of a depletion region and applicability of (1). It should be noted that $n_{e,c}$ values are close to $n_{e,H}$ ones. At

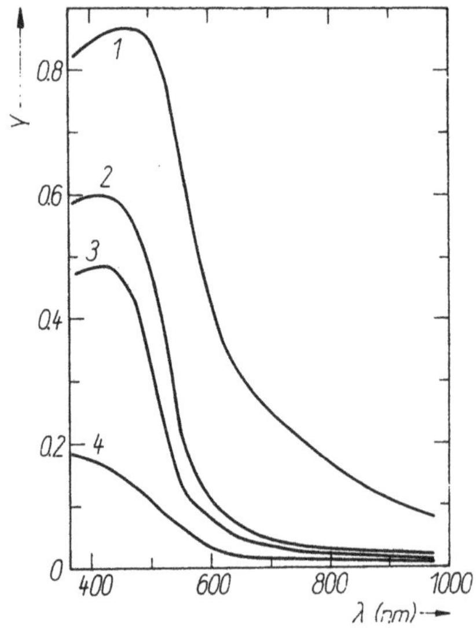

Fig. 3

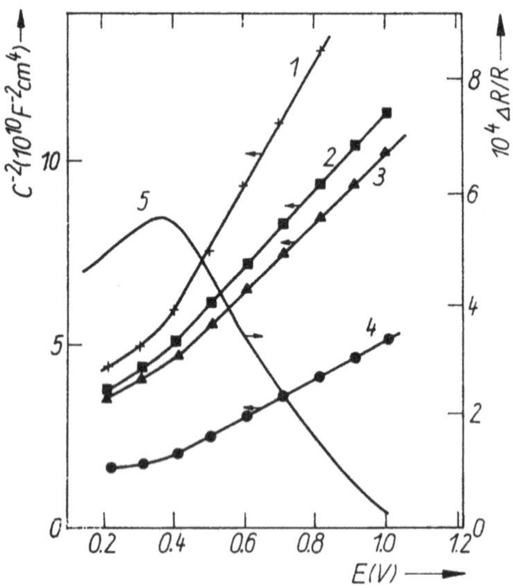

Fig. 4

Fig. 3. Dependence of quantum yield Y on wavelength in 0.2 M NaSCN, $E = 0.8$ V. (1) $T_a = 700$, (2) 800, (3) 900, (4) 1000 °C

Fig. 4. The reciprocal square of the measured capacity C^{-2} ((1) to (4)) and relative variation in the reflection coefficient $\Delta R/R$ (5) vs. CdO electrode potential E in 0.1 M NaOH ((1) to (4)) and 0.2 M NaSCN (5). (1), (5) $T_a = 700$, (2) 800, (3) 900, (4) 1000 °C

Table 1

Quantum yield (Y), relative position of Fermi level (η), indirect (E_g^I) and direct (E_g^{II}) optical energy gaps of CdO samples annealed at different temperature T_a

T_a (°C)	$\eta = (F - E_c)/kT$	Y (%) ($\lambda = 450$ nm)	E_g^I (eV)	E_g^{II} (eV)
700	2.0	80	0.90	2.06
800	5.3	55	1.06	2.22
900	7.1	43	1.10	2.26
1000	11.7	12	1.10	2.29

the same electrode potentials (i.e. in the vicinity of E_{on}) an essential electrooptical effect (at $\lambda = 500$ to 700 nm) is shown to appear becaues of the CdO reflection coefficient change caused by electrode potential modulation (Fig. 4).

All these facts together confirm that the linearity of the C^{-2}–E plot at $E > E_{on}$ is not incidental and a depletion region (i.e. space charge layer) is really formed at the degenerate CdO/electrolyte solution interface. In this case one can reasonably suppose that photocurrent generation in degenerate CdO is similar to the processes encountered in photoelectrochemical systems based on nondegenerate semiconductors. Thus, the CdO band gap may be determined from [19]

$$Y\hbar\omega = A_n(\hbar\omega - E_g)^n (L_p + L_{sc}) ,$$

where $L_{sc} = 2\varepsilon\varepsilon_0(E - E_{fb})^{1/2}/(en_e)^{1/2}$ [19] is the space charge layer width, L_p the diffusion length of holes, $\hbar\omega$ the irradiation energy, A_n a constant, $n = 1/2$ for direct and $n = 2$ for indirect transitions. CdO energy gaps determined from $(Y\hbar\omega)^{1/2} - \hbar\omega$ and $(Y\hbar\omega)^2 - \hbar\omega$ plots (Fig. 5) are given in Table 1 for indirect and direct optical

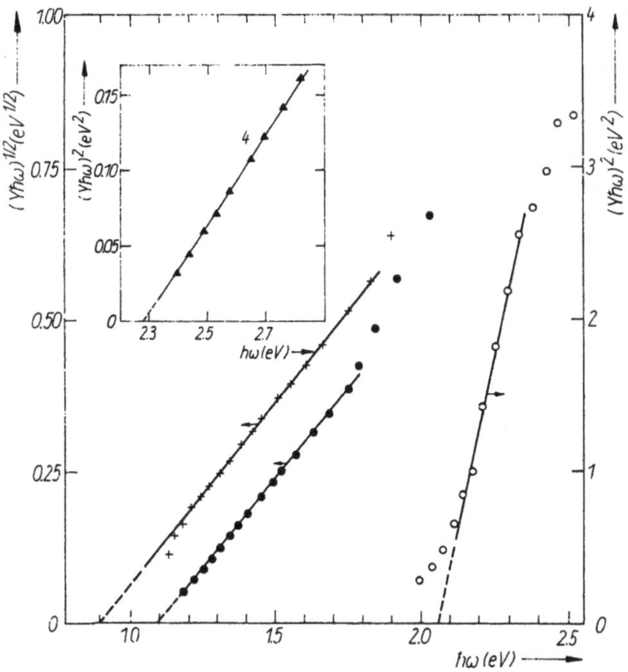

Fig. 5. $(Y\hbar\omega)^n$ vs. $\hbar\omega$ plots for CdO annealed at (1), (3) 700 °C and (2), (4) 1000 °C; (3), (4) $n = 2$ for direct and (1), (2) $n = 1/2$ for indirect transitions. $E = 0.8$ V, 0.2 M NaSCN (+ (1), ● (2) [× 5], ○ (3))

transitions, respectively. The differences in the band gap widths between the samples annealed at 700 and 1000 °C are close to the values that may be caused by the Burstein shift amounting to 0.18 to 0.20 eV at the same change of n_e [14]. It is noteworthy that the shift of the Fermi level calculated by (2) for these CdO samples equals the same value (0.24 eV).

Furthermore, the increase of external bias (at $E > E_{on}$) somewhat affects the width of CdO energy gaps both for direct and indirect transitions, the gaps being decreased by 0.03 to 0.06 eV. This effect can be attributed to the increasing contribution to the photocurrent of the charges photogenerated in the depletion region (which is expanded at higher anodic potentials) and the band gap in such region must be narrower than in the quasi-neutral bulk (because degeneracy of majority carriers and Burstein shift do not occur in the depletion region).

4. Discussion

Experimental facts demonstrate that in the degenerate CdO/electrolyte solution systems the current of majority carriers is suppressed as compared to the one at the CdO/metal contact and despite the semiconductor degeneracy, a space charge layer is formed and considerable photoresponse is observed with anodic polarization at $E > E_{on}$. If an appropriate electrolytic contact is chosen, an appreciable quantum yield (units or tens per cent) may be attained even at highly degenerate CdO.

The shift in the quantum yield spectrum coinciding in its value with the Burstein shift determined from independent reflectance data [14] evidences that photogeneration processes are mainly concentrated in the quasi-neutral bulk. Considerable quantum yields, when the depletion region is comparatively narrow (ca. units or tens nm), and moderate CdO absorption coefficient ($\alpha = 10^3$ to 10^5 cm^{-1} at 2.0 to 3.0 eV [11]) corresponding to deep penetration of light into the sample (an order-of-magnitude estimate gives 10^2 to 10^4 nm) also support an essential contribution to the photocurrent of the carriers photogenerated in the quasi-neutral bulk. All this may account for the discrepancy between quantum yields calculated by Koffyberg and Benko [11] from band gap absorption spectra neglecting the diffusion component of the photocurrent and quantum yields determined experimentally both in [11] and in the present work. The possibility of effective transport of minority carriers in degenerate CdO from the quasi-neutral bulk to the CdO/electrolyte solution interface is studied now as well as the significance of the diffusion mechanism in hole transport.

5. Conclusions

1. Anodic polarization of degenerate CdO contacted with aqueous solutions of electrolytes results in the formation of a depletion region and the occurrence of intrinsic photoresponse (at $\lambda < 1200$ nm).

2 Electron donor species (such as SCN^- ions) added to the solution increase the quantum yield to 0.8 ($\lambda = 450$ nm) for CdO with $n_e = (1$ to $4) \times 10^{19}$ cm^{-3}.

3. The quantum yield in SCN^--containing solutions reduces from 0.8 ($\lambda = 450$ nm) to 0.1 when the electron concentration changes from (1 to 4) $\times 10^{19}$ to (0.9 to 1.6) $\times$ $\times 10^{20}$ cm^{-3}.

4. The CdO energy gaps determined from quantum yield spectra amount to 0.90 eV for indirect and 2.06 eV for direct optical transitions and increase by 0.20 to 0.25 eV when the electron concentration changes from (1 to 4) $\times 10^{19}$ to (0.9 to 1.6) $\times$ $\times 10^{20}$ cm^{-3}.

References

[1] F. P. KOFFYBERG, J. Solid State Chem. **2**, 176 (1970).
[2] F. P. KOFFYBERG, Solid State Commun. **9**, 2187 (1971).
[3] A. CIMINO, and M. MAREZIO J. Phys. Chem. Solids **17**, 57 (1960).
[4] T. K. LAKSHMANAN, J. Electrochem. Soc. **110**, 548 (1963).
[5] J. S. CHOI, Y. H. KANG, and K. H. KIM, J. Phys. Chem. **81**, 2208 (1977).
[6] H. J. ENGEL, Z. Electrochem., Ber. Bunsenges. phys. Chem. **60**, 905 (1956).
[7] R. HAUL and D. JUST, J. appl. Phys. **33**, 487 (1962).
[8] K. MASCHKE and U. ROSSLER, phys. stat. sol. **28**, 577 (1968).
[9] A. BREEZE and P. G. PERKINS, Solid State Commun. **13**, 1031 (1973).
[10] S. TEWARI, Solid State Commun. **12**, 437 (1973).
[11] F. A. BENKO and F. P. KOFFYBERG, Solid State Commun. **57**, 901 (1986).
[12] F. P. KOFFYBERG, Phys. Rev. B **13**, 4470 (1976).
[13] J. KOCKA and C. KONAK, phys. stat. sol. (b) **43**, 731 (1971).
[14] H. FINKENRATH and M. VON ORTENBERG, Z. angew. Phys. **23**, 323 (1967).
[15] M. ALTWEIN, H. FINKENRATH, C. KONAK, J. STUKE, and G. ZIMMERER, phys. stat. sol. **29**, 203 (1968).
[16] F. P. KOFFYBERG, Canad. J. Phys. **49**, 435 (1971).
[17] M. S. GRILIKHES, A. M. SUKHOTIN, and A. YA. SHIK, Elektrokhimiya **23**, 1529 (1987).
[18] J. McDOUGALL and E. STONER, Phil. Trans. Roy. Soc. **123**, 67 (1938).
[19] YU. YA. GUREVICH and YU. V. PLESKOV, Photoelektrokhimiya Poluprovodnikov, Izd. Nauka, Moskva 1983 (p. 167).

(Received May 19, 1988; in revised form September 15, 1988)

phys. stat. sol. (a) 111, 201 (1989)

Subject classification: 71.55 and 73.30; S8.12

P. N. Lebedev Physical Institute, Academy of Sciences of the USSR, Moscow[1])

Relaxation of Trapping Sites in the ZnSe Films of Au–ZnSe–SiO_2–Si and Au–ZnSe–Ge Structures

By

S. L. VINOGRADOV, O. K. MALINOVSKAYA, and V. A. TOLOKONNIKOV

A thermodepolarization analysis is made of trapping sites in the ZnSe layer of multi-layer Au–ZnSe–SiO_2–Si and Au–ZnSe–Ge structures. A hole trapping site with a cross-section typical of neutral centers, is found. The TS activation energy is found to increase with TS filling, and not to decrease, as is always the case. This fact is explained in terms of a Schottky barrier model, where the semiconductor layer has a thickness comparable to the depth of the spatial charge region. According to the proposed model, the energy level, wich corresponds to the trapping site, is distanced from ZnSe valont band bottom by no less than 0.91 eV.

Проведен термодеполяризационный анализ центров захвата в слое ZnSe многослойных структур Au–ZnSe–SiO_2–Si и Au–ZnSe–Ge. Обнаружен дырочный центр захвата с сечением, характерным для нейтральных центров. Наблюдалось увеличение энергии активации центров захвата по мере их заполнения, а не уменьшение, как это происходит обычно. Этому факту дано объяснение в рамках модели, основанной на представлении о барьере Шоттки, когда полупроводниковый слой имеет толщину, сравнимую с глубиной области пространственного заряда. В соответствии с предложенной моделью энергетический уровень, отвечающий центру захвата, отстоит от потолка валентной зоны ZnSe не менее, чем на 0,91 eV.

1. Introduction

In modern practice there are two kinds of photoelectric solid-state devices that can solve the problem of recording an optical image and transforming it into electrical signal. They are: (i) charge-coupled devices (CCD) and (ii) "optical" MNOS structures. The CCD is an expensive device which has limited possibilities to process the images and had to be combined with buffer memory devices. The MNOS structures are characterized by low threshold sensitivity and spatial resolution.

For the optical data recording it is of interest to use structures which consist of two photosensitive layers with different band gaps. The wide band-gap layer can be used for optical image recording by means of trapping charge carriers photoinjected to deep level trapping sites (TS). The second one had to obtain optical reading of the stored image and transforming its information into an electric signal. These principles are realized in the photoelectric structures on the basis of high-resistance ZnSe, and "ordinary" Si or Ge. In this technique of image recording one can successfully combine high sensitivity with a high spatial resolution.

As shown in [1], an optical image can be recorded in a metal–zinc selenide–germanium structure by means of internal photoemission of charge carriers from the metallic layer to the ZnSe layer. After that charge carriers go to trapping sites, thus producing a potential relief, which repeats the distribution of illumination in the

---

[1]) Leninskii prospekt 53, SU-117924 Moscow, USSR.

projected image. The time of data storage in this structure, as well as in similar metal–$ZnSe$–SiO_2–Si systems [2], would be determined by thermogeneration processes. These are the thermogeneration of charge carriers in the ZnSe layer (with further trapping in TS) from the adjacent layers and the thermogeneration of carriers from TS. Which of these two processes is dominant, would depend on TS parameters and properties of the interface, through which the carriers are thermogenerated in the ZnSe layer. The aim of this paper is to clarify this problem. The peculiarities of TS relaxation in the region of a Schottky barrier space charge are considered.

2. Experimental Apparatus

To measure thermoactivation processes we employed the same experimental apparatus as in [1]. The thermally stimulated depolarization (TSD) current was measured by means of a VK2-16 electrometer. The input resistance was $10^8\ \Omega$, which was much less than the resistance of the samples studied.

The samples was photopolarized by the radiation of a halogen lamp KGM12-100. Passing through a monochromator ZMR-3 and a filter the radiation was focused onto the sample by a system of mirrors. Measurements were taken in a liquid-nitrogen Dewar. To make measurements in the range 80 to 420 K we used samples with indium-soldered contacts. At $T \geqq 420$ K a finger contact was used. It was a piece of 0.4 mm diameter gold wire with one end partially fused in order not to damage the sample. Another end was pressed in a thin-wall stainless steel capillary. The heat conductivity in such a contact is determined by the heat conductivity of the capillary, hence, it is not high. Test measurements of TSD currents showed that the results are independent of the way of electric contact to the sample.

A cold copper finger of the Dewar has a cavity with a heater element in it. The heat transfer from the element to the cold finger occurs by radiation. The heater was operated by a programmed switching-on and -off of the power supply.

The program used ensured a constant rate of sample heating.

3. Experimental Results

3.1 Metal-$ZnSe$–SiO_2–Si

By studying the process of recharging of trapping sites in this structure we observed a significant difference in the behaviour of samples which had been subjected to high-temperature annealing and those which had not. Investigating charge leakage from TS on unannealed samples by measuring the shift of the surface photoelectromotive force versus the applied voltage, we observed a fast component which was absent in the case of annealed samples. This peculiarity might be due to the field recharging of surface states of TS in zinc selenide. To clear up its mechanism, both annealed and unannealed samples were subjected to TSD analysis. It showed a noticeable difference in the temperature dependences for TSD currents.

Fig. 1 shows a typical curve of TSD currents in an unannealed Au–$ZnSe$–SiO_2–Si sample, where the thickness of gold layer is ≈ 10 nm, the thickness of ZnSe layer 350 nm, the thickness of SiO_2 layer 100 nm, and the area of the structure $1.5 \times \times 1.5\ mm^2$.

The sample was charged up to saturation by light of 0.465 μm wavelength (the intrinsic absorption edge in ZnSe) at an applied bias of $+1$ V on the metal (predominant hole injection) at room temperature. After that the voltage was removed and the sample cooled down to 80 K in darkness. The curve of TSD current obtained with

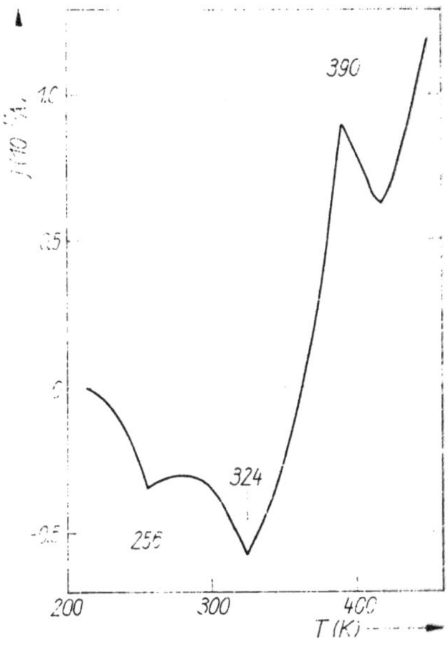

Fig. 1. TSD current curve for the unannealed Au–ZnSe–SiO_2–Si

linearly growing temperature at the rate $b =$ $= 0.323$ K/s, shows three distinct peaks with temperature maxima of 256, 324, and 390 K. The latter peak has been registered with a monotonously growing current, which starts at about 400 K. The behaviour of the curve in its high-temperature part at $T > 400$ K weakly depended on the charge state of TS. But it depends significantly on the material of the metallic layer. Therefore, we associate the current growth in this region with the beginning of thermoemission of charge carriers from metal to zinc selenide. On the contrary, the amplitude and position of peaks depend on the charge on TS in the ZnSe layer, and therefore, they are associated with thermogeneration processes caused by TS emptying.

This curve differs drastically from TSD current curves in annealed samples (see Fig. 2). The difference is in the presence of the first and third peaks, the third one being inverted relative to the first two peaks.

Perhaps, the fast component appearing in unannealed samples during the volume charge relaxation, can be explained by emptying a shallow TS, to which the peak with $T = 256$ K on the TSD current curve corresponds. The origin of this center can probably be associated with disordering of the ZnSe structure. The degree of disordering is diminished after annealing. This agrees well with the current view on the influence of annealing on shallow TS.

The appearance of an inverse peak can hardly be explained by depolarization of the initial state. Another process is much more probable, i.e. repolarization of a sample during its heating. It is known [3] that when the rates of outward flow and neutralization of the volume charge (due to thermogeneration from the Au layer, in our case) are compatible, the intrinsic electric field and TSD current are inverted. This follows from the fact that when a sample is heated, the intrinsic field of the initial volume charge may impose a thermostimulated polarization. As a result, a new polarization state appears. When this polarization state is destroyed, at further heating, an inverse peak of the TSD current appears. Such a situation seems to be the case with unannealed structures, and the peak with $T = 390$ K, which is absent on the TSD curves for the annealed samples, is caused by a destruction of a new polarization state.

We investigated the annealed samples in more detail. Below are the results for one of the studied structures, Au–ZnSe–SiO_2–Si, where the thickness of a gold layer is ≈ 10 nm, the thickness of a zinc selenide layer 160 nm, the thickness of silicon dioxide 100 nm, and the area of the structures 1.5×1.5 mm^2.

Normally, the TSD current curves in annealed samples had one peak, and the temperature position of its maximum depended significantly on the sample polarization. At the same time, the sample might be treated so that no peak would be observed on the current curve. For that purpose the sample was preliminarily heated to $T \leqq \leqq 400$ K (i.e. when thermoemission of charge carriers from gold to zinc selenide is negligible and thermoemptying of TS is completed) and then rapidly cooled down. To change the sample state so that the current curve had the peak, we had to realize hole injection into the ZnSe layer (for example, via internal photoemission from gold). The latter proves the assumption that we are dealing with trapping sites for holes.

The TS activation energy was determined by two methods, by varying the heating rate and by the method of initial rise, respectively. Here we also investigated the influence of the conditions for sample polarization.

Fig. 2 illustrates two curves for the TSD current in annealed samples obtained at different linear heating rates of the sample: $b_1 = 0.29$ K/s, $b_2 = 0.43$ K/s. Both curves have the same conditions of sample polarization: $U = +3.5$ V, $T = 300$ K; illumination by a beam of light with $\lambda = 0.465$ μm up to saturation; then cooling to 80 K in darkness without applied bias. As would be expected, the increase in the velocity of heating shifts the TSD peak towards higher temperatures and increases its amplitude. At first we estimated roughly the TS activation energy from the temperature position T_m of the peak maximum on the TSD current curve. The formula used was

$$E_t \approx ZkT_m \,, \tag{1}$$

where Z, with respect to the relaxation kinetics, would take values 15 to 30 [4]. It was found that E_t lies within the range of (0.4 to 0.9) eV (see the curves of Fig. 2).

The attempts to define more accurately the activation energy, by varying the heating rate (e.g., Booth [5]) led to unreal results, which were far beyong the range of values we derived from (1). From the curves of Fig. 2 it can be seen that the calculation for all possible types of relaxation processes gives very small values for the activation energy, 0.023 to 0.065 eV. This indicates an anomalously strong dependence of the temperature position of the TSD current peak on the heating rate. From this we can conclude about essential changes of the field conditions in a ZnSe layer during a TSD process. In such a situation, therefore, we should use only those methods which deal with the initial parts of the TSD current curve, when the conditions change negligibly.

We used the initial rise method (by Garlik and Gibson [6]) which consists in the following. The initial parts of TSD current curves, independent of the relaxation

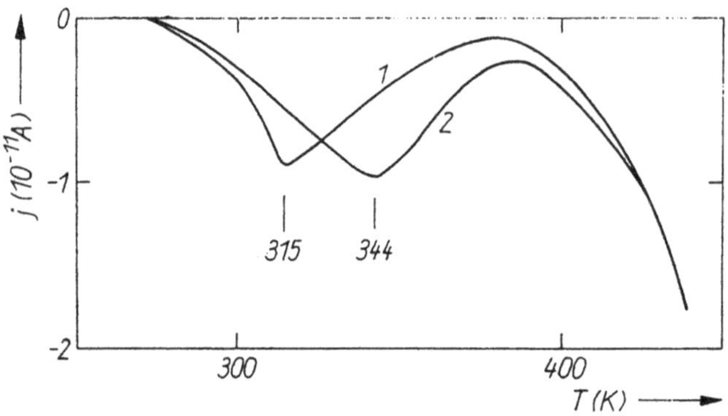

Fig. 2. Annealed Au–ZnSe–SiO_2–Si structure. Position of TSD peak vs. heating rate: (1) $b = 0.29$, (2) 0.43 K/s

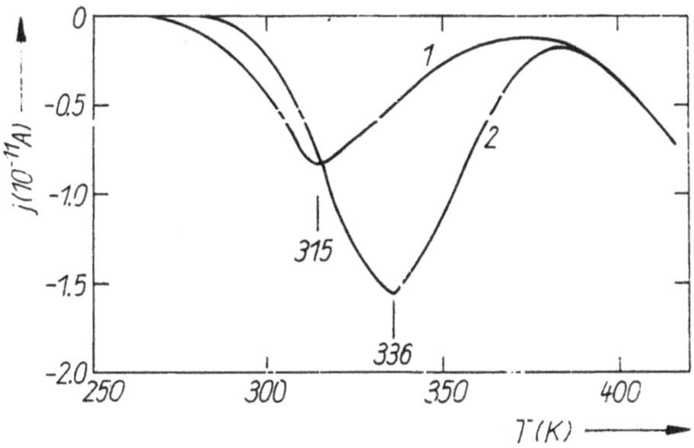

Fig. 3. Position and amplitude of TSD peak for Au–ZnSe–SiO_2–Si vs. value of positive bias: (1) 3.5 V, (2) no applied bias

kinetics, are known to have the form

$$j(T) = j_0 \exp\left(-\frac{E_t}{kT}\right), \tag{2}$$

where j_0 is some constant. From (2) it follows that the initial part of the curve is linearized in Arrhenius coordinates ($\ln j - 1/kT$). The slope of this straight line allows the calculation of the activation energy. We measured the activation energy on TS for different conditions of sample polarization.

Fig. 3 shows two curves for two samples with equal conditions of depolarization, heating rate of 0.29 K/s. The first and second experiments differ only by the values of the voltage on the sample during polarization. Curve 2 corresponds to the sample illumination without applied bias, and curve 1 is for +3.5 V voltage ($T = 300$ K, 0.465 μm wavelength) on the sample. The peak of curve 2 is greater and is shifted to higher temperatures. The results of the linearization of the initial part of these curves, in Arrhenius coordinates, are given in Fig. 4. The obtained values are 0.50 and

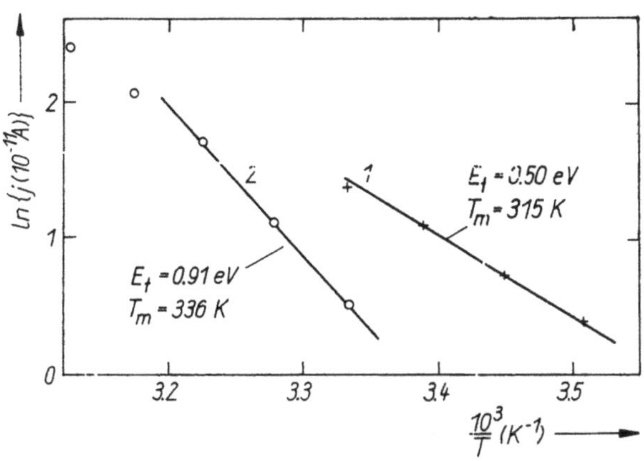

Fig. 4. Curves 1 and 2 of Fig. 3 in Arrhenius coordinates

0.91 eV, respectively. That mean that an increase in the concentration of charged centres caused an increase of activation energy, and not its diminishing, as is usually the case.

3.2 *Metal–ZnSe–Ge*

Experimental results for the internal photoemission current in Ni–ZnSe–Ge and Ni–ZnSe–SiO_2–Si structures are reported in [1, 2]. As follows from those results, the energy barriers for carriers at the Ni–ZnSe interface in those structures are not the same. That testifies to the fact that electrophysical properties of ZnSe film in our structures depend on the parameters of the surface. In fact, using a single-crystal Ge substrate, one can obtain a structure more ordered than an amorphous SiO_2 surface (the more so because the lattices constants of Ge and ZnSe are very close). This, perhaps, is the reason for the observed difference in electrophysical properties.

One could expect that this difference would also be manifested in a thermodepolarization analysis. We, therefore, investigated the trapping sites of ZnSe in the Au–ZnSe–Ge structure. A noticeable difference was found only between unannealed Au–ZnSe–SiO_2–Si and Au–ZnSe–Ge structures, and was practically not observed in annealed samples.

Fig. 5 illustrates the TSD current curve in unannealed Au–ZnSe–Ge structures. Its polarization conditions are: $T = 80$ K, $\lambda = 0.45$ µm, $U = -0.5$ V. Very well resolved are a low-temperature peak with $T = 230$ K and a high-temperature peak at 343 K, since the temperature interval between them increases (cf. Fig. 1). The low-temperature peak in the Au–ZnSe–Ge structure is significantly shifted towards lower temperatures as comparred to the Au–ZnSe–SiO_2–Si structure; it is more "spread" and has always a smaller amplitude. After annealing the structure it also disappeared. Perhaps, a more shallow level than in Au–ZnSe–SiO_2–Si corresponds to it. It is difficult to make an exact comparison, because one can hardly determine the depth of the level in the Au–ZnSe–SiO_2–Si structure because of the strong opposite influence of the high-temperature peak.

The peak with the maximum at $T = 187$ K is related to a relaxation process at the ZnSe–Ge interface. This is supported by the fact that the maximum temperature was shifted to 205 K in the Au–ZnSe–Si structure, i.e. when Ge was substituted for Si.

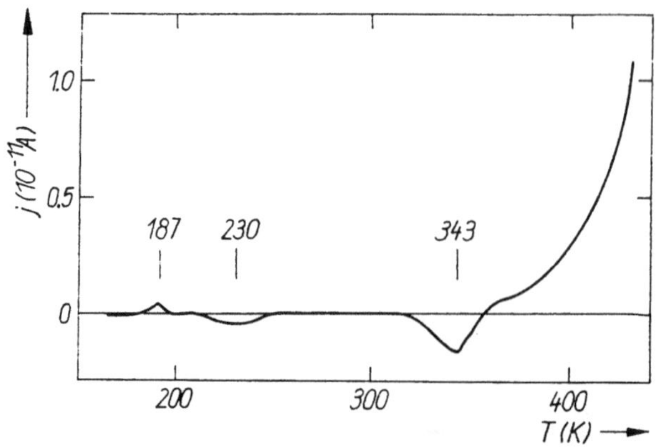

Fig. 5. TSD current for the unannealed Au–ZnSe–Ge structure

4. Discussion

4.1 Model

As seen from the experimental results presented in Section 3.1 the behaviour of TSD currents essentially depends on the degree of TS population. This leads to a strong deviation in the form of TSD current curves from the standard form of the curves. This is particularly true for the region of the peak maxima. It is natural that under such conditions the least reliable results are observed in the estimation of the activation energy by the method of heating rate variation [5], as this method utilizes the TSD current maxima T_m, and assumes unchangeable thermogeneration conditions in the depolarization process. The initial rise method [6] allows one to get reliable results irrespective of the depolarization process near the maximum. However, it should also be noted that linearization in Arrhenius coordinates in measuring the activation enerry takes place in a very small temperature range. And what is more, the activation energy itself strongly depends on the initial population of TS, as is seen from Fig. 3, 4. Thus, the combination of all the experimental data requires a detailed consideration of the effect of the volume charge and the corresponding strong intrinsic field on the TSD processes. At present we have no sufficient data for final conclusions. However, the present experimental data agree well with the following model. It says that band bending in a thin ZnSe layer (comparable with the depth of the space charge region) in case of a small concentration of charged trapping sites (Fig. 6a) is greater than in case of a high concentration (Fig. 6b). Hence, due to the Poole-Frenkel effect, the decrease in the activation energy, in the second case, proves to be less noticeable, so that $E_t'' > E_t'$. A maximum value E_t''' of the activation energy which determines more accurately the depth of the TS position, can be obtained after the final "flattening" of bands in the layer, Fig. 6c.

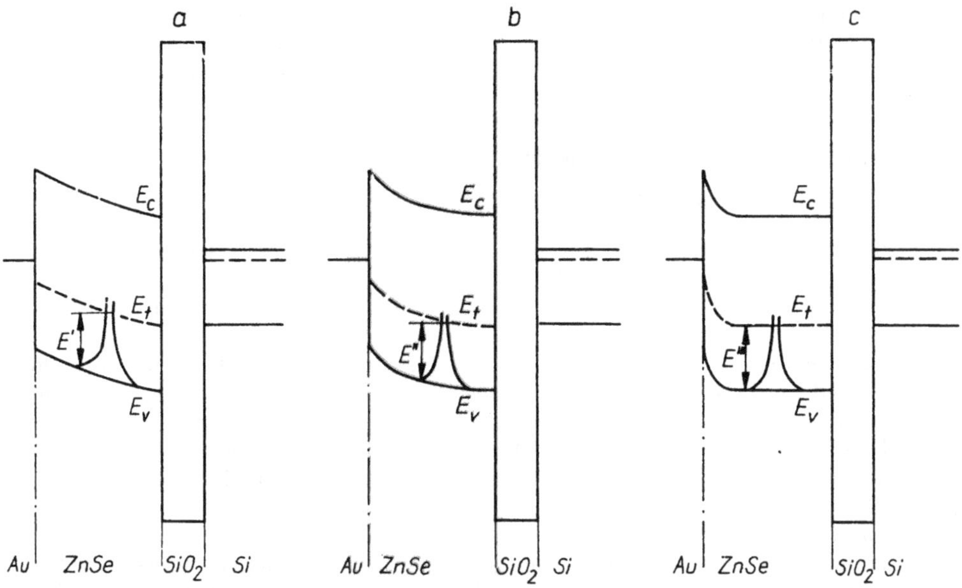

Fig. 6. Energy band diagram of the Au–ZnSe–SiO_2–Si structure vs. geometric coordinate normal to the layers with Poole-Frenkel distortion of the shape of the energy barrier for holes on TS; a), b), c) are different populations of TS in the volume of the ZnSe layer

Note that due to increasing concentration of trapping sites, the part of TS in ZnSe in the region adjacent to the interface with gold (Fig. 6b, c), gets into the stronger field, and its activation energy is further decreased. But a released charge entering the gold does not practically change its coordinate, and therefore it does not contribute significantly to the TSD current. But thermogeneration of charge carriers from the centres to the band increases the field intensity in the layer volume, which means a reduction of the activation energy in TS far away from the interface, which determine the behaviour of the TSD current curve.

We observed TSD currents when those processes compete, and therefore, perhaps, the temperature dependence of the peak on the heating rate is so strong.

The highest value of activation energy calculated from the experimental curves is 0.91 eV. So, according to our model, this value is closest to the depth of the energy level responsible for TS in the ZnSe forbidden gap. Curve 2 of Fig. 3 was obtained after polarization with $\lambda = 0.465\ \mu m$ light, without applied bias, which again supports our model (see Fig. 6c).

4.2 Time of data storage

From the data on the activation energy, we estimated an efficient frequency factor. We used the formula obtained in [7, 8], where it is assumed that $E_t \gg kTm$,

$$\nu_{\text{eff}} \approx \frac{bE_t}{kT_m^2} \exp\left(\frac{E_t}{kT_m}\right)\left[1 + (l - 1)\frac{2kT_m}{E_t}\right], \tag{3}$$

where l determines the kinetics of the relaxation process and takes values in the range 1 to 2. Note that $z \geqq 15$ in (1). Hence, the error in calculating ν_{eff} does not exceed 15%. To simplify calculations we omitted the second term in (3). The obtained value $\nu_{\text{eff}} = 1.5 \times 10^6\ s^{-1}$ corresponds to the trapping cross-section $S \approx 10^{-20}\ cm^2$, while the photocharging experiments yield a considerably higher value of $\approx 10^{-16}\ cm^2$, which is still typical for neutral centres. Since the TSD current curve has only one peak, and the electron injection into ZnSe reduces its amplitude, we can assume that the found centres are recombination centres and can either be neutral or positively charged.

The time of charge relaxation on TS from [9]

$$\tau' = \nu_{\text{eff}}^{-1} \exp(E_t/kT)$$

is ≈ 200 s, and corresponds to the lowest activation energy observed, $E_t = 0.5$ eV ($\nu_{\text{eff}} = 1.5 \times 10^6\ s^{-1}$ and $T = 300$ K).

At the same time, optical data recorded on a structure with Ni layer can be stored no longer than a minute. The data are destroyed in that case mostly due to a thermal injection of charge carriers from Ni to ZnSe. This would "smoothen out" a potential relief formed by TS photocharging, i.e. data recording. As it follows from the Richardson formula, the typical time τ'' for this process is one minute ($T = 300$ K). One should also take into account that the barrier for the hole injection from Ni is 0.96 eV [2]. The corresponding barrier in structures with Au layer is higher[2]), so that $\tau'' > \tau'$ ($E_t = 0.5$ eV), and the duration of data storage is determined by τ', which depends on the activation energy.

As shown experimentally in our case, the activation energy increases with the concentration of trapping sites where the holes are localized. To store information for

[2]) According to [10] the barrier for electron injection from Au to ZnSe is 1.36 eV, i.e. the Fermi level lies in the middle of the forbidden band of ZnSe.

a longer time, therefore, one should preliminarily increase the degree of TS filling. The TSD current was not investigated on a structure with nickel as the metallic layer. The peak was not observed, in that case, in its "pure" form, but was accompanied by a monotonously growing current caused by thermal injection from nickel into ZnSe.

5. Conclusions

1. Both structures are characterized by the current of TS depolarization and by the current of charge carrier thermoemission to the data carrier layer.

2. We observed hole trapping centres in ZnSe films. The calculation made on the base of TSD current curves gives reduced values for the trapping site cross-section. The electron injection into the layer reduces the concentration of charged centres.

3. The activation energy was found to increase with filling of TS. A qualitative model based on a Schottky barrier at the metal-semiconductor boundary, and the variation of activation energy due to the Poole-Frenkel effect has been proposed.

4 The effect of polarization conditions on the TS activation energy in ZnSe was studied. The maximal value of the activation energy is 0.91 eV. In the frames of our model this value determines most exactly the position of the energy level which characterizes TS in the forbidden band of ZnSe.

References

[1] N. A. Dovgaleva, A. F. Plotnikov, V. A. Tolokonniko, V. E. Shubin, V. I. Botsan, and V. A. Kasian, phys. stat. sol. (a) **100**, 527 (1987).
[2] O. K. Malinovskaya, A. F. Plotnikov, and V. A. Tolokonnikov, phys. stat. sol. (a) **98**, K191 (1986).
[3] Yu. A. Gorokhovatskii, Izv. Akad. Nauk SSR, Ser. fiz. tekh. Nauk, No. 1, 43 (1973).
[4] K. W. Boer, S. Oberländer, and J. Voigt, Ann. Phys. **2**, 130 (1958).
[5] A. H. Booth, Canad. J. Chem. **32**, 214 (1954).
[6] G. F. J. Garlik and A. F. Gibson, Proc. Phys. Soc. A **60**, 574 (1948).
[7] A. M. Andriesh, S. D. Shutov, V. G. Abashkin, and M. R. Chorny, Fiz. Tekh. Poluprov. **8**, 1936 (1974).
[8] G. A. Bodrovskii, V. G. Boitsov, and B. A. Demidov, Fiz. Tekh. Poluprov. **8**, 1918 (1974).
[9] Yu. A. Gorokhovatskii, Principles of thermodepolarization analysis, Izd. Nauka, Moskva 1981 (p. 176) (in Russian).
[10] C. M. Sze, Physics of Semiconductor Devices, John Wiley and Sons, Inc., New York 1981.

(Received August 12, 1987 in revised form May 10, 1988)

phys. stat. sol. (a) **111**, 211 (1989)

Subject classification: 71.20; S7.16

Laboratoire d'Etudes et de Recherches sur les Matériaux, associé au C.N.R.S., Institut des Sciences de la Matière et du Rayonnement, Université de Caen[1])

Etude par diffraction de R-X de la densité de charge de valence dans les deux semi-conducteurs tétraédriques $ZnSiAs_2$ et $ZnGeAs_2$

Par

M. LEVALOIS et G. ALLAIS

La répartition des électrons de valence dans $ZnSiAs_2$ et $ZnGeAs_2$ est décrite à l'aide de charges de liaison gaussiennes, centrées sur les axes des liaisons. Les résultats obtenus permettent de retrouver la différence, constatée précédemment lors de l'étude des paramètres structuraux, entre la liaison Zn–As et les liaisons Si–As ou Ge–As, plus fortement covalentes. En outre, la densité de charge de valence ainsi décrite est très voisine de celles qui résultent des calculs théoriques de structure de bandes.

Gaussian bond charge model is used to describe the valence charge density in $ZnSiAs_2$ and $ZnGeAs_2$. The main feature obtained is the significant difference, already pointed out in the structural parameters study, between the Zn–As bond and the Si–As or Ge–As bonds, which are more strongly covalent. Besides, the valence charge density appears to be close to that obtained from band structure calculations.

1. Introduction

La densité de charge $\varrho(\boldsymbol{r})$ joue un rôle primordial dans l'étude des milieux cristallins: pour l'état fondamental, elle détermine toutes les propriétés du cristal. La mesure du module des coefficients de Fourier de $\varrho(\boldsymbol{r})$ par diffraction de rayons X est une méthode directe d'approche de la densité de charge. La difficulté essentielle réside alors dans la précision limitée des mesures: pour pouvoir analyser la densité de charge, il est nécessaire de modéliser $\varrho(\boldsymbol{r})$ par des fonctions qui ne dépendent que d'un petit nombre de paramètres.

Les semi-conducteurs tétraédriques constituent un bon exemple de composés pour lesquels les calculs théoriques sont maintenant bien avancés, et ils ont souvent été utilisés du fait de la qualité des cristaux disponibles. Dans le cas du silicium, Aldred et Hart [1] ont obtenu des résultats très précis pour les facteurs de structure mesurés par la méthode des franges de solution pendulaire. Les résultats ont été utilisés avec divers modèles de la densité électronique: en particulier, Scheringer [2] a montré que la répartition des électrons de valence pouvait très bien être modélisée par des charges de liaison, situées entre les atomes de silicium, et prélevées sur les orbitales des atomes libres.

Dans le cas du composé binaire III–V GaAs, nous avons réalisé récemment [3] une étude de la densité de charge de valence à partir des mesures directes d'intensités de diffraction de R-X; nous avons montré que de telles measures, réalisées avec suffisamment de soin et sur des cristaux préparés par abrasion mécanique pour limiter

[1]) Boulevard du Maréchal Juin, F-14032 Caen Cédex, France.

les erreurs de mesure, permettaient une étude assez précise de la densité de valence, celle-ci étant représentée par un modèle de charges de liaison analogue à celui utilisé par Scheringer [2].

Les composées II–IV–V_2 sont les analogues ternaires des composés binaires III–V: nous avons réalisé une étude similaire sur les composés $ZnSiAs_2$ et $ZnGeAs_2$. La partie expérimentale, concernant l'enregistrement des spectres de diffraction de R-X dans ces composés, ainsi que la détermination des paramètres structuraux, a été rapportée précédemment [4]. Nous montrons dans ce second article qu'un modèle de charges de liaison permet, là encore, de décrire la densité de valence dans ces deux composés ternaires.

2. La densité de charge de valence

Dans le modèle de charges de liaison, chaque liaison est représentée par une répartition de charges centrée sur l'axe de la liaison, et prélevée sur les couches de valence des deux atomes constitutifs de la liaison. Dans les composés ternaires ABC_2, les liaisons AC et BC, qui relient des atomes différents, doivent être représentées par deux charges de valence de forme et d'amplitude différentes.

La densité électronique est la superposition de deux contributions: celle des atomes et celle des charges de liaison. Le nombre total d'électrons étant donné, les atomes ont une charge totale inférieure à celle des atomes libres. Pour un atome, la densité électronique est choisie à symétrie sphérique. La densité de valence est proportionnelle à la somme des densités des parties radiales des orbitales de l'atome libre: les déformations éventuelles sont prises en compte par les charges de liaison. Enfin, la densité électronique du coeur est la même que dans l'atome libre.

Nous choisissons une forme gaussienne pour les charges de liaison, mais en autorisant une extension de la charge dans la direction perpendiculaire à l'axe de la liaison, $V_\perp$, différente éventuellement de celle dans la direction de l'axe de la liaison, $V_\parallel$. La densité électronique correspondante, pour une amplitude Q de la charge, s'écrit alors

$$\varrho_Q(r, z) = \frac{Q}{\pi^{3/2} V_\parallel V_\perp^2} \exp\left[-\left(\frac{z^2}{V_\parallel^2} + \frac{r^2}{V_\perp^2}\right)\right],$$

r et z représentant les coordonnées perpendiculaire et parallèle à l'axe de la liaison. Le centre de la charge est sur la droite qui joint les centres des deux atomes: le facteur d'agitation thermique moyen de la charge est la moyenne des facteurs des deux atomes constitutifs de la liaison. Nous construisons de cette manière deux charges de liaison, l'une (type n° 1) pour la liaison AC, l'autre (type n° 2) pour la liaison BC.

Dans les deux composés $ZnSiAs_2$ et $ZnGeAs_2$, l'atome A est un atome de zinc, dans lequel les électrons de valence appartiennent aux orbitales s. Nous représentons le facteur de diffusion de l'atome par

$$f_A - 2A_s f_A^s ,$$

f_A étant le facteur de diffusion atomique de l'atome A, f_A^s le facteur de diffusion correspondant à une orbitale 4s seule, A_s le nombre d'électrons retirés à l'orbitale 4s et transférés aux charges de liaison ou aux autres atomes. Pour l'atome B (Si ou Ge), les électrons de valence appartiennent soit aux orbitales s soit aux orbitales p, et on a donc

$$f_B - 2B_s f_B^s - 2B_p f_B^p .$$

Enfin, pour l'atome C (As), le facteur de diffusion s'écrit

$$f_C - 2C_s f_C^s - 3C_p f_C^p .$$

Il y a quatre liaisons AC s'appuyant sur l'atome A, et quatre liaisons BC s'appuyant sur l'atome B; l'atome C voit deux liaisons de chaque type. Nous choisissons de séparer les contributions de l'atome C aux deux charges de liaisons. Avec $C_s = C_{1s} + C_{2s}$ et $C_p = C_{1p} + C_{2p}$ la conservation du nombre total d'électrons s'écrit

$$\frac{Q_1}{e} = \frac{1}{4}\,2A_s + \frac{1}{2}\,(2C_{1s} + 3C_{1p})\,,$$

$$\frac{Q_2}{e} = \frac{1}{4}\,(2B_s + 2B_p) + \frac{1}{2}\,(2C_{2s} + 3C_{2p})\,.$$

Il y a alors sept coefficients A_s, B_s, etc., mais il n'y a en fait que six coefficients indépendants: les quatre coefficients C_{1s}, C_{1p}, C_{2s}, C_{2p} n'interviennent dans les formules précédentes que par des combinaisons linéaires non indépendantes entre elles. L'utilisation de ces coefficients est commode pour les calculs, mais leur signification physique est limitée: l'affectation de composantes de la densité électronique à l'un ou l'autre des atomes est arbitraire, et seule compte en fait la répartition globale des électrons de valence. Ainsi, pour l'atome B, il y a $2(1 - B_s)$ électrons de valence dont la répartition spatiale est décrite par la partie radiale des orbitales s de l'atome B libre, et $2(1 - B_p)$ électrons par celle des orbitales p. Les électrons manquants (par rapport à l'atome libre) sont ailleurs, soit sur les charges de liaison, soit sur les autres atomes. Les coefficients A_s, B_s, etc. peuvent être négatifs, mais ils sont soumis à plusieurs contraintes: le nombre d'électrons enlevés à un atome ne doit pas excéder le nombre d'électrons de valence de l'atome libre, et le nombre d'électrons d'une charge de liaison doit être positif.

Le modèle comporte finalement douze paramètres: aux six paramètres précédents, il faut ajouter pour chaque liaison i ($i = 1, 2$) les deux largeurs $V^i_\perp$ et $V^i_\parallel$ ainsi que le paramètre de position de la charge U^i (les centres des charges sont situés sur les segments joignants les centres des deux atomes constitutifs de la liaison, et par conséquent leurs coordonnées ne dépendent que d'un seul paramètre U^i compris entre 0 et 1 et prenant la valeur 0,5 si la charge est placée rigoureusement au milieu de la liaison).

3. Conduite des affinements

Dans les deux composés, le calcul des paramètres a été conduit par une méthode d'essais–erreurs; l'accord entre les résultats expérimentaux et théoriques est caractérisé par le calcul des facteurs R et R_w et du facteur G_0F,

$$G_0F = \left[\frac{1}{n - m}\sum w(|F_0| - |F_c|)^2\right]^{1/2},$$

n est le nombre de réflexions indépendantes utilisées, et m le nombre de paramètres à affiner. Nous disposions de 87 réflexions indépendantes pour $ZnSiAs_2$, et 74 pour $ZnGeAs_2$; les pondérations w ont été évaluées à partir des écarts-types sur les intensités, comme indiqué précedemment [4]. La variance des divers paramètres a été évaluée par le calcul de la matrice variance–covariance.

Au départ, nous choisissons pour les paramètres de température β_{ij}, la coordonnée x de l'atome d'arsenic, et les paramètres E_X et E_C les valeurs obtenues lors des affinements des paramètres structuraux [4]; nous autoriserons des variations ultérieures, mais l'introduction des charges de liaison dans le calcul des facteurs de structure n'a apporté aucune modification significative à ces paramètres.

La séparation des largeurs longitudinale et transversale des charges des deux types de liaison ne s'est pas révélée significative dans les deux composés: les différences $(V_{\|} - V_{\perp})$ sont toujours restées très inférieures aux écarts-types calculés par moindres carrés. Nous avons donc retenu $V^i_{\perp} = V^i_{\|} = V^i$ $(i = 1, 2)$: les charges du modèle sont isotropes; nous avions relevé le même effet dans l'arséniure de gallium [3]. Par ailleurs, les calculs ont montré que les meilleurs résultats pour les facteurs R_w étaient obtenus lorsque les coefficients relatifs aux orbitales p étaient très faibles: pour les affinements définitifs, nous avons donc choisi de bloquer ces paramètres à des valeurs nulles: $B_p = C_{1p} = C_{2p} = 0$. Les résultats finaux ont donc été obtenus avec huit paramètres variables seulement: A_s, B_s, C_{1s}, C_{2s}, U^1, U^2, V^1, V^2. Dans le Tableau 1 figurent les valeurs finales de ces paramètres dans les deux composés, ainsi que leurs écarts-types. Nous y donnons également les différents facteurs R correspondant aux deux meilleurs affinements (sans charge de liaison, et avec charges de liaison).

La prise en compte des charges de liaison introduit une variation des facteurs de structure. Lorsque les réflexions sont faibles ou lorsque l'angle de Bragg θ est supérieur à 15°, cette variation est très inférieure à celle dûe aux erreurs de mesure. L'introduction dans le calcul des facteurs R de réflexions dont l'intensité est insensible à la répartition des électrons de valence limite artificiellement leur variation et peut

Tableau 1

Résultats des affinements des paramètres du modèle des charges de liaison utilisé

	$ZnSiAs_2$		$ZnGeAs_2$	
liaison 1: AC	$A_s = 0{,}15$ (0,04) $C_{1s} = 0{,}325$ (0,06) $Q_1 = 0{,}40$ (0,05) e $U^1 = 0{,}69$ (0,02) $V^1 = 0{,}050$ (0,004) nm		$A_s = 0{,}20$ (0,09) $C_{1s} = 0{,}25$ (0,09) $Q_1 = 0{,}35$ (0,05) e $U^1 = 0{,}71$ (0,03) $V^1 = 0{,}045$ (0,006) nm	
liaison 2: BC	$B_s = 0{,}35$ (0,06) $C_{2s} = 0{,}30$ (0,06) $Q_2 = 0{,}475$ (0,04) e $U^2 = 0{,}69$ (0,04) $V^2 = 0{,}0675$ (0,009) nm		$B_s = 0{,}20$ (0,08) $C_{2s} = 0{,}05$ (0,03) $Q_2 = 0{,}15$ (0,01) e $U^2 = 0{,}68$ (0,09) $V^2 = 0{,}060$ (0,013) nm	
facteurs R	sans charge de liaison	avec charges de liaison	sans charge de liaison	avec charges de liaison
1er ensemble de réflexions	$(n = 87)$ $R = 0{,}0076$ $R_w = 0{,}0127$ $G_0F = 3{,}52$	$(n = 87)$ $R = 0{,}0061$ $R_w = 0{,}0099$ $G_0F = 2{,}92$	$(n = 74)$ $R = 0{,}0113$ $R_w = 0{,}0099$ $G_0F = 3{,}51$	$(n = 74)$ $R = 0{,}0099$ $R_w = 0{,}0080$ $G_0F = 3{,}03$
2ème ensemble de réflexions	$(n' = 56)$ $R = 0{,}0074$ $R_w = 0{,}0115$ $G_0F = 3{,}43$	$(n' = 56)$ $R = 0{,}0056$ $R_w = 0{,}0071$ $G_0F = 2{,}32$	$(n' = 53)$ $R = 0{,}0127$ $R_w = 0{,}0117$ $G_0F = 3{,}94$	$(n' = 53)$ $R = 0{,}0104$ $R_w = 0{,}0082$ $G_0F = 3{,}08$

Les différents paramètres sont définis dans le texte, et les écarts-types figurent entre parenthèses. Nous donnons également les valeurs des facteurs R, R_w et G_0F calculés sur les ensembles de n réflexions de départ, puis sur les ensembles plus réduits de n' réflexions (cf. texte)

introduire un biais dans l'utilisation des critères de Hamilton [5]. Nous avons donc calculé un deuxième ensemble de facteurs R, dans lequel n'interviennent que les n' reflexions dont l'intensité est sensible à la répartition des charges de valence (cf. Tableau 1).

4. Discussion

L'introduction des charges de liaison provoque une diminution du facteur R_w. Les tests de Hamilton [5] appliqués à R_w montrent que cette variation est hautement significative, à un seuil de 0,5%, quel que soit le jeu de réflexions utilisé. Toutefois, les facteurs G_0F sont supérieurs à 2,3: ils sont plus élevés que ceux obtenus dans l'étude similaire de GaAs [3] ($G_0F \approx 1$). Ces tests statistiques montrent que, si le modèle de répartition de charges utilisé est moins bien adapté aux composés ternaires qu' à GaAs, il apporte néanmoins une amélioration significative de l'accord entre les facteurs de structure calculés et observés. Il doit donc rendre compte des caractéristiques principales de la répartition effective de la charge de valence.

Dans les résultats finaux, l'occupation des orbitales p est la même que dans les atomes libres: c'est cette situation qui correspond au meilleur accord entre les facteurs de structure observés et calculés. Lorsque les probabilités d'occupation des deux types d'orbitales (s et p) sont considérées comme des variables indépendantes, les variances et les coefficients de corrélation obtenus sont très élevés. La fixation de l'occupation des orbitales p à la valeur observée dans les atomes libres (fixation des coefficients B_p, C_{1p} et C_{2p} à des valeurs nulles) diminue fortement la variance des coefficients d'occupation des orbitales s (cf. Tableau 1). Tout se passe donc comme si les électrons des charges de liaison avaient été prélevés sur les orbitales s des atomes libres. Ces orbitales ont une extension radiale inférieure de 15% à celle des orbitales p. La composante de la densité électronique ayant la symétrie sphérique autour d'un atome a donc une extension sensiblement supérieure dans le cristal à ce qu'elle est dans l'atome libre. Ce résultat semble être une constante dans les semi-conducteurs à coordinence tétraédrique: Aldred et Hart [6] et Scheringer [2] l'ont observée dans le silicium, et nous l'avions retrouvée [3] dans l'arséniure de gallium.

Les largeurs des charges de liaison sont voisines de 0,055 nm dans tous les cas; dans GaAs, la charge de la liaison Ga–As avait une largeur de 0,048 nm. Dans tous ces composés dont les longueurs des différentes liaisons sont voisines (de l'ordre de 0,24 nm) les largeurs des charges restent voisines les unes des autres. Cette largeur moyenne est par ailleurs très supérieure à l'élargissement dû à l'agitation thermique (0,01 nm environ). La largeur de la liaison BC (Si–As ou Ge–As) est plus importante que celle de la liaison AC (Zn–As) dans les deux composés. Ce résultat doit être relié à la différence de nature entre les deux types de liaison: l'étude structurale antérieure [4] nous avait montré en effet que la liaison BC, plus rigide que la liaison AC, présentait un caractère plus fortement covalent.

Cette rigidité de la liaison BC fait que le tétraèdre BC_4 est régulier dans $ZnSiAs_2$ et $ZnGeAs_2$ (cf. [4]). Dans le Tableau 2 nous comparons les caractéristiques (charge

Tableau 2

liaison	Si–Si	Si–As	Ga–As	Ge–As
amplitude $Q(e)$	0,56	0,475	0,15	0,15
largeur V (nm)	0,068	0,0675	0,048	0,060
position U	0,50	0,69	0,56	0,68
longueur d (nm)	0,235	0,235	0,245	0,242

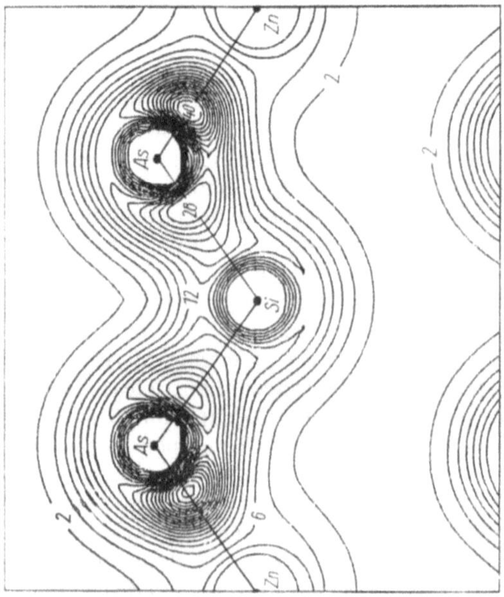

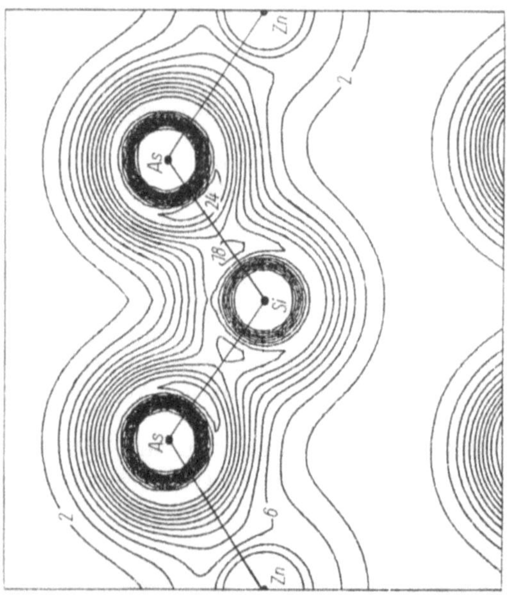

a

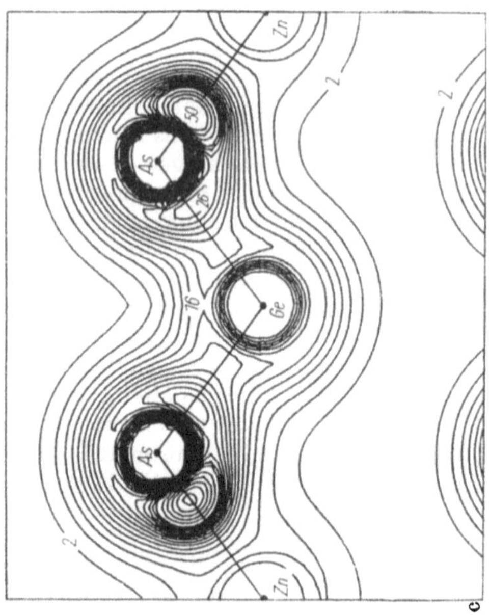

c

Fig. 1. Sections de la densité de charge de valence dans le plan (110). Les contours sont équidistants de $(2e/\Omega_c)$, et les régions des coeurs atomiques ont été laissées en blanc. a) $ZnSiAs_2$, densité sans charge de liaison, b) $ZnSiAs_2$, densité avec charges de liaison; c) $ZnGeAs_2$, densité avec charges de liaison totale Q, largeur V, position U et longueur d) des liaisons Si–As et Ge–As obtenues dans nos composés avec les résultats obtenus par Scheringer [2] sur la liaison Si–Si du silicium, ainsi qu'avec nos résultats précédents [3] sur la liaison Ga–As dans l'arséniure de gallium. (Dans ces deux derniers cas, les tétraèdres sont réguliers par symétrie.) On constate que les amplitudes Q des charges sont corrélées aux longueurs d des liaisons: pour les deux premières, qui font intervenir un atome de silicium au moins, la longueur d est plus courte (0,235 nm) et l'amplitude Q plus forte (0,5e environ); pour les deux autres (Ga–As et Ge–As), la charge est nettement moindre (0,15e), mais la longueur est plus grande. En ce qui concerne les positions des charges de liaison, on constate une différence notable entre la liaison Ga–As et les liaisons Si–As et Ge–As: dans les deux composés ternaires, les charges de liaison sont très décalées vers l'atome d'arsenic (U = 0,69 et 0,68). Les maximums des charges sont ainsi situés à une distance d'environ 0,075 à 0,080 nm du centre de l'atome d'arsenic; ils coincident avec le maximum de la densité de valence de ce dernier. Par contre, dans Ga–As, la charge de liaison reste assez voisine du milieu de la liaison (U = 0,56).

Ce décalage très net des charges dans les deux composés ternaires se retrouve également pour la liaison Zn–As (U = 0,69 et 0,71; cf. Tableau 1). Il faut souligner d'ailleurs les caractéristiques très voisines de cette liaison dans les deux composés (Q = 0,40 et 0,35e, V = 0,050 et 0,045 nm, U = 0,69 et 0,71): cette cohérence entre les résultats obtenus à partir des enregistrements de deux composés différents doit accréditer la fiabilité du modèle.

Finalement, la différence essentielle entre les deux composés ternaires et l'arséniure de gallium réside dans ce décalage net des charges de liaison vers l'atome d'arsenic: cela correspond sans doute à un caractère ionique plus marqué des liaisons dans les composés ternaires.

5. Représentations de la densité de charge de valence

Les Fig. 1a et b représentent les sections de la densité électronique de valence du composé $ZnSiAs_2$ dans le plan (110) qui contient les liaisons. La Fig. 1a correspond à la densité sans charge de liaison, elle utilise les densités calculées pour les atomes libres; la Fig. 1b correspond à la densité calculée avec charges de liaison. L'équidistance des courbes de niveau y est à chaque fois de $2e/\Omega_c$ ($\Omega_c = \frac{1}{8}$ du volume de la maille élémentaire), et nous n'avons pas représenté les oscillations de la densité de valence à l'intérieur des coeurs atomiques (dont l'extension moyenne est de 0,05 nm).

La comparaison des cartes a) et b) fait ressortir l'augmentation de la densité de charge de valence sur les liaisons; elle est plus importante pour la liaison Zn–As (maximum de $40e/\Omega_c$) que pour la liaison Si–As (maximum de $28e/\Omega_c$). En l'absence de charge de liaison (carte a)), les deux liaisons ne sont pratiquement pas différenciées, avec des maximums voisins de $24e/\Omega_c$ toutes les deux. Les centres des charges de liaison coïncident avec les positions des maximums de la densité de charge de l'atome d'arsenic: les maximums de la carte b) sont donc aussi sur ces positions, c'est-à-dire assez proches de l'atome d'arsenic (U = 0,69 environ).

La situation est très similaire dans $ZnGeAs_2$: la densité sans charge de liaison (non représentée) est extrêmement proche de la densité représentée sur la Fig. 1a; la densité avec charges de liaison est représentée sur la Fig. 1c (nous avons supprimé quelques courbes de niveau dans les régions où les pics de densité étaient trop pointus pour rendre plus lisible la figure). Les maximums sont, là aussi, situés assez près de l'atome d'arsenic (U = 0,69) et on retrouve la dissymétrie entre la liaison Zn–As (maximum de $50e/\Omega_c$) et la liaison Ge–As (maximum de $26e/\Omega_c$).

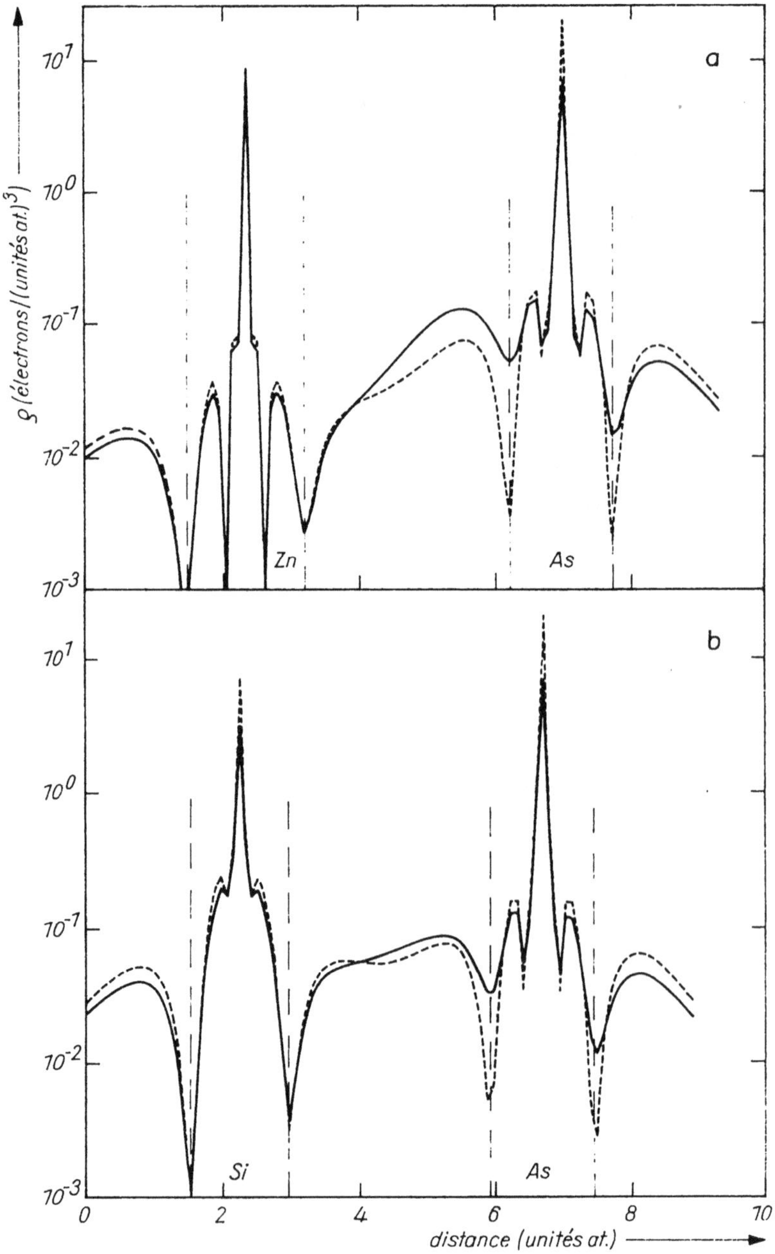
a
b
Zn
As
Si
As
10^1
10^0
10^{-1}
10^{-2}
10^{-3}
ρ (électrons/(unités at.)^3)
0
2
4
6
8
10
distance (unités at.)

Fig. 2

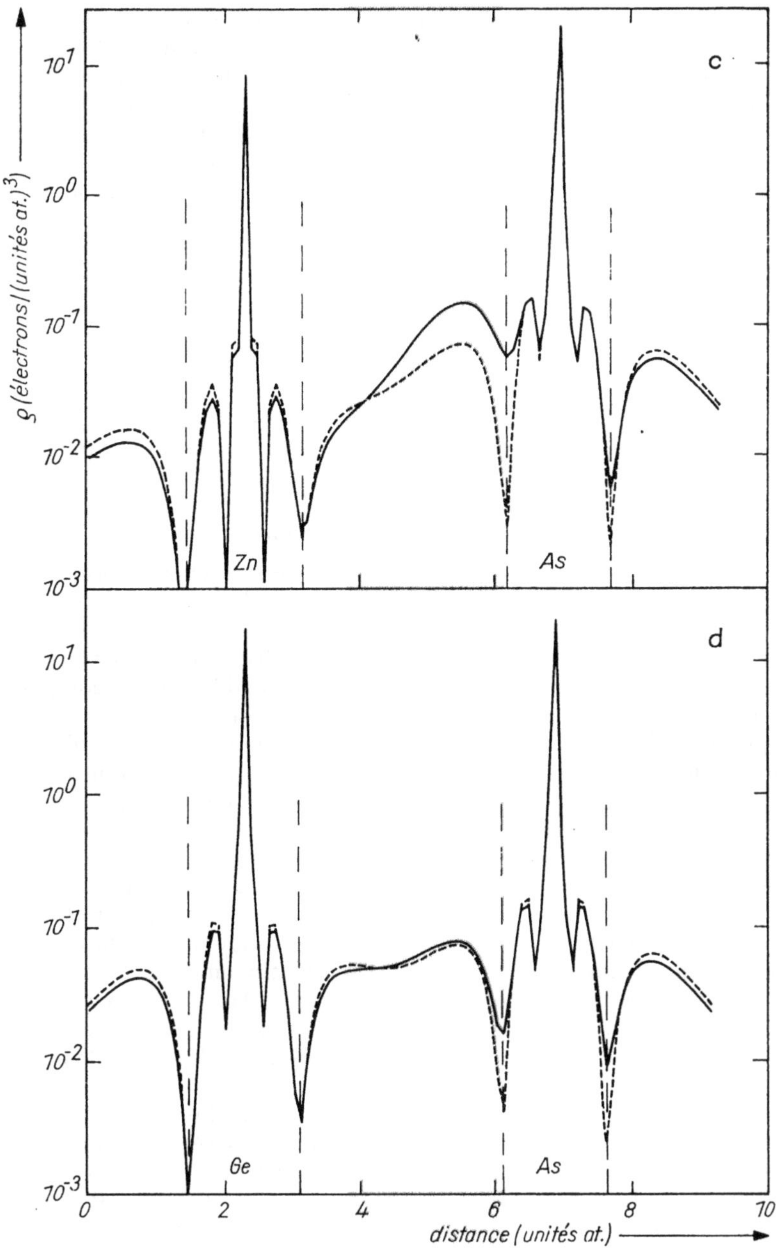

Fig. 2. Courbes représentant les variations de la densité de charge de valence le long des liaisons AC et BC pour les deux composés ABC_2 étudiés. Les distances en abscisses sont en unités at., et les densités en (c/(unités at.)^3), en échelle logarithmique. Les courbes en pointillé représentent la densité en l'absence de charge de liaison, et celles en trait plein la densité avec charges de liaison. Enfin, les lignes en tirets verticales délimitent les régions des coeurs atomiques. a) $ZnSiAs_2$, liaison Zn–As; b) $ZnSiAs_2$, liaison Si–As; c) $ZnGeAs_2$, liaison Zn–As; d) $ZnGeAs_2$, liaison Ge–As

Afin de mieux illustrer encore l'allure de la densité de charge dans la zone des liaisons, nous avons représenté les variations de la densité le long des liaisons, sur les Fig. 2a à d. Les distances sont exprimées en unités at. et les densités de charge en $e/(\text{unités at.})^3$, en échelle logarithmique, afin de pouvoir représenter les oscillations de la densité de charge dans les régions des coeurs atomiques (délimitées par des lignes en tirets verticales sur les figures). La densité de charge calculée sans charge de liaison est représentée en pointillé, et la densité avec charges de liaison de notre modèle en trait plein.

Les Fig. 2a et b représentent les liaisons Zn–As et Si–As dans $ZnSiAs_2$, les Fig. 2c et d les liaisons Zn–As et Ge–As dans $ZnGeAs_2$. Les variations de la densité de charge de valence sont similaires dans les deux composés. Les charges de liaison sont beaucoup plus visibles sur les liaisons Zn–As que sur les liaisons Si–As ou Ge–As, bien que les charges totales soient comparables. Cet effet résulte du fait que la liaison Zn–As a une extension notablement plus faible que les autres.

Le décalage du centre des charges vers l'atome d'arsenic est visible: seul l'atome d'arsenic voit ses minimums de densité de charge (situés aux limites du coeur) fortement modifiés par les charges de liaison.

6. Comparaison avec les résultats de calculs théoriques

Dans les semi-conducteurs ternaires II–IV–V_2, de nombreux calculs théoriques de structure de bandes ont été développés (par exemple [7 à 10]); les deux méthodes de calculs les plus employées utilisent la technique du pseudo-potentiel et celle de la fonctionnelle de densité. Plus rarement, les auteurs ont utilisé leurs résultats pour décrire la densité de charge de valence correspondante: il est possible alors de comparer cetted ensité théorique à la densité expérimentale que nous avons obtenue.

Nous avons reproduit (Fig. 3a) la carte de densité de charge de valence obtenue dans $ZnGeAs_2$ par Baldereschi et al. [8], à partir de calculs de pseudo-potentiel: elle est directement comparable à notre Fig. 1c. Sur la carte de Baldereschi et al., il n'apparaît pas de trous dans les régions des coeurs atomiques, ce qui est une conséquence directe de la méthode de calcul employée: les pseudo-fonctions d'onde obtenues ne représentent plus les fonctions d'ondes vraies dès que l'on s'approche des coeurs atomiques. Dans les autres régions, l'allure de la densité de charge est globalement la même sur les deux cartes; la différence essentielle réside dans le fait que la pseudo-densité de charge calculée (Fig. 3a) montre systématiquement des variations plus atténuées que la densité de charge expérimentale (Fig. 1c). Nous avions déjà constaté un tel phénomène dans l'arséniure de gallium [3], mais ici cette atténuation est telle qu'elle efface presque totalement les différences entre les deux types de liaisons.

Jaffe et Zunger [10] ont utilisé la technique de la fonctionnelle de densité pour étudier la structure électronique d'un certain nombre de composés ternaires. Nous avons reproduit sur la Fig. 3b l'allure de la densité de charge obtenue par ces auteurs dans le composé $ZnSiAs_2$, le long des liaisons Zn–As et Si–As; cette figure est directement comparable à nos Fig. 2a et b, où nous avons utilisé les mêmes échelles. On observe dans l'atome de zinc une différence notable, liée au fait que les auteurs ont inclu les électrons des orbitales 3d du zinc dans leur densité de valence. Cependant, les positions et les amplitudes des maximums sur les liaisons sont les mêmes sur les Fig. 2a, b et sur la Fig. 3b. De plus, on retrouve sur cette dernière le fait, constaté précédemment, que la charge de liaison est beaucoup plus visible sur la liaison Zn–As que sur la liaison Si–As.

Nos résultats expérimentaux sont donc en accord avec les résultats de calculs théoriques, la densité de valence calculée en utilisant la méthode de la fonctionnelle

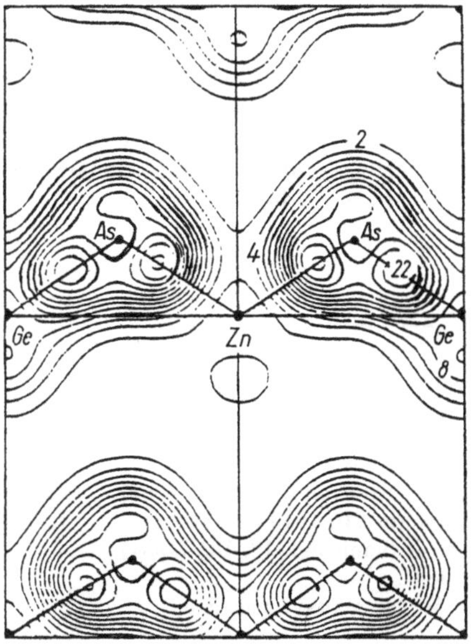

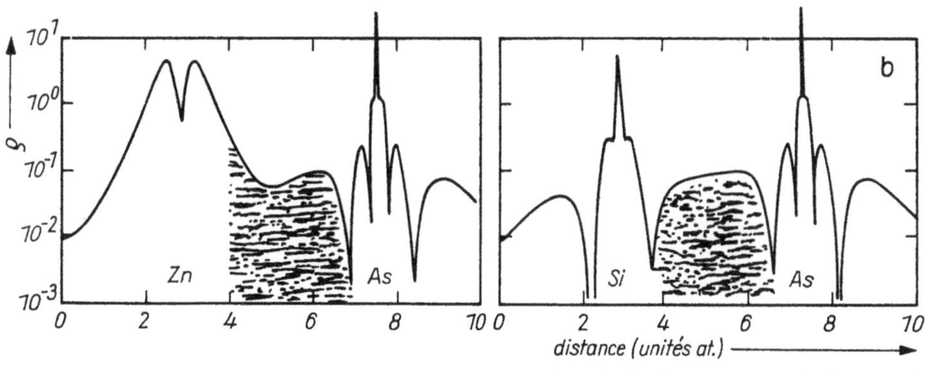

Fig. 3. Densité de charge de valence obtenue à partir de calculs théoriques. a) $ZnGeAs_2$ (d'après Baldereschi et al. [8]): Section de la densité de charge dans le plan (110) calculée par la méthode du pseudo-potentiel. L'équidistance des lignes de niveau y est de $(2e/\Omega_c)$. b) $ZnSiAs_2$ (d'après Jaffe et Zunger [10]): Variations de la densité de charge le long des liaisons Zn–As et Si–As, calculée en utilisant la méthode de la fonctionnelle de densité. Les notations sont analogues à celles de Fig. 2; les électrons 3d du zinc sont inclus dans la densité de valence, et les zones de charges de liaison sont représentées en gris sur la figure

de densité apparaissant plus proche de la réalité que celle calculée par la méthode du pseudo-potentiel.

7. Conclusion

Dans ces composés tétraédriques de poids atomique moyen, la mesure des facteurs de structure en diffraction de R-X permet d'avoir des indications sur la répartition de la densité de charge électronique des électrons de valence. En particulier, l'accumulation des électrons sur les liaisons est très visible, et la précision des mesures permet de mettre en évidence la distinction entre les liaisons AC et BC: ces dernières sont plus fortements covalentes; enfin, la liaison AC (Zn–As) a les mêmes caractéristiques dans les deux composés.

Les résultats expérimentaux sont très voisins de ceux obtenus par les calculs utilisant la méthode de la fonctionnelle de densité. Nous montrerons dans un prochain article [11] qu'ils diffèrent en revanche de ceux obtenus par des calculs utilisant la méthode du pseudo-potentiel: celle-ci conduit en effet à une "pseudo-densité de charge" qui semble peu sensible à la nature des liaisons considérées.

Bibliographie

[1] P. J. E. ALDRED et M. HART, Proc. Roy. Soc. **A332**, 223 (1973).
[2] C. SCHERINGER, Acta cryst. **A36**, 205 (1980).
[3] M. LEVALOIS et G. ALLAIS, Acta cryst. **B42**, 443 (1986).
[4] M. LEVALOIS et G. ALLAIS, phys. stat. sol. (a) **109**, 111 (1988).
[5] W. C. HAMILTON, Statistics in Physical Science, The Ronald Press Co., New York 1964.
[6] P. J. E. ALDRED et M. HART, Proc. Roy. Soc. **A332**, 239 (1973).
[7] N. A. ZAKHAROV et V. A. CHALDYSHEV, Soviet Phys. — Semicond. **16**, 763 (1982).
[8] A. BALDERESCHI, F. MELONI et M. SERRA, Nuovo Cimento **2D**, 1643 (1983).
[9] N. A. ZAKHAROV et V. A. CHALDYSHEV, Soviet Phys. — Semicond. **18**, 135 (1984).
[10] J. E. JAFFE et A. ZUNGER, Phys. Rev. B **30**, 741 (1984).
[11] A. HAIRIE, R. MADELON, E. PAUMIER, G. ALLAIS et M. LEVALOIS, à paraître.

(Received July 13, 1988)

phys. stat. sol. (a) 111, 223 (1989)

Subject classification: 72.20; 61.70; S8.11

Department of Applied Physics, Faculty of Engineering, University of Tokyo[1])

Microscopic Origin of Electric Conduction Anisotropy in Plastically Deformed CdS Single Crystals

By

Y. MERA, K. MAEDA, and S. HYODO

Extremely anisotropic results are reported previously regarding electrical conduction in plastically deformed CdS crystals. This effect is found to be substantially reduced if the distance between the electrodes for conductivity measurements decreases to the submillimeter range. It is revealed by transmission electron microscopy that the dislocations introduced are very heterogeneous and localized in slip bands. The very conspicuous anisotropy and the less conspicuous but still significant anisotropy observed in thin specimens are interpreted in terms of the slip bands which are considered to act as highly resistive layers.

Kürzlich wurden experimentelle Ergebnisse über die extrem anisotrope elektrische Leitung in plastisch verformten CdS-Kristallen berichtet. Es wurde gefunden, daß dieser Effekt wesentlich verringert wird, wenn der Abstand zwischen den Elektroden für die Leitfähigkeitsmessung auf den Submillimeter-Bereich reduziert wird. Mittels Transmissionselektronenmikroskopie wird gefunden, daß die eingeführten Versetzungen sehr heterogen sind und in Gleitbändern lokalisiert sind. Die sehr ausgeprägte Anisotropie und die wenig ausgeprägte aber immer noch deutliche Anisotropie, die in dünnen Proben beobachtet wird, lassen sich durch die Existenz der Gleitbänder interpretieren, die als hochohmige Schichten wirken.

1. Introduction

Plastically deformed CdS single crystals exhibit an anomalous anisotropy in the electrical conduction [1 to 3]: The conductivity measured along the Burgers vector of the primary slip system is extremely larger (by a factor of 10^9 at 200 K) than that measured in the directions perpendicular to the vector. To explain this, Elbaum [1] and Merchant and Elbaum [4] proposed that in the dislocations above stated conductive channels are formed with which one-dimensional energy bands are associated. Hutson [5] attributed this band conduction to an excess donor ionization that screens the piezoelectric polarization in the elastic strain field surrounding dislocations.

Döding and Labusch [6] also inferred the existence of one-dimensional dislocation bands from their elaborate experiments using electrical contacts of μm size. As for the mechanism of the anisotropy in bulk specimens, however, their conclusion was different [2]: Their view is that the deformed specimens should contain straight screw dislocations lying in parallel and that each dislocation is surrounded by a highly resistive cylinder produced by the compensation of shallow donors by acceptors somehow accumulated around the dislocation. The conductivity anisotropy is thus interpreted as percolation of insulating cylinders through a conductive matrix.

Although the Döding and Labusch model can explain the experimental results obtained so far to a satisfactory extent, no direct evidence has been reported for the dislocation arrangements assumed. In order to identify such microscopic origins of

[1]) Hongo, Bunkyo-ku, Tokyo 113, Japan.

the electrical anisotropy, we performed conductivity measurements on plastically deformed CdS single crystals together with the direct observation of their dislocation configuration by transmission electron microscopy (TEM).

2. Experiment

Samples were prepared from a high purity grade, undoped n-type CdS single crystal purchased from Eagle Pitcher Co., Ltd. Samples for deformation were cut from an ingot to rectangular bars of dimensions $2 \times 2 \times 5\ mm^3$ with edges in parallel to [0001], [1$\bar{3}$20], and [$\bar{5}$140] directions as shown in Fig. 1. This orientation, facilitating prismatic slips with a Burgers vector [11$\bar{2}$0] on a (1$\bar{1}$00) plane as a primary slip system, is the same as that used in previous studies [1, 2]. After being polished with concentrated HCl, these samples were plastically deformed at 400 °C by compression along the [$\bar{5}$140] axis in argon atmosphere at a crosshead speed of 0.02 or 0.3 mm/min.

Conductivity measurements were performed on specimens undeformed and those deformed to 2.7, 9.2, and 17.7% in strain. In order to discriminate the effects of heating history from those of deformation, we measured conductivity in a similar manner also for undeformed samples subjected to the same heat treatments. Some of the samples were sliced to thin plates (approximately 0.4 mm thick) with surfaces (0001), (1$\bar{1}$00), and ($\bar{1}\bar{1}$20) (see Fig. 2). Hereafter the specimens with thickness of 2 mm will be called "thick specimens", whereas those of 0.4 mm thickness "thin specimens", respectively.

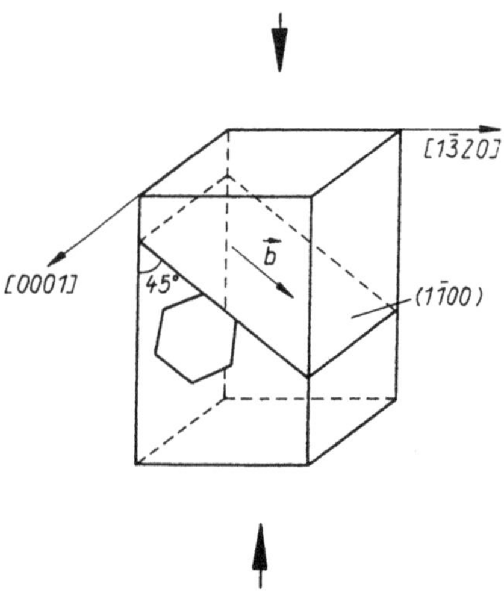

Fig. 1

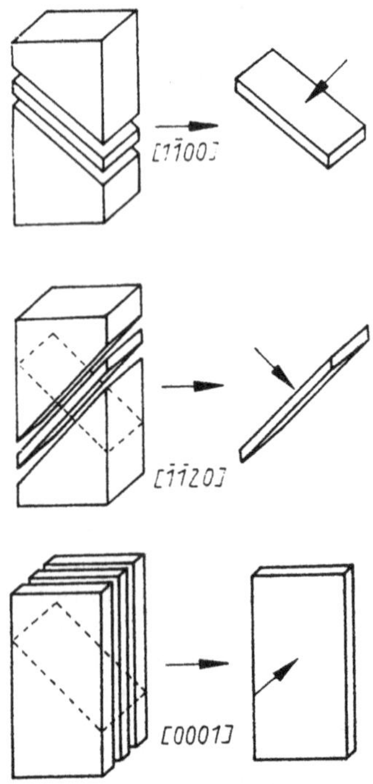

Fig. 2

Fig. 1. Crystal orientation for compression (marked by thick arrows)

Fig. 2. Specimen orientations for conductivity measureme ts and TEM observations

DC electrical conductivity was measured by a two-probe method at temperatures ranging from 2.0 to 290 K. After removal of damaged surface layers by chemical polishing using concentrated HCl, indium spots of 1 mm in diameter were formed on specimen surfaces by evaporation as electrical contacts. For thick specimens, two pairs of contacts were formed on a single specimen in two intersecting directions. These contacts were found to be ohmic in the temperature range used.

After conductivity measurements, the samples were thinned first mechanically, and then chemically using concentrated HCl. Finally they were etched by argon ions until they became thin enough for TEM observations of dislocation configuration with a JEOL-1250 operated at 1000 keV.

3. Results

3.1 Conductivity measurements

Fig. 3 a to c show electrical conductivity of the thick specimens strained to various degrees as a function of temperature; squares represent the data for undeformed specimens. In the case of the specimens undeformed and deformed up to 2.7%, with decreasing temperatures their conductivity increased above 100 K but begins to decrease below 100 K. No conductivity anisotropy was observed in these poorly deformed specimens.

For the specimens more heavily deformed, their conductivity decreased considerably with lowering temperature in the whole temperature range. For the specimens strained to 9.2%, the conductivity peak around 100 K almost disappeared and a conductivity anisotropy by a factor 10 became noticeable between the [1$\bar{3}$20] and [0001] directions. For the specimen most heavily deformed (17.7%), no anisotropy was observed; its conductivity in both (1$\bar{3}$20] and [0001] directions decreased with lowering temperature drastically to rather low levels with an activation energy of 770 meV. These results are, both qualitatively and quantitatively, similar to those obtained by

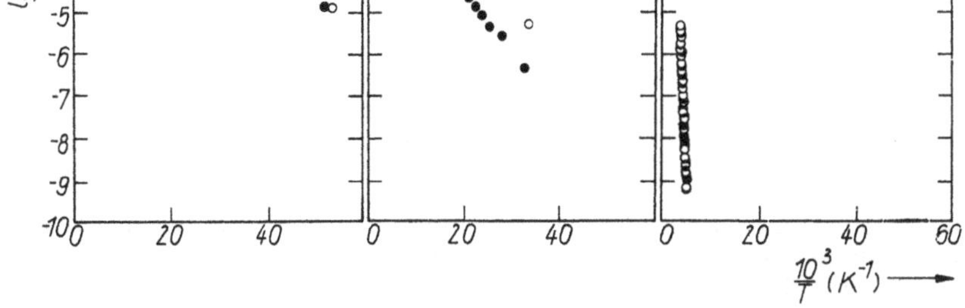

Fig. 3. Temperature variation of electric conductivity of CdS crystals deformed to a) 2.7, b) 9.2, and c) 17.7% in strain. Symbols □, ○, and ● represent data for undeformed reference sample showing isotropic conduction, measurement along [1$\bar{3}$20], and measurement along [0001], respectively. All specimens are "thick"

Döding and Labusch [2], although their samples were considerably different from ours in the carrier concentration, the dopants, and the growth process. The value of activation energy ≈ 0.7 eV in deformed crystals was also reported previously [3].

Fig. 4 shows conductivity versus temperature relations obtained from the most heavily deformed samples. While the data measured in the [0001] direction were obtained from a thick specimen, those in [$\bar{1}\bar{1}20$] from a thin one. One may see that these two sets of data are very different, i.e., by a factor of 10^6 at 250 K. Fig. 4 suggests further that, if one pays no heed to the effect of specimen thickness, one should believe the presence of a conspicuous anisotropy in electrical conduction in deformed CdS single crystals, as was previously reported. However, it should be noted that the conductivity of thin specimens we measured in the [0001] direction was much larger than that of thick ones. This indicates that the conductivity values actually determined can depend on the sample thickness or on the distance between the electrodes. Although no reference was made to the specimen thickness in the previous report [2], it seems to us highly probable that the anisotropy proposed can be ascribed mostly to a difference in the specimens thickness, rather than to the current flow direction.

Fig. 5 shows the conductivity versus temperature relation regarding their directions of the most heavily deformed crystals, which were approximately of the same thickness (≈ 0.4 mm). Although the extraordinary anisotropy has disappeared, there still remains an appreciable anisotropy, i.e., a conductivity difference by a factor 10 between [$1\bar{1}00$] and the other two directions. The very cause of this anisotropy will be clarified in the next section through TEM observations of dislocation configurations in deformed crystals.

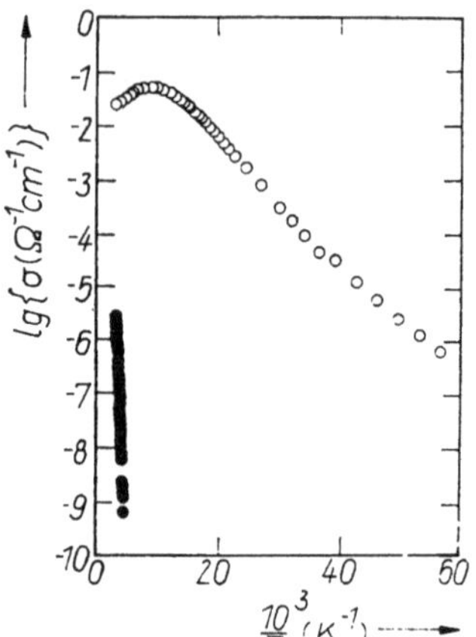

Fig. 4

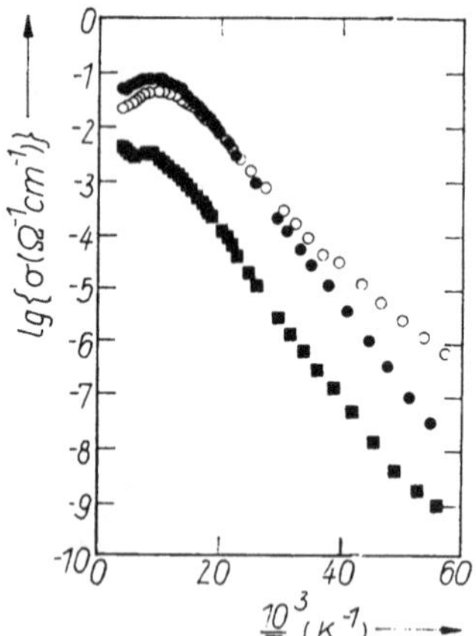

Fig. 5

Fig. 4. Temperature variation of electric conductivity of CdS crystals most heavily deformed up to 17.7%. ○ marks indicate conductivity measured along [$\bar{1}\bar{1}20$] obtained for a thin specimen and ● marks along [0001] for a thick specimen

Fig. 5. Inconspicuous but still significant anisotropy observed in thin specimens. Note that conductivity along ● [0001] and that along ○ [$\bar{1}\bar{1}20$] are of nearly the same magnitude and both higher than that measured along ■ [$1\bar{1}00$]

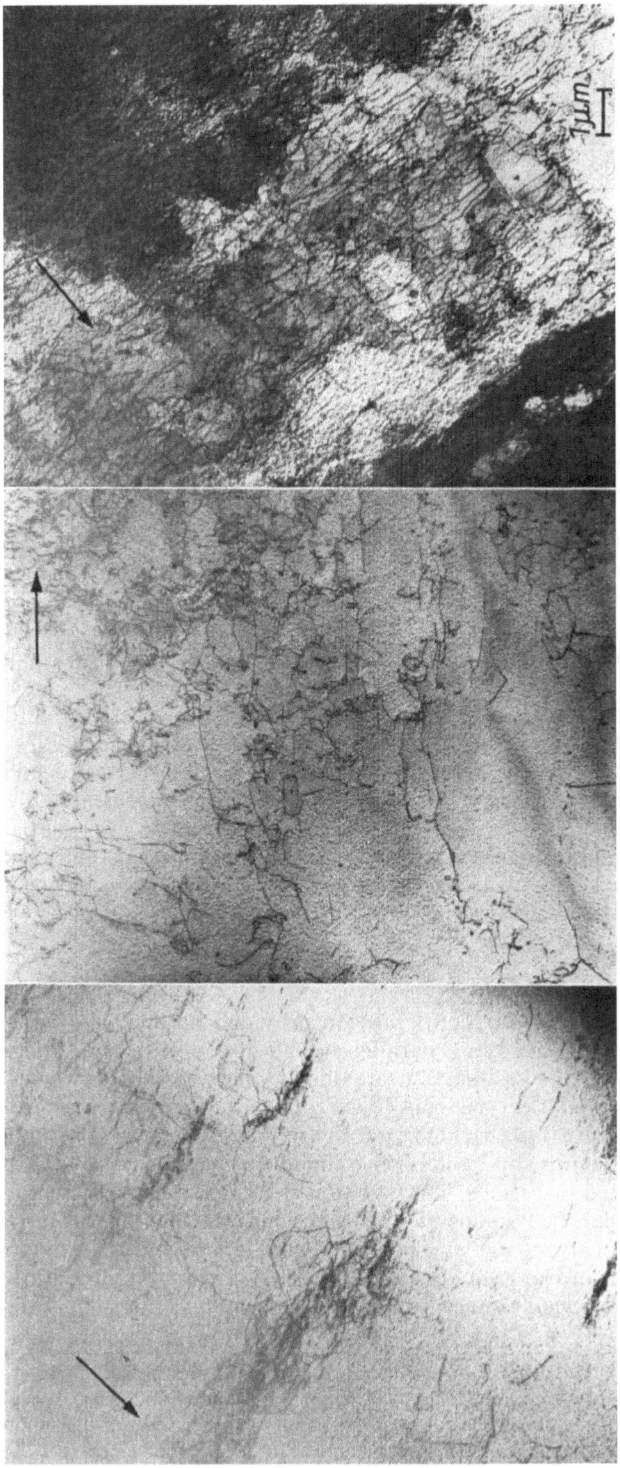

a b c

Fig. 6. Typical dislocation configurations in the primary $(\bar{1}100)$ slip plane. a) 4.3, b) 7.8, c) 17.6%. (The arrows indicate the direction of the Burgers vector)

3.2 *TEM observations*

Fig. 6 represents typical transmission electron micrographs regarding the specimens deformed to 4.3, 7.8, and 17.6%. Dislocations in the primary slip planes $(1\bar{1}00)$ are seen to be in general rather tangled. From contrast experiments under various diffraction conditions, the Burgers vectors of the main dislocations introduced were identified; they lay in $[11\bar{2}0]$ direction and had no component in c-axis. This identification is in agreement with the fact that no expansion of the samples along the c-axis was induced by compression. The slip planes were found to be mainly $(1\bar{1}00)$, as expected also from the orientation of the compression system used. Although both screw and edge components were always observed, there was no parallel array of a single type of dislocations in any specimens.

Fig. 7a shows a light field image of a specimen deformed to 17.7%, viewed from a direction parallel to the primary slip plane. The stripe contrasts indicate that the crystal is slightly bent alternately between two successive slip bands. The right-hand side of Fig. 7a represents the selected-area diffraction patterns corresponding to the regions in the left-hand side as indicated by arrows. From these patterns the crystal was found to be corrugated around [0001] axes parallel to the stripe direction and to have corrugation angles of a few degrees. Fig. 7b, another light field image of the same specimen but of a different region, shows that edge dislocations on the secondary slip system $[\bar{2}110]$ $(0\bar{1}10)$ pile up against the primary slip bands observed in black contrast. Hence, the corrugation is considered to arise from the pile-up of edge dislocations; this mechanism is illustrated in Fig. 8. It is suggested from these results that in the case of the CdS single crystals the used deformation must have proceeded quite heterogeneously.

Fig. 9 shows the dislocation configuration in (0001) basal planes in a sample deformed to 14.5%. Although the deformation axis permitted no dislocation glides in the basal plane, one can notice in Fig. 9 many dislocations actually introduced in those planes. According to our view, this unexpected result is due to a small misorientation (which is inevitable developed as deformation proceeds) and to cross slips of screw dislocations in the primary prismatic slip system to the basal system. Fig. 9 shows also that the dislocations in the basal planes are distributed more uniformly than those in the secondary prismatic system (Fig. 7 a and b).

4. Discussion

Our TEM observation of the dislocation configuration provided no evidence for the dominant presence of screw segments lying parallel to the direction of the Burgers vector. This is inconsistent with the assumption previously made to explain the conductivity anisotropy in deformed CdS crystals [2]. From our observation, the dislocation configuration in the crystal is characterized by irregular and severely tangled networks which are localized within slip bands (the bands being separated by dislocation-free layers), as is illustrated in Fig. 10. Thus we consider that such an anisotropic texture of deformed crystals can be responsible for their conductivity anisotropy.

If, as often presumed, dislocations act as acceptors or collect acceptor impurities around them, carrier (electron in the case of n-type crystals) depletion cylinders will form surrounding the dislocations. Otherwise, heavy deformation may introduce

Fig. 7. TEM images viewed from $[\bar{1}\bar{1}20]$ direction parallel to the primary slip plane. a) Selected-area diffractions indicate that the specimen is slightly corrugated around [0001] axis parallel to the stripe direction. b) The regions between the primary slip bands are almost free of dislocations except for a few edge dislocations in the secondary slip system $[\bar{2}110]$ $(0\bar{1}10)$ (deformed to 17.7%) ▶

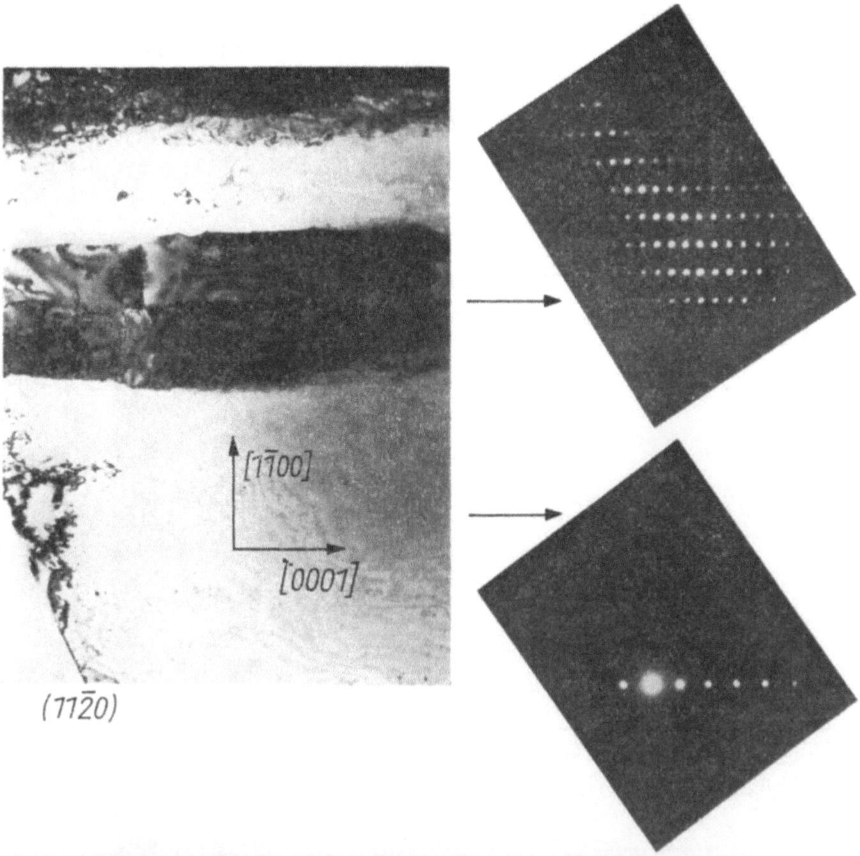

a

b

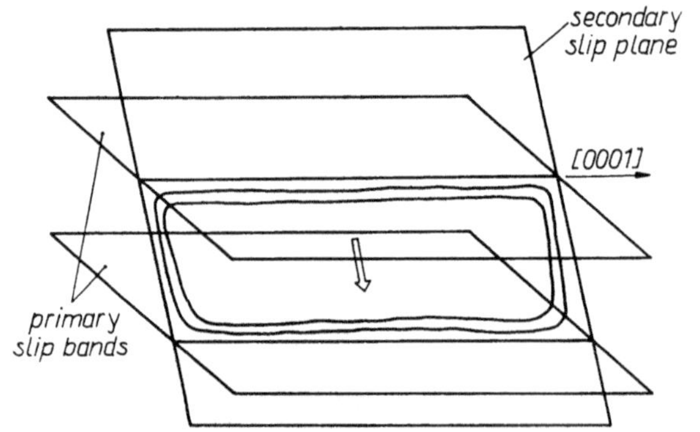

Fig. 8. Edge dislocations in the secondary slip plane piling up against the primary slip bands give rise to the corrugation shown in Fig. 7

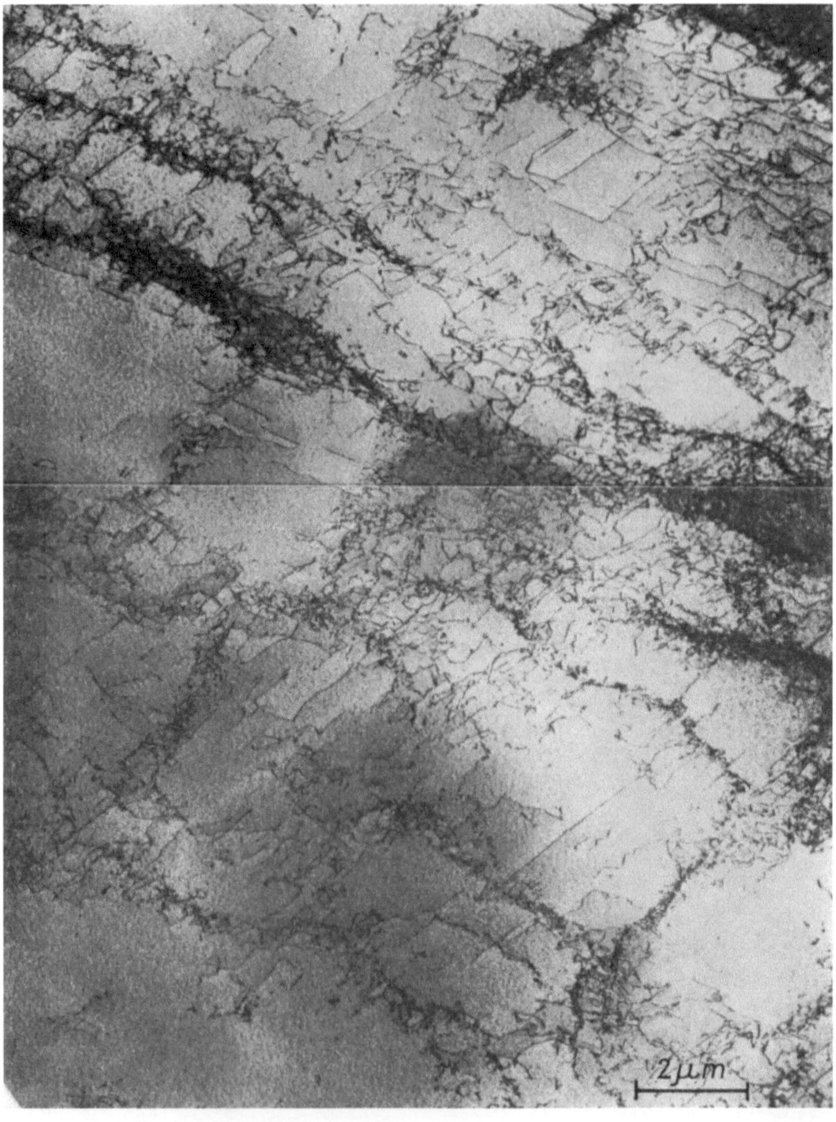

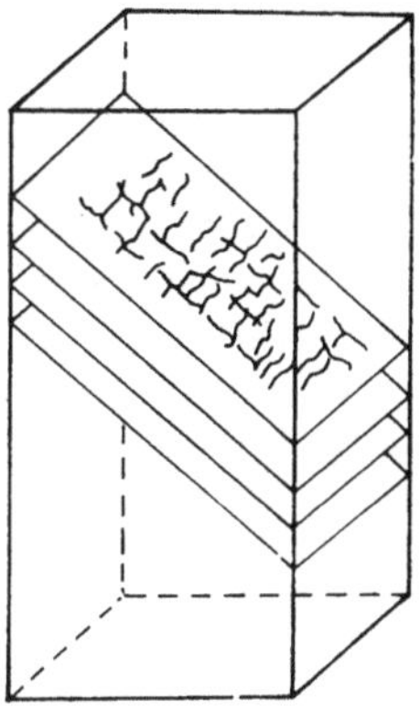

Fig. 10. Irregular and severely tangled networks of dislocations localized within the primary slip bands are separated by dislocation-free layers

native point defects that act as acceptors. Whatever the mechanism may be, slip bands will then form highly resistive layers in the conductive matrix. It is natural that, in the presence of such resistive layers, current flow across the layers should be blocked while that along the layers should be rather free. The anisotropic results shown in Fig. 5 can be interpreted in terms of the current direction to the orientation of the glide plane rather than to that of the Burgers vector: The $[1\bar{1}00]$ direction exhibiting low conductivity is perpendicular to the $(1\bar{1}00)$ primary slip plane whereas both [0001] and $[\bar{1}\bar{1}20]$ directions exhibiting comparatively high conductivity are parallel to the main slip plane.

As shown in Fig. 3b, a conductivity difference by a factor 10 was observed between the [0001] and $[1\bar{3}20]$ directions in thick specimens deformed to 9.2%. Obviously this result is incompatible with the above explanation which predicts higher conductivity in [0001] than in $[1\bar{3}20]$, since the carrier paths in $[1\bar{3}20]$ are blocked by the resistive slip bands.

Contrary to our expectation, many dislocations were introduced in the basal planes. The thickness dependence of conductivity perpendicular to the basal plane (Fig. 4) implies that the deformed crystals are not homogeneous but had some oriented structure at an intermediate scale. Since the dislocations in the basal planes are distributed rather uniformly, they may overlap between the electrodes in the case of thick specimens. If resistive layers parallel to (0001) planes arise also from the basal dislocations, carrier flow across the planes will be considerably obstructed. Thus we attributed the anisotropy observed in thick specimens to the resistive layers formed along basal planes; the layers are considered to be separated at a sub-millimeter distance. In this way, the fact that anisotropy was not observed in the thick and most heavily deformed specimens can be explained by a complete blocking of current in the direction of [0001] by the basal slip bands and in $[1\bar{3}20]$ by the prismatic ones.

Vyvenko and Tulev [3] showed that significant anisotropy is observed only when the initial carrier density of the deformed crystal is within a relatively narrow range of 9×10^{15} to 5×10^{16} cm^{-3} which covers that of the present crystal ($\approx 10^{16}$ cm^{-3}). However, they observed pronounced anisotropy between [0001] and $[\bar{1}\bar{1}20]$ in thin specimens (≤ 1 mm) whereas we obtained a small and rather contrary result in the present study (Fig. 5). Microscopic observations of the dislocation configuration may serve for clarifying the cause of this discrepancy.

◀Fig. 9. Configuration of dislocations unexpectedly introduced in the (0001) basal planes in a specimen deformed to 14.5%

5. Summary

The conspicuous anisotropy of electric conduction in plastically deformed CdS crystals was found to be substantially reduced, when two electrodes for conductivity measurements were close to each other at a distance of sub-millimeter scale. TEM observations revealed that the dislocation configuration in each slip plane was very irregular but nearly all of the dislocations were localized in slip bands. With the assumption that the slip bands are electrically resistive, the inconspicuous anisotropy in thin specimens can be interpreted by the resistive slip bands belonging to the primary prismatic slip system: Conduction paths perpendicular to the bands are intercepted by resistive layers while they are left unaffected in directions parallel to them. We consider that the conspicuous anisotropy in thick specimens is also due to highly resistive basal slip bands with larger separations.

Acknowledgements

This work was supported partly by a Grant-in-Aid for Scientific Research from the Ministry of Education, Science, and Culture and by the Miyashita Foundation for Material Scientific Research.

References

[1] C. Elbaum, Phys. Rev. Letters **32**, 376 (1974).
[2] G. Döding and R. Labusch, phys. stat. sol. (a) **68**, 143 (1981).
[3] O. F. Vyvenko and A. V. Tulev, Soviet Phys. – Solid State **29**, 487 (1987).
[4] P. Merchant and C. Elbaum, Phys. Rev. B **19**, 2992 (1979).
[5] A. R. Hutson, Phys. Rev. Letters **46**, 1159 (1981).
[6] G. Döding and R. Labusch, phys. stat. sol. (a) **68**, 469 (1981).

(Received July 14, 1988)

phys. stat. sol. (a) **111**, 233 (1989)

Subject classification: 72.60; 72.20; S1.61

Department of Electrical Engineering, Salesian Polytechnic, Tokyo[1])

Semiconducting and Thermoelectric Properties of Sintered Iron Disilicide

By

T. KOJIMA

The electrical resistivity, thermoelectric power, and Hall coefficient of sintered iron disilicide, $Fe_{1-x}Mn_xSi_2$, and $Fe_{1-y}Co_ySi_2$ are measured over the temperature range from 77 to 1400 K. Regardless of the purity of raw materials, $Fe_{1-x}Mn_xSi_2$ and $Fe_{1-y}Co_ySi_2$ show semiconducting properties below the semiconductor-to-metal transition temperature T_c. Mn and Co atoms in iron disilicide act as acceptor and donor, respectively. The energy gap of $FeSi_2$ at 0 K is 1.00 eV, deduced from the temperature dependence of the Hall coefficient in the intrinsic region. The energy gap of Mn-doped $FeSi_2$ decreases from 0.95 to 0.83 eV with increasing Mn amount. Above 150 K the lg ϱ versus $1/T$ dependence of Mn-doped $FeSi_2$, exhibits an S-like decaying curve. This cannot be explained completely by ordinary band conduction, but can be done by small polaron conduction with crystalline distortions.

An den gesinterten Eisendisiliziden $Fe_{1-x}Mn_xSi_2$ und $Fe_{1-y}Co_ySi_2$ werden spezifischer Widerstand, Thermospannung und Hallkoeffizient im Temperaturbereich zwischen 77 und 1400 K gemessen. Unabhängig von der Reinheit der Ausgangsmaterialien zeigen $Fe_{1-x}Mn_xSi_2$ und $Fe_{1-y}Co_ySi_2$ unterhalb der Temperatur T_c beim Halbleiter–Metall-Übergang halbleitende Eigenschaften. Mn- bzw. Co-Atome im Eisendisilizid wirken als Akzeptor bzw. Donator. Die Energielücke von $FeSi_2$ beträgt bei 0 K 1,00 eV, was aus der Temperaturabhängigkeit des Hallkoeffizienten im Eigenleitungsgebiet ermittelt wird. Die Energielücke von $FeSi_2$ vermindert sich von 0,95 auf 0,83 eV mit der Zunahme der Mn-Dotierung. Oberhalb 150 K zeigt die $1/T$-Abhängigkeit von lg ϱ einen S-förmigen Verlauf. Dieses Verhalten des spezifischen Widerstandes ist im Rahmen des gewöhnlichen Bändermodells schwer verständlich, es kann jedoch durch kleine Polaronen und Kristalldeformationen gedeutet werden.

1. Introduction

Stoichiometric iron disilicide, β-$FeSi_2$, possesses stable semiconducting properties below 1250 K. Above this temperature, however, it transits to the metallic phase [1 to 4] which is a eutectic alloy composed of ε-FeSi [7] and α-$Fe_{1-\square}Si_2$ ($\square$ means Fe vacancy concentration) [5, 8]. As shown in the phase diagram of Fig. 1 [6], the silicon content in the eutectic alloy at the eutectic point corresponds to that of $FeSi_2$, and the semiconducting phase is obtained by annealing the eutectic alloy below 1259 K. The semiconductor becomes p-type by doping with Al [9 to 11] or Mn [12], and n-type by doping with Co [9 to 11]. The sintered $FeSi_2$ has higher mechanical strength and heat resistance than chalcogenides, and also it is chemically stable at high temperatures, so it proves to be a very useful thermoelectric material for thermoelectric generators operating in high temperature atmosphere without any protection. $FeSi_2$ therefore has a great potential for commercial use.

On the other hand, $FeSi_2$ is interesting from the physical point of view as well, and its thermoelectric properties have been reported by several authors [10 to 14].

---

[1]) 2-35-11, Igusa, Suginami-ku, Tokyo 167, Japan.

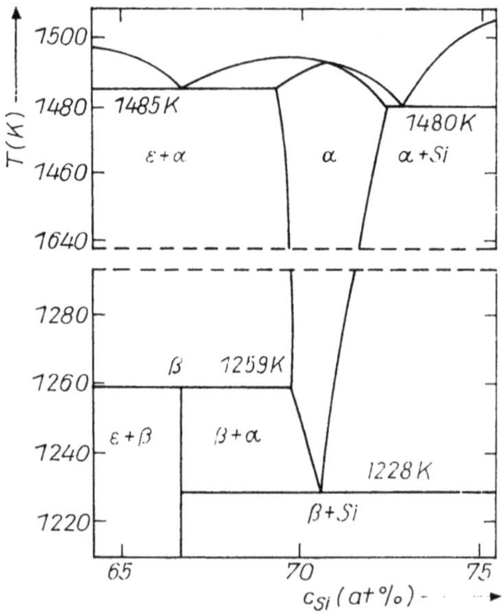

Fig. 1. Equilibrium phase diagram of the Fe–Si system [6]

The conduction mechanism of $FeSi_2$ doped with Al or Co was investigated by Birkholtz and Schelm [10] who concluded that p-type and n-type $FeSi_2$ were governed by band and polaron conductions, respectively. Moreover, it was noted that the thermoelectric power of p-type $FeSi_2$ doped with Mn is rather high, and its conductivity is interesting especially near the semiconductor-to-metal transition temperature [12, 15]. However, there are no detailed investigations concerning the conduction mechanism of this material over the temperature range from 300 to 1000 K, which is important for a thermoelectric material. Therefore, it is essential to clarify in detail the electrical properties of $FeSi_2$.

The present work was undertaken in order to collect intrinsic data on the thermoelectric properties of $FeSi_2$, which are useful when applying this material in thermoelectric generators. Undoped, Mn-doped, and Co-doped $FeSi_2$ were prepared from two series of raw materials of high and low purities, and the electrical resistivity, thermoelectric power, and Hall coefficient were measured as functions of temperature. We will discuss the semiconducting properties of Mn-doped specimens and clarify the hole conduction mechanism of this semiconductor.

2. Experimental

High purity specimens were synthesized from zone-refined iron (99.99% pure), n-type silicon single crystal (99.9999% pure), electrolytic manganese (99.99% pure), and electrolytic cobalt (99.99 pure). These materials were weighed to compose $FeSi_2$, $MnSi_2$, and $CoSi_2$ and then arc-melted in argon atmosphere. The specimens of $Fe_{1-x}Mn_xSi_2$ and $Fe_{1-y}Co_ySi_2$ were obtained by remelting $FeSi_2$ with $MnSi_2$ or $CoSi_2$ in the arc-melting furnace. The remelted ingots were crushed into powder with a particle size of about 1.8 μm and hot-pressed in argon atmosphere at 1425 K for 8 min under a pressure of 29.4 MPa. These sintered specimens were encapsuled in an evacuated quartz tube and homogenized at 1433 to 1453 K for 10 h and then annealed at 1073 K for 100 h. Each of them was confirmed to be single phase by an X-ray powder diffractometer. The content of Mn and Co in the specimens is shown in Table 1.

Low purity specimens were prepared from electrolytic iron, manganese, cobalt, and metallic silicon with industrially low purity (better than 98%). The starting material

Table 1
Composition of $Fe_{1-x}Mn_xSi_2$ and $Fe_{1-y}Co_ySi_2$

specimen	composition		doping concentration (10^{20} cm^{-3})		purity*)	symbol
No.	x	y	Mn	Co		
1	0.00		0.00		h	○
2	0.03		7.97		h	△
3	0.06		15.9		h	□
4	0.00		0.00		l	●
5	0.05		13.3		l	▲
6	0.08		21.2		l	■
7		0.05		13.3	h	▽
8		0.10		26.5	h	◇
9		0.05		13.3	l	▼
10		0.10		26.5	l	◆

*) h and l denote high and low purities, respectively.

was weighed to compose $FeSi_{2.05}$ and then melted in an induction furnace. At this point, some more Si than the stoichiometric composition was added to compensate for contaminations in Fe (2 mass%), which cannot be avoided in the powdering process. Specimens doped with Mn or Co are made by mixing the appointed quantities of $MnSi_2$ or $CoSi_2$ in the vacuum melting process. The ingots thus made were crushed by a stamp-mill and then powdered to the size of 2 to 3 μm in diameter by an iron ball-mill. Polyvinyl alcohol (PVA) and colloidal paraffine of 1 mass % each was added to these powder raw materials as binder and lubricant. After granulation, they were cold-pressed under a pressure of 2.94×10^2 MPa. The mean packing density of the green samples ($7 \times 10 \times 35$ mm^3) was 3.14×10^3 kg/m^3. The green samples were annealed in air from 300 to 700 K at a heating rate of 80 K/h to remove the binder and the lubricant, and then sintered in vacuum (1.33×10^{-1} Pa) at (1428 ± 3) K for 10.8×10^3 s (3 h). Since as-sintered specimens are of the high temperature metallic phase, each of them was encapsuled in a quartz tube and annealed at 1073 K for 3.6×10^5 s (100 h).

The electrical resistivity, thermoelectric power, and Hall coefficient of both high and low purity specimens, were measured in the temperature range from 100 to 1400 K, using the dc method as reported by Sakata and Tokushima [16]. The Hall coefficient was measured under an applied magnetic field of 0.5 T.

3. Experimental Results

3.1 Temperature dependence of electrical resistivity

The temperature dependence of the electrical resistivity ϱ is shown in Fig. 2 a and b for high and low purity specimens, respectively. For undoped specimens of both high and low purities the resistivity changes lie on the same curve. It indicates that even such low purity material possesses semiconducting properties. An abrupt resistivity drop of one order in magnitude is observed near 1259 K, at which the semiconductor-to-metal transition occurs [12, 15]. The resistivities of the specimens above 500 K fall quite on the same curve, and the plot of lg ϱ against the reciprocal of absolute temperature $1/T$ is a straight line in the temperature range from 500 to 1000 K. The energy gap E_g obtained from the slope is 1.0 eV. Though it is difficult to judge only from the

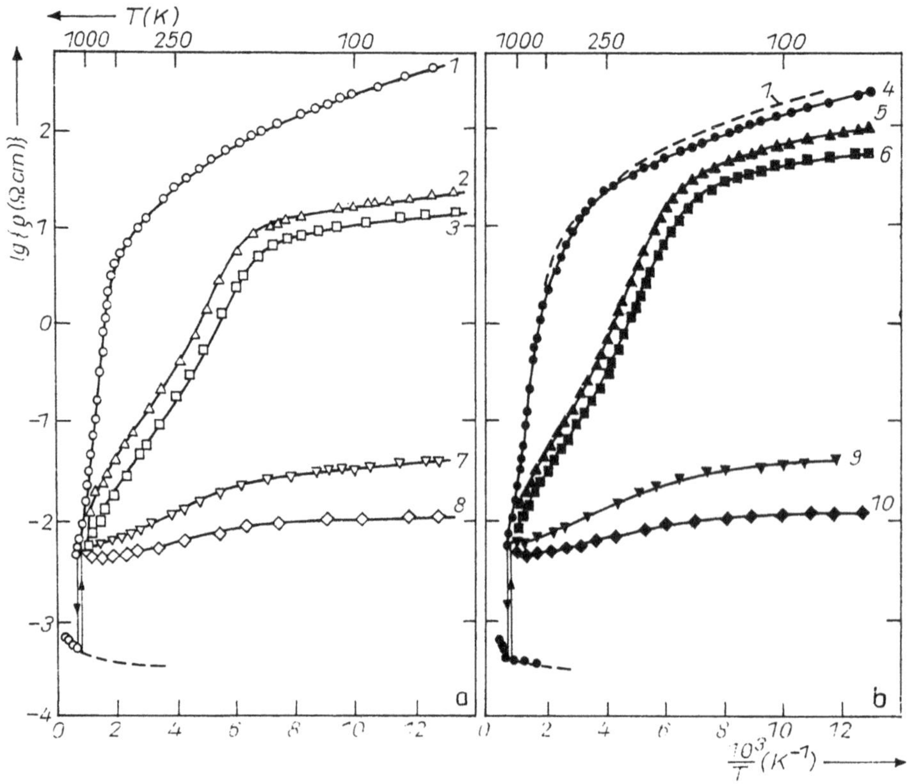

Fig. 2. Temperature dependence of the electrical resistivity for $Fe_{1-x}Mn_xSi_2$ and $Fe_{1-y}Co_ySi_2$ a) with high purity and b) with low purity. (1) $x = 0.00$, (2) $x = 0.03$, (3) $x = 0.06$, (7) $y = 0.05$, and (8) $y = 0.10$. b) (4) $x = 0.00$, (5) $x = 0.05$, (6) $x = 0.08$, (9) $y = 0.05$, and (10) $y = 0.10$

figure, the energy gap tends to decrease slightly above 1000 K, and it becomes 0.98 eV around the semiconductor-to-metal transition temperature T_c. Below 500 K, the resistivity of the low purity specimen is less than that of the high purity specimen because of extrinsic conduction resulting from unknown impurities contained in the starting materials.

Although extrinsic conduction due to unknown impurities with a shallow level of 0.010 to 0.016 eV is observed for Mn-doped specimens below 150 K, the changes of ϱ of both high and low purity specimens are different from the undoped specimen in the temperature range 150 to 900 K. The resistivity curve shows an S-like decaying behavior (indicated by 2 and 3 in Fig. 2a and by 5 and 6 in Fig. 2b), but this tendency is less predominant for high purity specimens. The ordinary band theory cannot explain this resistivity behavior sufficiently. Some additional conduction mechanism must be considered. Above 1000 K (in the intrinsic region) the slope of the lg ϱ versus $1/T$ plot decreases with increasing Mn content, resulting in an energy gap E_g of 0.95 to 0.83 eV.

The ϱ behavior of the Co-doped specimen is found to be independent of the unknown impurity content in the starting material as shown in Fig. 2a (curves 7, 8) and 2b, (curves 9, 10). The influence of Co doping is present in the temperature region from 150 to 900 K accompanying the S-like decaying behavior of ϱ which is less severe than in the case of Mn doping. Donor ionization energies E_d deduced from the slopes of lg ϱ versus $1/T$ plots are 0.0245 eV for $y = 0.05$ and 0.013 eV for $y = 0.10$, and these energies decrease also with increasing Co content. In the same temperature region the slope of lg ϱ for Co-doped specimens is comparatively gentler than that of Mn-doped specimens. Thus the ionization energy of the donor is smaller than that of the acceptor.

3.2 Temperature dependence of the Hall coefficient

Fig. 3 shows plots of the Hall coefficient R against the reciprocal temperature for undoped, Mn-, and Co-doped specimens. R of undoped specimen is independent of the purity of starting materials from 500 to 1000 K. The plot of lg $RT^{3/2}$ versus $1/T$ fits well a straight line in that temperature region. The energy gap obtained from the slope is in good agreement with that given by the temperature dependence of ϱ. On the other hand, R values below 500 K depend on the purity of the starting material. These impurities act as acceptor for the high purity specimen and donor for the low purity one.

The Hall coefficient of the Mn-doped specimens is positive over the temperature range measured, and the plot of lg $RT^{3/2}$ against $1/T$ is a straight line in the region from 200 to 500 K as shown in Fig. 4. From this slope, the activation energy E_a of the acceptor level is estimated to be 0.102 and 0.100 eV for high purity specimens of $x = 0.03$ and 0.06, and 0.123 and 0.118 eV for low purity specimens of $x = 0.05$ and 0.08, respectively. These acceptor energies of each specimen decrease with increasing

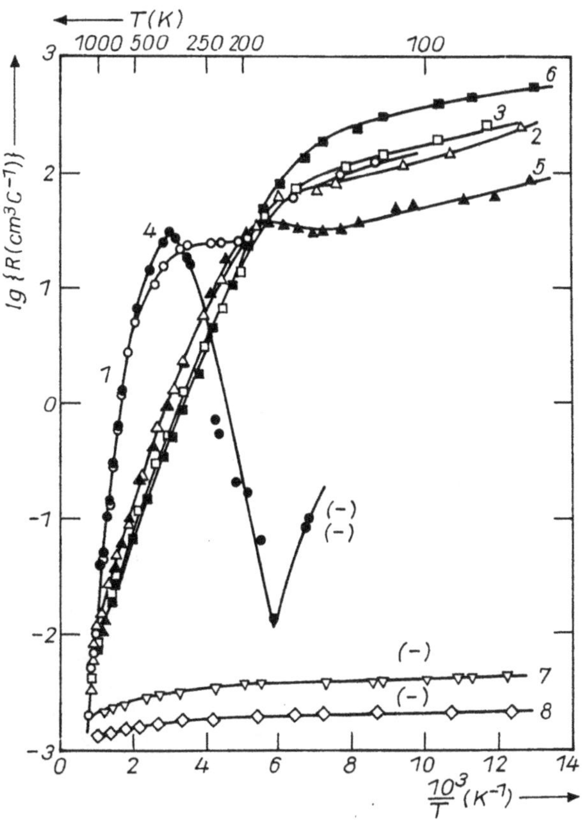

Fig. 3. Temperature dependence of the Hall coefficient for $Fe_{1-x}Mn_xSi_2$ and $Fe_{1-y}Co_ySi_2$ (open and closed symbols show high and low purity specimens, respectively). (1) $x=0.00$, (2) $x=0.03$, (3) $x=0.06$, (4) $x=0.00$, (5) $x=0.05$, (6) $x=0.08$, (7) $y=0.05$, and (8) $y = 0.10$

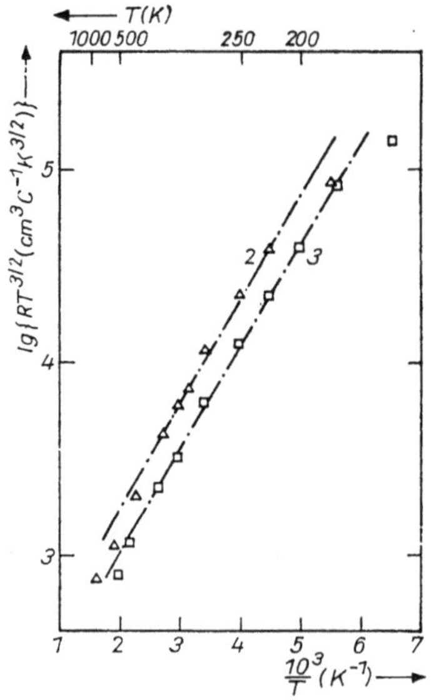

Fig. 4. Plots of lg $RT^{3/2}$ against reciprocal temperature for $Fe_{1-x}Mn_xSi_2$ (high purity). (2) $x = 0.03$ and (3) $x = 0.06$

Mn content. The hole mobility μ_h estimated from R and ϱ is 8.0 cm^2/Vs at room temperature, which is four times larger than that of Al-doped p-type specimens studied by Birkholtz and Schelm [10].

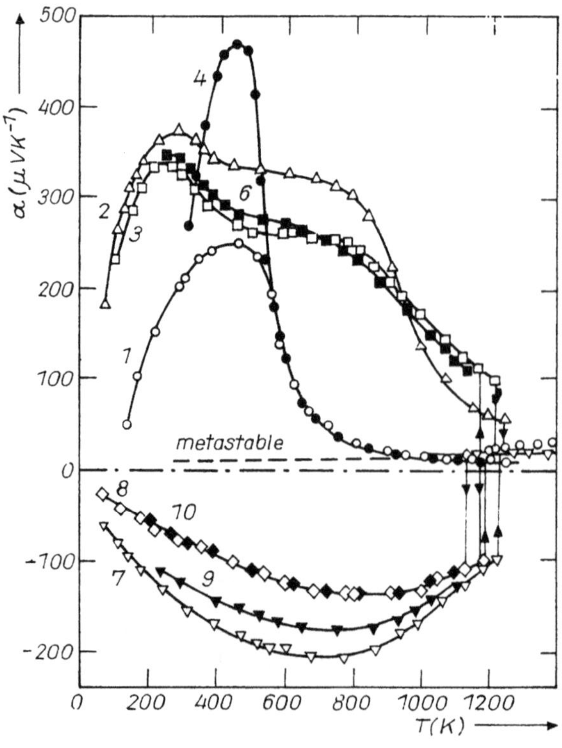

Fig. 5. Temperature dependence of the thermoelectric power for $Fe_{1-x}\cdot Mn_xSi_2$ and $Fe_{1-y}Co_ySi_2$ (open and closed symbols show high and low purity specimens, respectively). (1) $x = 0.00$, (2) $x = 0.03$, (3) $x = 0.06$, (4) $x = 0.00$, (6) $x = 0.08$, (7) $y = 0.05$, (8) $y = 0.10$, (9) $y = 0.05$, and (10) $y = 0.10$

On the other hand, R of Co-doped specimens is negative over the whole temperature range investigated, and the electron mobility μ_e is 0.31 cm^2/Vs at 300 K, which is 1.2 times larger than that reported by Birkholtz and Schelm [10]. The mobility ratio μ_e/μ_h is 0.04 at 300 K, and has a large temperature dependence. This is in good agreement with previous results reported by Nishida [12].

3.3 Temperature dependence of thermoelectric power

Fig. 5 shows the temperature dependence of the thermoelectric power α for undoped, Mn-doped, and Co-doped specimens. α of the undoped specimen increases as the temperature rises up to 450 K, and above this temperature it rapidly decreases. Although the maximum value of α for the low purity specimens is twice as large as that for the high purity specimens, α of specimens with high and low purities produces the same characteristic curve above 500 K, which corresponds to the intrinsic conduction range of the electrical resistivity ϱ. A very slight "drop" of α occurs with small hysteresis near 1260 K. This is also attributable to the semiconductor-to-metal transition. α of Mn-doped $FeSi_2$ is larger than that of undoped specimens over the temperature range from 150 to 900 K where an S-like ϱ curve was observed. The magnitude of the "drop" of α increases with Mn content, resulting in a decrease in T_c [12]. $|\alpha|$ of Co-doped specimens monotonously increases with rising temperatures up to a maximum value and then decreases. With increase of the Co content, the maximum value decreases and shifts to higher temperature.

4. Discussion

Both ϱ and R of $FeSi_2$ decrease with rising temperature over the temperature range studied, which is characteristic of semiconducting behavior. The resistivity ϱ in the intrinsic region at high temperatures, however, cannot be explained by the general two-band model, and E_g decreases with increasing carrier concentration. In the analysis of ϱ, this fact should be taken into consideration. The change of E_g in the temperature region below T_c is represented by the relationship between Jahn-Teller effect and carrier concentration. The resistivity ϱ_i [12, 17] is expressed by

$$\varrho_i = \varrho_0 \exp\left[\frac{E_{g0}}{2kT}\left(1 - \frac{2}{1-\delta}\right)\frac{n_p}{N}\right], \tag{1}$$

where E_{g0} is the energy gap at 0 K, n_p the hole concentration, N the state density, and δ a variable which takes values from 1/4 to 1/2. δ is a function of the ratios E_{g0}/kT_c and n_p/N at the transition temperature T_c.

In the temperature range from 150 to 900 K, the resistivity curve does not show a simple straight line but an S-like decaying behavior, which is affected by doping with Mn. In order to analyze theoretically the temperature dependence of this resistivity, the small polaron model was applied in which the mobility increases with rising temperature. The polaron mobility μ_{pol} [16, 18, 19] is given by

$$\mu_{pol} = \mu_{p_0} \exp\left(-\frac{E_h}{kT}\right) = \frac{M}{T} \exp\left(-\frac{E_h}{kT}\right), \tag{2}$$

where E_h is the hopping energy, M a constant factor independent of temperature. In this region the hole concentration aslo increases in proportion to $\exp(-E_a/kT)$ as the temperature is raised. Thus the electrical resistivity should be given as follows

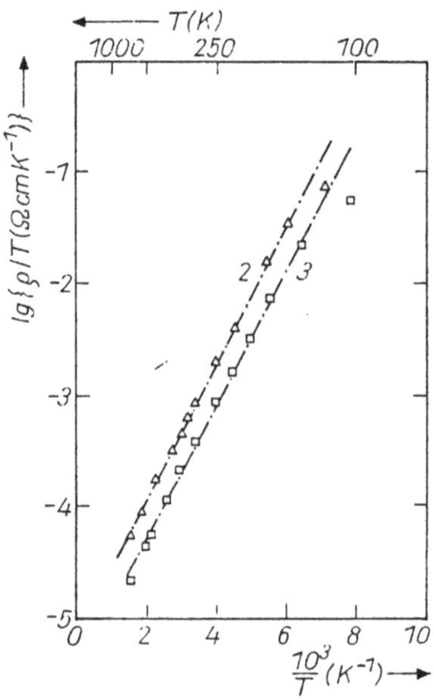

Fig. 6. Plots of lg (ϱ/T) against reciprocal temperature for $Fe_{1-x}Mn_xSi_2$ (high purity). (2) $x = 0.03$ and (3) $x = 0.06$

for $E_a \gg kT$ [19]:

$$\varrho_{pol} = \varrho_{p_0} T \exp [(E_a + E_h)/kT] . \tag{3}$$

Using experimental $\varrho(1/T)$ data, lg (ϱ/T) versus $1/T$ plots for $Fe_{1-x}Mn_xSi_2$ samples (high purity specimens with $x = 0.03$ and 0.06) are shown in Fig. 6. They fit well to straight lines. It is obvious from (3) that the slope indicates the sum of E_a and E_h. The energies $E_a + E_h$ obtained from the slopes in Fig. 6 are 0.126 and 0.122 eV for $x = 0.03$ and 0.06, respectively. The hopping energies can be estimated to be 0.024 and 0.022 eV by subtracting the hole activation energy, and the mean value is 0.023 eV.

The temperature dependence of the electrical resistivity below 150 K due to the unknown impurity in the starting material, can be written as

$$\varrho_u = \varrho_{u_0} \exp (E_u/kT) . \tag{4}$$

The total resistivity ϱ is expressed from (1), (3), and (4) as

$$1/\varrho = 1/\varrho_i + 1/\varrho_{pol} + 1/\varrho_u , \tag{5}$$

and the resistivities at various temperatures were calculated using (5).

In the calculation, a fixed value of 0.023 eV for E_h was assumed, ϱ_0, ϱ_{p0}, ϱ_{u0}, and E_u ware determined experimentally from resistivity data and E_a was evaluated from the slope of log $RT^{3/2}$ against $1/T$. Furthermore, the energy gaps, $E_g = E_{g0}[1 - 2/(1 - \delta)] n_p/N$, were calculated using the same method as reported by Nishida [12]. The calculated results are shown by solid lines in Fig. 7a for high purity specimens of $x = 0.03$ and 0.06 and in Fig. 7b for low purity specimens of $x = 0.05$ and 0.08. The dahsed line in each figure indicates band conduction which was calculated assuming that acoustic lattice scattering is predominant in the range from 150 to 900 K. As shown in Fig. 7a, the theoretical curves are in good agreement with the observed values. The agreement for low purity specimens in Fig. 7b is worse, therefore it is suggested that the hopping energy is slightly larger than 0.023 eV. Nevertheless,

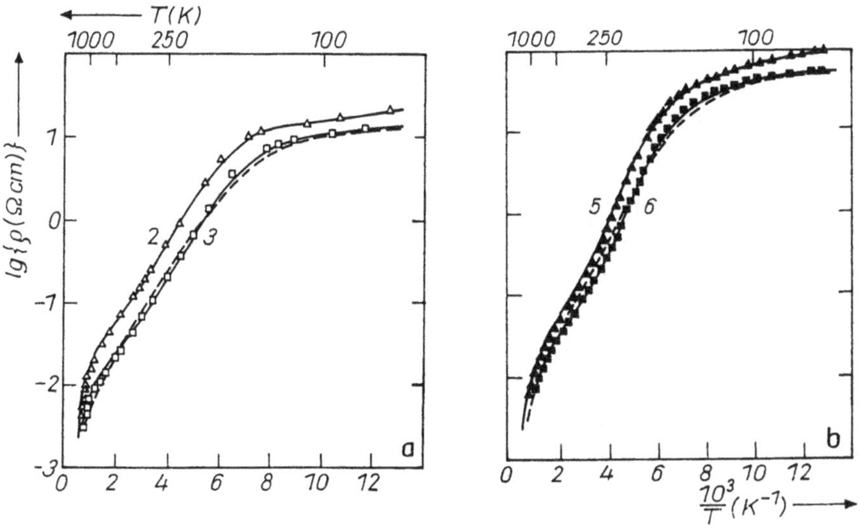

Fig. 7. Comparison of experimental and theoretical electrical resistivity for Mn-doped $FeSi_2$. The solid lines represent values calculated on assuming small polaron conduction and the broken line assuming band conduction. a) High purity (2) $x = 0.03$ and (3) $x = 0.06$; b) low purity (5) $x = 0.05$, (6) $x = 0.08$

fundamentally, the variation of resistivity with temperature of Mn-doped $FeSi_2$ is interpreted by small polaron conduction with crystalline distortion.

The temperature dependence of α for Mn-doped p-type specimens ($x = 0.03$ to 0.08) shows the same behavior as that of α for Co-doped n-type specimens ($y = 0.0028$ to 0.028) which was also reported by Birkholtz and Schelm [10]. Although the resistivity considerably decreases in the region from 200 to 800 K in Fig. 2 a and b, the thermoelectric power does not decrease extremely and its large value is maintained. In order to analyze the temperature variation of the thermoelectric power of thermoelectric materials, the general equation given by Ioffe [20] is usually employed. According to this analysis, the thermoelectric power should decrease gradually with increasing carrier concentration [21]. But in Fig 6, the thermoelectric power of Mn-doped $FeSi_2$ is almost constant with rising temperature from 400 to 800 K. This temperature behavior of α is assumed to be influenced by small polarons, however, this cannot yet be interpreted quantitatively by the conventional model. In conclusion, the temperature dependence of resistivity of Mn-doped $FeSi_2$ could be explained theoretically by applying the small polaron model with crystalline distortion.

Acknowledgements

The author would like to thank Dr. T. Sakata for his encouragement and advice and Dr. I. Nishida of the National Research Institute for Metals for his useful discussions.

References

[1] P. Y. Dusausoy, J. Protas, R. Wandji, and B. Roques, Acta cryst. **B27**, 1209 (1971).
[2] R. Bucksch, Z. Naturf. **22a**, 2124 (1967).
[3] R. Wandji, C.R. Acad. Sci. (France) **C267**, 1587 (1968).
[4] R. Wandji, C.R. Acad. Sci. (France) **C269**, 907 (1969).

[5] F. A. SIDRENKO, P. V. GELD, and L. B. DUBROVSKAYA, Fiz. Metallov i Metallovedenie 8, 735 (1959).
[6] J. P. PITON and M. F. FAY, C.R. Acad. Sci. (France) **C266**, 514 (1968).
[7] L. PAULING and A. M. SOLDATE, Acta cryst. **1**, 212 (1948).
[8] G. V. SAMSONOVE, Refractory Transition Metal Compounds, Academic Press, New York/London 1955 (p. 178).
[9] R. M. WARE and D. J. MCNEIL, Proc. Inst. Electr. Engng. **111**, 178 (1964).
[10] U. BIRKHOLTZ and J. SCHELM, phys. stat. sol. **27**, 413 (1968).
[11] J. HESSE and R. BUCKSCH, J. Mater. Sci. **5**, 272 (1970).
[12] I. NISHIDA, Phys. Rev. B **7**, 2710 (1973).
[13] J. HESSE, Z. angew. Phys. **28**, 652 (1969).
[14] J. HESSE, Z. Metallk. **60**, 652 (1969).
[15] U. BIRKHOLTZ and J. SCHELM, phys. stat. sol. **34**, K177 (1969).
[16] T. SAKATA and T. TOKUSHIMA, Trans. Nat. Res. Inst. Metals **5**, 34 (1963).
[17] D. ADLER and H. BROOKS, Phys. Rev. **155**, 826 (1967).
[18] R. R. HEIKES and W. D. JOHNSTON, J. chem. Phys. **26**, 582 (1957).
[19] D. ADLER and F. APPEL, Solid State Phys. **21**, 58 (1968).
[20] A. F. IOFFE, Physics of Semiconductor, Infoseach, London 1960 (p. 305).
[21] I. NISHIDA and T. SAKATA, J. Phys. Chem. Solids **39**, 499 (1978).

(Received September 21, 1988)

phys. stat. sol. (a) 110, 243 (1988)

Subject classification: 75.50; S1.1; S1.2

Institut für Physik, Max-Planck-Institut für Metallforschung, Stuttgart[1])

The Kinetics of the Magnetically Induced Anisotropy in Amorphous Ferromagnetic $Co_{71}Fe_1Mo_1Mn_4Si_{14}B_9$ and $Fe_{41}Ni_{41}B_{18}$ Alloys

By

T. Reininger, N. Moser, A. Hofmann, and H. Kronmüller

Analysing the kinetics of the induced anisotropy at various annealing temperatures (200, 230, 240, 260, 300, and 320 °C) it is possible to determine the spectrum of activation enthalpies for reversible processes in a nearly nonmagnetostrictive amorphous $Co_{71}Fe_1Mo_1Mn_4Si_{14}B_9$ alloy. A magnetostrictive amorphous $Fe_{41}Ni_{41}B_{18}$ alloy is preannealed by means of a short current pulse (0.5 s, 240 °C) while applying a tensile stress in order to induce internal stresses with a well-defined stress axis. After this treatment it is possible to measure the kinetics of the uniaxial anisotropy induced by a magnetic field perpendicular to the internal stresses at a temperature $T_a = 250$ °C. The formation of an induced anisotropy is attributed to the thermally activated alignment of mobile atom pairs. The thermodynamic and micromagnetic properties of these atom pairs are discussed on the basis of the model of two-level-systems [1 to 3].

Für eine nahezu magnetostriktionsfreie amorphe $Co_{71}Fe_1Mo_1Mn_4Si_{14}B_9$ Legierung wird aus der Kinetik der induzierten Anisotropie bei verschiedenen Anlaßtemperaturen (200, 230, 240, 260, 300 und 320 °C) das Spektrum der Aktivierungsenthalpien für reversible Prozesse bestimmt. Eine magnetostiktive amorphe $Fe_{41}Ni_{41}B_{18}$-Legierung wird bei angelegter Zugspannung durch einen elektrischen Strompuls (0,5 s; 240 °C) kurz getempert, um innere Spannungen mit einer definierten Richtung zu induzieren. Nach dieser Vorbehandlung ist es möglich die Kinetik der uniaxialen Anisotropie, induziert durch ein Magnetfeld senkrecht zu den inneren Spannungen bei einer Anlaßtemperatur $T_a = 250$ °C zu messen. Die Ausbildung der induzierten Anisotropie wird einer thermisch aktivierten Ausrichtung von beweglichen Atompaaren zugeschrieben und im Rahmen eines Zwei-Niveau-Modells behandelt [1 bis 3].

1. Introduction

Amorphous ferromagnetic alloys show remarkable physical properties like large magnetic permeability, small hysteresis losses, low electrical conductivity, etc. However, as is well known nearly all these properties degenerate with time and temperature.

As mentioned in various publications [4 to 8] there exists a strong correlation between the time dependence of the properties of the hysteresis loop and the structural stability. Accordingly one can use macroscopic properties for investigating microscopic processes. In the present paper magnetic annealing was used to stabilize the physical properties. Depending on the value of the induced anisotropy either rotational processes or wall movements govern the main magnetization process. In the case of prevailing rotational processes hysteresis losses will decrease. With the kinetics of the induced anisotropy it was possible to monitor relaxation phenomena as well as to obtain information on mobile defect structures in amorphous alloys.

[1]) Heisenbergstr. 1, Postfach 800665, D-7000 Stuttgart 80, FRG.

The formation of an induced anisotropy in amorphous alloys is supposed to be connected with the directional ordering of mobile atom pairs in the neighbourhood of free volumes.

The spectrum of activation enthalpies for reversible relaxation processes in amorphous $Co_{71}Fe_1Mo_1Mn_4Si_{14}B_9$ is obtained and discussed on the basis of a micromagnetic theory of the induced anisotropy based on the thermodynamics of two-level systems.

Furthermore it was possible to measure an isothermal annealing curve of the induced anisotropy also in a magnetostrictive $Fe_{41}Ni_{41}B_{18}$ amorphous alloy, after inducing a well-defined internal stress axis by a special pretreatment.

2. Theory

In binary crystalline alloys the induced anisotropy is due to the alignment of atom pairs parallel to the direction of the spontaneous magnetization by means of migrating vacancies. The activation enthalpy for this process is therefore governed by the self-diffusion enthalpy of these materials.

In amorphous alloys the free spaces between the atoms play a similar role as vacancies in crystalline materials. As described in various publications (e.g. [9]) there exist excess free volumes in as-quenched amorphous materials. The irreversible relaxation effects are connected with mobile atom pairs in the environment of these free volumes which change their configuration and become immobile during a preannealing treatment by an annealing of the excess free volumes.

For preannealed samples the reversible reorientation of atom pairs may be considered as an isomorphic transition (see Fig. 1), where each mobile atom pair reorients between two structurally equivalent configurations. These two equivalent configurations are assumed to be separated by a potential barrier which has to be overcome by the mobile atom either by a thermally activated jump or by a tunneling process. As proposed previously this atomic model may be described by a two-level system [1 to 3] (see Fig. 2). The magnetic interaction energy $\varepsilon_{i,j}$ of the i-th atom pair in orientation j with the spontaneous magnetization is given by

$$\varepsilon_{i,j} = \varepsilon_{0,i}^{(j)} \cos^2 \theta_{i,j} , \tag{2.1}$$

where $\varepsilon_{0,i}^{(j)}$ corresponds to a local anisotropy constant of the atom pair and $\theta_{i,j}$ is the angle between the i-th atom pair in orientation j and the magnetization.

In order to determine the occupation numbers f_j of these different energy levels in thermal equilibrium we neglect transitions between configurations of different splitting energies 2Δ, i.e., we assume isolated double wells. Using Boltzmann statistics and a rate equation for the number of mobile atom pairs occupying the two different

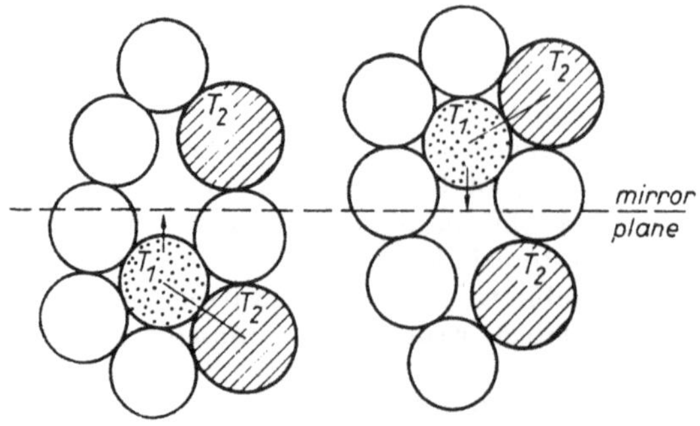

Fig. 1. Isomorphic reorientation of an atom in the vicinity of a free volume. The relaxation process is reversible

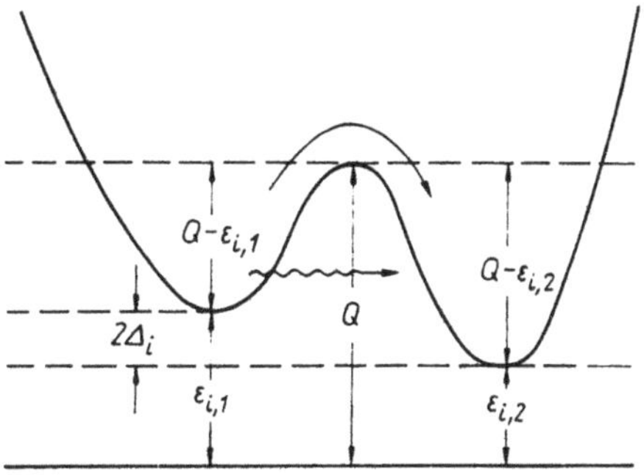

Fig. 2. Model of a two-level system describing the reorientation of an atom pair axis

orientations within their double-well potential [2] we obtain

$$f_j(\Delta, t) = 1/2\,[1 \pm \Delta/kT(1 - e^{-\nu_r t})] \tag{2.2}$$

for the occupation number. The relaxation frequency, ν_r, is assumed to obey an Arrhenius equation $\nu_r = 2\nu_0\, e^{-Q/kT}$, where ν_0 corresponds to the atomic attempt frequency and Q denotes the activation enthalpy. The total magnetic interaction energy of the mobile atom pairs with the spontaneous magnetization after the magnetic annealing treatment for some time t is given by

$$\Phi_0(t) = \sum_{i,j} f_j(\Delta, t)\, \varepsilon_{i,j}\,. \tag{2.3}$$

After an arbitrary rotation of the spontaneous magnetization the total induced anisotropy energy is equal to the change of the total magnetic interaction energy [2]

$$\Phi_u = \frac{c_p}{15kT} \langle \varepsilon_{0,i}^2 \rangle\, (1 - e^{-\nu_r t}) \sin^2 \theta\,, \tag{2.4}$$

where an averaging over the statistical parameters of the pair axis orientation has been performed and θ denotes the angle between M_s and the easy axis of the induced anisotropy, $\langle \varepsilon_{0,i}^2 \rangle$ corresponds to the average of the squares of the local anisotropy constants. The induced anisotropy therefore is given by

$$K_{\text{ind}} = \frac{c_p}{15kT} \langle \varepsilon_{0,i}^2 \rangle\, (1 - e^{-\nu_r t})\,, \tag{2.5}$$

where c_p corresponds to the density of mobile atom pairs. For amorphous alloys we expect a spectrum of activation enthalpies for the reversible processes, therefore the general time dependence of the induced anisotropy may be written as

$$K_{\text{ind}} = \frac{c_p}{15kT} \langle \varepsilon_{0,i}^2 \rangle\, G_{\text{rev}}(t, T) = A(T)\, G_{\text{rev}}(t, T) \tag{2.6}$$

with the amplitude factor $A(T)$ and the relaxation function

$$G_{\text{rev}}(t, T) = \int P_r(Q_r)\, [1 - e^{-t/\tau_r(Q_r)}]\, dQ_r\,. \tag{2.7}$$

$P_r(Q_r)$ denotes the distribution function of the activation enthalpies Q_r. For reversible annealing effects c_p in (2.6) denotes the hypothetical equilibrium density of mobile atom pairs.

The irreversible effects are connected with those atom pairs which change their configuration and become immobile during the annealing treatment by annealing of excess free volumes.

3. Experimental Procedure

For our experiments we used a magnetostrictive ($\lambda = 3 \times 10^{-5}$) $Fe_{41}Ni_{41}B_{18}$ alloy in order to compare our results with radio tracer diffusion measurements done by Horvath, Pfahler and Frank. The second material of interest for technical applications because it is nearly nonmagnetostrictive ($\lambda \ll 3 \times 10^{-7}$) was a Co-based alloy with a composition of $Co_{71}Fe_1Mo_1Mn_4Si_{14}B_9$. Both alloys were produced by the melt-spinning technique. The dimensions of the specimens were: thickness 40 μm, width 2 mm, and length 50 mm.

The magnetostrictive specimens were pretreated by a short current pulse of 0.5 s while simultaneously a tensile stress (2.5×10^7 N/m^2) was applied parallel to the ribbon. The maximum temperature on the surface was detected with a fast infrared spectropyrometer. After each pretreatment with different surface temperatures the domain structure was observed and the coercive field H_c, and the virgin curve were measured. All magnetization curves and coercive fields were measured by a Förster coercimeter [10] in an earth-field-compensated solenoid.

After this pretreatment the specimens were annealed in a transverse magnetic field of 0.12 T parallel to the surface of the specimens (magnetic annealing). After certain magnetic annealing times the virgin magnetization curves were measured at room temperature and the induced anisotropy was derived from the magnetization work required for saturation.

The nearly nonmagnetostrictive $Co_{71}Fe_1Mo_1Mn_4Si_{14}B_9$ alloys were preannealed above the Curie temperature at 370 °C for 2 h in order to decrease the excess free volume and to approach an equilibrium density of mobile atom pairs. Thereafter the as-quenched and preannealed samples were magnetically annealed at $T_a = 200$, 230, 240, 260, 300, 320 °C. For this alloy a special pretreatment was not necessary because internal stresses do not influence the magnetization. All the annealing treatments were performed in vacuum ($p = 10^{-5}$ Torr).

4. Experimental Results and Analysis

After annealing the specimens in a transverse (parallel to the surface) magnetic field, the induced anisotropy was derived from the relation $K_{ind} = H_k M_s/2$, where H_k denotes the saturation field determined by an extrapolation of the linear magnetization curve to the ferromagnetic saturation when the magnetic field was applied parallel to the ribbon axis, i.e., perpendicular to the induced easy axis.

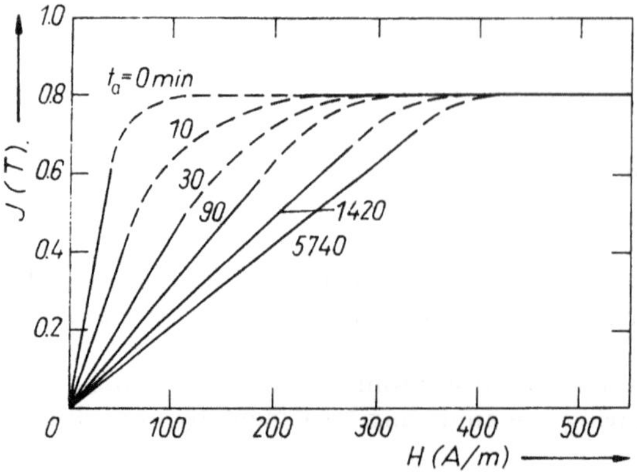

Fig. 3. The virgin magnetization curves of an as-quenched sample after different magnetic annealing times at $T_a = 200$ °C (– – – schematically)

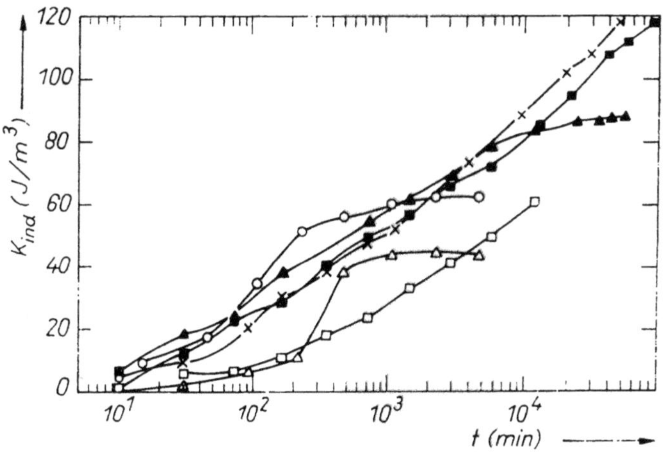

Fig. 4. Isothermal annealing curves of K_{ind} of preannealed samples (370 °C, 1.5 h) after magnetic annealing: □ T_a = 200, × 230, ■ 240, ▲ 260, ○ 300, △ 320 °C

4.1 *Measurements on $Co_{71}Fe_1Mo_1Mn_4Si_{14}B_9$*

Fig. 3 shows the magnetization curves of a $Co_{71}Fe_1Mo_1Mn_4Si_{14}B_9$ sample obtained after different magnetic annealing times t_a at T_a = 200 °C. The main contribution to the magnetization of the sample are rotation processes of the spontaneous magnetization

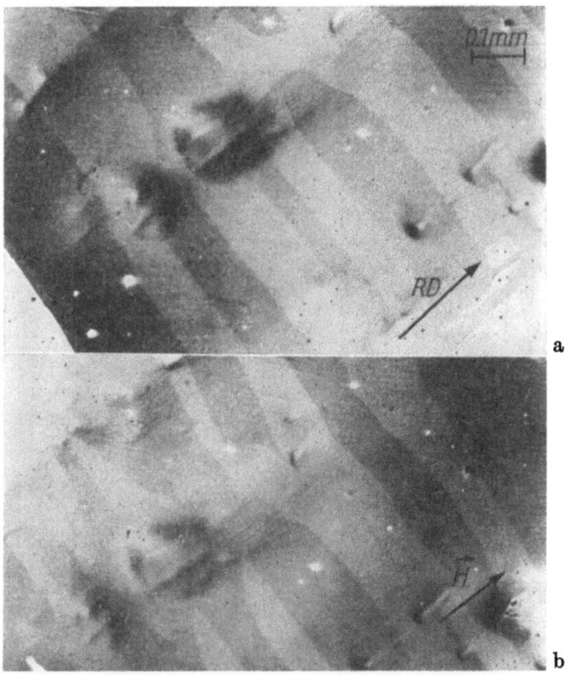

Fig. 5. Domain structure of a magnetically annealed $Co_{71}Fe_1Mo_1Mn_4Si_{14}B_9$ sample at T_a = 300 °C and t_a = 15 min. a) Without magnetic field; b) with magnetic field H = 40 A/m parallel to the rolling direction (RD)

within the domains. Therefore we can determine the induced anisotropy from the inverse slope of the virgin curve [9]. Fig. 4 shows the isothermal magnetic annealing curves of K_{ind} between 200 and 320 °C. The equilibrium amplitude decreases with increasing annealing temperature T_a. Large deviations of the measured isotherms from the theoretical curves occur above $T_a = 260$ °C for small magnetic annealing times ($t_a < 3$ h). They are due to curved Bloch walls which are responsible for magnetization processes by wall movements instead of rotation processes (see Fig. 5) during a magnetization of the sample. For lower annealing temperatures ($T_a < 260$ °C) the Bloch walls are better oriented and less curved, even for short annealing times ($t_a = 15$ min).

A quantitative analysis of the experimental results by a fit of the function $A(T)\,G_{rev}(t, T)$ to the data points gives a distribution of the activation enthalpies between 1.3 and 2.1 eV and a pre-exponential factor of $\tau_0 = 10^{-13\pm2}$ s as shown in Fig. 6.

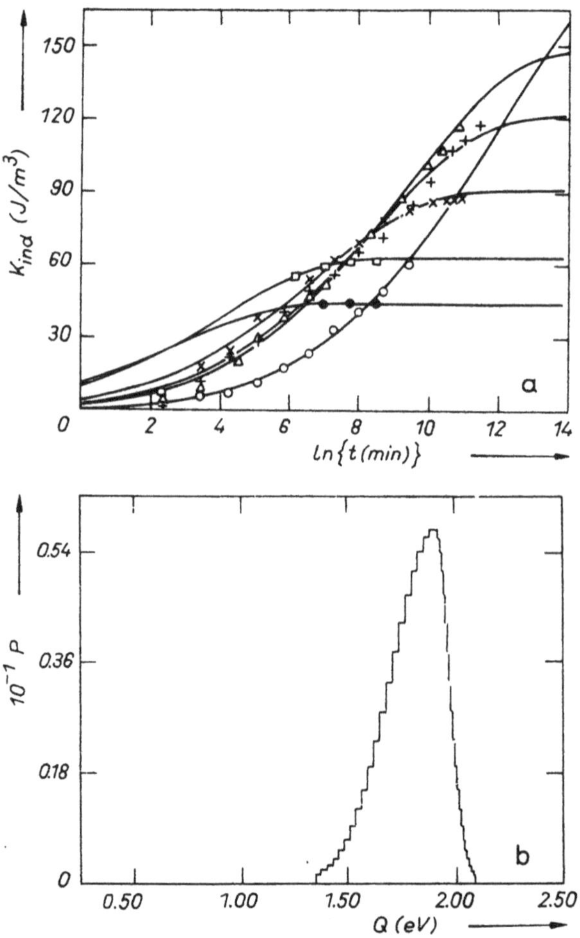

Fig. 6. Analysis of isothermal relaxation curves of K_{ind}. a) A fit of the function $A(T)\,G_{rev}(t, T)$ to the data points. ○ $T = 200$, △ 230, + 240, × 260, □ 300, ● 320 °C; b) the resulting distribution of the activation enthalpies ($\tau_0 = 10^{-13\pm2}$ s)

4.2 Crossover effect

The crossover effect was studied by annealing the sample for the time t_a at a temperature T_a in a magnetic field until $K(t_a, T_a^{(1)})$ is equal to the saturation amplitude $K_{ind}(\infty, T_a^{(2)})$ of another isotherm with $T_a^{(2)} > T_a^{(1)}$. At the crossover point a temperature change from $T_a^{(1)}$ to $T_a^{(2)}$ is performed. As shown in Fig. 12 we observe for small times ($t_a < 30$ min) a rapid decrease of $K_{ind}(t_a, T_a^{(2)})$. At larger times ($t_a > 60$ min) $K_{ind}(t_a, T_a^{(2)})$ increases again towards the equilibrium value $K_{ind}(\infty, T_a^{(2)})$.

4.3 Measurements on $Fe_{41}Ni_{41}B_{18}$

The virgin magnetization curves of as-quenched as well as preannealed and magnetically annealed $Fe_{41}Ni_{41}B_{18}$ samples show no linear range because domain wall movements contribute to the magnetization of the sample. The domain pattern is mainly governed by internal stresses and was not remarkably influenced by a magnetic annealing treatment (see Fig. 7). The reduction of the internal stresses by an annealing treatment gives only a stabilization of the domain pattern because of a structural relaxation.

Therefore we applied a uniaxial stress ($\sigma = 25$ MPa) parallel to the ribbon leading to a laminar domain structure oriented parallel to the ribbon axis. At the same time

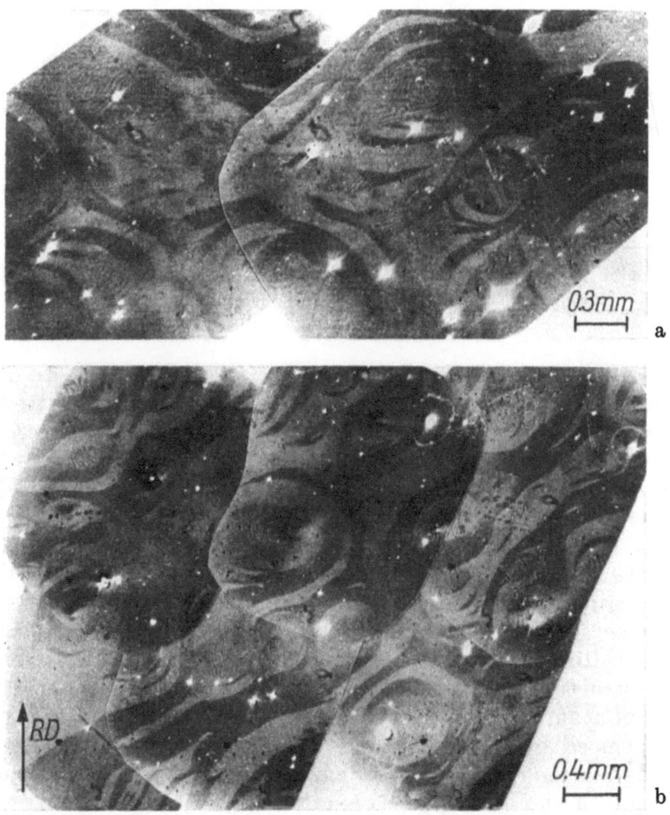

Fig. 7. Domain structure of a $Fe_{41}Ni_{41}B_{18}$ sample a) as-quenched; b) after magnetic annealing at $T_a = 250$ °C, $t_a = 50$ min and $H = 9.5 \times 10^4$ A/m

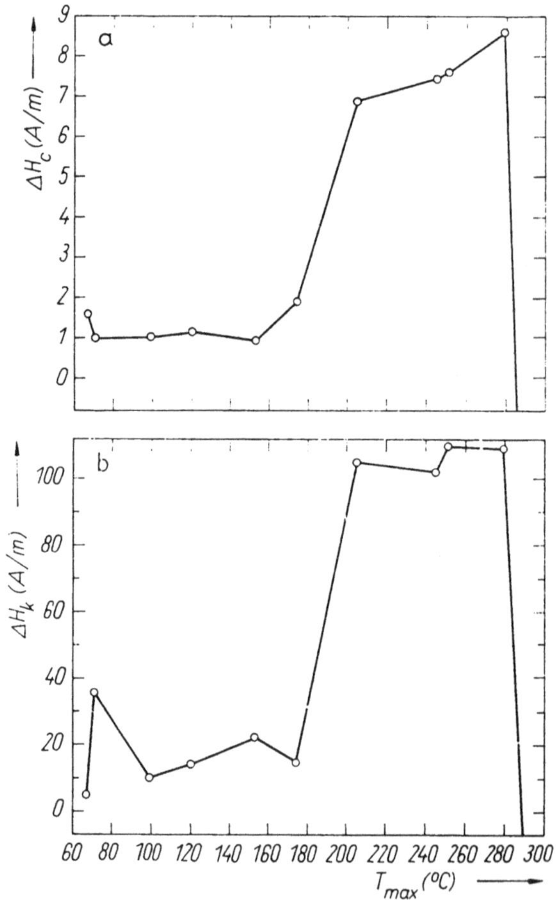

Fig. 8. Change in coercivity, ΔH_c, and in the saturation magnetic field, ΔH_k, for different short time preannealing temperatures for a $Fe_{41}Ni_{41}B_{18}$ sample

a short current pulse of 0.5 s flew through the sample raising the temperature to T resulting in a structural relaxation which causes a stabilization of the domain pattern which was externally forced. The annealing temperature T_a achieved was detected by an infrared spectropyrometer.

Fig. 8 shows the change in coercivity, ΔH_c, and in the saturation magnetic field, ΔH_k, for different short time preannealing temperatures. At temperatures lower than 170 °C no changes could be observed. At temperatures 210 °C $\leqq T_a \leqq$ 240 °C the coercive field decreased to 10%, the virgin curve became steep and linear and the domains changed from fine structured domains to wide laminar domains parallel to the applied tensile stress (see Fig. 9a). At higher temperatures the coercive field increased drastically. On annealing such a sample, the well-defined uniaxial internal stresses will decrease due to structural relaxations. Simultaneously an anisotropy will be induced by an alignment of atom pairs parallel to the direction of the spontaneous magnetization which is influenced by internal stresses [9, 11]. The principle shapes of the measured virgin magnetization curves caused by a distinct domain pattern and the corresponding distribution of the atom pair axes after short-time pulse current annealing treatments under an external stress are schematically represented in Fig. 10. Magnetic annealing of such pretreated samples will induce a second anisotropy

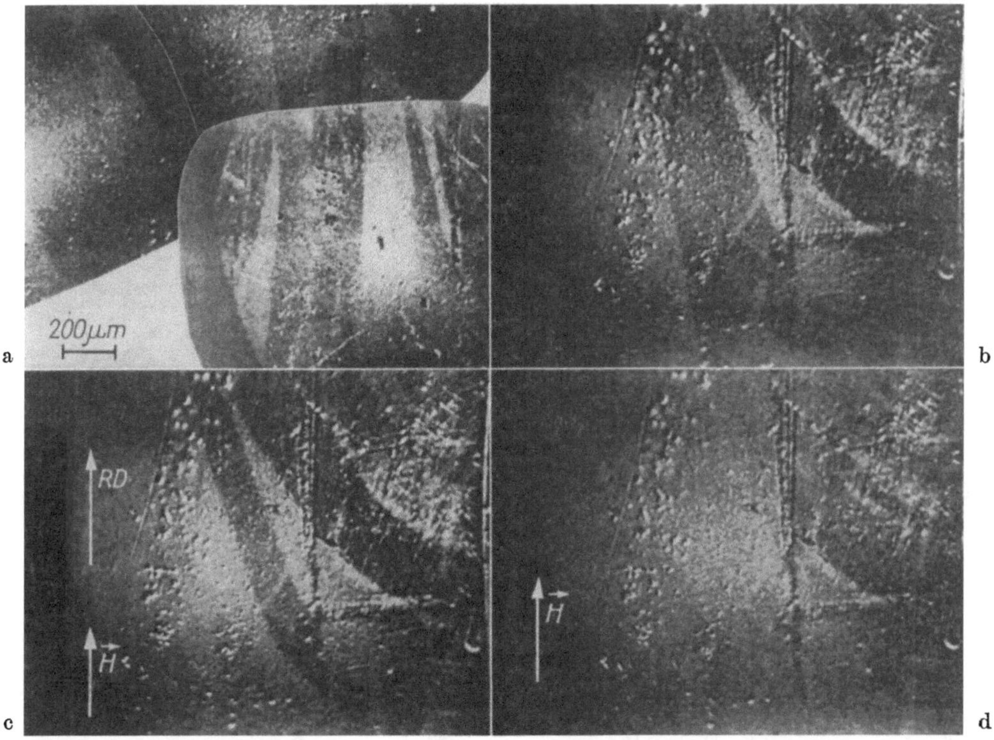

Fig. 9. Domain structure of a $Fe_{41}Ni_{41}B_{18}$ sample a) after annealing under applied uniaxial stress parallel to the rolling direction by means of a short current pulse at $T_a = 247$ °C and $t_A = 0.5$ s. b) Same as a) but after magnetic annealing at 250 °C for 2 h. c) Same as b) but with additional magnetic field ($H < 20$ A/m) parallel to the ribbon axis. d) Same as b) but with additional magnetic field ($H > 20$ A/m) parallel to the ribbon axis

axis perpendicular to the ribbon axis while the stress induced anisotropy along the ribbon axis will decrease. This means a reorientation of atom pairs from a parallel to a perpendicular orientation to the ribbon axis. This process is the origin of the two linear ranges in the virgin magnetization curve as schematically shown in Fig. 10.

The first linear range in the magnetization curve (Fig. 10) results from Bloch wall movements aligned parallel to the anisotropy axis along the ribbon axis as produced by internal stresses (see Fig. 9 b and c).

The second linear range is governed by rotation processes (see Fig. 9d) caused by the magnetically induced anisotropy axis perpendicular to the ribbon axis. With increasing magnetic annealing time the second linear range increases as well and the slope decreases. From these second slopes of the magnetization curves we determined the 250 °C-K_{ind}-isotherm represented in Fig. 11.

5. Discussion

5.1 Activation enthalpy spectra

On the basis of the two-level model the kinetics of the induced anisotropy may be described quantitatively. From a fit of isothermal annealing curves to the experimental data we got a spectrum of activation enthalpies $P_r(Q_r)$ for the reversible an-

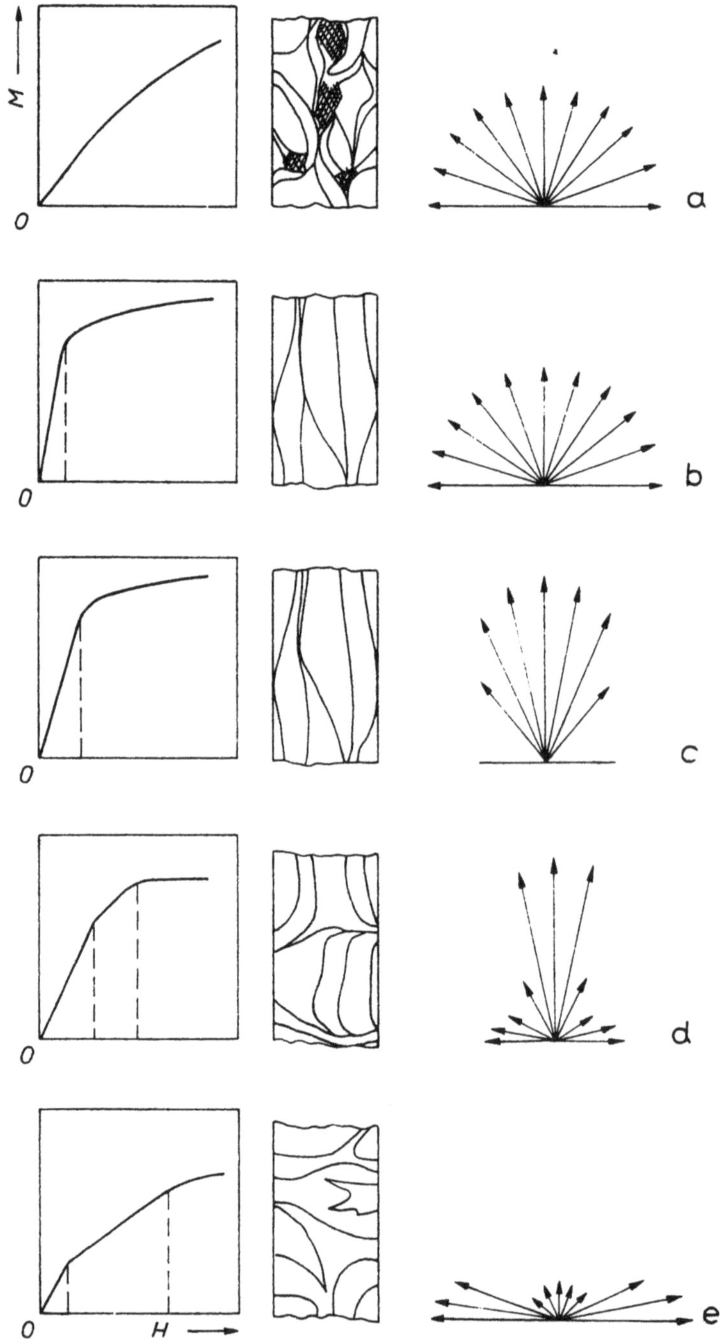

Fig. 10. Relations between virgin magnetization curves, domain structure, and distribution of atom pair axes after different treatments for the $Fe_{41}Ni_{41}B_{18}$ sample. H applied parallel to the ribbon axis. a) As-quenched $Fe_{41}Ni_{41}B_{18}$ alloy. b) After short-time pulse current annealing and simultaneously applied stress. c) After additional annealing (285 °C, 1.5 h). d) After magnetic annealing ($t_a = t_a^{(1)}$). e) After magnetic annealing ($t_a = t_a^{(2)}$; $t_a > t_a^{(1)}$)

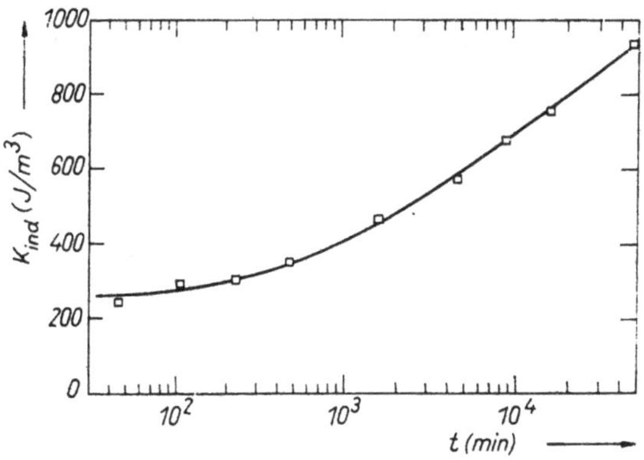

Fig. 11. Isothermal annealing curve of K_{ind} of a specially pretreated $Fe_{41}Ni_{41}B_{18}$ sample (pulse current annealed under applied stress parallel to the ribbon axis with $T_a = 247$ °C, $t_a = 0.5$ s and $\sigma = 25$ MPa

nealing processes with a maximum at $Q = 1.9$ eV $\pm$ 0.3 eV and a pre-exponential factor of $\tau_0 = 10^{-13\pm2}$ s for the nearly nonmagnetostrictive $Co_{71}Fe_1Mo_1Mn_4Si_{14}B_9$ alloy. This result is in good agreement with the work done by Kronmüller [2] and Guo et al. [12]. Magnetic after-effect experiments with short annealing times ($t_a \leqq 180$ s) performed for similar alloys lead to activation enthalpies between 0 eV $< Q_r \leqq 1.3$ eV and a pre-exponential factor of $\tau_0 = 10^{-17}$ s [13]. According to these results we assume two reversible processes:

1. A local rearrangement of atom pairs in the neighbourhood of free volumes. This is the main process for short annealing times and is governed by low energy barriers.

2. A long-range diffusion of free volumes leading to a further alignment of atom pair axes parallel to the magnetization. This diffusion process requires higher activation enthalpies, as radio-tracer diffusion experiments show [14] and becomes important at higher annealing times.

According to this model the system is first passing all the possible low energy barriers, while for longer annealing times the system has to overcome also higher energy barriers.

The activation enthalpy of $Q = 1.9$ eV determined as the most probable activation enthalpy indicates a long-range diffusion process. Also the pre-exponential factor $\tau_0 = 10^{-13\pm2}$ s indicates that long-range rearrangements are present because for short-time experiments $\tau_0 \approx 10^{-17}$ s has been determined.

The annealing time scale for the magnetically induced anisotropy goes from 15 min to several weeks, that means we detect both reversible processes described above. For short annealing times the diffusion plays a subordinate role and the essential part is the local rearrangement of mobile atom pairs. While for long annealing times the alignment of atom pairs takes place by a diffusional process.

Also the analysis of long-time isotherms of magnetic after-effect measurements shows that the activation enthalpy spectrum approaches the self-diffusion enthalpy for increasing annealing time [15].

5.2 Analysis of crossover experiments

The crossover experiment delivers a verification of the reversibility and the existence of a spectrum of activation enthalpies in amorphous alloys. A quantitative model for the crossover process is given in Fig. 12. Here we used a schematic representation for the distribution of the mobile atom pairs in an ensemble of two-level systems with equal splitting energy Δ. The most simple way to explain this crossover process is to

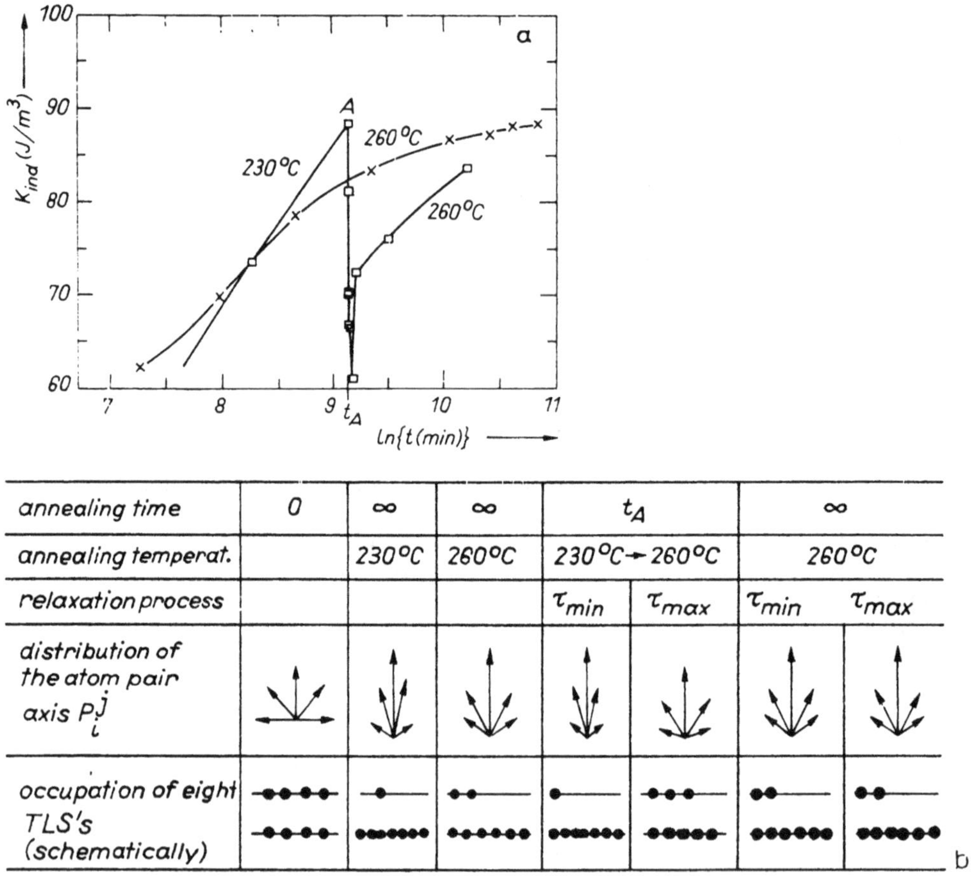

Fig. 12. a) Crossover behaviour of K_{ind} of preannealed $Co_{71}Fe_1Mo_1Mn_4Si_{14}B_9$ from the 240 and 260 °C isothermal annealing curves. Point A denotes the crossover point of the asymptotic curve for $T_a = 240$ °C with the equilibrium value K_{ind}^{gg} of the 260 °C isotherm. b) A qualitative model for the crossover process in the case of spectra of relaxation times

assume two relaxation processes with different relaxation times, $\tau_{min} \ll \tau_{max}$. We start at $t = 0$ s with randomly oriented atom pair axes at a temperature T_1. After an annealing time, $t = t_A$, the amplitude of $K_{ind}(t_A, T_1)$ is equal to the saturation amplitude $K_{ind}(\infty, T_2)$ of the isotherm with temperature $T_2 (T_1 < T_2)$. At $t = t_A$ the fast relaxation processes are already in their thermodynamic equilibrium whereas the slow processes have not even reached the equilibrium configuration of T_2. After a temperature change from T_1 to T_2 both processes proceed to reach the new equilibrium configuration of T_2. The fast processes, i.e. the low activation energies determine the behaviour for short times. Therefore the previously reached order for $t < t_A$ now changes to a lower order corresponding to the higher annealing temperature T_2. This leads to a rapid decrease of K_{ind} for short times immediately after the temperature change. For longer times the slow processes become more and more important in reaching the equilibrium configuration of T_2. This manifests itself in an asymptotic increase towards the equilibrium amplitude $K_{ind}(\infty, T_2)$.

References

[1] M. R. J. Gibbs, J. E. Evetts, and J. A. Leake, J. Mater. Sci. **18**, 278 (1983).
[2] H. Kronmüller, phys. stat. sol. (b) **118**, 661 (1983).
[3] H. Kronmüller, Phil. Mag. **B48**, 127 (1983).
[4] H. Fujimori, H. Morita, Y. Obi, and S. Ohta, Amorphous Magnetism 2, Ed. R. A. Levy and R. Hasegawa, Plenum Press, 1977 (p. 393).
[5] H. Fujimori, S. Ohta, T. Masumoto, and K. Nakamoto, Rapidly Quenched Metals 3, Ed. B. Cantor, Vol. 2, The Metal Society, London 1978 (p. 232).
[6] F. E. Luborsky, P. G. Frischmann, and L. A. Johnson, J. Magnetism magnetic Mater. **8**, 318 (1978).
[7] N. Moser and H. Kronmüller, J. Magnetism magnetic Mater. **19**, 275 (1980).
[8] H. Kronmüller, Rapidly Quenched Metals 4, Ed. T. Masumoto and K. Suzuki, Japan Institute of Metals, Sendai 1982 (p. 971).
[9] H. Kronmüller, H. Q. Guo, W. Fernengel, A. Hofmann, and N. Moser, Crystal Lattice Defects **11**, 135 (1984).
[10] F. Förster, Z. Metallk. **46**, (1955).
[11] L. Köszegi and H. Kronmüller, Appl. Phys. A **34**, 95 (1984).
[12] H. Q. Guo, H. Kronmüller, N. Moser, and A. Hofmann, Scripta metall. **20**, 185 (1986).
[13] F. Rettenmeier and H. Kronmüller, phys. stat. sol. (a) **93**, 221 (1986).
[14] J. Horvath, K. Pfahler, W. Ulfert, W. Frank, and H. Kronmüller, Internat. Conf. Vacancies and Interstitials in Metals and Alloys, Berlin 1986 (p. 148).
[15] M. Bourrous and H. Kronmüller, to be published.

(Received October 5, 1988)

phys. stat. sol. (a) **111**, 257 (1989)

Subject classification: 61.70; 76.30; S5.11

Institute of Semiconductor Physics, Academy of Sciences of the USSR, Siberian Branch, Novosibirsk[1])

The EPR Study of Inhomogeneous Deformations in Neutron-Irradiated Silicon

By

A. A. Karanovich and A. V. Dvurechenskii

The tensosensitivity parameters of the EPR response of negatively charged divacancies (G7) to uniaxial compression are studied. The data obtained are used in the analysis of the EPR spectra of divacancies in silicon doped with elements of group V (P, As, Sb) within $N_i = 2 \times 10^{18}$ to 1×10^{19} cm^{-3} concentration range and neutron-irradiated by a reactor dose of $D = 5 \times 10^{17}$ to 4×10^{19} cm^{-2}. On the assumption of the uniaxial character of local deformations, the average deformation of lattice outside the disordered regions (DR) is estimated as $\varepsilon \approx 10^{-4}$. The distribution of divacancies is given by absolute values of deformations acting on them in the DR($(dN/d\varepsilon) = N \cdot F_N(\varepsilon)$). The average value of the uniaxial deformation in DR $\langle\varepsilon\rangle$ is 1.1×10^{-3}.

На основании исследования отклика спектра ЭПР отрицательно заряженных дивакансий (G7) на одноосное сжатие определены параметры тензочувствительности данного дефекта. Полученные данные использовались при анализе спектров ЭПР дивакансий, наблюдаемых в кремнии, легированном элементами V группы (P, As, Sb) в интервале концентраций $N_i = 2 \times 10^{18}$ до 1×10^{19} cm^{-3} и облученном нейтронами реактора дозой $D = 5 \times 10^{17}$ до 4×10^{19} cm^{-2}. В предположении об одноосном характере локальных деформаций получена оценка средней величины деформации решетки вне разупорядоченных областей (PO): $\varepsilon \approx 10^{-4}$. Построено распределение дивакансий по абсолютным значениям действующих на них деформаций в PO ($(dN/d\varepsilon) = N \cdot F_N(\varepsilon)$). Среднее значение одноосных деформаций в PO $\langle\varepsilon\rangle$ составляет $1{,}1 \times 10^{-3}$.

1. Introduction

Fast neutron irradiation ($E > 0.1$ MeV) produces in silicon crystals elastic deformations caused by a high concentration of irradiation defects [1 to 3]. It has been found that the bulk distribution of defects is in this case rather inhomogeneous: within the disordered regions (DR) their concentration is considerably higher than in the rest of the crystal (matrix) [4]. This, in turn, leads to an inhomogeneity of the deformation distribution [5]. The existing experimental procedures (X-ray diffraction analysis [2], infra-red spectroscopy [1]) allow us to obtain only the whole crystal, bulk averaged values of deformations.

The present investigation was aimed at obtaining data on the inhomogeneity of lattice deformation distributions by using a negatively charged divacancy as a tensosensitive paramagnetic probe in neutron-irradiated silicon. This problem statement is justified, first, by a high concentration of divacancies in neutron-irradiated silicon, and, second, by a strong dependence of the observed EPR spectrum of this defect upon lattice deformations.

---

[1]) Prospekt Akad. Lavrenteva 13, SU-630090 Novosibirsk, USSR.

2. Tensosensitivity of Divacancies

A divacancy in silicon has three configurations of the defect electron subsystem per pair of vacancies [6]. In the unperturbed lattice all three divacancy configurations are equal in energy and, hence, equally populated. The activation energy of the transition (or of re-orientation) from one configuration to another is small ($E_a \approx 0.06$ eV), so already at $T \gtrsim 80$ K the re-orientation frequency becomes so high that the EPR spectrum is converted into a so-called "motionally-averaged" (MA) spectrum [7].

The uniaxial deformation of the lattice created in divacancy localization removes the energy degeneracy of various defect orientations (unlike hydrostatic expansion or concentration that do not produce such an effect) and, therefore, changes the populations of these orientations.

In the low-temperature EPR spectrum of divacancies it is seen in changed line intensities, in the MA spectrum, accordingly, in changed line positions.

The investigation of divacancy response to uniaxial compression [6] has shown that the change of energy of a certain orientation of the defect depends only on the value of the uniaxial deformation component of the lattice along a certain direction, i.e., the tensosensitivity axis of the defect (this axis is directed along $\langle 110 \rangle$, and couples the atoms between which a closed bond in the defect exists). In addition, this energy change varies linearly, within measurement error limits, with the deformation value, i.e.

$$E = E_0 + M\varepsilon_t\,, \tag{1}$$

where E_0 is the divacancy energy in the unperturbed lattice, M the tensosensitivity coefficient, ε_t the deformation component along the tensosensitivity axis.

In the present study, the dependence of line positions in the MA spectrum of divacancies (in electron-irradiated silicon) on the value of applied compression at $\boldsymbol{P} \parallel [1\bar{1}0]$, $\boldsymbol{B} \parallel [111]$ (Fig. 1a) was studied in an attempt of finding the coefficient M. For a given orientation of $\boldsymbol{P}$ and $\boldsymbol{B}$ and assuming that the populations of different defect orientations are described by the Boltzmann relationship, the position of line c of the spectrum (Fig. 1a), as shown by the analysis, is defined as

$$B_c(P, T) = 2\,\frac{1 - e^{-M\frac{P}{kT\lambda}}}{2\,e^{-M\frac{P}{kT\lambda}} + 1}\,; \qquad \frac{1}{\lambda} = \frac{1}{4}\left(\frac{\lambda_1}{(\lambda_1 - \lambda_2)(\lambda_1 + 2\lambda_2)} + \frac{1}{\lambda_3}\right), \tag{2}$$

where $B_c(P, T)$ is the value of the magnetic field for line c, k the Boltzmann coefficient; $\lambda_1, \lambda_2, \lambda_3$ are the rigidity coefficients of crystal silicon [8].

Fig. 1b shows that the position of line c is indeed well described by (2) for the parameter $M = (52 \pm 6)$ eV. This value seems more credible than 32 eV obtained in [6] because our accuracy in measuring the compression-dependent line position was much higher than that of [6] measuring the line intensity under conditions of low spectral resolution.

3. Divacancies in Neutron-Irradiated Silicon

In neutron-irradiated slightly doped silicon the Fermi level within DR occupies a certain limiting position under which divacancies are non-paramagnetic [9]. In order to shift the Fermi level to the upper half of the forbidden region, we used a material with a high content of n-type impurities (P [5], As, Sb). The following experimental results were obtained:

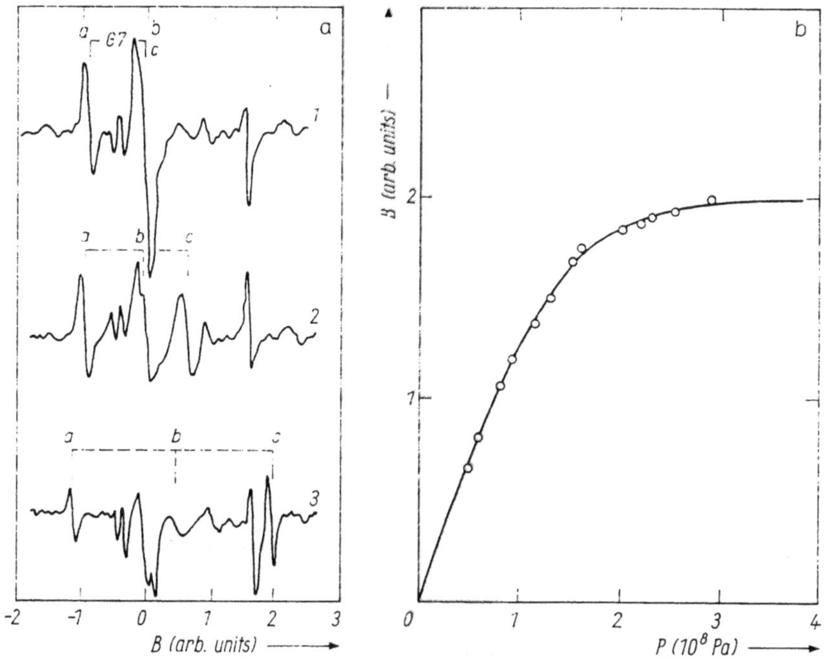

Fig. 1. a) The EPR spectra of divacancies (lines a, b, and c) in electron-irradiated silicon ($E = 1$ MeV, $D = 10^{19}$ cm^{-2}) after application of uniaxial compression: (1) $P = 0$; (2) 5×10^7 (3) 3×10^8 Pa. $\boldsymbol{P} \parallel [110]$, $\boldsymbol{B} \parallel [111]$, $T = 140$ K, $\nu = 35$ GHz. b) The position of line c of the G7 spectrum as a function of the applied uniaxial compression. Points refer to the experiment, solid line to computation

1. Within the range of irradiation doses of 5×10^{17} to 1×10^{18} cm^{-2} in silicon with impurity content of $N_i = 2 \times 10^{18}$ cm^{-3} the spectrum of negatively charged divacancies G7 is dominant. Besides, a weak signal from non-reorientable divacancies of H9 is observed [5]. A higher content of doping impurities at these exposure doses increases the conductivity of samples thus making EPR measurements impossible. A comparison with electron irradiation (Fig. 2) shows that the spectrum G7 is distorted: one of its lines is of abnormal width and low intensity.

2. At radiation doses from 5×10^{18} to 1×10^{19} cm^{-2} in silicon with impurity content $N_i = 2 \times 10^{18}$ cm^{-3} the spectra G7 and H9 have about the same intensity, but when the dose increases, the intensity of H9 is increasing and that of G7 decreasing [5]. The increased content of doping impurity (P, As, Sb) within the range of 2×10^{18} to 1×10^{19} cm^{-3} considerably increases (5 to 7 fold) the intensity of H9. In this case the G7 intensity remains almost unchanged. The experiment has shown that the shape of the spectrum (intensity, concentration ratio of G7 to H9 centres) depends on the concentration of the doping impurity and not on its type.

3. In silicon exposed to the maximum neutron doses ($D = (2$ to $4) \times 10^{19}$ cm^{-2}) the G7 spectrum is absent (Fig. 3). The H9 intensity is growing with the increase of the doping impurity reaching the respective spin concentration of $\approx 7 \times 10^{18}$ cm^{-3}. The absorption spectrum obtained through numerical integration of the experimental spectrum H9 (in all experiments the first derivative of the SHF power absorption spectrum was recorded) is marked by the absence of individual lines (Fig. 3). A tem-

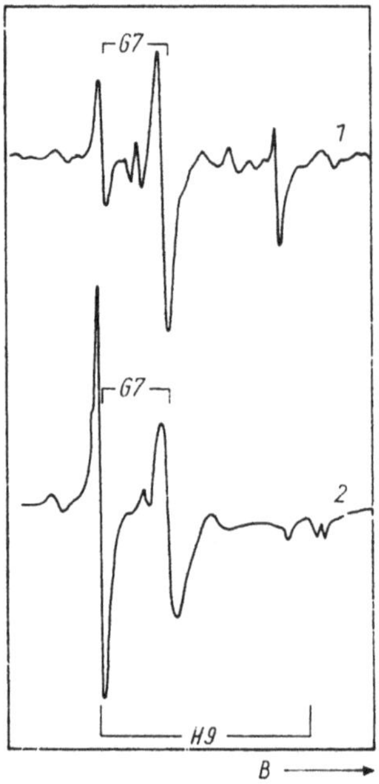

Fig. 2

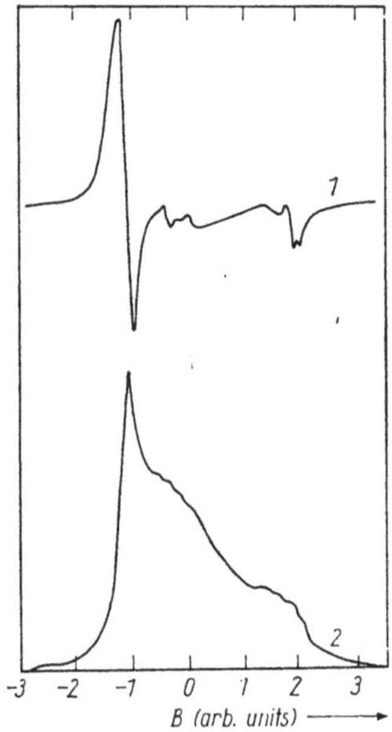

Fig. 3

Fig. 2. The EPR spectrum of G7 divacancies in silicon exposed a) to electron-irradiation of an energy of $E = 1$ MeV by $D = 10^{19}$ cm^{-2} dose and b) to neutron-irradiation by $D = 5 \times 10^{17}$ cm^{-2}. $T = 140$ K, $\nu = 35$ GHz

Fig. 3. The experimental spectrum of H9 divacancy (1) and the result of its numerical integration (2). $D = 2 \times 10^{19}$ cm^{-2}, $T = 125$ K, $\nu = 35$ GHz

perature increase within 90 to 300 K essentially modifies the spectrum H9: a broad line appears the position of which coincides with one of the G7 lines.

4. Discussion of the Results

The obtained experimental data are well explained by the idea of inhomogeneous lattice deformations the value of which in DR is much higher than in the matrix. Divacancies localized in the matrix are yielding a normal, only slightly distorted MA spectrum G7, and the H9 spectrum corresponds to divacancies localized in DR [5].

The deformations induced by the doping impurity atoms (because their covalent radii differ from those of silicon atoms) are small compared to the deformations in the neutron-irradiated crystal. This follows directly from the fact that, first, the shape of spectral divacancies is independent of the type of the doping impurity (though the covalent atom radii of the used impurities are substantially different [10, 11]), and, second, electron irradiation ($E = 1$ MeV, $D = 10^{19}$ cm^{-2}) of the used material ($N_i = 2 \times 10^{18}$ cm^{-3}) does not lead to the emergence of non-reorientable divacancies (see Fig. 2).

So the effect of the impurity (in the investigated range of concentrations) was reduced in our experiments only to the Fermi level shift. While in silicon with n-type impurity concentration of 2×10^{18} cm^{-3} only divacancies are observed localized in the matrix and in some periphery of DR, in a maximum-doped material ($N_i = 1 \times$

$\times 10^{19}$ cm^{-3}) the divacancies are paramagnetic in the whole crystal bulk which follows from the coincidence (within measurement error limits) of their observed and actual number, the latter measured by IR absorption [1]. This explains the H9 intensity growth observed in the experiment when the donor impurity concentration was increased.

The dependence of the relative intensities of the G7 and H9 spectra on the radiation dose is caused by the relationship of the volumes of DR and matrix in the crystal [5].

The absence of the lines of the H9 absorption spectrum seems to evidence that this spectrum is enveloping signals from divacancies found in different deformation fields. This also explains the temperature variation of the H9 spectrum observed in the experiment: with a temperature increase, the condition $\Delta E > kT$ (ΔE is the energy difference for different defect configurations defined by (1)) begins to hold. The signal from such defects is shaped like the spectrum G7 of unperturbed divacancies, which causes, with temperature increase, the emergence of a broad line in H9 coinciding by its position with a G7 line. However, the line typical of the H9 spectrum remains, too, which indicates the presence of divacancies that are in stronger deformation fields.

5. Lattice Deformations in Neutron-Irradiated Silicon

Using the width of line c in G7 (Fig. 1, 2) in silicon exposed to neutron irradiation doses in the 5×10^{17} to 1×10^{18} cm^{-2} range (if the crystal volume outside DR is large enough), the obtained data on divacancy tensosensitivity allow us to estimate the average value of uniaxial deformations in the matrix: $\varepsilon \approx 10^{-4}$.

To obtain the data on lattice deformations in DR, the H9 spectrum was analyzed. Since the shape of H9 does not depend on the doping impurity concentration (within the range under investigation), we used higher radiation doses when G7 is absent (because the crystal volume outside DR is negligibly small) and the maximum level of doping when divacancies are paramagnetic over the whole crystal bulk.

The analysis of lattice deformations shaping the H9 spectrum was made on the following assumptions:

1. the lattice deformation near a divacancy is either axial contraction or axial expansion;
2. all directions of axial deformations are equally plausible. We introduce a system of coordinates so that the z-axis is along the line coupling two vacancies. Obviously, the divacancy spectrum shape is determined only by the component ε_2 of the deformation lying in the xy-plane (because in this plane the axes of tensosensitivity fitting all three configurations of the electronic subsystem of the defect are lying). Then the absorption spectrum shape of the divacancy $S(B)$ will be described by the following relationship:

$$S(B) = N \int_0^\infty F_{N2}(\varepsilon_2)\, \sigma(\varepsilon_2, B)\, \mathrm{d}\varepsilon_2\,, \tag{3}$$

where B is the value of the magnetic field, N the total number of divacancies, $F_{N2}(\varepsilon_2)$ the function of divacancy distribution by the absolute values of deformation component ε_2 of the lattice in the region of their localization, $\mathrm{d}N/\mathrm{d}\varepsilon_2 = NF_{N2}(\varepsilon_2)$, $\sigma(\varepsilon_2, B)$ is the spectrum which would be observed in case $F_{N2}(\varepsilon_2) = \delta(\varepsilon_2' - \varepsilon_2)$, i.e., if all divacancies were under the action of axial deformation equal in absolute value to ε_2 and uniformly distributed along all directions in the xy-plane.

In order to find the distribution $F_{N2}(\varepsilon_2)$, the values of the function $\sigma(\varepsilon_2, B)$ were computerized on the basis of the tensosensitivity of divacancies. The function $F_{N2}(\varepsilon_2)$ was presented as a histogram $\left(F_{N2}(\varepsilon_2) = C_i \text{ at } \varepsilon_i < \varepsilon < \varepsilon_{i+1}\right)$, after which the values

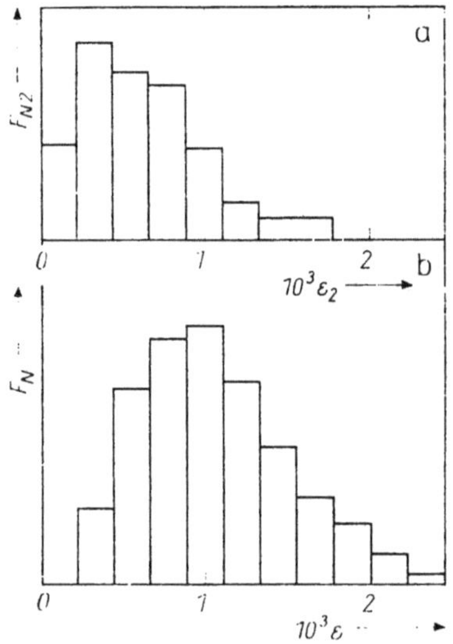

Fig. 4. The distribution of divacancies by absolute values of (1) deformation component ε_2 of the lattice and (2) the total uniaxial lattice deformation ε in the region of their localization

of parameters C_i varied until the best fit of the spectrum $S(B)$ calculated by (3) with the experimental spectrum (Fig. 3) was obtained. The distribution $F_{N2}(\varepsilon_2)$ corresponding to the best approximation is given in Fig. 4, curve 1. The calculations of the spectra recorded at other temperatures yield similar results. Note that since $\sigma(\varepsilon_2, B) = \sigma(-\varepsilon_2, B)$, we do not discriminate between deformations of contraction and expansion.

If all directions of uniaxial deformation in the crystal are equally plausible, we can, using the function $F_{N2}(\varepsilon_2)$, estimate the distribution function of divacancies by absolute value of total uniaxial deformation $F_N(\varepsilon)$,

$$F_N(\varepsilon) = \int\limits_0^{\varepsilon} F_{N1}(\varepsilon)\, F_{N2}(\varepsilon - \varepsilon_z)\, d\varepsilon_z\,, \tag{4}$$

where $F_{N1}(\varepsilon_z)$ is the distribution function of divacancies by the value of the deformation component along the z-axis (ε_z),

$$F_{N1}(\varepsilon_z) = \frac{1}{4\pi\varepsilon_z} \int\limits_{\varepsilon_z}^{\infty} \frac{F_{N2}(\varepsilon_2)}{\sqrt{(\varepsilon_2/\varepsilon_z) - 1}}\, d\varepsilon_2\,. \tag{5}$$

The distribution function $F_N(\varepsilon)$ computed by $F_{N2}(\varepsilon_2)$ (see Fig. 4, curve 1) from (4), (5) is supplied in Fig. 4, curve 2.

Therefore, through the analysis of the spectrum of divacancies in neutron-irradiated silicon, the average value of uniaxial lattice deformation in the matrix $\varepsilon \approx 10^{-4}$ was found, and a distribution function of divacancies by absolute values of uniaxial deformations in DR was derived. Generally speaking, for the most complete description of a disorderly deformed material, it is necessary to know the distribution $F_V(\varepsilon)$ which at given ε determines the relative portion of the crystal volume with lattice deformation equal to ε: $dV = VF_V(\varepsilon)\, d\varepsilon$. At inhomogeneous distribution of divacancies in the volume, the functions $F_N(\varepsilon)$ and $F_V(\varepsilon)$ are obviously different. However, in case of high radiation doses ($D \gtrsim 2 \times 10^{19}$ cm^{-2}), as the result of multiple DR overlapping, a statistically homogeneous distribution of divacancies seems to form; then $F_N(\varepsilon)$ and $F_V(\varepsilon)$ must coincide. In particular, the average deformation value $\langle\varepsilon\rangle = 1.1 \times 10^{-3}$ computed by $F_N(\varepsilon)$ seems to be also the average value of deformation in the crystal bulk (or in the DR bulk for radiation doses $< 10^{19}$ cm^{-2}).

References

[1] V. D. AKHMETOV, V. V. BOLOTOV, L. S. SMIRNOV, and V. A. KHARCHENKO, phys. stat. sol. (a) **75**, 601 (1983).
[2] T. O. BALDWIN, J. T. THOMAS, and J. K. HALDEMAN, J. appl. Phys. **50**, 3748 (1979).
[3] M. L. SWANSON, J. R. PARSONS, and C. W. HOELKE, in: Radiation Effects in Semiconductors, Ed. J. W. CORBETT and G. D. WATKINS, L.-N.Y.-Paris, Gordon and Breach 1971 (p. 359).
[4] L. S. SMIRNOV (Ed.), Questions of Radiation Technology of Semiconductors, Izd. Nauka, Novosibirsk 1980 (in Russian).
[5] A. V. DVURECHENSKII and A. A. KARANOVICH, Fiz. Tekh. Poluprov. **19**, 1944 (1985).
[6] G. D. WATKINS and J. W. CORBETT, Phys. Rev. **138**, A543 (1965).
[7] J. C. BOURGOIN and M. LANNO, Point Defects in Semiconductors, Izd. Mir, Moscow 1985 (in Russian).
[8] L. D. LANDAU and E. M. LIFSHITS, Theory of Elasticity, Izd. Nauka, Moscow 1987 (in Russian).
[9] R. C. NEWMAN, Rep. Progr. Phys. **45**, 1163 (1982).
[10] F. COTTON and J. WILKINSON, Fundamental of Nonorganic Chemistry, Izd. Mir, Moscow 1979 (in Russian).
[11] A. GORDON and R. FORD, Handbook of Chemist, Izd. Mir, Moscow 1976 (in Russian).

(Received July 6, 1988)

phys. stat. sol. (a) **111**, 265 (1989)

Subject classification: 75.60; S1.2

Physics Department, Faculty of Science, Ain-Shams University, Cairo[1])

Effect of Rolling and Heat Treatment on the Irreversible Magnetization in Nickel–Chromium Wire

By

A. H. WAFIK

The mean potential difference (U_B) of Barkhausen jumps and the frequency of the cluster of jumps (F_B) are measured as a function of the percentage of thickness reduction. An increase of the percentage of thickness reduction up to 55% causes a decrease of U_B and F_B. U_B vanishes at certain values of the magnetizing frequencies. The vanishing magnetizing frequency depends on the percentage of thickness reduction. The vanishing of U_B is attributed to the produced eddy current in the sample. Heat treatment has a considerable effecton both U_B and the vanishing magnetizing frequency.

Die mittlere Potentialdifferenz (U_B) von Barkhausensprüngen und die Frequenz der Sprungcluster (F_B) werden als Funktion der prozentualen Dickenreduzierung gemessen. Der prozentuale Anstieg der Dickenreduzierung bis zu 55% verursacht den Abfall von U_B und F_B. U_B verschwindet bei gewissen Werten der Magnetisierungsfrequenz. Die verschwindende Magnetisierungsfrequenz hängt von der prozentualen Dickenreduzierung ab. Das Verschwinden von U_B wird den Wirbelströmen in der Probe zugeordnet. Wärmebehandlung hat einen beträchtlichen Einfluß sowohl auf U_B als auch auf die verschwindende Magnetisierungsfrequenz.

1. Introduction

The Barkhausen jumps (BJ) are related to the irreversible motion of domain walls [1]. The motion of domain walls is affected by dislocation density, grain boundaries, eddy currents, etc., B. J. are strongly affected by the different kinds of mechanical deformations [2].

The aim of this work is to study the effect of rolling process and heat treatment on B. J. in Ni–Cr wire and to compare the obtained results with the results of a previous work [3] on the effect of plastic tensile deformation on B.J. in the same kind of material.

2. Experimental

The examined material is a nickel–chromium wire (Nichrome) of composition 60 Ni–16 Cr–24 Fe. The mean diameter of the wire is 0.4 mm. The mean potential difference of Barkhausen jumps (U_B) and the frequency of the cluster of jumps (F_B) were measured by an induction technique. The block diagram of the measuring circuit and the technical data of both the magnetizing and pick-up coils are given in [4, 5]. U_B and F_B of eight samples of equal length (150 mm) were measured in the as-received state. The samples showed nearly the same values. The average U_B is considered to represent the as-received state. Then five samples were subjected to five different percentages of thickness reduction (using a high-mill). The percentages of

[1]) Abbassia, Cairo, Egypt.

thickness reduction were 15, 20, 25, 35, and 55%, respectively. Each individual percentage of thickness reduction was established in one pass of an as-received sample through the mill. The percentage of thickness reduction was calculated as the ratio of $(h_0 - h_f)/h_0$, where h_0 and h_f are the initial and final thickness, respectively.

One as-received sample was annealed at 500 °C and another one was annealed at 1100 °C. The annealing time was one hour. The samples were left to cool slowly in the oven. The heat-treated samples are abbreviated hereafter as sample-500 °C and sample-1100 °C, respectively.

3. Effect of Rolling on U_B

Fig. 1 shows the behaviour of U_B as function of the magnetizing current for samples rolled to different percentages of thickness reduction. U_B of the as-received sample is represented on the same figure. The sample of 15% thickness reduction shows the highest values of U_B over the whole range of the used magnetizing field. The increase of the percentage of thickness reduction up to 55% causes the decrease of U_B. Fig. 2a shows the relation between the percentage of thickness reduction (percentage of deformation) and U_B measured at constant magnetizing current of 0.15 A ($H \approx \approx 0.20 \times 10^4$ A m^{-1}). The corresponding values of F_B are represented on Fig. 2b.

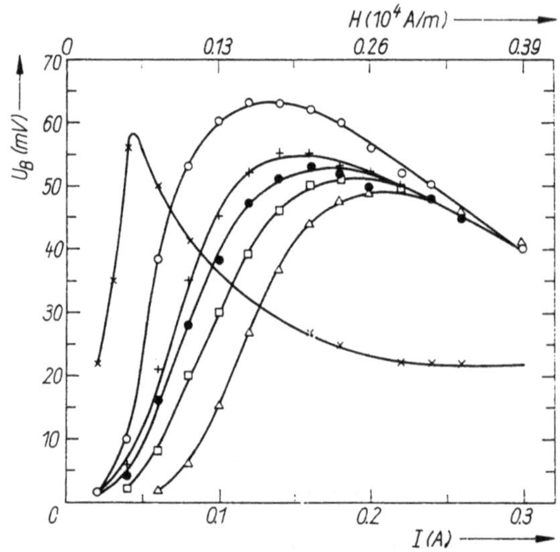

Fig. 1

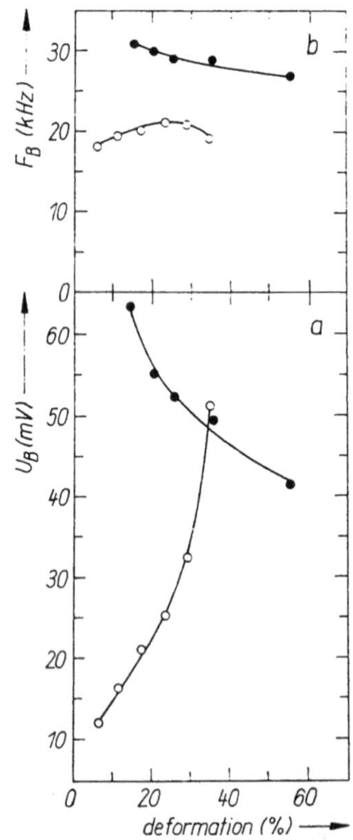

Fig. 2

Fig. 1. U_B of as-received and rolled samples as function of the magnetizing current; × as-received, percentage of thickness reduction ○ 15, + 20, ● 25, □ 35, △ 55%

Fig. 2. Dependence of a) U_B and b) F_B on the percentage of deformation at constant magnetizing current of 0.15 A; ● rolling, ○ tension

F_B is considered to be a function of the number of jumps per unit time. The higher the frequency of cluster the larger is the number of jumps [4]. To compare between the effect of both rolling and plastic tensile deformation on B. J., the previous results [3] showing the change of U_B and F_B (at constant mangetizing current of 0.15 A) with plastic tensile deformation of the same kind of wire are represented on Fig. 2a and b, respectively. In case of rolling the increase of the deformation percentage causes the decrease of U_B, while the increase of deformation percentage due to plastic tensile strains (5.4, 10.8, 17.0, 23.0, 28.8, and 34.0%) causes the increase of U_B. Also U_B and F_B of the rolled samples are higher than those of the plastically tensiled samples (except U_B which is corresponding to 34% tensile deformation because this is the sample which showed complete fracture [3]). This means that the rolled samples show a higher number of jumps and consequently higher values of U_B than the tensiled samples. Dislocations have a pinning effect on the motion of the domain walls. More pinning sites would lead to more Barkhausen jumps [2]. Thus, one can conclude that the dislocation network formed during the rolling process are of higher density than that formed by plastic tensile strain.

Fig. 3 shows U_B at constant magnetizing current of 0.15 A plotted as function of the magnetizing frequency for the as-received and the rolled samples. U_B increases with the increase of the magnetizing frequency till it reaches a maximum value, then it begins to decrease till it vanishes at a certain value of the magnetizing frequency (the vanishing magnetizing frequency). U_B of samples of 15, 20, 25, 35, and 55% thickness reduction vanishes at 200, 180, 170, 150, and 115 Hz, respectively, while U_B of the as-received sample vanishes at 400 Hz. The decrease of the vanishing magnetizing frequency with the increase of the percentage of thickness reduction is represented in Fig. 4, curve a, the deformation percentage of the as-received sample is considered to be zero relative to the other rolled samples. In previous works it was considered that the higher the vanishing magnetizing frequency, the lower is the produced eddy current in the sample [3, 4]. Consequently, the increase of the percentage of thickness reduction causes the increase of the produced eddy current in the sample. To represent the increase of the eddy current graphically with the increase of the percentage of thickness reduction, the reciprocal of the vanishing

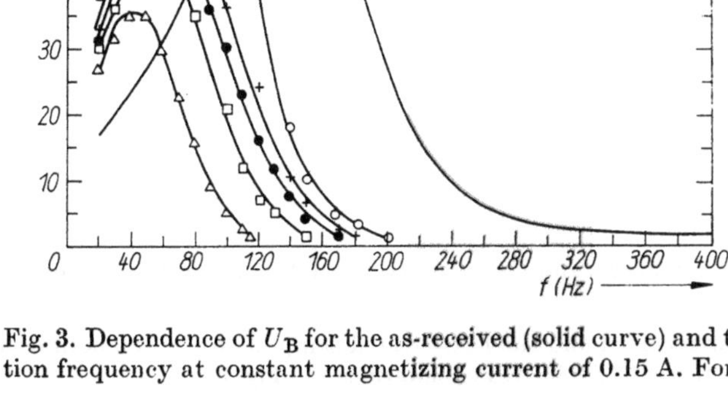

Fig. 3. Dependence of U_B for the as-received (solid curve) and the rolled samples on the magnetization frequency at constant magnetizing current of 0.15 A. For notations see Fig. 1

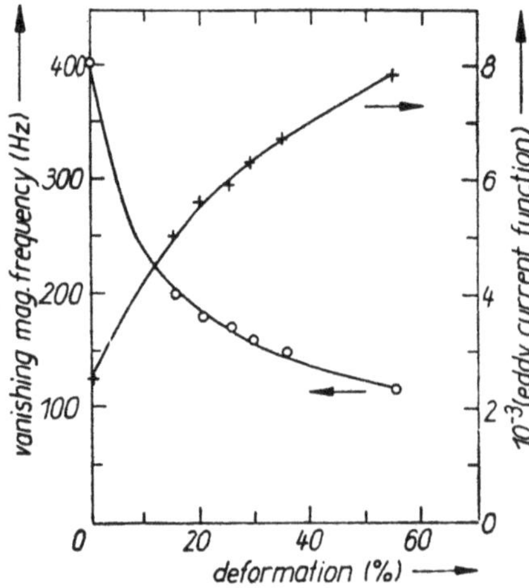

Fig. 4. Dependence of the vanishing magnetizing frequency (○, curve a) and the eddy current (+, curve b) on the percentage of thickness reduction (percentage of deformation)

magnetizing frequency is considered to be a function (in arbitrary units) of the produced eddy current. The increase of the produced eddy current function with the increase of the percentage of thickness reduction is shown of Fig. 4, curve b.

4. Heat Treatment

Fig. 5 shows the effect of heat treatment at 500 and 1100 °C on the behaviour of U_B for the examined Nichrome wire. The heat-treated samples show lower U_B values than the as-received sample over the whole range of the used magnetizing field. The sample-1100 °C shows higher values of U_B than sample-500 °C. In the equilibrium

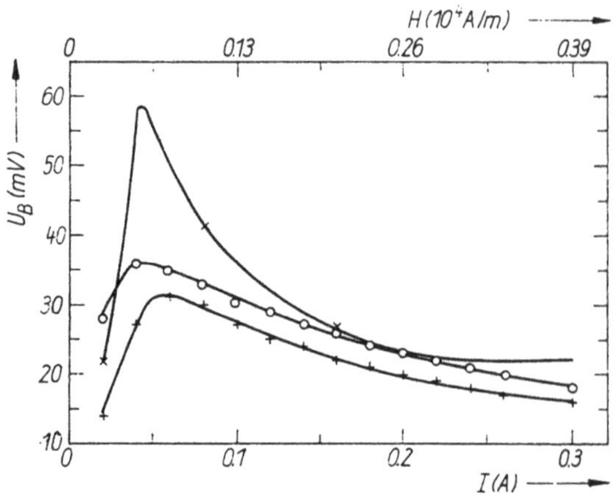

Fig. 5

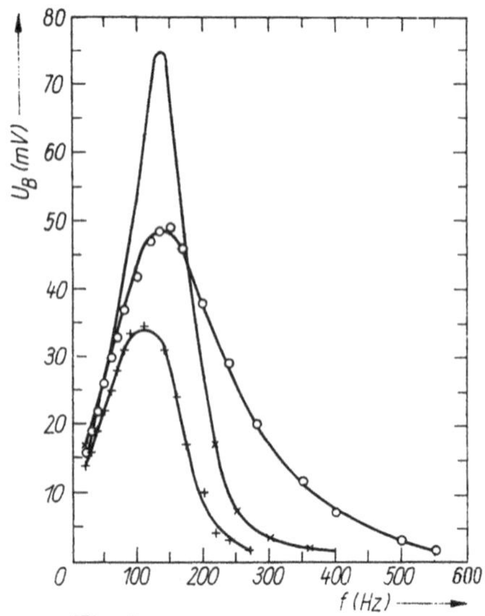

Fig. 6

Fig. 5. U_B of as-received and heat-treated samples as function of the magnetizing current; × as-received, ○ 1100 °C, + 500 °C

Fig. 6. Dependence of U_B for the as-received and the heat-treated samples on the magnetization frequency at constant magnetizing current of 0.15 A. For notations see Fig. 5

phase diagrams of binary chromium–nickel alloy [6] and iron–nickel alloy [7], the structure of the alloy and the amount of each phase present are dependent on the nickel content, heat treatment, cooling rate, and amount of cold working. The heat treatment dependence of U_B in the examined ternary alloy could be related to the formation of different alloy phases.

Fig. 6 shows U_B, at constant magnetizing current of 0.15 A plotted as function of the magnetizing frequency for the as-received and heat-treated samples. The vanishing magnetizing frequencies of sample-1100 °C and sample-500 °C are nearly at 550 and 270 Hz, respectively, while U_B due to the as-received sample vanishes at intermediate value of 400 Hz. This means that the produced eddy current in sample-1100 °C is of lower magnitude than in the as-recived sample, while the produced eddy current in sample-500 °C is of higher magnitude than in the as-received sample. The magnitude of the produced eddy current could be related to the microstructure of the sample [8]. Thus, the change of the produced eddy current values could be attributed to the formation of two different phases as a result of heat treatment.

References

[1] K. Stierstadt, Der magnetische Barkhausen-Effekt, Springer Tracts in Modern Physics, Vol. 40, Springer-Verlag, 1966 (p. 1).
[2] A. H. Wafik, J. Magnetism magnetic Mater. **42**, 23 (1984).
[3] A. H. Wafik and I. A. Ibrahim, phys. stat. sol. (a) **98**, K141 (1986).
[4] A. H. Wafik and S. A. Mazen, Solid State Commun. **61**, 523 (1987).
[5] A. H. Wafik, phys. stat. sol. (a) **91**, K133 (1985).
[6] M. Hansen and K. Anderko, Constitution of binary alloys, McGraw-Hill Publ. Co., 1958 (p. 541).
[7] Metals Handbook, Amer. Soc. Metals, Metals Park, Ohio, 1948 (p. 1211).
[8] A. H. Wafik, phys. stat. sol. (a) **104**, K133 (1987).

(Received August 16, 1988)

phys. stat. sol. (a) **111**, 271 (1989)

Subject classification: 76.30; 61.80; 66.70; S11

Institute of Molecular Physics, Polish Academy of Sciences[1]) *(a), and Institute of Physics, A. Mickiewicz University*[2]) *(b), Poznań*

EPR Studies of $LiNaSO_4$ Cr^{6+} Doped Monocrystals after X-Ray Irradiation

By

S. Waplak (a) and T. Krajewski (b)

The EPR spectra of the paramagnetic centres SO_4^-, SO_3^-, and CrO_4^{3-} are used to probe the local dynamics of $LiNaSO_4$ crystal in the 100 to 500 K temperature range. It is shown that high mobility of Li^+ ions takes place well below the phase transition temperature to the fast-ion-conducting phase. The motions of Li^+ ions and SO_4^{2-} anions are briefly considered in relation to the phase transition mechanism.

Die EPR-Spektren der paramagnetischen Zentren SO_4^-, SO_3^- und CrO_4^{3-} werden benutzt, um die lokale Dynamik von $LiNaSO_4$-Kristallen im Temperaturbereich von 100 bis 400 K zu sondieren. Es wird gezeigt, daß die hohe Beweglichkeit der Li^+-Ionen bereits unterhalb der Phasenübergangstemperatur zur Leitungsphase der schnellen Ionen einsetzt. Die Bewegungen der Li^+-Ionen und SO_4^{2-}-Anionen werden kurz bezüglich des Phasenübergangsmechanismus diskutiert.

1. Introduction

Lithium sodium sulphate crystallizes in a hexamolecular unit cell of the dimensions $a_0 = 0.76270$ nm, and $c_0 = 0.98579$ nm [1] in space group P31c. Three crystallographically different sulphate groups are situated on two different threefold axes. Lithium and sodium ions are surrounded by irregular arrangements of four and eight oxygen atoms, respectively. At about 790 K there is a transition into a body-centred cubic phase [2] that exhibits high ionic conductivity [3].

One of the major problems in the study of fast ion conducting materials is to understand the dynamical behaviour of the mobile ion species. Teeters and Frech [4 to 6] studied the influence of temperature on Raman spectra of $LiNaSO_4$ and found a significant broadening of the bands corresponding to external optical modes of lithium ions. The modes get broadened and disappear at temperatures significantly lower than that of the transition to the superionic phase. To explain the observed dynamics of the Raman spectra two alternative mechanisms have been proposed [4 to 6]:

(i) thermally activated local disorder in the sublattice of Li ions;

(ii) the loss of correlation among sulphate ions in an elementary cell in result of anharmonic vibrations of oxygen atoms.

To obtain new information about local dynamics of tetrahedral groups in $LiNaSO_4$, we subjected those crystals to EPR investigations in the temperature range 100 to 500 K. We studied $LiNaSO_4$ crystals pure and doped with Na_2CrO_4, after they had

[1]) Smoluchowskiego 17/19, PL-60-179 Poznań, Poland.

[2]) Grunwaldzka 6, PL-60-780 Poznań, Poland.

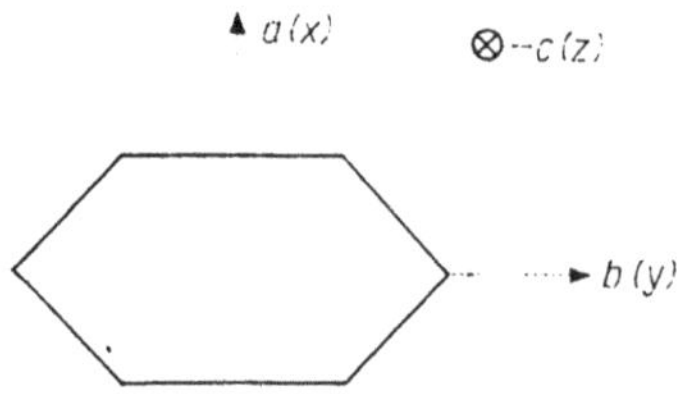

Fig. 1. The reference orthogonal system used for anisotropy studies

been irradiated with X-rays. As a consequence of irradiation we obtained the radicals SO_4^- and SO_3^- in the pure compound and paramagnetic CrO_4^{3-} as well in the doped crystals.

2. Experimental

$LiNaSO_4$ crystals were grown by the dynamical method under isothermal conditions at 315 K from aqueous solutions containing the initial salts in stoichiometric ratio. As the Na_2SO_4 salt was admixed with Na_2CrO_4 at a ratio 0.02 wt% the solution also contained CrO_4^{2-} ions. In order to obtain the paramagnetic centres, the pure and doped crystals were irradiated for 3 to 15 h with X-rays generated in a tungsten X-ray valve operating at 30 kV and 20 mA. The irradiation was performed at room temperature.

In order to find g-tensors and directional cosines we assumed the reference orthogonal system defined on Fig. 1. The EPR measurements were carried out on the X-band spectrometer, manufactured by Radiopan, for temperatures ranging from 100 to 500 K. The sample was maintained at a constant temperature with an accuracy of ± 0.5 K. The necessary cooling and heating was ensured by a flow of cold nitrogen vapour throughout the heating system.

3. EPR Spectra at RT

Single crystals of $LiNaSO_4$ X-ray irradiated at the temperature of liquid nitrogen have been previously studied by Alyabakov et al. [7]. They found the presence of the following centres: SO_4^-, SO_3^-, SO_2^-, and O_3^-. In our samples, which had been annealed at 470 K for 24 h prior to the irradiation, we did not find the centres SO_2^-

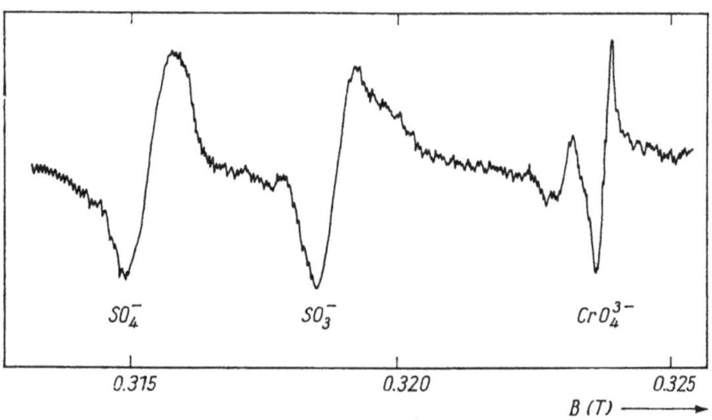

Fig. 2. EPR spectra of SO_3^-, SO_4^-, and CrO_4^{3-} centres for $\mathbf{c} \parallel \boldsymbol{B}_0$

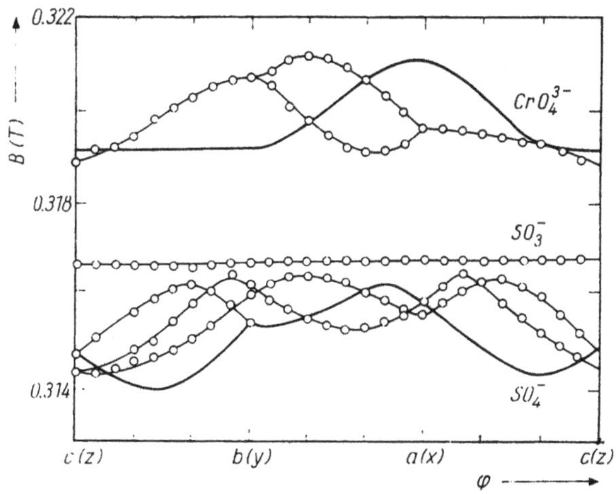

Fig. 3. Anisotropies of SO_4^-, SO_3^-, and CrO_4^{3-} centres for *zy*, *xy*, and *xz* planes

and O_3^-. A comparison of the EPR line intensities of SO_4^- radicals obtained at the same irradiation dose and in the same conditions for pure and for Na_2CrO_4 doped $LiNaSO_4$ samples, showed a significantly higher concentration of SO_4^- radicals in the latter. This may be due to the effect of electron trapping by CrO_4^{3-} centres.

In the $LiNaSO_4$ samples doped with Cr^{6+} ions the paramagnetic centres SO_4^- and CrO_4^{3-} are formed in the processes

$$SO_4^{2-} \rightarrow SO_4^- + e^-$$

and

$$CrO_4^{2-} \rightarrow CrO_4^{3-} - e^- .$$

A similar relation was reported in our paper [8] concerning the investigation of the irradiated pure and Cr^{6+} doped $LiKSO_4$ crystals. Fig. 2 presents the EPR spectra of SO_4^-, SO_3^-, and CrO_4^{3-} centres for the axis $\mathbf{c} \parallel \mathbf{B}_0$. Anisotropies of EPR spectra of these centres in the planes *zy*, *xy*, and *xz* are presented in Fig. 3. The main values of the *g*-tensor and their direction cosines for the centres marked in Fig. 3 with a thick line, are given in Table 1. The other centres SO_4^- and CrO_4^{3-} are arranged at the positions determined by C_6 symmetry operation for a rotation about the *c*(*z*) axis.

Table 1

centre	principal *g*-values	directional cosines with respect to axes		
		a	*b*	*c*
SO_4^-	$g_1 = 2.0045$	−0.58133	0.45372	0.67542
	$g_2 = 2.0111$	−0.66874	−0.73929	−0.07896
	$g_3 = 2.0420$	0.46351	−0.49758	0.73319
SO_3^-	$g = 2.0055$	isotropic		
CrO_4^{3-}	$g_\perp = 1.9735$			
	$g_\parallel = 1.9480$	1	0	0

4. Temperature Dependence of EPR Spectra of SO_4^-, SO_3^-, and CrO_4^{3-} Centres

EPR spectra of the above-mentioned paramagnetic centres show a series of anomalies in their temperature dependences. The amplitude of the EPR line of the SO_4^- centre tends to zero at a temperature of about 480 K while the amplitude of EPR line of the SO_3^- centre reaches the maximum at this temperature. The widths of EPR lines of both these centres do not change. The most interesting and, on the other hand, that which provides the most information on dynamics of Li ions are the changes that occur in the EPR spectrum of CrO_4^{3-} centre. The temperature evolution of this spectrum is shown in Fig. 4. At about 480 K the EPR spectrum of CrO_4^{3-} disappears. Further decreasing temperature, we obtained an EPR spectrum of quite a different structure. A characteristic superhyperfine structure appears in the spectrum as a result of the interaction between the unpaired electron of a Cr^{5+} ion on the orbital $d_{x^2-y^2}$ and the ^7Li nucleus of the spin $I = 3/2$. Anisotropy of this spectrum in the zy plane is presented in Fig. 5. When we again increased the temperature from 100 to 480 K, we could only record the spectrum of the SO_3^- centre and the above described $LiCrO_4^{2-}$ centre. Thus, the transformation process of CrO_4^{3-} to the centre $LiCrO_4^{2-}$ is irreversible. This transformation process is accompanied by a significant change in the main directions of the crystal field tensor. Fig. 6 presents the positions of the $g_{\parallel}$ axis of the centres 1, 2, 3, 4, 5, 6 of CrO_4^{3-} with respect to the $xyz(abc)$ laboratory frame and the positions of the $g_{\parallel}$ axis of the corresponding $LiCrO_4^{2-}$ centres (1', 2', 3', 4', 5', 6'). As a result of the interaction of Cr^{5+} ion with Li^+ nucleus the directions of $g_{\parallel}$ of particular $LiCrO_4^{2-}$ centres deviate by 30° from the xy plane. A centre of $LiCrO_4^{2-}$ is described by the following parameters of the spin-Hamiltonian: $g_{\perp} = 1.9751$, $g_{\parallel} = 1.9593$, and a superhyperfine structure constant $A = 8.9$ mT. Temperature changes in the EPR line amplitude of individual paramagnetic centres measured in the cycles of heating and cooling the sample, are presented in Fig. 7 in arbitrary units disregarding the differences in particular EPR linewidths of individual centres. The temperature changes of EPR line amplitude obtained for the centres SO_3^- and CrO_4^{3-} show the effect of shortening of spin–lattice relaxation time T_1. Since there are no temperature changes in the spin–spin interaction (no temperature chan-

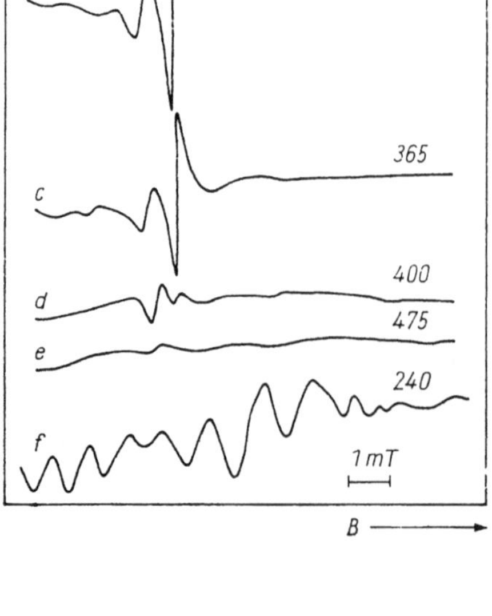

Fig. 4. Temperature evolution of CrO_4^{3-} spectra in heating and cooling processes (5 h X-ray)

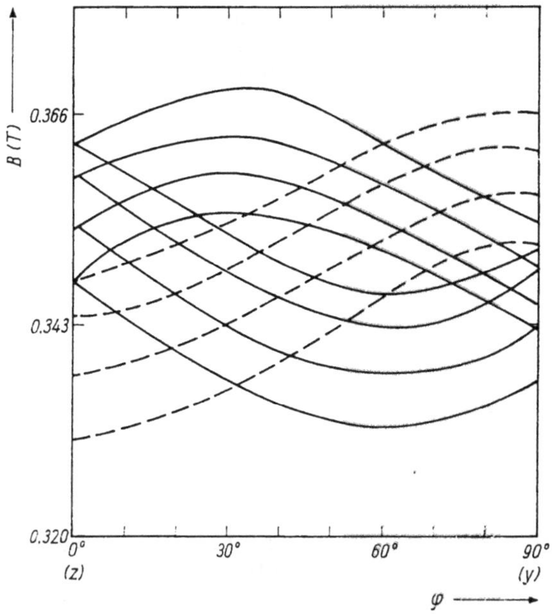

Fig. 5. The spectrum anisotropy of LiCrO$_4^{2-}$ centres in zy plane (x-rotation), the dashed lines correspond to smaller intensity

ges in the width of the line), we may assume that the imaginary part of susceptibility, and, thus, the intensity of EPR line as well, depends on T_1 in the following way [9]:

$$\chi'' \sim \frac{\chi_0 \omega}{(1 + \gamma^2 H_1^2 T_1^2)^{1/2}}, \tag{1}$$

where H_1 is the amplitude of the microwave field and γ the gyromagnetic ratio. The shortening of T_1 at high temperature is probably a consequence of Raman relaxation process due to the coupling of the SO_3^- spin system with the lattice mode A_1 (314 cm^{-1}). A detailed analysis of the relaxational process will be the subject of a separate paper.

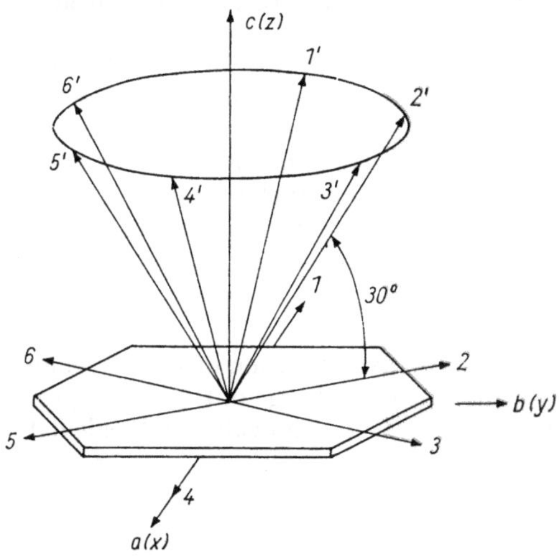

Fig. 6. The relative orientation of $g_{||}$ for CrO$_4^{3-}$ (1 to 6) and LiCrO$_4^{2-}$ (1' to 6') centres in LiNaSO$_4$

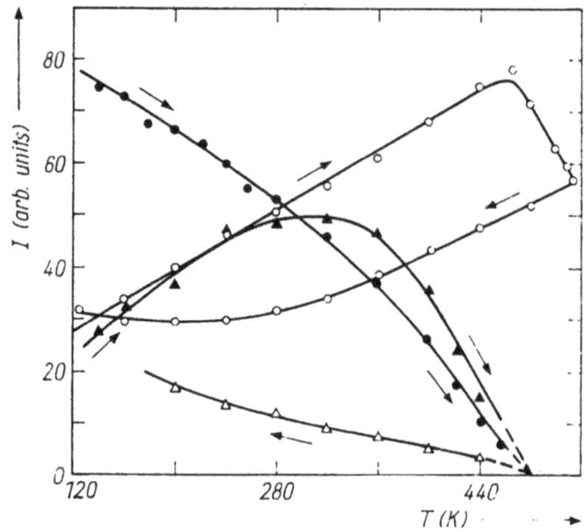

Fig. 7. Temperature dependences of line intensities for SO_4^- (●), SO_3^- (○), CrO_4^{3-} (▲), and $LiCrO_4^{2-}$ (△) centres in $LiNaSO_4$

5. Discussion

Similar to resistivity measurements in $LiNaSO_4$ by Bręczewski et al. [10] for the temperature range 300 to 900 K, three regions characterized by different slope coefficients can be distinguished in the slope $\ln \varrho_{ij} = f(1/T)$. An abrupt change in the slope of $\ln \varrho_{ij}$ at 790 K is related with a change in the symmetry of the crystal from hexagonal to cubic in which the crystal shows superionic properties. The change of slope coefficient of $\ln \varrho_{ij}$ at about 440 K indicates, in their opinion [10], an increase in the lithium ion contribution to ionic conductivity of the crystals.

Our results show that Li^+ ions in the doped $LiNaSO_4$ are coupled with Cr^{5+} ions at a temperature of about 480 K. This coupling is not dipole–dipole as evidenced by the temperature evolution of the spectrum with superhyperfine structure and its isotropic character. Thus we conclude that the presence of the superhyperfine structure due to Li nuclei demands the contact interaction between Li nuclei and Cr^{5+} ion and most probably occurs through the hopping of Li^+ ion to the nearest of the oxygens of the tetrahedral group CrO_4^{3-}. As a consequence a molecular bond Li–O–Cr is formed on the orbital $1s2pd_{x^2-y^2}$ which results in a significant change in the crystal field gradient direction at the centre $LiCrO_4^{2-}$ with respect to its direction at the centre CrO_4^{3-} (Fig. 6).

Such a hopping of Li^+ ion should take place over a significant distance. However, a large amplitude of Li^+ oscillation and local disorder like Brownian motion are strongly suggested by Raman data [4].

This hopping in chromium doped $LiNaSO_4$ is additionally enhanced by the internal electric field gradient evoked in the lattice by CrO_4^{3-} anion replacing SO_4^{2-}.

Consider the case of ionic motion in which Li^+ ions are made to move under an internal electric field E. As a result the height of the barrier for Li–O_1 jump will be diminished to [11]

$$\Delta V = \Delta V_0 - (1/2) qaE\,, \tag{2}$$

where $\Delta V = 480\ \text{K} \triangleq 0.04$ eV, $\Delta V_0 = 0.5$ eV is energy activation of an Li^+ removal from the site position into a vacant site [6], q is the ionic charge of the CrO_4^{3-} group, a the interionic distance, and E the internal electric field. The value of E calculated from (2),

$$E = \frac{2(\Delta V - \Delta V_0)}{qa}\,, \tag{3}$$

is reciprocally proportional to the a-value and will be increased with increasing Li^+ oscillation amplitude. The E value for $a = 0.187$ nm (this is Li–O_1 distance in rigid lattice) is equal 1.4×10^7 V/cm. As a result Li^+ ion is captured to CrO_4^{3-} group. There arises also question about the motion of SO_4 tetrahedra in the temperature region under investigation. Because the linewidths of SO_4^- and SO_3^- are temperature independent in the region (100 to 480) K i.e. $\Delta\omega\tau_c \gg 1$ we conclude that SO_4 tetrahedra motions are still correlated. Unfortunately EPR signals disappears far below the phase transition temperature to the fast-ion conducting phase and we lost possibility of direct EPR investigation of phase transition mechanism. We could only suggest that similar to Li_2SO_4 [12] which belongs to the same family as $LiNaSO_4$, reorientation of sulphate tetrahedra above 790 K is a significant factor in enhancing of Li^+ ion diffusion rate.

6. Conclusion

(i) Superhyperfine contact interaction of Li^+ ion with CrO_4^{3-} proves a large amplitude of Li vibration at 480 K.

(ii) In the EPR limit ($\omega_0 = 6 \times 10^{10}$ s^{-1}) SO_4 groups behave like a correlated sublattice.

(iii) Li^+ ions at first undergo oscillation with increasing amplitude far below the transition temperature to the fast-ion-conducting phase.

Acknowledgement

This work was supported by the Institute of Low Temperature and Structure Research, Polish Academy of Sciences, Wrocław, Poland, Grandt No. CPBP 01.12.

References

[1] B. Morosin and D. L. Smith, Acta cryst. **22**, 906 (1967).
[2] K. Schroeder, A. Kvist, and H. Ljungmark, Z. Naturf. **27a**, 1252 (1972).
[3] A. M. Josefson and A. Kvist, Z. Naturf. **24a**, 466 (1969).
[4] D. Teeters and R. Frech, Solid State Ionics **5**, 437 (1981).
[5] D. Teeters and R. Frech, Phys. Rev. B **26**, 4132 (1982).
[6] D. Teeters and R. Frech, Phys. Rev. B **26**, 5897 (1982).
[7] A. A. Alybakov, V. M. Bujko, and K. Sharsheev, Cryst. Res. Technol. **16**, 611 (1981).
[8] S. Waplak, J. Minge, and T. Krajewski, phys. stat. sol. (a) **99**, K27 (1987).
[9] R. W. Kedzie, J. R. Shane, and M. Kestigian, Phys. Letters **11**, 286 (1964).
[10] T. Bręczewski, B. Mróz, and P. Piskunowicz, to be published.
[11] Suresh Chandra, Superionic Solids, North-Holland Publ. Co., 1981.
[12] L. Nilson, J. O'Thomas, and B. C. Tofield, J. Phys. C **13**, 6441 (1980).

(Received April 25, 1988; in revised form October 3, 1988)

phys. stat. sol. (a) 111, 279 (1989)

Subject classification: 75.80; S1.1; S1.2

Departamento de Electricidad y Electrónica, Facultad de Ciencias, Universidad del Pais Vasco, Bilbao[1]) *(a) and*
Department of Electrophysics, The Technical University of Denmark, Lyngby[2]) *(b)*

Magnetoelastic Properties of Some Fe-Rich Fe–Co–Si–B Metallic Glasses

By

J. GUTIÉRREZ (a), J. M. BARANDIARÁN (a), and O. V. NIELSEN (b)

Magnetoelastic measurements are performed in metallic glasses of composition $(Fe_{0.79}Co_{0.21})_{75+x} \cdot Si_{15-1.4x}B_{10+0.4x}$ ($x = 0, 2, 4, 6, 8, 10$) by means of the resonance–antiresonance method. High values of k (magnetoelastic coupling coefficient) and of the ΔY effect are observed in some compositions, reaching 0.7 for k and 50% for ΔY in the as-quenched state. Annealing treatments further improve these values.

An amorphen Metallen der Zusammensetzung $(Fe_{0,79}Co_{0,21})_{75+x}$ $Si_{15-1.4x}B_{10+0.4x}$ ($x = 0, 2, 4, 6, 8, 10$) werden die magnetoelastischen Größen unter Verwendung der Resonanz–Antiresonanzmethode bestimmt. Hohe Werte von k (der magnetoelastische Koeffizient) sowie vom ΔY-Effekt werden für einige der Legierungen gefunden, mit Werten bis zu 0,7 für k und 50% für ΔY im abgeschreckten Zustand. Noch höhere Werte werden durch Ausglühen erreicht.

1. Introduction

Amorphous ferromagnetic alloys containing Fe, Co, and Ni as metallic components show outstanding magnetic and magnetoelastic properties of great interest in technological applications. This fact is due to the absence of magnetocrystalline anisotropy and structural defects like grain boundaries.

The magnetoelastic coupling coefficient k [1], defined as the fraction of elastic energy generated when the material is magnetically excited, gives us a good idea of the utility of the alloy in technological applications, such as sensors and transducers. This is intimately related to the change in Young's modulus, the ΔY effect, under the influence of applied magnetic fields. In this paper we discuss Young's modulus, Y, the magnetoelastic coupling coefficient, k, and the damping factor, Q^{-1}, from measurements performed in as-quenched samples of composition $(Fe_{0.79}Co_{0.21})_{75+x}Si_{15-1.4x}B_{10+0.4x}$ ($x = 0, 2, 4, 6, 8, 10$).

These samples show the high values of the saturation magnetostriction, λ_s, and spontaneous magnetization, J_s [2], needed for a good magnetoelastic response.

The sample with composition $x = 10$ is commercially available from Allied Co. as Metglas 2605CO and has shown a good value of k after transverse field annealing [3], whereas the other compositions show a good response to stress-annealing which induces a transverse magnetic anisotropy. The sample with $x = 0$ has $Si_{15}B_{10}$ as a glass former, the same glass former used by a large number of investigators in the Fe/Co system.

[1]) Leioa, Aptdo. 644, E-48080 Bilbao, Spain.
[2]) Building 322, DK-2800 Lyngby, Denmark.

2. Experimental Method

Samples with 0.5 mm width and 20 to 30 μm thickness were obtained by the single roller quenching technique in the form of long ribbons. Pieces of about 10 cm in length were cut for the measurements. The measuring method used in this work is based on the magnetic excitation of the sample and consequent detection of the magneto-elastic resonances [4]. To expedite the measurement, we built the fully automatized experimental set-up shown in Fig. 1.

The sample lies freely along the axis of an assembly of three coaxial solenoids with the following functions: (i) generation of the bias field H, (ii) generation of the excitation ac field h, (iii) pick-up and compensation for obtaining directly the sample susceptibility.

A lock-in amplifier measures the real and imaginary components of the response of the material. A multimeter continuously monitors the bias field H applied to the sample. The system is controlled by a microcomputer. The resonance and antiresonance frequencies, ω_r and ω_a, respectively, are obtained from the modulus of the magnetic susceptibility, χ. The frequencies of the maximum ω_M, and minimum, ω_m, of the real part of the susceptibility, χ', are also stored by the microcomputer. Noting that $\chi(\omega_a) = 0$, we obtain from [4] and [5] the following expressions for the magnetoelastic coupling coefficient, k, the Young's modulus, Y^H, and the quality factor, Q, at a given bias field:

$$k^2 = \frac{\pi^2}{8}\left(1 - \frac{\omega_r^2}{\omega_a^2}\right),$$

$$Y^H = \left(\frac{l}{\pi}\right)^2 \omega_r^2 \varrho ,$$

$$Q = \frac{\omega_r}{\omega_m - \omega_M},$$

where l and ϱ are the length and density of the sample, respectively. Measurements were performed around the first harmonic, $n = 1$, which appears at about 20 kHz.

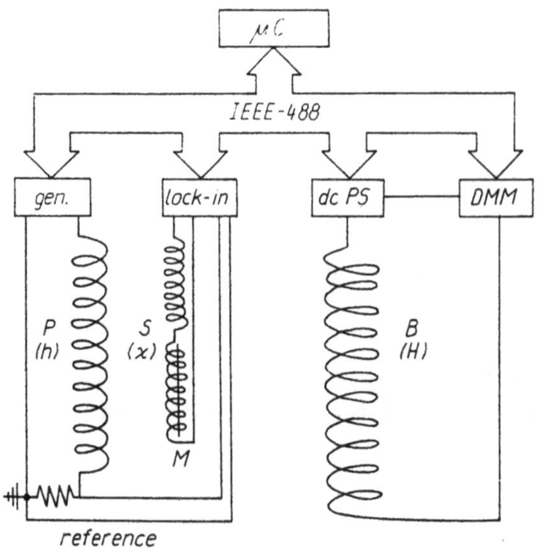

Fig. 1. Experimental set-up: μC microcomputer, gen sine wave generator, dc PS dc power supply (current regulated), DMM digital multimeter; P primary, S secondary, and B bias coils, respectively; M magnetic sample

Table 1

Magnetic and magnetoelastic parameter of the series $(Fe_{0.79}Co_{0.21})_{75+x}Si_{15-1.4x}B_{10+0.4x}$

x	$10^6\lambda_s$	J_s (T)	Y_s (GPa)	$\Delta Y/Y_s$ (%)	k_{max}	$Q(k_{max})$	$H(k_{max})$ (A/m)
0	42	1.5	160	50	0.52	15	40
2	38	1.6	179	42	0.66	47	56
4	49	1.6	160	15	0.22	253	440
6	52	1.65	155	16	0.29	42	640
8	43	1.7	165	28	0.43	63	240
10	28	1.75	147	19	0.29	222	224

3. Results

In Table 1 we show the results obtained for the magnetoelastic parameters Y_s (Young's modulus at magnetic saturation), $\Delta Y/Y_s$ (relative magnitude (in %) of the ΔY effect), k_{max} (maximum magnetoelastic coupling coefficient), and $Q(k_{max})$ (quality factor corresponding to k_{max}) together with some magnetic parameters of the samples taken from [2].

In all the samples studied, the maximum ΔY effect takes place in the presence of low external magnetic fields H, of the order of a few hundred A/m. Moreover, the maximum values of the magnetoelastic coupling coefficient k and the damping factor Q^{-1} are also observed in this range. Fig. 2 shows the typical behaviour of these parameters as a function of the bias field. Values of ΔY and correspondingly k_{max} are quite high in some compositions.

The behaviour of λ_s and $H(k_{max})$ as a function of composition shows a certain correlation (see Fig. 3). We assume that the field $H(k_{max})$ is about the value of the anisotropy field, $H = 2K/M_s$, determined by the internal stress. The magnitude of the anisotropy is $\frac{3}{2}\lambda_s\sigma$, where σ is the local value of the internal stress. If this stress is almost equal in all the compositions studied in this work, the bias field for k_{max} should follow the value of the saturation magnetostriction, λ_s.

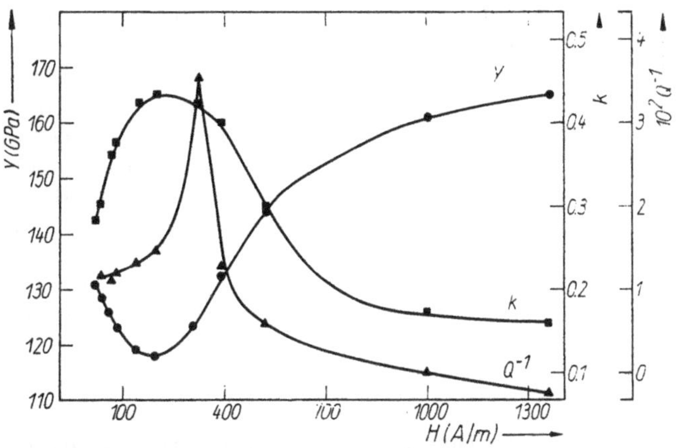

Fig. 2. Typical behaviour of the magnetoelastic parameters as a function of the bias field, H, for the composition $Fe_{65.57}Co_{17.43}Si_{3.8}B_{13.2}$ ($x = 8$)

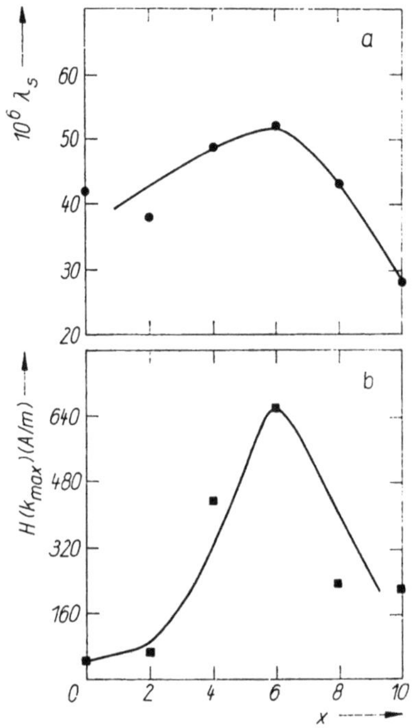

Fig. 3

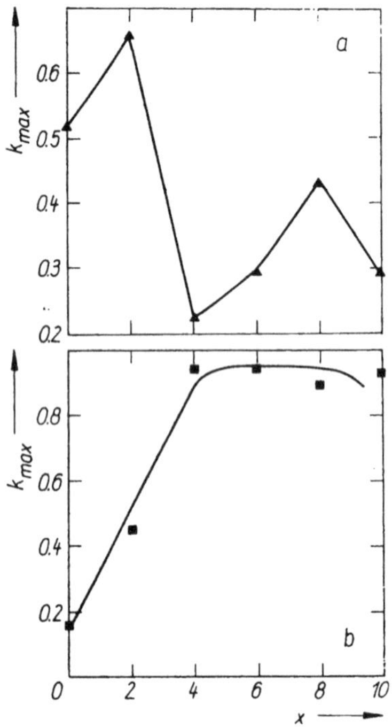

Fig. 4

Fig. 3. a) Variation of λ_s and b) $H(k_{max})$ vs. composition

Fig. 4. a) Experimental and b) theoretical values obtained for k_{max} vs. composition. The latter assumes a uniform transverse anisotropy in all the samples

A rough estimate of the validity of this assumption may be carried out by means of the theoretical expression for the magnetoelastic coupling coefficient k which was derived in [6 to 8]. We anticipate that the assumption made in [6 to 8], i.e. a homogeneous anisotropy perpendicular to the ribbon axis, is not fulfilled in our case. As an estimate, however, we assume that the anisotropy originates from a uniform stress perpendicular to the ribbon axis. Under this assumption we compare in Fig. 4 the theoretical and the experimental values for k_{max}.

The poor agreement between theoretical calculations and experimental results shows that it is not possible to consider the internal stress uniform and perpendicular to the longitudinal axis of the ribbon. It is more probable that the fabrication procedure introduces an irregular distribution of internal stresses. In a more realistic model, this distribution could be Gaussian [9], with equal magnitudes of the mean value of the internal stress for all compositions, but with different widths. We consider the values of k to be rather high for unannealed samples. We have therefore started a series of annealing treatments to homogenize the samples. Preliminary results show that transverse field annealing for 2 h, for the composition $x = 0$, showed a change from 0.52 to 0.81 in the value of k_{max}.

4. Conclusions

The samples studied in this work, belonging to the series $(Fe_{0.79}Co_{0.21})_{75+x}Si_{15-1.4x}B_{10+0.4x}$, $x = 0, 2, 4, 6, 8, 10$, have high values of λ_s, J_s, and Y_s. The maximum experimental values of k are found at very low values of H and near $x = 0$. As-quenched samples do not show a clear correlation between magnetoelastic parameters and composition, due mainly to the inhomogeneity of the internal stress distribution.

Acknowledgements

One of us, J.G., wishes to thank to the Basque Government for the concession of a grant. This work has been supported by the University of the Basque Country, with charge to the project No. 310.05-7/86. We wish to thank M.L. Fdez-Gubieda for the annealing treatments in some of the samples. We would like also to thank Dr. H. T. Savage for his careful reading of the manuscript.

References

[1] A. Hernando, Los nuevos materiales: vidrios metálicos, Eudema, Madrid 1987 (p. 145).
[2] M. L. Fdez-Gubieda, O. V. Nielsen, and P. Voitanik, Proc. Symp. Magnetic Properties of Amorphous Metals, ed. A. Hernando, V. Madurga, M. C. Sánchez Trujillo, and M. Vázquez, Elsevier Science Publishers, 1987 (p. 38).
[3] C. Modzelewski, H. T. Savage, L. T. Kabacoff, and A. E. Clark, IEEE Trans. Magnetics **17**, 2837 (1981).
[4] A. Hernando, V. Madurga, M. Liniers, and H. T. Savage, Rev. Real Acad. Ci. exact., físic. natur. (Madrid) **75**, 949 (1981).
[5] H. T. Savage and R. Abbundi, IEEE Trans. Magnetics **14**, 545 (1978).
[6] M. L. Spano, K. B. Hathaway, and H. T. Savage, J. appl. Phys. **53**, 2667 (1982).
[7] J. D. Livingston, phys. stat. sol. (a) **70**, 591 (1982).
[8] A. E. Clark, H. T. Savage, and M. L. Spano, Physica (Utrecht) **119B**, 187 (1983).
[9] M. Vázquez, W. Fernengel, and H. Kronmüller, phys. stat. sol. (a) **80**, 195 (1983).

(Received September 19, 1988)

phys. stat. sol. (a) **111**, 285 (1989)

Subject classification: 75.50; S1.2; S1.4

Department of Metal Physics, Charles University, Prague[1]) *(a) and*
Department of Physics, Ural State University, Sverdlovsk[2]) *(b)*

Magnetic Behaviour of $TmCu_2$ below T_N

By

P. SVOBODA (a), Z. SMETANA (a), and A. V. ANDREEV (b)

Magnetisation isotherms below Néel temperature in magnetic fields up to 5 T are studied on $TmCu_2$ single crystal. Besides a large magnetocrystalline anisotropy (with easy *b*-axis) four antiferromagnetic phases are observed in the narrow temperature region.

Ниже температуры Нееля проведены исследования намагничености монокристалла $TmCu_2$ в магнитных полях до 5 Т. Кроме большой магнитокристаллической анизотропии найдены четыре антиферромагнитной фазы (с легким направлением вдоль оси *b*) в узкой температурной области.

1. Introduction

The RCu_2 intermetallic compounds (R heavy rare earth) crystallize in the orthorhombic $CeCu_2$-type structure [1].The bulk magnetic properties have been studied both in polycrystalline samples [2] and single crystals [3] and an antiferromagnetic ordering with a large magnetocrystalline anisotropy has been revealed in the sequence from Gd to Er. The competition between the weak indirect antiferromagnetic exchange via conduction electrons, the magnetocrystalline anisotropy, the magnetoelastic interactions, and dipole–dipole interactions leads to various types of magnetic structure formed subsequently within a relatively narrow temperature range [4].

The $TmCu_2$ compound, which has been already investigated in a polycrystalline form with respect to the magnetic phase transitions below T_N, displays the lowest ordering temperature of this series, namely $T_N = 6.3$ K [5]. This fact indicates a relatively weak antiferromagnetic coupling between Tm ions. There are several experimental evidences for the existence of four antiferromagnetic phases between 2.5 and 6.3 K [4, 5, 8]. In order to obtain more reliable information the investigation of single crystals is desirable. In this paper we compare the results of magnetic measurements of $TmCu_2$ in the low-temperature range performed both on monocrystalline and polycrystalline samples.

2. Experimental Details

The single crystal of $TmCu_2$ has been prepared by remelting polycrystalline material [5] in an alumina crucible under helium atmosphere at temperature of about 20 K above the melting point with subsequent slow cooling down. The single crystal of dimensions $2 \times 2 \times 1.5$ mm³ has been cut out from the ingot. The X-ray analysis proved the whole sample to be a single crystal with the lattice parameters $a = 0.425(1)$ nm, $b = 0.676(1)$ nm, and $c = 0.724(1)$ nm.

---

[1]) Ke Karlovu 5, CS-121 16 Prague 2, Czechoslovakia.

[2]) Lenin av. 51, SU-620083, Sverdlovsk, USSR.

Magnetisation measurements have been performed by means of a vibrating sample magnetometer in the temperature range from 2.5 to 20 K and in magnetic fields up to 5 T. The single crystal of $TmCu_2$ has been measured in fields applied parallel to all three principal crystallographic directions. The experimental uncertainty in orientation of the sample with respect to the applied field is at maximum 5°.

3. Results and Discussion

In Fig. 1 we compare the magnetisation isotherms obtained at $T = 4.2$ K on the single crystal and on the polycrystalline sample. An anisotropy in magnetic ordering of $TmCu_2$ is clearly visible. The *b*-axis appears to be the easy magnetisation axis; in this direction the magnetisation process consists of two steps. We can consider both steps as the metamagnetic transition of possible spin flip type with critical fields $H_{c1} = 0.06$ T and $H_{c2} = 0.36$ T. Above the second step the magnetisation quickly reaches the saturated value of $(6.9 \pm 0.2)\ \mu_B$ per Tm atom which is in good agreement with the value for a free Tm^{3+} ion. The magnetisation along *a*- and *c*-axes is an order of magnitude lower (does not exceed 1 μ_B/Tm atom) than that along the *b*-axis. The M versus H dependences consist of two approximately linear parts and do not saturate up to the maximum field measured (3.5 T). The corresponding values of the differential susceptibility are $\chi_{a1} = 2.28 \times 10^{-8}\ m^3\ mol^{-1}$ (0 to 1 T), $\chi_{a2} = 1.11 \times 10^{-8}\ m^3\ mol^{-1}$ (1 to 3.5 T), $\chi_{c1} = 3.45 \times 10^{-8}\ m^3\ mol^{-1}$ (0 to 1.5 T) and $\chi_{c2} = 1.11 \times 10^{-8}\ m^3\ mol^{-1}$ (1.5 to 3.5 T). The type of the magnetisation process suggests, moreover, that the ground state is probably an antiferromagnet of Ising type with magnetic moments along the *b*-axis.

In order to study the different magnetic phases which appear when cooling down between 6.3 and 2.5 K [5] (which we denote as AF1, AF2, AF3, and AF4 corresponding to the temperature intervals of existence 6.3 to 4.3 K, 4.3 K to 3.9 K, 3.9 to 3.3 K, and below 3.3 K, respectively) the magnetisation curves along the easy axis at selected temperatures and the temperature dependence of magnetisation in different magnetic fields have been measured. In Fig. 2 the temperature dependences of magnetisation obtained in the field of 0.06 T on the single crystal along each principal crystallographic axis and on polycrystalline sample can be compared. In all these dependences we can see more or less sharp features (maxima or shoulders), which are probably connected with temperatures of magnetic phase transitions denoted as T_N, T_1, T_2, and T_3,

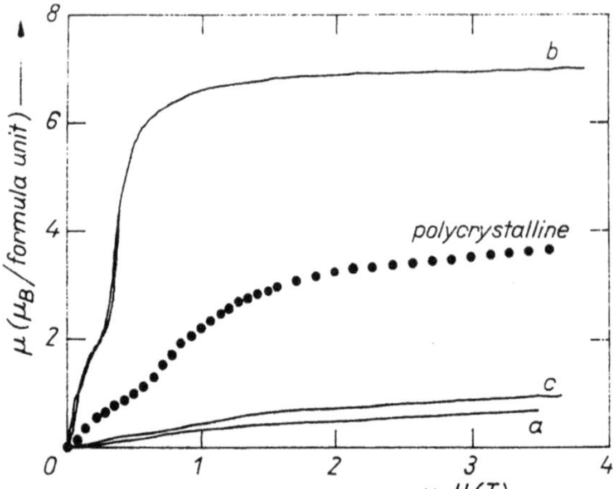

Fig. 1. Magnetisation curves of $TmCu_2$ along each principal crystallographic axis (full lines) in comparison with the polycrystalline one (dotted line) obtained at $T = 4.2$ K

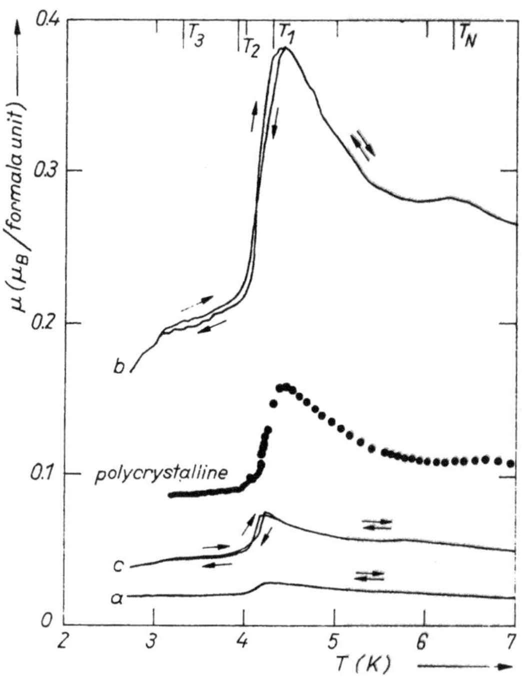

Fig. 2. Temperature dependence of magnetisation of $TmCu_2$ obtained on single crystal (full lines) and polycrystalline sample (dotted line) in external field $\mu_0 H = 0.06$ T. The marks indicate the phase transition temperatures obtained from specific heat measurement [5]. The arrows show the temperature course

obtained from specific heat and ac susceptibility measurements [5, 8]. The shape of the anomalies is strongly field dependent, namely that of a tiny sharp maximum at $T_2 = 3.9$ K in the polycrystalline M versus T dependence, which is depressed with increasing field and moves towards the large peak at T_1. For a field of 0.1 T the tiny maximum disappears completely in the background of the large peak. The M versus T dependences obtained for the single crystal show similar anomalies but their positions slightly differ along each axis.

The shapes of the M versus T dependence along b- and c-axes are also different from the polycrystalline one. The hysteresis appears below T_1 and we can say that the phase transition connected with T_1 probably is the first-order transition. The hysteresis increases with increasing applied field and appears along the a-axis below T_1 in the field of 0.1 T. These effects are influenced by internal stresses due to magnetoelastic interactions — mainly at temperatures around 3 K [8].

From the shape of the magnetisation curves taken along the easy axis at various temperatures (Fig. 3) we can observe differences in magnetisation processes characteristic of different magnetic phases. Below T_N the magnetic structure described in [4] (phase AF1) is established and with decreasing temperature the easy-axis magnetisation curve gradually transforms from a curve with one step to a two-step curve accompanied by an increase of hysteresis. Between T_1 and T_2 (phase AF2) the shape of the easy axis magnetisation curve is similar to that taken at 5 K, but almost no hysteresis was observed. In the AF3 phase ($T = 3.5$ K) the hysteresis reappears and the shape of the curve changes (the magnetisation after the first step is higher than in previous phases). Below T_3, the curve changes even from the two-step type to the three-step type.

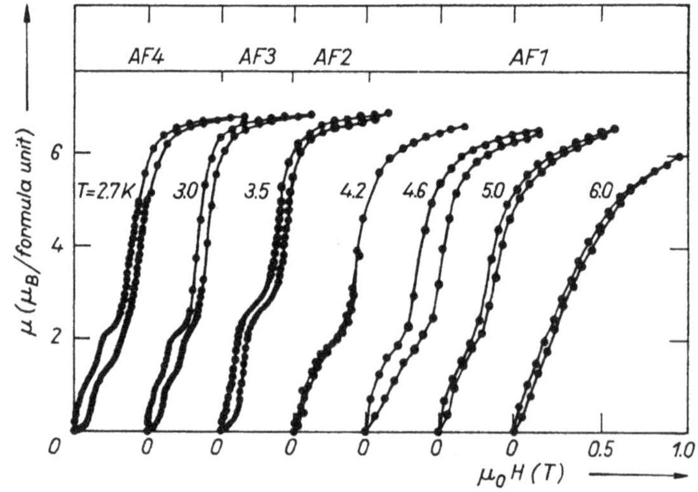

Fig. 3. Magnetisation curves of $TmCu_2$ along the b-axis at different temperatures below T_N. The horizontal bar above the curves indicates the phase from which each curve was taken

4. Concluding Remarks

According to the relatively simple twelve sublattice model described by Kimura [6, 7] we have tried to compare our measured curves with the calculated ones. The neutron diffraction experiment at $T = 5$ K showed the non-zero component of magnetic moment parallel to the a-axis [4]. This component may exist due to the dipole–dipole interaction [8] and probably disappears at the beginning of the magnetisation process. (The magnetisation curve along the a-axis is the lowest one.) Therefore, we can suggest that in the AF1 phase (just below T_N), the magnetic structure of type I_1 [6, 7] is established after the first step of the magnetisation process. The resultant moment in the AF1 phase is equal to zero according to [4] in zero external field and the first step of magnetisation process is probably the reorientation from the structure described in [4] to the structure of Ising type I_1 [6, 7]. On the other hand, the second step in magnetisation is most probably the spin flip transition from I_1 to F [7].

The situation in the temperature range below T_1 is more complicated. There is no direct experimental evidence about the magnetic structure yet and a similar comparison with the model of Kimura would be unexplicit.

References

[1] A. R. STORM and K. E. BENSON, Acta cryst. **16**, 701 (1963).
[2] R. C. SHERWOOD, H. J. WILLIAMS, and J. H. WERNICK, J. appl. Phys. **35**, 1049 (1964).
[3] Y. HASHIMOTO, H. FUJII, H. FUJIWARA, and T. OKAMOTO, J. Phys. Soc. Japan **47**, 73 (1979).
[4] B. LEBECH, Z. SMETANA, and V. SIMA, J. Magnetism magnetic Mater. **70**, 97 (1987).
[5] Z. SMETANA, V. SIMA, J. BISCHOF, P. SVOBODA, S. ZAJAC, L. HAVELA, and A. V. ANDREEV, J. Phys. F **16**, L201 (1986).
[6] I. KIMURA, Bull. Fukuoka Univ. Educ. **35**, Part III, 67 (1986).
[7] I. KIMURA, Bull. Fukuoka Univ. Educ. **36**, Part III, 43 (1987).
[8] V. SIMA, Z. SMETANA, M. DIVIS, P. SVOBODA, S. ZAJAC, J. BISCHOF, B. LEBECH, and F. KAYZEL, J. Physique, to be published.

(Received August 19, 1988)

phys. stat. sol. (a) **111**, 289 (1989)

Subject classification: 78.20; S1.63; S8.16

A. F. Ioffe Physico-Technical Institute, Academy of Sciences of the USSR, Leningrad[1])

Fundamental Optical Absorption Edge in $MnIn_2Te_4$ Single Crystals

By

G. A. Medvedkin, Yu. V. Rud, and M. A. Tairov

The optical properties are studied in the region of the fundamental absorption edge of $MnIn_2Te_4$ single crystals ($I\bar{4}2m$, defect stannite) oriented along (100) and (001). The width of the energy gap is measured to be $E_g = 1.55$, 1.425, and 1.41 eV, for the temperatures 77, 300, and 370 K, respectively. The $\alpha(\hbar\omega)$ spectra are established to follow Urbach's rule throughout the temperature range studied, the steepness of the absorption edge remaining constant. Crystal annealing followed by fast cooling shifts the absorption edge longward by 50 meV, the exponential character of $\alpha(\hbar\omega)$ persisting within the whole range $T = 77$ to 370 K. An analysis shows the optical absorption in the region of the fundamental edge to be governed by the sum of the effects of the density-of-states tails associated with a high concentration of lattice vacancies and of the electron–phonon interaction. The optical linear dichroism of the absorption edge of $MnIn_2Te_4$ crystals with defect stannite structure is discovered and studied. It is found that the crystals belonging to the tetragonal syngony, such as $MnIn_2Te_4$ ($I\bar{4}2m$) and $CuInSe_2$, $CuInTe_2$ ($I\bar{4}2d$), and exhibiting a weak tetragonal lattice deformation $\tau \approx 0$, reveal, in contrast to most chalcopyrites with $\tau > 0$, a positive linear dichroism of transmittance close to E_g. Minimal direct transitions in $MnIn_2Te_4$ are allowed in the $\boldsymbol{E} \perp \boldsymbol{c}$ polarization, the crystal splitting of the valence band at the Γ point being positive.

Исследованы оптические свойства в области фундаментального края поглощения ориентированных в кристаллографических направлениях (100) и (001) монокристаллов $MnIn_2Te_4$ ($I\bar{4}2m$ — дефектный станнит). Определена ширина запрещенной зоны $E_g = 1{,}55$; 1,425 и 1,41 eV при температурах 77, 300 и 370 K. Установлено, что спектры $\alpha(\hbar\omega)$ во всем температурном диапазоне следуют правилу Урбаха, причем крутизна края поглощения сохраняется неизменной. Отжиг кристаллов с последующим быстрым охлаждением приводит к смещению края поглощения в длинноволновую область на 50 meV с сохранением экспоненциального характера $\alpha(\hbar\omega)$ во всем диапазоне $T = 77$ до 370 K. Анализ результатов показывает, что оптическое поглощение в области фундаментального края определяется суммарным эффектом: хвостами плотности состояний, связанных с высокой концентрацией вакансий решетки и электрон–фононным взаимодействием. Обнаружен и изучен оптический линейный дихроизм края поглощения монокристаллов $MnIn_2Te_4$ с решеткой дефектного станнита. Установлено, что монокристаллы, принадлежащие тетрагональной сингонии-$MnIn_2Te_4$ ($I\bar{4}2m$), также как $CuInSe_2$, $CuInTe_2$ ($I\bar{4}2d$) и имеющие слабую тетрагональную деформацию решетки $\tau \approx 0$, в отличие от большинства халькопиритов с $\tau > 0$, обнаруживают положительный линейный дихроизм пропускания вблизи энергии E_g. Минимальные прямые переходы в $MnIn_2Te_4$ разрешены в поляризации $\boldsymbol{E} \perp \boldsymbol{c}$, а кристаллическое расщепление валентной зоны в точке Г имеет положительный знак.

1. Introduction

The ternary compounds $Mn–III_2–VI_4$ belong to a class of magnetic semiconductors attracting an ever increasing interest in connection with an unusual combination of optical, magnetic, and semiconducting properties [1 to 4]. Compounds of this class

[1]) Politekhnicheskaya 26, SU-194021 Leningrad, USSR.

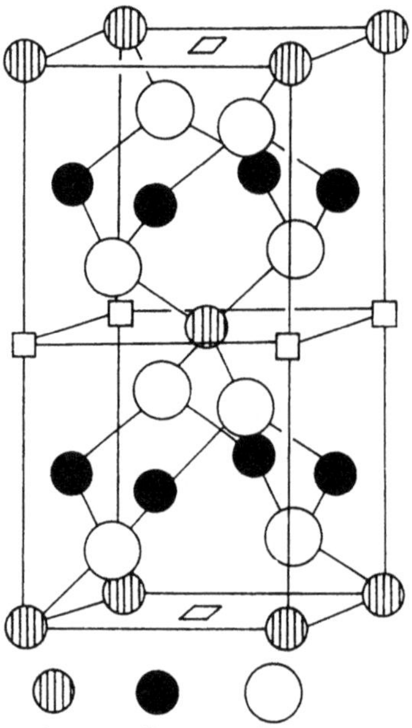

Fig. 1. Lattice structure of $MnIn_2Te_4$

crystallize typically in a number of structures, including tetragonal (stannite) and cubic (spinel). $MnIn_2Te_4$ is an adamantine semiconductor with tetrahedral coordination, belongs to the spatial group I$\bar{4}$2m and crystallizes in the defect stannite lattice shown in Fig. 1. A specific feature of the $MnIn_2Te_4$ lattice is an ordered arrangement not only of the atoms of the three chemical elements, Mn, In, and Te, but of the vacancies as well, which represent essentially virtual states without a nucleus and electron shells (in Fig. 1 the vacancies are specified by squares). The defect stannite lattice has a fourth order axis and assumes uniaxial optical anisotropy, as is the case, for example, with the chalcopyrite group (I$\bar{4}$2d) [5]. However, the tetragonal lattice deformation producing a strong anisotropy may not manifest itself ($c/a = 2$), in which case quasi-isotropic structure prevails, as is observed, for instance, with $ZnGa_2Se_4$, $ZnIn_2Se_4$, $ZnIn_2Te_4$, $CdIn_2Te_4$, $HgIn_2Te_4$, and $HgGa_2Te_4$ [3].

We have already reported on some optical and optoelectronic properties of $MnIn_2Te_4$ [6 to 8]. The present paper discusses the dependence of the fundamental optical absorption edge of single crystals and films of $MnIn_2Te_4$ on the linear polarization of the incident radiation, the temperature, and heat treatment of the crystals used.

2. Experimental

$MnIn_2Te_4$ single crystals were grown by directed cooling of a stoichiometric melt of Mn, In, Te of purity not below 99.999 wt%. No doping was carried out. An analysis performed on a JXA-5A electron microprobe analyzer showed the composition of the compound obtained to correspond to $MnIn_2Te_4$. X-ray diffraction studies showed the crystals to belong to the I$\bar{4}$2m spatial group (defect stannite) with the following lattice parameters: $a = (6.196 \pm 0.005) \times 10^{-1}$ and $c = (12.402 \pm 0.010) \times 10^{-1}$ nm at $T = 300$ K. These parameters correspond to a very weak tetragonal distortion of the lattice, $\tau = (1 - c/2a) \times 100\% = 0.081\%$, the lattice being extended (rather than compressed) along a fourth-order axis, just as is the case with some $CuInVI_2$ compounds with chalcopyrite structure. The single crystals were oriented by means of the X-ray diffraction and optical polarization techniques. We studied plane-parallel plates 2 mm to 10 μm thick oriented along (100) and (001). The **c**-axis was oriented to within $\pm 1°$. The surfaces of the specimens used in the measurements were lapped, etched in an alcohol solution of Br and subsequently washed in CCl_4 and ethanol. The $Cu/MnIn_2Te_4$ Schottky barriers were prepared by chemical deposition of a copper layer from an aqueous solution of blue vitriol: $CuSO_4 + In \rightarrow In(SO_4)_3 + Cu\downarrow$. The current leads were soldered with pure indium. Thin polycrystalline films of $MnIn_2Te_4$ were prepared by vacuum deposition on optical glass and other substrates. The chemical composition of the films was checked with the electron microprobe analyzer, their thickness being measured with a МИИ-4 microinterferometer.

Spectral measurements of the optical transmission T and reflection R of crystals and films, as well as of the photoresponse of the Schottky barriers were carried out on a computerized spectrometer КСБУ-3М using a code СПЕКТР-КСБУ. A mirror reflection attachment with an incidence angle $\theta \approx 15°$ was employed to obtain $R(\hbar\omega)$ spectra. Optical and photoelectric spectra were also measured on a spectrometer based on a prism monochromator SPM-2. The optical absorption was calculated by the relation

$$\alpha = \frac{1}{d} \ln \left(A + \sqrt{A^2 + R^2}\right), \tag{1}$$

where $A = (1 - R)^2/2T$, and d is the specimen thickness: 2 mm, 388, 220, 96, 72, 55, 15, and 10 μm.

Temperature dependences were studied in a Dewar flask with a mixture of liquid nitrogen with ethanol serving as cryogenic liquid, the temperature being varied by properly adjusting the composition of the liquid. A film IR polaroid ($\Delta\lambda = 700$ to 2500 mm) was used as a polarizing element. The axis of the maximal polaroid transmission and the tetragonal axis of the crystal were aligned to within ±1°. The spectral resolution of the instrumentation employed was 1 to 2 meV.

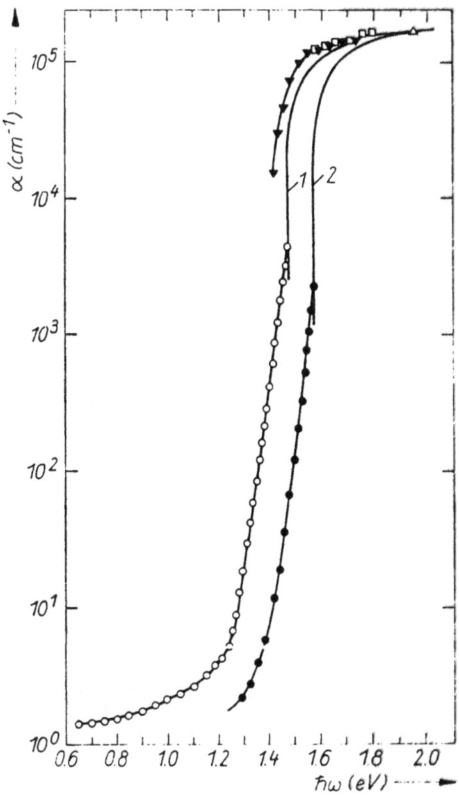

Fig. 2. Edge absorption of $MnIn_2Te_4$ at T = 300 K (○, □, △, ▲, and curve 1) and 77 K (● and curve 2). Experiment: ○, ● single crystal, □ thin polycrystalline film, ▼ photo-e.m.f. U_{ph} (in arb. units) of Cu/$MnIn_2Te_4$ Schottky barrier, △ ellipsometric data; curves 1, 2 calculations by parabolic approximation

3. Absorption and Reflection in Natural Light

Fig. 2 presents spectral responses of the optical absorption coefficient and photo-e.m.f. taken at 300 and 77 K. The long-wavelength tail of the fundamental absorption edge of $MnIn_2T_4$ single crystals (points ●, ○) is described well by Urbach's rule [10]

$$\alpha = \alpha_0 \exp\left[-\frac{\sigma}{kT}(E_0 - \hbar\omega)\right], \tag{2}$$

where α_0 and E_0 are constants of the semiconductor material used. The Urbach parameter, $\sigma = 0.5$ and 0.1 for $T = 300$ and 77 K, respectively, corresponds to direct optical transitions in the crystal. Relation (2) approximates well the absorption edge over more than two decades, $\alpha = 10^1$ to 5×10^3 cm^{-1}. Since the concentrations of free carriers (both electrons and holes) in $MnIn_2Te_4$ do not usually exceed 10^{12} to 10^{13} cm^{-3} [6], the absorption of light in these compounds can be described as this is done for doped semiconductors with a nondegenerate carrier gas having Gaussian fluctuations of the r.m.s. potential γ [11]. A remarkable feature of the crystalline structure of $MnIn_2Te_4$ (Fig. 1) is the presence of a high concentration of intrinsic lattice defects, namely, vacancies (4×10^{21} cm^{-3}) and probable antistructural interchange of cations, Mn $\rightleftarrows$ In, as well as of cations with vacancies, Mn(In) $\rightleftarrows$ V. The existence of such a complex ensemble of defects in the lattice of $MnIn_2Te_4$ creates a random Coulomb field of charged local centres, $\varepsilon = Ze/\varkappa r^2$. For the contribution to the absorption coefficient of the concentration fluctuations Z of the atoms Mn, In, and structural vacancies V (Fig. 1) in an n-type semiconductor one can write, as this is done in [11],

$$\alpha \sim \exp\left(-\frac{Z^2}{N_t r^3} - \frac{r\sqrt{m_e(E_g - \hbar\omega)}}{\hbar}\right), \tag{3}$$

where $N_t = N_D + N_A$ is the total concentration of the charged donors and acceptors, and r is the linear dimension of the fluctuation.

The long-wavelength tail of $\alpha(\hbar\omega)$ deviates for photon energies <1.25 eV from an exponential law (Fig. 2) and approaches a Gaussian resembling the density-of-states function for a heavily doped semiconductor [11],

$$G_0(X) = \frac{1}{\sqrt{\pi}} \int_{-\infty}^{X} e^{-y^2} (X - y)^{1/2}\, dy\,, \tag{4}$$

where X has a meaning of the quantity E_g/γ.

To obtain the spectral response of absorbance of the ternary compound for $\alpha > 6 \times 10^3$ cm^{-1}, thin vacuum-deposited polycrystalline films and Schottky barriers on bulk crystals were used (Fig. 2, points □, ▼). The optical absorption coefficient of $MnIn_2Te_4$ films in the region 1.5 to 1.8 eV, $\alpha = (1.2$ to $1.5) \times 10^5$ cm^{-1}, approximates well to the relation $\alpha \sim \sqrt{E_g - \hbar\omega}$ characteristic of a direct gap semiconductor [12]. When the surface barrier structure Cu/$MnIn_2Te_4$ is illuminated from the side of the metal, the spectral response of the photo-e.m.f. corresponds to photoactive absorption in the semiconductor. The long-wavelength fall off of $U_{ph}(\hbar\omega)$ reveals an exponential pattern, just as $\alpha(\hbar\omega)$ does, (curves ○, ●), saturation being reached for $\hbar\omega > 1.5$ eV (Fig. 2, curve ▲).

To ascertain the absolute value of α for $\hbar\omega > E_g$, we measured also the optical constants of $MnIn_2Te_4$ at $\lambda = 632.8$ nm with a ЛЭФ-3M laser ellipsometer. By analogy with $CuInSe_2$ [13], a model of an isotropic absorbing crystal with a transparent oxide film ($n = 2.12$) was chosen. For $MnIn_2Te_4$ crystals with a polished specular surface the complex index is $N = n - ik = 3.18 - i \cdot 0.81$ for an oxide layer thickness of 3.8 nm. The optical absorption coefficient calculated with the expression $\alpha = 4\pi k/\lambda$ was used to construct the $\alpha(\hbar\omega)$ curves in the region $\hbar\omega > E_g$.

For a direct gap semiconductor and a parabolic approximation for allowed optical transitions at the centre of the Brillouin zone the following relation is valid [12]:

$$\alpha = A' \, |\boldsymbol{P}_{m_0}|^2 \, \frac{(\hbar\omega - E_g)^{1/2}}{\hbar\omega}\,, \tag{5}$$

where

$$A' = \frac{2\pi^2(2m_r)^{3/2}}{3m_0^2 n\varepsilon_0 ch^2}.$$

A quantitative analysis of $\alpha(\hbar\omega)$ cannot presently be carried out because of a lack of information concerning some fundamental parameters of $MnIn_2Te_4$ among them the matrix element of momentum $|\boldsymbol{P}_{m_0}|$, and the electron and hole effective masses. Therefore, the term $A' |\boldsymbol{P}_{m_0}|^2$ in (5) was used as a fitting parameter. The experimental data were fitted with the values of $\alpha(\hbar\omega)$ derived from the spectral and ellipsometrical data (points □ and △ in Fig. 2).

The best fit of (5) to the experimental data is obtained with the value $E_g^{opt} = 1.45$ eV for 300 K, and 1.55 eV for 77 K. As seen from Fig. 2, the simple parabolic approximation describes well the experiment in the region of strong optical absorption, $\alpha > > 10^5$ cm^{-1}, i.e. deep in the fundamental absorption band of $MnIn_2Te_4$. The long-wavelength tail of $\alpha(\hbar\omega)$ deviates from the theoretical curve and can be approximated with an exponential form of the type (3) or (2). The value of the optical width of the gap obtained from Kane's parabolic approximation is somewhat smaller than that of E_g derived from van Hove's singular points, e.g. from the reflectance spectra.

Fig. 3 presents spectra of the optical reflectance of $MnIn_2Te_4$ obtained in unpolarized light at $T = 300$ K. For undoped single crystals of $MnIn_2Te_4$ (curve 1) the behaviour of the $R(\hbar\omega)$ spectrum in the vicinity of the singular point at $\hbar\omega = 1.48$ eV resembles that which is known, e.g., for GaAs [12, 13]. Therefore, this maximum of reflectance can be attributed to strong interband transitions in $MnInTe_4$. For comparison, Fig. 3 shows a spectrum of the reflection coefficient obtained on a thin film of $MnIn_2Te_4$ on glass (curve 2). The maximum of $R(\hbar\omega)$ for the film near E_g is blurred because of its polycrystalline microstructure. In the long-wavelength region one observes interference oscillations of $R(\hbar\omega)$ originating from the small thickness of the film. The long-wavelength maximum at $\hbar\omega = 1.04$ eV in the reflectance spectrum of single crystal $MnIn_2Te_4$ can be accounted for by the existence of levels due to intrinsic lattice defects of the compound in question [6]. The energy position of the peak in the $R(\hbar\omega)$ may be compared with the deep level activation energy $E_{act} = 0.4$ eV. In the region of the fundamental absorption band of the crystal the amplitude of R increases smoothly with photon energy up to 35 to 40% (Fig. 3, curve 1) and exhibits a normal spectral dispersion within the range 1.5 to 3.5 eV. The value of R at $\hbar\omega = 1.96$ eV derived from an independent ellipsometric measurement (point ▽) agrees well with the spectral response $R(\hbar\omega)$ (Fig. 3).

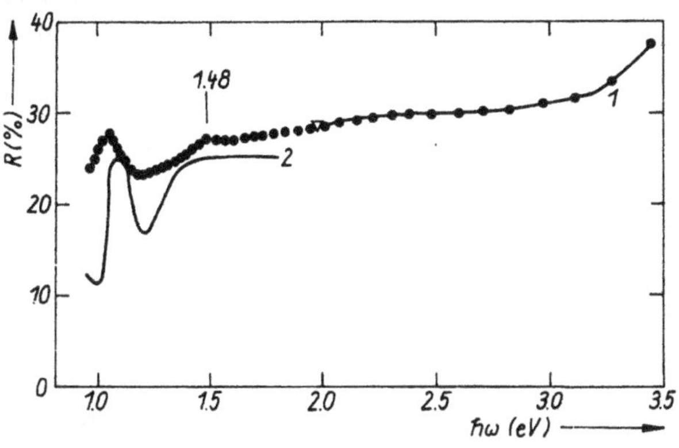

Fig. 3. $MnIn_2Te_4$ reflectance spectra, $T = 300$ K. (1) Single crystal, (2) thin ($d = 0.3$ μm) polycrystalline film, ellipsometric data

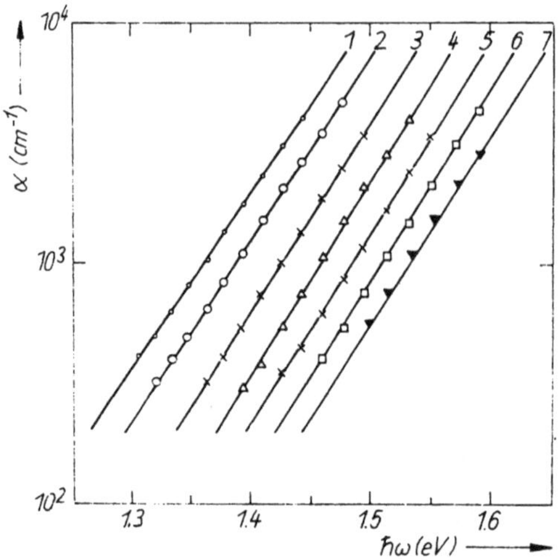

Fig. 4. Shift of $MnIn_2Te_4$ exponential absorption edge in the temperature range $T = 370$ to 4.2 K (curves 1 to 7); crystal thickness 15 μm

4. Temperature Behaviour of Edge Absorption

As the temperature is varied in the range $T = 370$ to 4.2 K, the fundamental absorption edge of $MnIn_2Te_4$ crystals shifts toward shorter wavelengths following Urbach's rule (2). Fig. 4 illustrates precision measurements of the temperature dependence of $\alpha(\hbar\omega)$ made on 15 μm thick crystals. The experimental points can be rectified to form a straight line on a log scale (Fig. 2) for temperatures from 370 down to 4.2 K. The lines shift with temperature nearly parallel to one another, the focal point lying at very large values of the constants $\alpha_0 = 10^{20}$ cm^{-1}, $E_0 = 3.7$ eV. A similar temperature shift of the absorption edge was obtained by us on samples prepared in different ways and varying in thickness from 96 to 6 μm. Changing the growth conditions of deliberately undoped $MnIn_2Te_4$ crystals in various modifications of the directed crystallization technique, results in differences in the steepness of the absorption edge, $s = d(\ln \alpha)/d(\hbar\omega)$ within 17 to 25 eV^{-1}. Note that the major features of the spectral response $\alpha(\hbar\omega)$ persist, as seen from Fig. 2 and 4. This suggests that the dominant contribution to the energy spectrum of the ternary semiconductor comes from inherent lattice defects which loosen the crystal lattice and produce density-of-state tails in the energy gap of $MnIn_2Te_4$.

The temperature dependences of the energy parameters of single crystal $MnIn_2Te_4$ derived from the spectra $\alpha(\hbar\omega)$ are presented in Fig. 5. The values of E_g and E_1

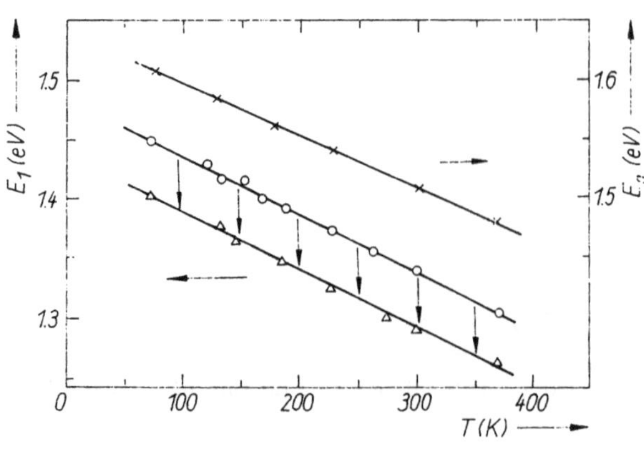

Fig. 5. Temperature behaviour of $MnIn_2Te_4$ gap width. From parabolic approximation (×), from extrapolation of $\alpha^2(\hbar\omega)$ to zero (○, △) (○ before △ after heat treatment)

obtained by extrapolation of the dependences $\alpha^2(\hbar\omega)$ to $\alpha \approx 10^5\ cm^{-1}$ (Fig. 2) and to zero vary linearly with temperature as is seen from Fig. 5. The temperature coefficient of the energy gap width for crystals from different technological lots lies in the range $dE_g/dT = -(4.5 \text{ to } 5.3) \times 10^{-4}$ eV/K.

In contrast to the crystals $CuInS_2$, $CuInSe_2$, and $CuInTe_2$ having a small tetragonal extension $\tau < 0$ [14 to 16], $MnIn_2Te_4$ crystals, which likewise have a small τ, exhibit a high temperature coefficient of the energy gap width which is comparable with those for the compounds III–V, II–IV–V_2, and II–III_2–VI_4 [3, 4, 17]. Substitution of manganese for copper ($CuInTe_2 \rightarrow MnIn_2Te_4$) increases dE_g/dT by about a factor three.

To study the effect of inherent defects on the energy spectrum of $MnIn_2Te_4$ crystals, the samples were thermally annealed in vacuum at $T = 680$ °C for 30 min with a subsequent rapid cooling ($v \approx 10^3$ K/min). When compared with as-grown samples, such samples revealed a parallel shift of the absorption edge by ≈ 50 meV without any change in the spectral responses $\alpha(\hbar\omega)$. Fig. 5 presents a temperature dependence of E_1 for $MnInTe_4$ crystals cooled rapidly after the high temperature treatment (△). As seen from Fig. 5, the character of the optical transitions persists after the additional thermal defects frozen in during the fast cooling have been introduced in the $MnIn_2Te_4$. One may thus conclude that the predominant contribution to the formation of edge absorption in $MnIn_2Te_4$ comes from inherent defects in the lattice of the ternary semiconductor.

Urbach's parameter describing the exponential long-wavelength tail in $\alpha(\hbar\omega)$ of $MnIn_2Te_4$ obeys the relation [9]

$$\sigma = \sigma_0(2kT/\hbar\omega_p)\ \mathrm{th}\ (\hbar\omega_p/2kT)\,, \tag{6}$$

where σ_0 is the maximal value of Urbach's parameter, $\hbar\omega_p$ the energy of the phonons involved in the interaction with the electrons. Over a broad temperature range σ depends linearly on T (Fig. 6, lowes curve). Introducing thermal defects into $MnIn_2Te_4$ crystals (points △) does not affect this relation. The theoretical curve (6) describes well the experiment (points ×, △) for $\sigma_0 = 0.9$, $\hbar\omega_p = 90$ meV. Since thermal treatment does not result in a qualitative change in the mechanism of optical edge absorption and only reduces E_g, one may conclude that the effect of inherent structural defects in $MnIn_2Te_4$ dominates over the electron–phonon mechanism.

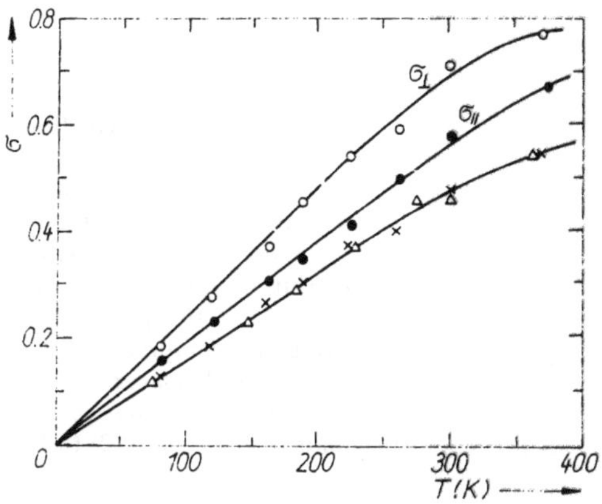

Fig. 6. Temperature dependence of Urbach's parameter for unpolarized light (× before, △ after heat treatment) and for two polarizations, $\boldsymbol{E} \perp \boldsymbol{c}$ and $\boldsymbol{E} \parallel \boldsymbol{c}$

5. Polarization Studies of Optical Absorption

The principal contribution to the polarization dependence of the coefficient R comes from the transmittance T,

$$T_{\parallel,\perp} = \frac{(1 - R)^2 \exp(-\alpha_{\parallel,\perp} d)}{1 - R^2 \exp(-2\alpha_{\parallel,\perp} d)}, \tag{7}$$

where the subscripts "$\parallel$" and "$\perp$" identify the $\boldsymbol{E} \parallel \boldsymbol{c}$ and $\boldsymbol{E} \perp \boldsymbol{c}$ orientations, respectively. Anisotropy in R affects the amplitude of α in a much lesser degree. Therefore the possible anisotropy in the reflectance was disregarded in our polarization measurements.

The spectral responses of $\alpha_{\parallel}$ and $\alpha_{\perp}$ calculated by (1) from the experimental values of $T_{\parallel}$, $T_{\perp}$, and R are presented in Fig. 7. The exponential form of the $\alpha(\hbar\omega)$ curves persists in both natural and linearly polarized radiation (○ and ●). In the $\boldsymbol{E} \perp \boldsymbol{c}$ polarization, the absorption edge turns out to be shifted relative to that for $\boldsymbol{E} \parallel \boldsymbol{c}$ towards longer wavelengths. Since the $\alpha(\hbar\omega)$ is steeper in the perpendicular polarization, the minimal direct transitions in $MnIn_2Te_4$ should occur predominantly for $\boldsymbol{E} \perp \boldsymbol{c}$. Despite the quasi-isotropic lattice of the ternary compound, the difference in the transmittances $(\alpha_{\perp} - \alpha_{\parallel})$ in this spectral region is as high as 1800 cm^{-1} implying a high linear dichroism of the crystals. The inset in Fig. 7 shows a quasi-cubic model of the band structure obtained for the defected stannite ($I\bar{4}2m$), the lattice in which $HgIn_2Te_4$ likewise crystallizes [18]. The natural optical anisotropy in these two crystals is of the same origin and is determined by ordered vacancies in the lattice and local electrostatic field effects.

Fig. 6 presents Urbach's parameters calculated from the $\alpha(\hbar\omega)$ curves for linearly polarized light as a function of T. The difference between $\sigma_{\perp}$ and $\sigma_{\parallel}$ (curves ○ and ●) is seen to be largest at high temperatures, which suggests an increase of the optical anisotropy with temperature. Calculation by (6) yields the limiting value $\sigma_0 = 1.5$ which is the same for $\boldsymbol{E} \parallel \boldsymbol{c}$ and $\boldsymbol{E} \perp \boldsymbol{c}$. However, the photon energy $\hbar\omega_p$ which describes satisfactorily the experiment is different for the two polarizations: $\hbar\omega_{p,\perp} =$

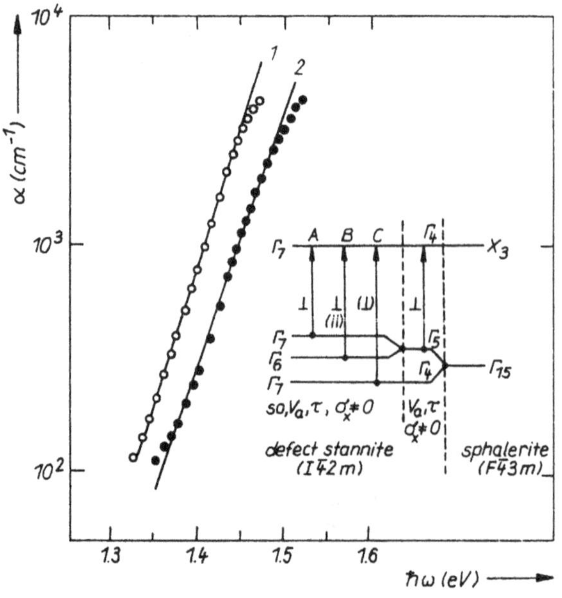

Fig. 7. Splitting of $MnIn_2Te_4$ fundamental absorption edge in linearly polarized light. Inset: model of $MnIn_2Te_4$ band structure with selection rules for direct optical transitions. So spin–orbit interaction, V_a antisymmetric part of the pseudopotential, τ tetragonal compression of the lattice, σ_x anion displacement from ideal tetrahedral positions. Curve (1) corresponds to $\boldsymbol{E} \perp \boldsymbol{c}$ and (2) to $\boldsymbol{E} \parallel \boldsymbol{c}$

$= 80$ meV and $\hbar\omega_{p,\parallel} = 100$ meV. This difference can be attributed to an anisotropy of the phonon spectrum which contributes to the mechanism of formation of the anisotropic absorption edge in $MnIn_2Te_4$ single crystals. Thus, one may conclude from the results depicted in Fig. 6 that the fundamental absorption edge in $MnIn_2Te_4$ is formed by two mechanisms, namely, electron–phonon interaction and electronic transitions to the density-of-states tails.

6. Optical Linear Dichroism

In contrast to $CuInVI_2$ crystals where τ is small and the maximum dichroism in absorption $|\mathcal{P}_\alpha^{max}| \leqq 20\%$, in the ternary telluride of manganese and indium the amplitude of $\mathcal{P}_\alpha$ turns out to be substantially higher, reaching $\approx 60\%$. There is no optical anisotropy of $MnIn_2Te_4$ single crystals in the isotropic direction [001] thus implying a uniaxial character of optical anisotropy in this compound.

Fig. 8 presents linear dichroism spectra of the optical transmittance of $MnIn_2Te_4$ crystals having different thickness. The maximum of $\mathcal{P}_T$ attains 95 to 97% for all thicknesses (curves 1 to 4) and shifts from 1.34 to 1.48 eV as d is reduced from 388 to 10 μm. The spectral profile of the linear dichroism of T is symmetrically bell-shaped only for thick samples (Fig. 8, curves 1 and 2). Therefore in the corresponding spectral region the curves $\alpha_\parallel(\hbar\omega)$ and $\alpha_\perp(\hbar\omega)$ have an equal slope, no features in the form of local levels being revealed. By transferring to small thicknesses, one can reach the energies corresponding to the minimal direct transitions in $MnIn_2Te_4$. As follows from Fig. 8, curve 3 consists of two $\mathcal{P}_T(\hbar\omega)$ bands, with the dominant maximum lying at 1.425 eV, and the short-wavelength shoulder at 1.485 eV. For the thinnest crystal ($d = 10$ μm, curve 4) the position of the absolute maximum of $\mathcal{P}_T$ corresponds to the short-wavelength shoulder in the $\mathcal{P}_T$ spectrum for the $d = 15$ μm crystal (curve 3). The sharp fall-off of $\mathcal{P}_T(\hbar\omega)$ in the short-wavelength region turns out to be the same for the two samples. The latter implies that the experiment has revealed deeper interband transitions differing from the minimal transitions in the selection rules for polarized light. These data have been used to identify for $MnIn_2Te_4$ crystals the following direct transitions: $A = 1.425$ eV ($\perp$), $B = 1.485$ eV ($\perp$, $\parallel$), and $C = 1.55$ eV ($\perp$) for $T = 300$ K. The selection rules for these transitions are in agreement with the quasi-cubic model of the band structure at the Γ point (see inset in Fig. 7).

Thus the minimal direct transitions occurring in $MnIn_2Te_4$ are observed in the light

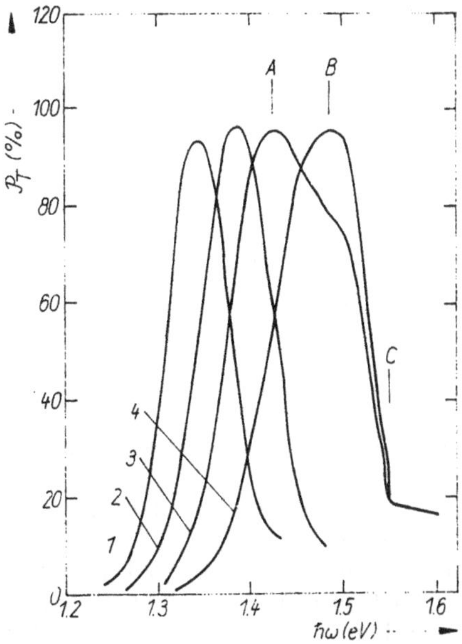

Fig. 8. Dependence of $\mathcal{P}_T(\hbar\omega)$ spectra on crystal thickness. (1) $d = 388$, (2) 72, (3) 15, (4) 10 μm; $T = 300$ K

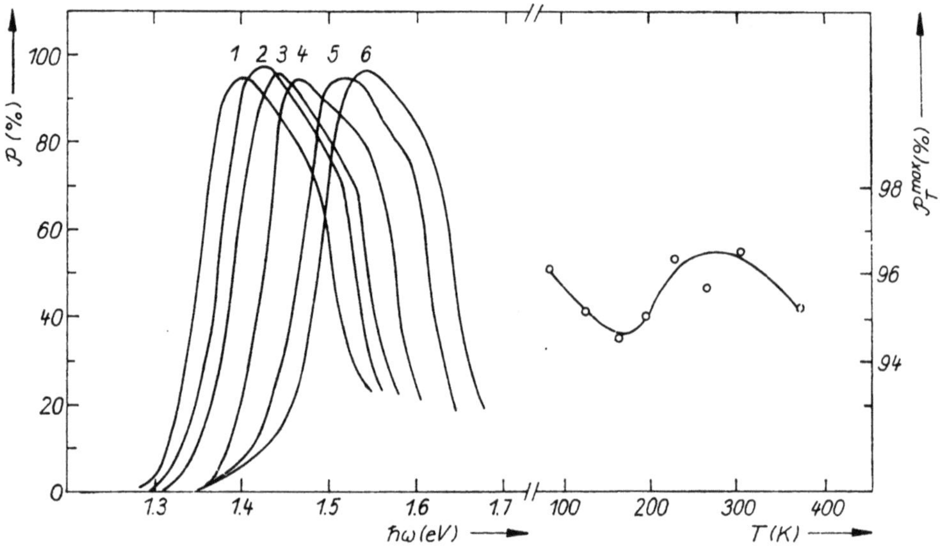

Fig. 9. Temperature-induced shift of $\mathcal{P}_T(\hbar\omega)$ spectra in the range $T = 370$ to 77 K (curves 1 to 6) and temperature dependence of the maximum amplitude $\mathcal{P}_T^{max}$; $d = 15\ \mu m$. At the ordinate read $\mathcal{P}_T$ instead of $\mathcal{P}$

polarized perpendicular to the c-axis. Although the lower minimum in the conduction band of the manganese-based compound is the same as in $HgIn_2Te_4$ and is associated with the X-point of the Brillouin zone in sphalerite, the minimum transitions in $MnIn_2Te_4$ turn out to be not pseudodirect as is the case with chalcopyrite crystals [5], since the matrix element of the $\Gamma \rightarrow X$ transition does not vanish [18, 19]. Hence in $MnIn_2Te_4$ there is a positive crystalline splitting of the valence band, which corresponds to the band model for the defect stannite structure. This disagrees with the sign of Δ_{cr} for most chalcopyrites for which $\tau > 0$, but coincides with the similar splitting in $CuInSe_2$ and $CuInTe_2$ with $\tau \leqq 0$ [14 to 16].

As the temperature is varied from 370 to 77 K, the spectral profile of the transmittance dichroism shifts towards higher photon energies with a coefficient dE_g/dT. As seen from Fig. 9, the profile has a complex shape for all temperatures and is made up of two bands corresponding to direct optical transitions in $MnIn_2Te_4$. Hence the valence band splitting in $MnInTe_4$ is observed at all temperatures in the range 77 to 370 K.

The values of Δ_{cr} and Δ_{so} were calculated by the expression [5]

$$E_{1,2} = -\frac{1}{2}\left(\Delta_{so} + \Delta_{cr}\right) \pm \frac{1}{2}\left[\left(\Delta_{so} + \Delta_{cr}\right)^2 - \frac{8}{3}\,\Delta_{so}\Delta_{cr}\right]^{1/2}. \tag{8}$$

Here the values of $E_{1,2}$ determine the energies of the Γ_7 levels relative to the Γ_6 level (Fig. 7).

The valence band splittings thus obtained and the minimum direct transitions in $MnIn_2Te_4$ for three temperatures are given in Table 1.

Fig. 9 presents the temperature dependence of the maximum dichroism in the transmittance of the crystals. In accordance with the relation between $\mathcal{P}_T^{max}(T)$ and $\tau(T)$ established by us earlier for $CuInTe_4$ [15] one may assume that for $MnIn_2Te_4$ the non-uniform temperature behaviour of the maximum dichroism reflects the temperature dependence of τ. The minimum in the $\mathcal{P}_T^{max}$ curve (Fig. 9) is observed to lie at $T \approx$

Table 1

Band parameters of $MnIn_2Te_4$ (in eV)

T (K)	A	B	C	Δ_{cr}	$-\Delta_{so}$
77	1.55	1.61	1.68	0.075	0.085
300	1.425	1.485	1.55	0.07	0.08
370	1.41	1.475	1.54	0.08	0.08

$\approx$ 160 K, suggesting that τ for the $MnIn_2Te_4$ lattice passes through an extremum at this temperature. No X-ray diffraction data are presently available on the temperature variation of the lattice parameters of this compound, so that the above hypothesis requires experimental verification.

To demonstrate the polarizing action of $MnIn_2Te_4$ crystals with a quasi-isotropic lattice ($\tau = 0$), two (100) oriented plates of this compound were crossed. Fig. 10 displays the spectral response of the degree of polarization for the polarizer-analyzer system made of $MnIn_2Te_4$. The FWHM of the spectrum is $\Delta\hbar\omega = 38$ meV for 300 K. The short-wavelength falloff of the $\mathcal{P}_{cr}(\hbar\omega)$ curve is accounted for by the onset of inherent absorption in the crystal, so that for $d = 55$ μm the spectrum peaks at 1.405 eV. Hence crystal plates of the ternary compound $MnIn_2Te_4$ can be used as optical polarizers for a narrow spectral interval with the maximum varying from 1.35 to 1.45 eV depending on their thickness.

Thus the fundamental optical absorption edge in $MnIn_2Te_4$ is formed due to the density-of-states tails associated with a high concentration of inherent lattice defects, and electron–phonon interaction. Single crystals of this ternary compound, despite its quasi-isotropic lattice structure, represent a strongly anisotropic uniaxial material with a positive linear dichroism of optical transmittance. Unlike chalcopyrite crystals with $\tau > 0$, the valence band of $MnIn_2Te_4$ reveals a positive crystal splitting, so that the minimum optical transitions occur in the $\boldsymbol{E} \perp \boldsymbol{c}$ polarization over the temperature range $T = 77$ to 370 K.

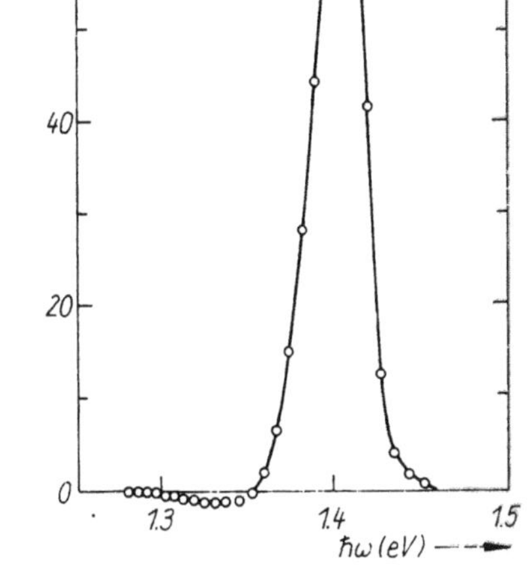

Fig. 10. Linear dichroism spectra of optical transmittance for the polarizer–analyzer system made of $MnIn_2Te_4$, (100) orientation, $d = 55$ μm

References

[1] E. L. Nagaev, Physics of Magnetic Semiconductors, Izd. Nauka, Moscow 1979 (in Russian).
[2] Magnetic Semiconductors, Proc. FIAN, Vol. 139, Izd. Nauka, Moscow 1982 (in Russian).
[3] A. V. Novoselova and V. B. Lazarev (Ed.), Physicochemical Properties of Semiconductor Materials (Hdb), Izd. Nauka, Moscow 1979 (in Russian).
[4] Landolt-Börnstein, Vol. 17 h, Physics of Ternary Compounds, Ed. O. Madelung, Springer-Verlag, 1985.
[5] J. L. Shay and J. H. Wernick, Ternary Chalcopyrite Semiconductors: Growth, Electronic Properties and Applications, Pergamon Press, New York 1975.
[6] R. N. Bekimbetov, A. A. Vaipolin, N. N. Konstantinova, L. V. Kradinova, G. A. Medvedkin, V. D. Prochukhan, Yu. V. Rud, and M. A. Tairov, Crystal Res. Technol. **22**, K238 (1987).
[7] R. N. Bekimbetov, G. A. Medvedkin, V. D. Prochukhan, Yu. V. Rud, and M. A. Tairov, Zh. tekh. Fiz. Pisma **13**, 1040 (1987).
[8] G. K. Averkieva, R. N. Bekimbetov, N. N. Konstantinova, L. V. Kradinova, V. D. Prochukhan, Yu. V. Rud, and M. A. Tairov, Izv. Akad. Nauk SSSR, Ser. neorg. Mater. **24**, 597 (1988).
[9] M. V. Kurik, phys. stat. sol. (a) **8**, 9 (1971).
[10] B. L. Gelmont, V. I. Perel, and I. N. Yassievich, Fiz. tverd. Tela **25**, 727 (1983).
[11] B. I. Shklovskii and A. L. Efros, Electronic Properties of Doped Semiconductors, Izd. Nauka, Moscow 1979 (in Russian).
[12] T. S. Moss, G. J. Burrel, and B. Ellis, Semiconductor Opto-Electronics, Butterworth, London/Washington 1973.
[13] Yu. I. Ukhanov, Optical Properties of Semiconductors, Izd. Nauka, Moscow 1977 (in Russian).
[14] G. A. Medvedkin, Yu. V. Rud, and M. A. Tairov, phys. stat. sol. (b) **144**, 809 (1987).
[15] G. A. Medvedkin, Yu. V. Rud, and M. A. Tairov, phys. stat. sol. (b) **110**, 631 (1988).
[16] G. A. Medvedkin, Yu. V. Rud, and M. A. Tairov, Optika i Spektroskopiya **66**, No. 1 (1989).
[17] A. N. Georgobiani, S. I. Radautsan, and I. M. Tiginyanu, Fiz. Tekh. Poluprov. **19**, 193 (1985).
[18] A. MacKinnon, A. Miller, and G. Ross, Proc. 3rd Internat. Conf. Ternary Compounds, Edinburgh 1977, Inst. Phys. Conf. Ser. **35**, 171 (1977).
[19] A. Miller, A. MacKinnon, and D. Weaire, Solid State Phys. **36**, 119 (1981).

(Received May 25, 1988)

phys. stat. sol. (a) **111**, 301 (1989)

Subject classification: 78.20 and 78.55; 78.60; S8.11

Institute of Microelectronics Technology and Superpure Materials, Academy of Sciences of the USSR, Chernogolovka[1]) *(a)*
and P. N. Lebedev Physical Institute, Academy of Sciences of the USSR, Moscow[2]) *(b)*

Change of the Optical Properties of CdS Single Crystals upon High Dose Indium and Gallium Implantation

By

A. N. GEORGOBIANI (b), A. N. GRUZINTSEV (a),
A. V. SPITSYN (b), and I. M. TIGINYANU (b)

Results of a complex study of photoluminescence, cathodoluminescence, and reflection spectra of CdS single crystals irradiated with indium and gallium ions at doses of 10^{15} and 10^{16} cm^{-2} are presented. It is shown, that after annealing of the CdS:In$^+$ and CdS:Ga$^+$ samples at low temperatures the spectral characteristics of the implanted side are similar to the spectra of $CdIn_2S_4$ and $CdGa_2S_4$ compounds, respectively. The influence of annealing on the properties of the crystals is investigated.

Приводятся результаты комплексного исследования спектров фотолюминесценции, катодолюминесценции и отражения монокристаллов CdS, облученных ионами индия и галлия при дозах 10^{15} и 10^{16} cm^{-2}. Показано, что после низкотемпературного отжига образцов CdS:In$^+$ и CdS:Ga$^+$ спектральные характеристики имплантированных сторон аналогичны спектрам соединений $CdIn_2S_4$ и $CdGa_2S_4$, соответственно. Исследовано влияние отжига на свойства кристаллов.

1. Introduction

It is known that a great amount of III-group atoms in $A^{II}B^{VI}$ compounds may be responsible for the formation of a phase of the $A^{II}C_2^{III}B_4^{VI}$ new compound [1, 2]. The process of ternary compound formation depends on many factors and first of all on the reaction ability of a dopant. For example, in CdS ($E_g \approx 2.58$ eV at 0 K [3]) gallium is more active than indium. In accordance with it high temperature annealing ($T = 640$ to 940 °C) of CdS single crystals in gallium melt leads to the formation of the $CdGa_2S_4$ ternary compound [2], while thermal treatment of CdS in indium melt does not lead to the formation of a new phase [4]. It is explained by the fact that the indium solubility in cadmium sulphide is comparatively small (10^{-2} mol% [5]). At the same time in CdS high indium solubility as ternary compound $CdIn_2S_4$ is observed (15 mol% [6]).

Interaction of $A^{II}B^{VI}$ compounds with III-group elements in case of thermodynamic non-equilibrium doping (ion implantation) can occur quite differently. In connection with it implantation of a large amount of indium can lead to the formation of a phase of the new compound $CdIn_2S_4$ in CdS. It should be noted, that synthesis of new compounds has been already performed by ion implantation, for example by implanting some elements in Si, GaAs, and ZnS [7 to 9].

The present paper investigates the possibility to form layers of $CdIn_2S_4$ ($E_g = 2.62$ eV at $T = 300$ K [10]) and $CdGa_2S_4$ ($E_g = 3.77$ eV at 10 K [11]) ternary semiconductor compounds by implantation of In$^+$ and Ga$^+$ ions in CdS single crystals, respectively.

[1]) SU-142432 Chernogolovka, Moscow District, USSR.
[2]) Leninskii prospekt 53, SU-117924 Moscow, USSR.

The starting cadmium sulphide single crystals were obtained by resublimation in a stream of inert gas. They were of n-type conductivity, the resistivity being equal to $\varrho \approx 10^2\ \Omega$ cm at 300 K. Faces (11$\bar{2}$0) of CdS plates were irradiated by In^+ and Ga^+ ions with energy 250 keV (at doses of $D = 10^{15}$ and 10^{16} cm^{-2}) using a heavy ion accelerator (High Voltage Engineering Europa). The target temperature under irradiation depends on the current of implanted ions and did not exceed 100 °C. After implantation CdS single crystals were annealed in a stream of argon for 10 to 20 min at different temperatures 410, 520, and 750 °C. $CdIn_2S_4$ and $CdGa_2S_4$ single crystals, which were used for comparison, were obtained by chemical transport reactions. All samples of ternary compounds are of n-type conductivity. At room temperature $CdIn_2S_4$ samples have resistivity $\varrho \geqq 10^6\ \Omega$ cm and $CdGa_2S_4$ $\varrho \geqq 10^8\ \Omega$ cm.

To excite photoluminescence (PL), light with $\lambda = 313$ nm from a 500 W high pressure mercury discharge lamp was employed. To investigate reflection spectra, radiation of a 100 W iodine incandescent lamp modulated by a chopper with a frequency of 30 Hz was employed. Cathodoluminescence (CL) was excited by electron fluxes of different energy with current densities on the sample equal to 10^{-4} A/cm^2. Radiation was detected by cooled photomultipliers PhM-62 and PhM-79. Then the signal was amplified by a selective nanovoltmeter (Unipan). A monochromator MDR-3 (Lomo, Leningrad) was used as a spectrometer. The energy position of spectral features was determined with an accuracy not worse than 0.01 eV.

2. Results

2.1 Properties of CdS:In+ single crystals

Fig. 1 illustrates PL spectra of CdS single crystal sides undoped and implanted with indium at doses 10^{15} and 10^{16} cm^{-2}. On the undoped side at $T_a = 410$ °C one can ob-

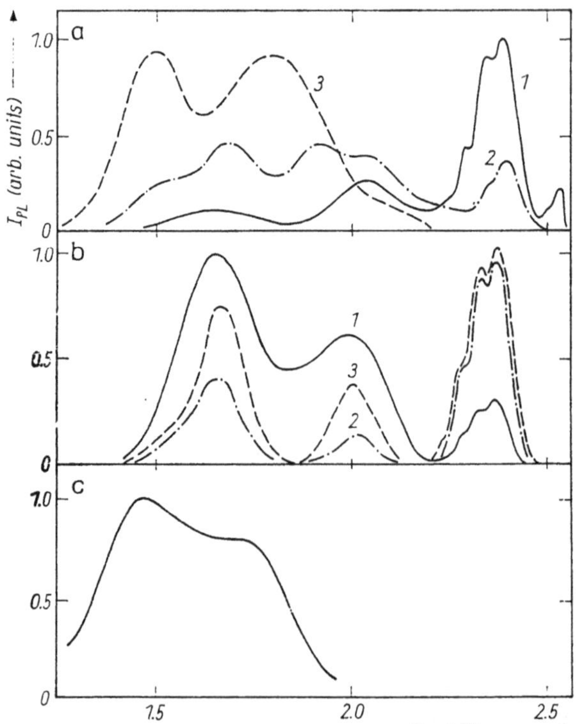

Fig. 1. PL spectra of the CdS samples undoped (1) and implanted with indium ions ((2) $D = 10^{15}$, (3) 10^{16} cm^{-2}) after annealing at a) 410 and b) 750 °C; c) PL spectrum of $CdIn_2S_4$ single crystals. $T = 80$ K

serve luminescence bands typical for cadmium sulphide [12]. Implantation of indium at $D = 10^{15}$ cm^{-2} causes in the PL spectrum of the implanted sides the appearance of peculiarities of the spectrum in the form of shoulders at energies ≈ 1.5 and 1.8 eV. An increase in the dose of implanted ions leads to the appearance of clearly defined maxima in the given region of the PL spectrum (curve 3 in Fig. 1a). In this case the typical luminescence bands of CdS are practically absent. The PL spectra of the implanted sides ($D = 10^{16}$ cm^{-2}) after annealing at 410 °C are similar to the spectra of $CdIn_2S_4$ compound single crystals (Fig. 1c).

An inrcease of annealing temperature in inert atmosphere up to 750 °C greatly changes the type of the PL spectra of CdS:In^+ samples (Fig. 1b). In this case in the spectra of the undoped sides the luminescence bands of CdS at energies 1.65 and 2.0 eV predominate, these bands are usually associated with intrinsic defects. On the implanted sides one can observe an intensive green band at 2.41 eV. Its intensity increases with increasing concentration of implanted indium.

If we assume that low temperature annealing of implanted layers ($D = 10^{16}$ cm^{-2}) leads to the formation of a new compound, it will be interesting to investigate its spatial distribution with the help of cathodoluminescence (CL). CL was measured at the sample temperature $T = 80$ K, the electron energy varying from 5 to 20 keV. According to the data in [13] the electron penetration depth into the sample varied from 0.2 μm for 6 keV and up to 1.8 μm for 21 keV. Fig. 2 shows the CL spectra of the CdS side implanted with indium ($D = 10^{16}$ cm^{-2}) after annealing at 410 °C. At low electron energy (curve 1) only the implanted layer is excited (the layer is ≈ 0.1 μm thick). Here, in the CL spectrum radiation bands typical for $CdIn_2S_4$ are well observed at energies 1.5 and 1.8 eV. A gain in electron energy increases the contribution to the luminescence of the undoped CdS substrate situated under the implanted layer. In the CL spectrum at the electron energy of 21 keV the green 2.41 and blue exciton 2.54 eV luminescence bands of CdS predominate.

Since In^+ implanted ions are situated in the thin surface layer, the investigation of reflection spectra (Fig. 3) may give some information on the band structure of the forming compound. The dependence of the reflection spectra on the dose of implanted indium is shown in Fig. 3a. On the undoped sides in reflection one can observe two exciton peaks of CdS with a maximum corresponding to the interband electron transition at 2.58 eV. After implantation of a maximum indium dose the structure associated with excitons disappeared and a new reflection peak appeared at an energy

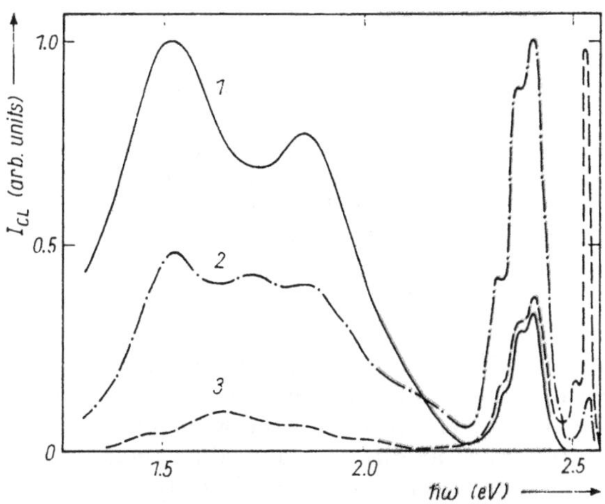

Fig. 2. CL spectra of the CdS sample sides implanted with indium ions ($D = 10^{16}$ cm^{-2}) after annealing at 410 °C with the following electron energies: (1) 6, (2) 13, and (3) 21 keV, $T = 80$ K

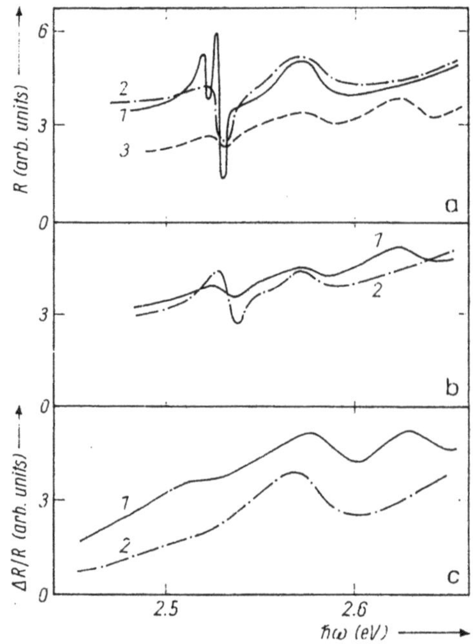

Fig. 3. a) Reflection spectra of the starting CdS single crystals (1) and the CdS:In$^+$ sample sides implanted with (2) $D = 10^{15}$ and (3) 10^{16} cm^{-2} after annealing at 410 °C; b) reflection spectra and c) electroreflection spectra of the sides implanted with indium ions $D = 10^{16}$ cm^{-2} after annealing at (1) 410 and (2) 750 °C. $T = 80$ K

of 2.63 eV close to the energy of direct interband transitions of the $CdIn_2S_4$ compound at the given temperature [10]. The presence of this reflection peak also depends on annealing temperature. It is seen that annealing at 750 °C (Fig. 3 b, c) of the implanted layers ($D = 10^{16}$ cm^{-2}) leads to the disappearance of the given peak. As a confirmation we note that in the spectrum of electroreflection of the implanted sides (Fig. 3c) the peak is also observed at 2.63 eV.

2.2 Properties of CdS:Ga+ single crystals

The formation of the $CdGa_2S_4$ compound upon implantation of gallium ions into cadmium sulphide even arouses greater interest. The fact is that in this case the ternary compound has a wider band gap than CdS and ion implantation can lead to optical transparency of the thin surface layer and to the appearance of luminescence bands with an energy greater than the width of the CdS band gap.

Let us consider the spectral characteristics of the CdS:Ga$^+$ system obtained by gallium ion implantation at doses 10^{15} and 10^{16} cm^{-2}. After implantation the samples were annealed in a stream of argon at 410, 520, and 750 °C. Fig. 4 illustrates the PL spectra of the implanted and undoped sides of CdS:Ga$^+$ samples at various implantation doses and annealing temperatures. It is seen that at low temperature annealing at 410 °C a gain in the dose of implanted gallium results in the appearance of absolutely new, broad luminescence bands in the red ≈ 1.9 and blue ≈ 2.9 eV spectral regions (Fig. 4a). Exactly these bands are typical for the PL spectra of $CdGa_2S_4$ compound single crystals (Fig. 5a).

For high temperature annealing at $T = 750$ °C the PL spectra of the implanted and undoped sides differ only in the increasing intensity of green luminescence at 2.39 eV with the dose of implanted gallium. The difference between luminescence spectra of the implanted sides ($D = 10^{16}$ cm^{-2}) at different annealing temperatures testifies the formation of a phase of the new ternary compound $CdGa_2S_4$ in the im-

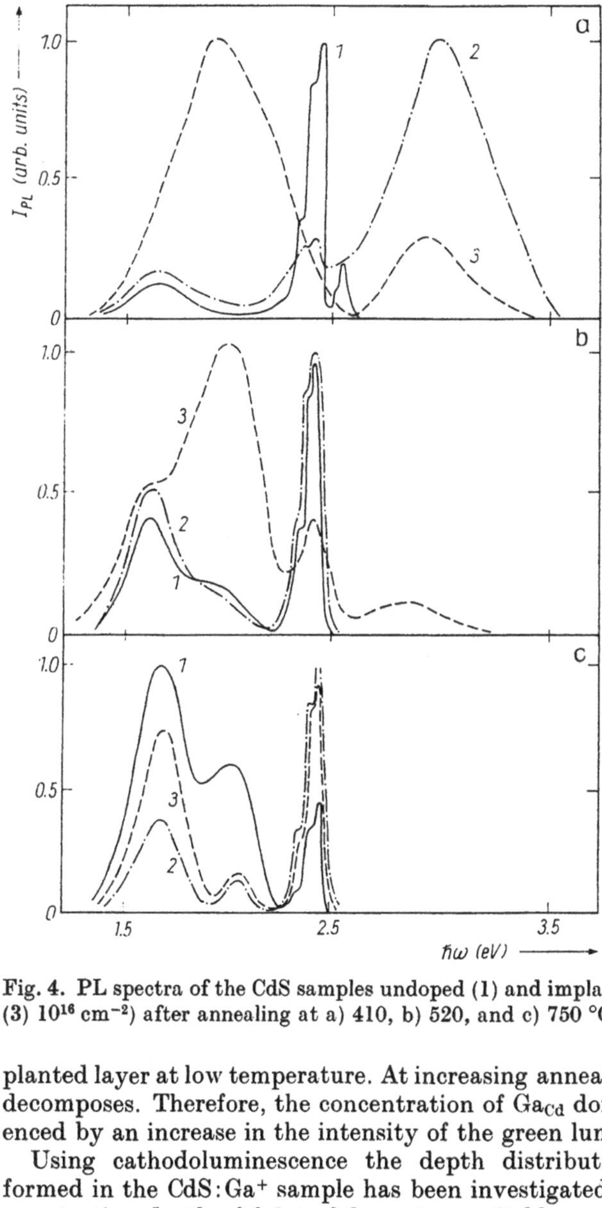

Fig. 4. PL spectra of the CdS samples undoped (1) and implanted with gallium ions ((2) $D = 10^{15}$, (3) 10^{16} cm^{-2}) after annealing at a) 410, b) 520, and c) 750 °C; $T = 80$ K

planted layer at low temperature. At increasing annealing temperature the compound decomposes. Therefore, the concentration of Ga_{Cd} donor centres grows which is evidenced by an increase in the intensity of the green luminescence (Fig. 4c).

Using cathodoluminescence the depth distribution of the ternary compound formed in the CdS:Ga^+ sample has been investigated. It is seen that at the electron penetration depth of 0.1 to 0.2 μm (curve 1) blue ≈ 2.9 and red ≈ 1.9 eV luminescences of $CdGa_2S_4$ predominate (Fig. 5b) in the CL spectrum. With increasing electron penetration depth (energy) the contribution of green and exciton luminescences of deeper undoped layers of CdS to CL increases.

The reflection study of the CdS:Ga^+ implanted side at the dose of implanted gallium $D = 10^{16}$ cm^{-2} (Fig. 6) permits revealing, after low temperature annealing, a peculiarity at 3.77 eV. A similar maximum is obrsrved in the electroreflection spectra of Ga^+ implanted CdS samples at maximum implantation dose and annealing temperature of 410 °C (Fig. 6b). The energy position of this maximum is close to the

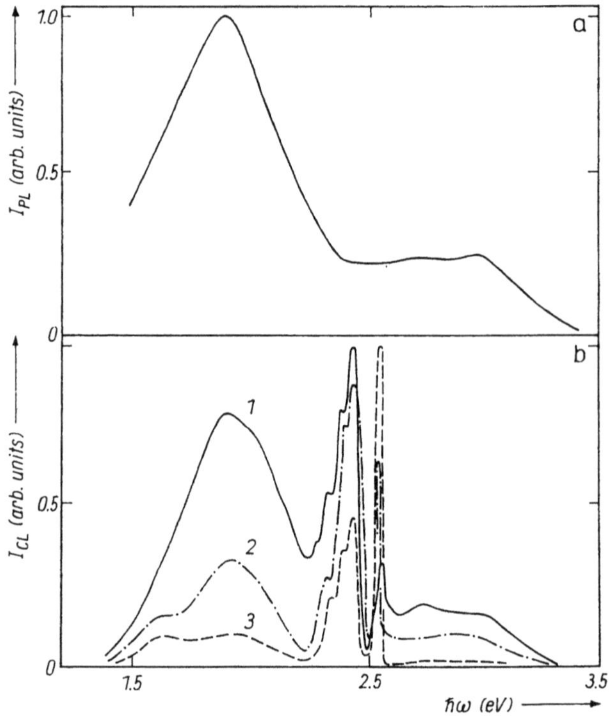

Fig. 5. a) PL spectra of $CdGa_2S_4$ single crystals; $T = 80$ K. b) CL spectra of the CdS sample side implanted with gallium ions ($D = 10^{16}$ cm^{-2}) after annealing at 410 °C with the following electron energies: (1) 5, (2) 13, and (3) 21 keV; $T = 80$ K

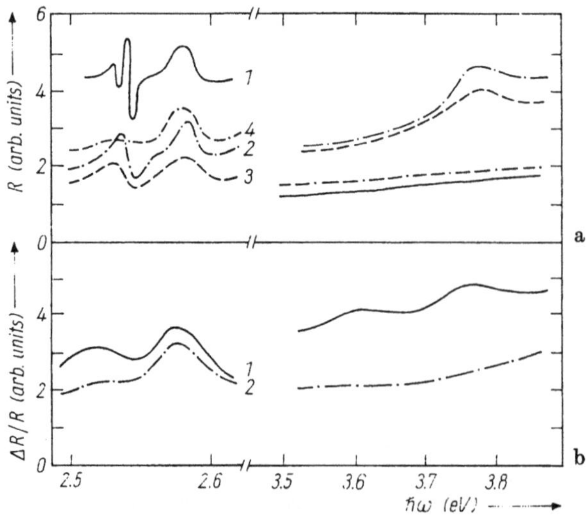

Fig. 6. a) Reflection spectra of the CdS single crystals implanted with gallium ions, $D = 10^{16}$ cm^{-2}, after annealing at (2) 410, (3) 520, (4) 750 °C and of the starting CdS single crystals (1); b) electroreflection spectra of the CdS:Ga^+ samples, $D = 10^{16}$ cm^{-2}, after annealing at (1) 410 and (2) 750 °C. $T = 80$ K

band gap of the ternary semiconductor $CdGa_2S_4$ ($E_g \approx 3.77$ eV at 10 K). In the electroreflection spectrum there also exists a maximum at 3.62 eV which is likely to be associated with the formation of free excitons in the $CdGa_2S_4$ compound [11]. An increase in annealing temperature up to 750 °C leads to decomposition of the new phase and the reflection spectrum of the implanted side of the CdS:Ga sample becomes similar to the reflection spectrum of the undoped side (Fig. 6a, curve 1 and 4). But on the implanted side the peculiarities in the exciton region at 2.54 eV are absent.

3. Discussion

The results of the investigations indicate the formation of new phases of the ternary semiconductor compounds $CdGa_2S_4$ and $CdIn_2S_4$ upon implantation of high doses ($D = 10^{16}$ cm^{-2}) of gallium and indium into CdS. At smaller dose ($D = 10^{15}$ cm^{-2}) the number of regions of the new compounds decreases. This is evidenced by a decrease in the contribution of the corresponding peculiarities to the luminescence and reflection spectra (Fig. 2 and 3) of the implanted sides. It can be also noted that at a dose of 10^{15} cm^{-2} the blue luminescence (at 2.9 eV) of the new phase of $CdGa_2S_4$ is more intensive than at a dose of 10^{16} cm^{-2} (Fig. 4), in agreement with the results of [14] where the blue luminescence of $CdGa_2S_4$ is explained by lack of gallium as compared to stoichiometry.

The concentration of indium and gallium atoms in the corresponding stoichiometric ternary compounds is according to [15] $N_T \approx 1.25 \times 10^{22}$ cm^{-3}. Their concentration in the implanted layer of cadmium sulphide can be evaluated by the formula [16]

$$D = 2.5\, \Delta R_p\, N_{max}\,,$$

where D is the does of implanted ions, ΔR_p the root-mean-square scatter of the implanted ion projected path, and N_{max} the concentration of ions in the maximum of the distribution profile. Using $\Delta R_p \approx 3 \times 10^{-6}$ cm typical for implanted ions of 250 keV energy we obtain $N_{max} = 1.3 \times 10^{21}$ cm^{-3} for $D = 10^{16}$ cm^{-2}. The distribution profile of ions has Gaussian shape, but even in the maximum of the distribution profile the concentration of implanted indium and gallium ions in CdS differs from the value N_T by nearly an order. Here it should be noted that as shown in [17] the formation of the new phase by ion implantation does not require accumulation of the implanted impurity to the concentration determined by the stoichiometry of the new compound. The transition from one compound to another includes a stage of cluster formation (microregions with a structure of the new phase) which collect the implanted element from the surrounding bulk. Thus microcrystals of the new phase compounds are formed. They have a few possible orientations in the CdS lattice. The number of these orientations is the greater the nearer the symmetry classes of the new and old phases are. Electron diffraction patterns of the implanted sides also give evidence of a polycrystalline structure of the implantation formed new compounds $CdIn_2S_4$ and $CdGa_2S_4$.

The disappearance of the reflection peculiarities and PL bands (typical for ternary compounds) after annealing of the samples at 750 °C is in agreement with [17], where it was shown that the new phase nuclei formed by ion implantation can exist up to a definite critical temperature T_{crit} the value of which is determined by the surface tension on the interface between new and old phases. In our case at annealing temperatures above T_{crit} (520 °C $< T_{crit} <$ 750 °C) a reconstruction throughout the implanted layers of the CdS phase should be expected. However, the near surface region of the samlpe remains enriched in group-III atoms in the form of a substitutional solution which is confirmed by an increased intensity of green luminescence.

4. Conclusion

Thus the performed investigations of CdS single crystals irradiated by indium and gallium ions at doses 10^{15} and 10^{16} cm^{-2} indicate that after annealing the samples in an argon stream at 410 and 520 °C the luminescence spectra and reflection spectra of the implanted side are similar to the spectra of $CdIn_2S_4$ and $CdGa_2S_4$ compounds, respectively. It testifies the formation of phases of new compounds during ion implantation in the form of clusters with different crystal orientations. An increase in annealing temperature leads to the destruction of the clusters with subsequent diffusion of group-III elements through the whole volume of the CdS single crystal.

References

[1] A. N. Georgobiani, Z. P. Ilyushina, S. I. Radautsan, and I. M. Tiginyanu, Preprint No. 259, P. N. Lebedev Phys. Inst., Moscow 1981 (p. 47).
[2] E. D. Jones and H. Mykura, J. Phys. Chem. Solids **41**, 1261 (1980).
[3] Zh. Pankov, Opticheskie protsessy v poluprovodnikakh, Izd. Mir, Moskva 1973 (p. 456).
[4] E. D. Jones and H. Mykura, J. Phys. Chem. Solids **39**, 11 (1978).
[5] N. I. Vitrikhovskii and I. B. Mizetskaya, Izv. Akad. Nauk SSSR, Ser. fiz. **28**, 1316 (1964).
[6] R. Kasik and Kh. Koppel, Izv. Akad. Nauk SSSR, Ser. khim. geol. **22**, No. 1, 42 (1973).
[7] N. P. Akimchenko, V. S. Vavilov, V. V. Galkin, V. S. Ivanov, V. V. Krasnopevtsev, and Yu. V. Malyutin, Fiz. Tekh. Poluprov. **6**, 1182 (1972).
[8] M. I. Belii, G. A. Gumanskii, V. I. Caras, V. M. Lomako, I. S. Tashlykov, and V. S. Tishkov, Fiz. Tekh. Poluprov. **9**, 2027 (1975).
[9] A. N. Georgobiani, Z. P. Ilyushina, A. V. Spitsyn, and I. M. Tiginyanu, Short Commun. P. N. Lebedev Phys. Inst., Ser. Phys. No. 8, 31 (1981).
[10] H. Nakanishi, Japan. J. appl. Phys. **19**, 103 (1980).
[11] A. N. Georgobiani, Yu. V. Ozerov, S. I. Radautsan, and I. Tiginyanu, Fiz. tverd. Tela **23**, 2094 (1981).
[12] L. S. Pedrotti and D. C. Reynolds, Phys. Rev. **119**, 1897 (1960).
[13] F. J. Bryant, Crystal Growth and Charact. **6**, No. 2, 191 (1983).
[14] A. N. Georgobiani, S. I. Radautsan, and I. M. Tiginyanu, phys. stat. sol. (a) **69**, 513 (1982).
[15] F. G. Donika, V. F. Zhitar, and S. I. Radautsan, Poluprovodniki sistemy $ZnS-In_2S_3$, Izd. Shtiintsa, Kishinev 1980 (p. 148).
[16] A. F. Burenkov, F. F. Komarov, and M. A. Kumakhov, Tablitsy parametrov prostranstvennogo raspredeleniya ionno-implantirovannykh primesei, Izd. Belorus. Gos. Univ., Minsk 1980 (p. 395).
[17] P. V. Pavlov, E. I. Zorin, and D. I. Tetelbaum, Proc. Internat. Conf. Ion Implantation of Semiconductors, Izd. Budapest Central Res. Inst. for Physics, Budapest 1975 (p. 107).

(Received March 14, 1988)

phys. stat. sol. (a) **111**, 309 (1989)

Subject classification: 77.80; S12

National Physical Laboratory, New Delhi[1])

Polarization Reversal Current Behaviour for Parallel and Antiparallel Rubbed Surface Stabilized Ferroelectric Liquid Crystals

By

S. S. BAWA, K. SAXENA, and S. CHANDRA

The polarization reversal current response for parallel and antiparallel rubbed samples of a 3 μm thick sample of DOBAMBC are studied by applying triangular pulses of varying voltage. The antiparallel rubbed sample shows two peaks and parallel rubbed sample shows three peaks of polarization reversal. The third peak appears before the peak of bulk switching and the zero field point of the applied pulse. It is shown that the selective pretilt angles with the surface results in tilted or bent smectic layer to give the multipeak response of the polarization reversal current.

Das Polarisationsumkehrstromverhalten für parallel und antiparallel geschliffene Proben einer 3 μm dicken DOBAMBC-Probe werden mit dreieckförmig anwachsenden Spannungsimpulsen untersucht. Die antiparallel geschliffene Probe zeigt zwei Maxima und die parallel geschliffene Probe zeigt drei Maxima von Polarisationsumkehr. Das dritte Maximum erscheint vor dem Maximum des Volumenverhaltens und dem Nullfeldpunkt des angelegten Impulses. Es wird gezeigt, daß die ausgewählten Vorneigungswinkel mit der Oberfläche zu einer gekippten oder gebogenen smektischen Schicht führen, die den Mehrfachmaximumrespons des Polarisationsumkehrstroms liefert.

1. Introduction

In recent years, surface stabilized ferroelectric liquid crystals (SSFLC) have been extensively studied [1 to 6] regarding applications to bistable, μs switching light valves. In spite of these studies, one does not yet clearly understand the molecular configuration, switching mechanism, and the temporal experimental evolutions for different thicknesses, surface treatments [7], pretilts or tilts of surface layers, material constants, inherent chirality, the negative dielectric anisotropy, the polarization electric field [8], etc. Clark and Lagerwall [9] and Handschy and Clark [10] first suggested a model to describe the switching mechanism in a thin sample. According to this model there are two switching regimes for different fields. At moderate field, the polarization switching takes place through the motion of walls and the switching time depends on the nucleation rate and the domain wall velocity. At high electric fields, the polarization reversal is accompanied by the simultaneous reversal of all molecules from one state to another. Takezoe et al. [6] suggested another model for a slightly thicker cell with disclinations both in the bulk of the cell and on the surface. Later Yamada et al. [7] performed switching simulations in terms of the model of Handschy and Clark by taking strong surface anchoring forces into account.

Recently Nakagawa et al. [8] studied the dynamic response of SSFLC on the basis of continuum theory by including a polarization electric field through Poisson's

---

[1]) Dr. K. S. Krishnan Marg, New Delhi 110012, India.

equation and including a polarization charge accompanied by elastic deformation. From numerically calculated results, it has been found that the polarization current and hysteresis curves are crucially affected by the inherent chirality, negative dielectric anisotropy, and the polarization electric field resulting in a doubly or triply peaked polarization reversal current by an alternating electric field.

Recently Clark and Lagerwall [11] by means of X-ray measurements have shown that smectic layers are inclined with the bounding glass plates or apparently bent between them. Ouchi et al. [12] also proposed a model for the existence of inclined or bent layers by means of stroboscopic and optical response measurements. The origin of bent or inclined smectic layers is mainly due to selective surface pretilt angles.

In this paper we report the results of our polarization current measurements for SSFLC cells prepared with parallel and antiparallel rubbed directions for the same material and for the same thicknesses of the cells. These configurations result in different orientations of the smectic layers, i.e. inclined or bent due to different surface pretilts. The polarization reversal current response shows multi-peaks for parallel (↑↑) and antiparallel (↑↓) rubbed samples and is described here.

2. Experimental

The sample cell comprises two ITO-coated etched glass plates separated by 3 μm thick mylar spacer. The alignment of the sample is obtained by initially treating the plates with adhesion promoter [13] and then with a polyamide nylon solution [14]. The glass plates are baked at 130 °C for one hour and then rubbed unidirectionally. DOBAMBC was filled between ITO-coated glass plates by capillary action at its isotropic (120 °C) temperature. Samples were then slowly cooled (10 K/h) to the SmA phase and further down to the SmC* phase to obtain an excellent monodomain cell. The temperature of the sample is kept constant at 76 °C for all measurements and is observed under a polarizing microscope.

The bridge circuit used in the present studies is the same as used previously [15]. Triangular-shaped fields at low frequencies (5 to 50 Hz) were applied from a pulse generator to a 3 μm thick sample. Pulses generated have the perfect linearity up to 20 V_{pp} signal. The applied frequency was chosen such that the multipeaks may appear separately on decreasing the voltage from 20 V_{pp} to about 5 V_{pp}.

The circuit used in the present studies has been modified slightly to balance out the resistive equivalent but to retain the capacitive and the nonlinear polarization currents of the FLC sample cell. This is achieved by subtracting a proportional current of the applied signal (CH2) from the overall current response (CH1) and recording the signal (CH1–CH2) on the storage oscilloscope (ECIL OS 768S). This ensures the occurrence of the resistively balanced polarization current response to occur horizontally on the oscilloscope. Thus it becomes more convenient to identify the occurrences of the polarization reversal current peaks.

3. Results and Discussions

Fig. 1 illustrates the schematic representation of the current response to a triangular pulse. In an antiparallel rubbed sample, two peaks of polarization reversal occur due to bulk and surface molecular reorientation as shown in Fig. 1, curve b. In thin cells, the resistive component becomes nonlinear due to the collection of ionic charge at the electrodes [16]. The small nonlinear contribution has been shown in Fig. 1,

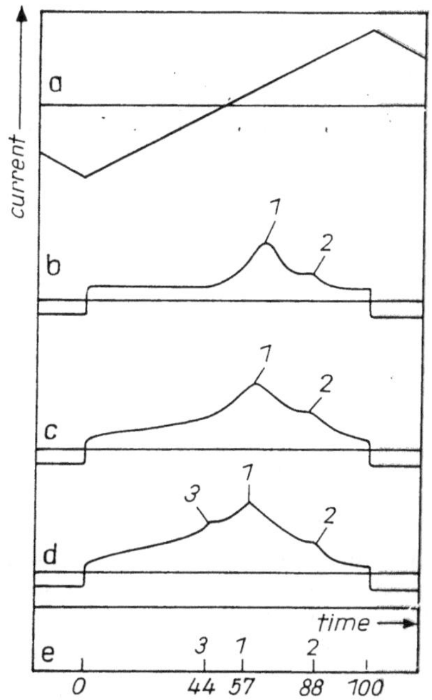

Fig. 1. Schematic representation of the current response to a triangular pulse: (a) applied pulse, (b) current response of an antiparallel rubbed sample neglecting ionic charge. (c) current response showing the nonlinear contribution due to the collection of ionic charge at the electrodes for an antiparallel rubbed sample. (d) current response of a parallel rubbed sample including the ionic charge contribution, (e) scale indicating the occurrence of first, second, and third peaks at 57%, 88%, and 44% of the half pulse, respectively

curve c. This nonlinear part is more evident at low frequencies, high voltages, and thin cells [16]. Fig. 1, curve d illustrates schematically the occurrence of three peaks in a parallel rubbed sample at low voltages (12 V) for a 3 μm thin sample.

Fig. 2 shows the current response of a 3 μm thick antiparallel rubbed sample of DOBAMBC at 8 Hz and 76 °C at different applied voltages of triangular pulse shape. Clearly there exist two peaks of polarization reversal current due to bulk and surface molecular reorientations. The occurrence of the peak due to surface reorientation appears at the trailing end of the first peak indicating the different response of surface and the bulk. With the decrease in electric field both peaks are separated and the peak due to surface switching shifts away faster indicating a much lower switching speed of surface than of the bulk. Similar behaviour, i.e. the shift of second peak, has also been observed with the increase of frequency [17].

Fig. 3 shows the current response of a parallel rubbed sample under identical operating condition as for an antiparallel rubbed sample. At high applied voltage (19 V_{pp}) there appear two peaks of polarization reversal. As the voltage is decreased

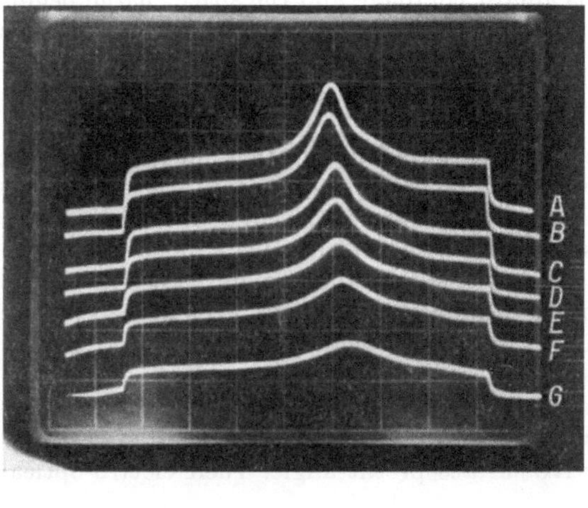

Fig. 2. Oscillograph of polarization reversal current for antiparallel rubbed 3 μm thick sample at 76 °C, 10 Hz and at different applied voltages: (A) 20, (B) 18, (C) 16, (D) 14, (E) 12, (F) 9, (G) 7 V. Ordinate scale is arbitrary in each case

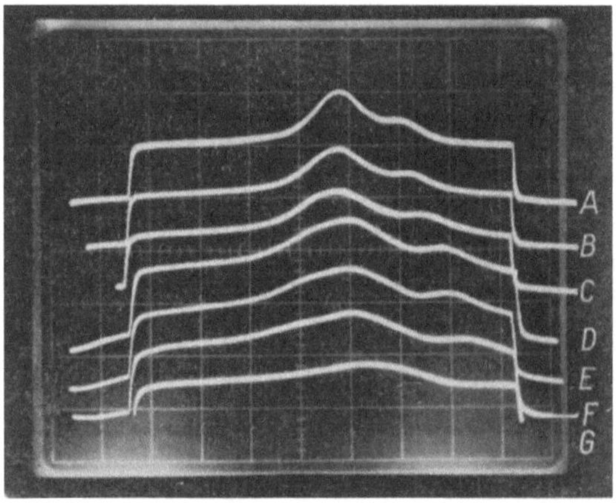

Fig. 3. Oscillograph of polarization reversal current for parallel rubbed 3 μm thick sample at 76 °C, 10 Hz and at different applied voltages: (A) 19, (B) 17, (C) 15, (D) 13, (E) 11.5, (F) 10, (G) 7.5 V

(17 V_{pp}) the third peak begins to appear (Fig. 3, curve B). The third peak occurs before the peak of bulk polarization reversal does. As the voltage is further decreased (15 to 10 V_{pp}) the third peak appears clearly separated. However, it should be noted that the appearance of the third peak is before the zero field point and occurs at about 44% of the applied half pulse (Fig. 2, curve B). For very low applied voltage the response of the second peak slows down and crosses the end point of the applied triangular pulse. It should be noted that the behaviour of the first two peaks for parallel and antiparallel rubbed sample with decreasing voltage is almost similar.

To check whether the occurrences of a double or triple peak in the polarization reversal current is mainly due to the parallel or antiparallel rubbing, a cell consisting of two segments with parallel and antiparallel rubbed configuration were made on a highly polished glass plates. The polarization reversal current response for both types of alignment was also studied. Clearly there appear three peaks for parallel and two peaks for antiparallel assembled configurations.

The effect of ionic charge accumulation at the electrodes has also been observed in both parallel and antiparallel rubbed samples. In both cases the overall response is nonlinear [16] in the background current. The slope of the trace of the current versus time at the beginning of the trace remains linear, indicating Ohm's law to be satisfied. However, the current quickly becomes nonlinear indicating that the ionic flow from one side of the electrode to the other is almost complete and the space charge begins to accumulate at the electrodes. It should be noted that the third peak which appears in parallel rubbed samples separates out from the first on decreasing the voltage and is of different nature than the effect of ionic charge accumulation in double charge layers for both electrodes.

The different behaviour for parallel and antiparallel configurations may be explained by considering the different possible deformations with a surface selective pretilt angle. It is well known that there is a selective pretilt of molecules with the surfaces in SSFLC's. The pretilts of molecules in parallel and antiparallel rubbed configurations are shown in Fig. 4 a and b. For parallel rubbed configuration the pretilts at the surfaces form mirror images of each other, while for the antirubbed surface the tilts are parallel. The smectic layers in such cases may be bent or tilted, respectively, as shown in Fig. 4.

Ouchi et al. [12] have also proposed a similar model of tilted and bent layers by means of stroboscopic and optical response measurements. They suggested two possible deformations based on the selective surface pretilt angle. Their electro-optical

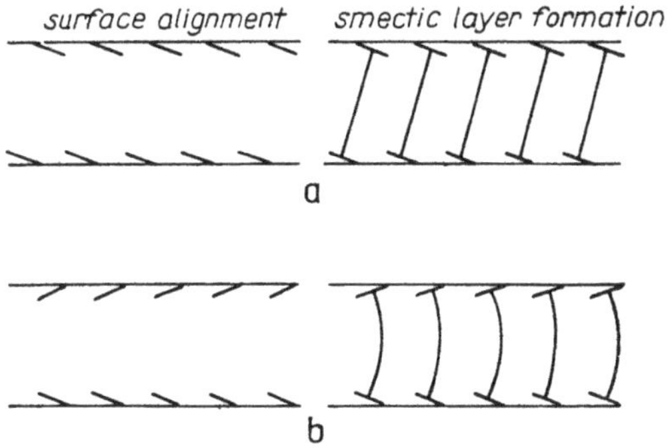

Fig. 4. Conceivable surface alignment and smectic layer formation for a) antiparallel rubbed and b) parallel rubbed samples

observation implied that one of the processes is accomplished by nucleation and growth of a loop disclination when the applied electric field is reversed and the other is similar to the Nambu-Goldstone mode and is much faster than the previous one. Here in the present case it is quite possible that the third peak for a parallel rubbed sample corresponds to the Nambu-Goldstone mode. As discussed before the third peak which appears at 44% of the applied triangular field is much faster than the other two peaks.

The theory put forward to explain the dynamic behaviour of SSFLC by Nakagawa et al. [8] is also based on the basic postulates suggested by Ouchi et al. [12]. The occurrence of three peaks due to the change in inherent chirality, dielectric anisotropy, and polarization electric field is also indirectly derived from the tilted and bent surface layers. Nakagawa et al. [8] have also found that the multi-peak switching current disappears with a decrease in the applied voltage or the frequency. But in the present case we have found that the decrease in voltage from 19 to 10 V_{pp} makes the third peak to appear. However, the second peak moves further away and may cross the peak of the applied triangular pulse due to the increase in the response time of switching ($\tau = \eta/PE$). Further the decrease in frequency (0.5 Hz) suppresses the polarization reversal surface current peak due to reduced peak area and not due to vanishing of the switching current. The increase in frequency of the applied triangular pulses shifts the polarization reversal current peak along the abscissa axis and may cross the peak of the applied pulse due to slower response of surface switching than the quarter period of the applied triangular pulse.

The appearance of the third peak before the occurrence of the peak of bulk switching indicates a much faster response than in the bulk. Ouchi et al. have suggested that the process resembling the so-called Nambu-Goldstone mode is much faster. This mode seems to be responsible for the occurrence of the third peak of polarization reversal. The occurrence of the third peak at 44% of the applied pulse (i.e. below the zero-field point) can be explained as follows: The process responsible for the third peak has a threshold near -2.0 V (say) and possesses some finite response time. At high applied voltages (± 20 V_{pp}) the region -2.0 to 0V is scanned at a much faster speed ($T/20$ s, where T is the period of the applied triangular pulses) than the response time and merges with the bulk switching peak. At low voltage (10 V_{pp}) of the applied pulses, the scanning time ($T/10$) from -2 to 0 V may be larger than the response time and hence the peak appears before the zero-field point.

4. Conclusions

We have obtained information on the polarization reversal current for parallel and antiparallel rubbed samples of surface stabilized ferroelectric liquid crystals. Tilted and bent smectic layer structures may result for antiparallel and parallel rubbed samples resulting in a double or triple peak of polarization reversal current. Such an experiment is a powerful tool in designing ferroelectric liquid crystal devices.

Acknowledgements

The authors sincerely thank Dr. S. K. Joshi, Director, National Physical Laboratory for continuous encouragement. The help provided by Liquid Crystal Group and Mr. M. S. Rannot of CSIR is gratefully acknowledged.

References

[1] R. B. MEYER, Mol. Cryst. liquid Cryst. **40**, 33 (1977).
[2] PH. MARTINOT-LAGARDE, R. DUKE, and G. DURAND, Mol. Cryst. liquid Cryst. **75**, 249 (1981).
[3] N. A. CLARK, M. A. HANDSCHY, and S. T. LAGERWALL, Mol. Cryst. liquid Cryst. **94**, 213 (1983).
[4] S. T. LAGERWALL and I. DAHL, Mol. Cryst. liquid Cryst. **114**, 151 (1984).
[5] M. A. HANDSCHY, N. A. CLARK, and S. T. LAGERWALL, Phys. Rev. Letters **51**, 471 (1983).
[6] H. TAKEZOE, Y. OUCHI, R. ISHIKAWA, and A. FUKUDA, Mol. Cryst. liquid Cryst. **139**, 27 (1986).
[7] Y. YAMADA, T. TSUGE, N. YAMAMOTO, M. YAMAWAKI, H. ORIHARA, and Y. ISHIBASHI, Japan. J. appl. Phys. **26**, 1811 (1987).
[8] M. NAKAGAWA, M. ISHIKAWA, and T. AKAHANE, Japan. J. appl. Phys. **27**, 456 (1988).
[9] N. A. CLARK and S. T. LAGERWALL, Appl. Phys. Letters **36**, 899 (1980).
[10] M. A. HANDSCHY and N. A. CLARK, Ferroelectrics **59**, 69 (1984).
[11] N. A. CLARK and S. T. LAGERWALL, Proc. 6th Internat. Display Res. Conf., September 30 to October 2, 1986, Tokyo 1986 (p. 456).
[12] Y. OUCHI, H. TAKEZOE, and A. FUKUDA, Japan. J. appl. Phys. **26**, L21 (1987).
[13] S. S. BAWA, A. M. BIRADAR, and S. CHANDRA, phys. stat. sol. (a) **102**, 829 (1987).
[14] J. S. PATEL, T. M. LESLIE, and J. W. GOODBY, Ferroelectrics **59**, 137 (1984).
[15] S. S. BAWA, A. M. BIRADAR, and S. CHANDRA, Ferroelectrics **76**, 69 (1987).
[16] J. S. PATEL and J. W. GOODBY, Chem. Phys. Letters **137**, 91 (1987).
[17] S. S. BAWA, A. M. BIRADAR, K. SAXENA, and S. CHANDRA, Rev. Sci. Instrum. **59**, 2023 (1988).

(Received June 8, 1988; in revised form September 6, 1988)

phys. stat. sol. (a) **111**, 315 (1989)

Subject classification: 78.55 and 78.60; 71.55; S11.2

Institut für Anorganische Chemie der Universität Tübingen[1])

Cathodo- and Photoluminescence in the Mn^{2+} Activated Garnets $Cd_{3-x}Ca_xAl_2Ge_3O_{12}$

By

J. WIEHL, R. HIRRLE, W. WISCHERT, and S. KEMMLER-SACK

Cathode ray irradiation of the pure Cd garnet ($x = 0$) yields an intense blue emission; activation with manganese results in an intense long-lived yellow emission due to an energy transfer from defect centres to the activator. UV excitation of $Cd_3Al_2Ge_3O_{12}$:Mn results in an yellow emission as well. In the system $Cd_{3-x}Ca_xAl_2Ge_3O_{12}$ a complete series of solid solutions is formed. For higher Ca content ($x > 1$) the intensity of the defect and also the Mn emission is reduced considerably. The luminescence properties are studied as a function of the Cd/Ca ratio.

Bei Kathodenstrahl-Anregung zeigt das reine Cd-Wirtsgitter ($x = 0$) eine intensive blaue Emission; Aktivierung mit Mangan führt zu einer intensiven, gelben, langlebigen Emission, welche auf eine Energieübertragung zwischen Defektzentren und dem Aktivator zurückzuführen ist. Unter UV-Anregung zeigt $Cd_3Al_2Ge_3O_{12}$:Mn ebenfalls eine gelbe Emission. Im System $Cd_{3-x}Ca_xAl_2Ge_3O_{12}$ wird die Bildung einer vollständigen Mischkristallreihe beobachtet. Für höhere Ca-Gehalte ($x > 1$) geht sowohl die Intensität der Defekt- als auch der Mangan-Emission beträchtlich zurück. Die Lumineszenzeigenschaften werden in Abhängigkeit vom Cd/Ca-Verhältnis untersucht.

1. Introduction

The activation of garnet hosts with rare earth and/or transition metal ions may result in an intense cathodoluminescence (CL), photoluminescence (PL), or stimulated emission [1]. For this purpose garnets are increasingly studied.

This work deals firstly with the CL and PL properties of the garnet $Cd_3Al_2Ge_3O_{12}$ and its activation with manganese. The influence of the substitution of Ca for Cd on the luminescent properties is studied as well (system $Cd_{3-x-y}Ca_xMn_yAl_2Ge_3O_{12}$) by means of the CL, PL, and diffuse reflectance spectra.

2. Experimental

The pure hosts and the manganese activated compounds of the system $Cd_{3-x-y}Ca_x \cdot Mn_yAl_2Ge_3O_{12}$ were prepared via solid state reactions by using high purity starting materials (CdO (specpure; Johnson Matthey); $CaCO_3$ (99.9975%; Ventron); $Al(OH)_3$ (p.A.; Merck); GeO_2 (99.99%; Fluka); $MnCO_3$ (99.999%; Ventron)). After grinding in an agate mortar the samples were fired in alumina crucibles (Degussit Al23) in air at 700, 800, 900 (1 h), and between 1100 and 1200 ($x \leqq 1.5$) or 1300 °C ($x \geqq 2.5$), respectively, for 20 h to 40 h. The observed weight losses are 1 to 2%. X-ray checks (Philips powder diffractometer; CuK_α radiation) showed all samples to be homogeneous. The lattice parameter of the pure hosts ($x = 0$: $a = 12.068(5) \times 10^{-1}$ nm; $x = 3$: $a = 12.111(5) \times 10^{-1}$ nm) are in agreement with the data in [2]. In the system $Cd_{3-x}Ca_xAl_2Ge_3O_{12}$ complete miscibility is observed; the lattice constants obey

[1]) Auf der Morgenstelle 18, D-7400 Tübingen, FRG.

Végards law. The values of the corresponding activated compounds did not vary considerably. The densities (determined by using n-octane as buoyancy fluid) are in agreement with the theoretical values ($x = 0$: $\varrho_{obs} = 5.97(5)$ g cm^{-3}; $\varrho_{calc} = 6.05$ g cm^{-3}; $x = 1.5$: $\varrho_{obs} = 5.15(5)$ g cm^{-3}; $\varrho_{calc} = 5.19$ g cm^{-3}; $x = 3$: $\varrho_{obs} = 4.37(5)$ g cm^{-3}; $\varrho_{calc} = 4.37$ g cm^{-3}). The pure garnet hosts are white; samples with low manganese concentration show a slight pink touch, which intensifies with increasing y.

The spectroscopic equipment for the measurement of the CL [3], PL [4], and diffuse reflectance spectra [5, 6] has been described elsewhere; SiO_2 was used as standard in the range from 25000 to 50000 cm^{-1}.

For CL decay measurements the electron beam of the CRT is magnetically deflected. The current density of the electron beam reaches approximately 1 μA/cm². The emitted light is detected by a photomultiplier (R636; Hamamatsu). A fast preamplifier (Model 115, EG & G, Princeton Applied Research Corporation) amplifies the phototube signal. The decay event is displayed on a digital storage oscilloscope (VP 5730P; Panasonic). Lifetime is the time of decay to 37% of the CL emission at cessation of excitation.

3. Results

3.1 Diffuse reflectance spectra

The optical properties of the pure hosts $Cd_3Al_2Ge_3O_{12}$ ($x = 0$) and $Ca_3Al_2Ge_3O_{12}$ ($x = 3$) vary considerably. In the diffuse reflectance spectra (Fig. 1) for the Ca compound an optical absorption edge is just visible around ≈ 210 nm; its position is in agreement with the value for other compounds with GeO_4 tetrahedra (e.g. Gd_2GeO_5: absorption edge at ≈ 210 nm [7]). In addition a weaker, broad absorption band is observed at ≈ 270 nm; the strong dependence of its intensity on the conditions of preparation indicates that it is due to lattice defects. In the pure Cd host $Cd_3Al_2Ge_3O_{12}$ the absorption edge is situated at considerably lower energies (≈ 270 nm). The substitution of Ca for Cd leads to a uniform shift to higher energies (Fig. 1).

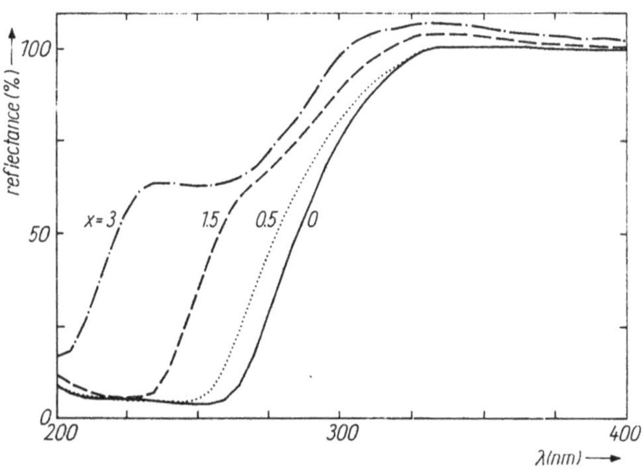

Fig. 1. Diffuse reflectance spectra of $Cd_{3-x}Ca_xAl_2Ge_3O_{12}$

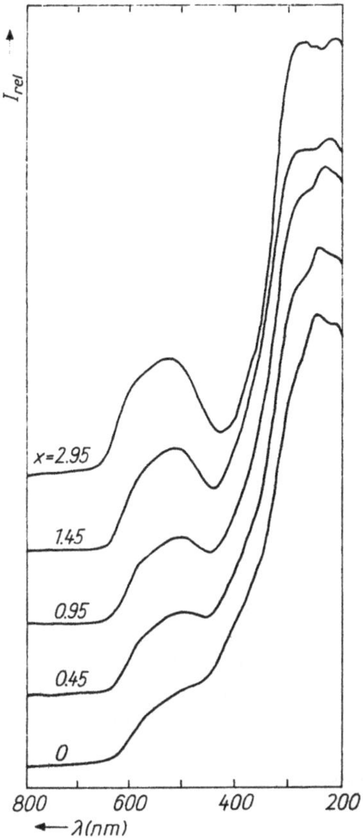

Fig. 2
Diffuse reflectance spectra of $Cd_{2.95-x}Ca_xMn_{0.05}Al_2Ge_3O_{12}$

The incorporation of manganese generates for $x = 0$ an absorption tail, getting more pronounced at higher Mn concentrations ($y \geqq 0.05$) and increasing Ca content (Fig. 2). The broad band at ≈ 500 nm points at the presence of higher valent manganese impurities (Mn^{4+}: $^4A_2 \rightarrow$ $\rightarrow {^4T_2}, {^2T_2}$; Mn^{3+}: $^5E \rightarrow {^5T_2}$ [8]).

3.2 Cathodoluminescence

Cathode ray excitation of $Ca_3Al_2Ge_3O_{12}$ yields no measurable luminescence. On the contrary, the pure Cd host $Cd_3Al_2Ge_3O_{12}$ shows an intense blue emission (broad maximum at ≈ 430 nm; Fig. 3a) which decays non-exponentially with a decay time $\tau \approx 1$ ms. This luminescence is not due to any ions in regular lattice positions and points at the presence of defect centres (cf. Section 4). With increasing Ca content the barycentre of the emission slightly shifts to higher energies, due to the formation of a new, higher energetic maximum (Fig. 3a), mainly present for $x \geqq 1.5$. The efficiency increases between $x = 0$ and 0.75; for higher Ca concentrations a marked decrease takes place (Table 1).

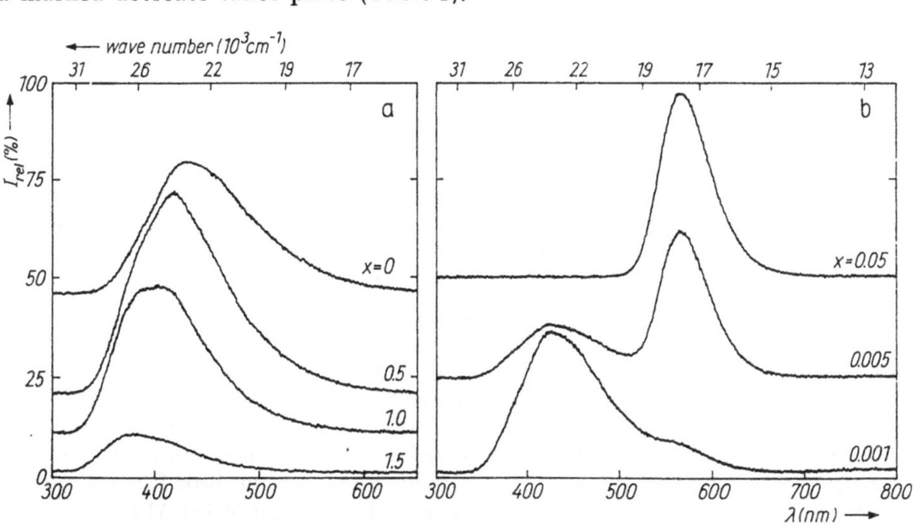

Fig. 3. CL spectra of a) $Cd_{3-x}Ca_xAl_2Ge_3O_{12}$ and b) $Cd_{3-y}Mn_yAl_2Ge_3O_{12}$

Table 1

CL efficiency (S_{rel}) and PL emission intensity (I_{rel}) for $Cd_{3-x-y}Ca_xMn_yAl_2Ge_3O_{12}$

x	y	S_{rel} a)	x	y	S_{rel} b)	I_{rel} c)	
						d → d	UV
0	0	72	0	0.2	15	15	7
0.25	0	89	0	0.05	100	25	44
0.5	0	100	0	0.01	86	14	100
0.75	0	97	0.45	0.05	68	19	18
1.0	0	74	0.95	0.05	62	29	6
1.5	0	18	1.45	0.05	45	20	<1
2.5	0	<1	2.95	0.05	5	12	<1
3.0	0	<1	2.99	0.01	12	<1	<1

a) Normalized to $S_{rel} = 100$ for $x = 0.5$; $y = 0$.
b) Normalized to $S_{rel} = 100$ for $x = 0$; $y = 0.05$.
c) Normalized to $I_{rel} = 100$ for $x = 0$; $y = 0.01$.

Activation with manganese results in a yellow emission at about 570 nm, ascribed to the $^4T_1(G) \rightarrow {}^6A_1(S)$ spin flip transition of the Mn^{2+} ($3d^5$) ion. Its spectral position is independent of x and y; its decay time is about 12 ms ($y = 0.05$), followed by a long persistent yellow phosphorescence ($x < 1.45$) with a much higher decay time in the range of seconds. For $x = 0$ and $y = 0.05$ the best efficiency is reached (Table 1), which is about 80% of the commercial phosphor P 46 ($Y_3Al_5O_{12}:Ce^{3+}$). With increasing Ca content the efficiency of the Mn^{2+} emission goes down.

A reduction of the activator concentration leads for $x = 0$ to the simultaneous observation of the Mn^{2+} emission at ≈ 570 nm, and the defect emission at ≈ 430 nm. From Fig. 3b it is clear that an increase in activator emission intensity correlates strongly with a decrease in defect emission. Correspondingly, the emission colour shifts from yellow ($y \geqq 0.01$) via white ($y \approx 0.005$) to blue ($y \leqq 0.001$). For low Mn concentration the CR pulse creates a luminescence turning in colour from white or blue to yellow, due to the marked differences in the decay time of the Mn^{2+} phosphorescence and the host emission.

3.3 Photoluminescence

The luminescence of the pure Ca host $Ca_3Al_2Ge_3O_{12}$ is difficult to excite with our instrumentation. In the diffuse reflectance spectra the optical band edge at 300 K is situated at relatively high energies (about 210 nm; cf. Section 3.1). Excitation with this wavelength (the spectral limit of our instrumentation) as well as in the absorption band at ≈ 270 nm yields neither at room temperature (RT) nor at 77 K a measurable luminescence. The same holds for the solid solution $Cd_{3-x}Ca_xAl_2Ge_3O_{12}$ by exciting with excitation wavelengths between 210 and 270 nm.

In agreement with the CL results Mn activation yields in PL at RT a yellow Mn^{2+} emission. In the RT excitation spectrum of $^4T_1(G) \rightarrow {}^4A_1(S)$ (≈ 570 nm) for $Ca_3Al_2Ge_3O_{12}$:Mn the forbidden d → d transitions of Mn^{2+} are present only (Fig. 4). On the opposite, for $Cd_3Al_2Ge_3O_{12}$:Mn an additional strong broad maximum at ≈ 260 nm is observed, corresponding to the absorption edge of $Cd_3Al_2Ge_3O_{12}$ in the diffuse reflectance spectra (cf. Section 3.1) and thus indicating an energy transfer

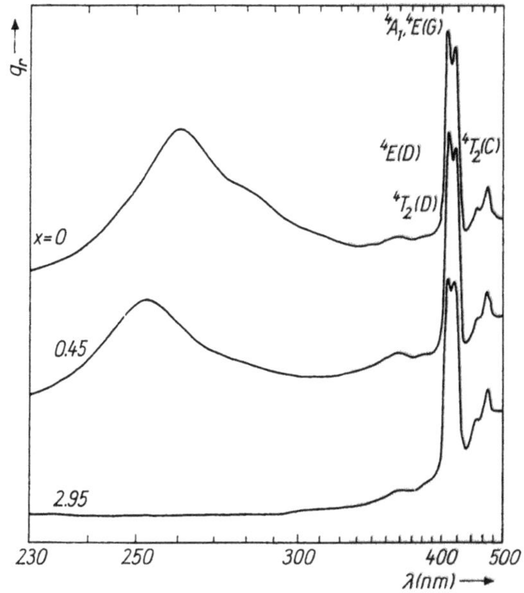

Fig. 4. PL excitation spectra of $Cd_{2.95-x}Ca_xMn_{0.05}Al_2Ge_3O_{12}$ at RT ($\lambda_{em} = 570$ nm); ground state $^6A_1(S)$ for Mn^{2+} (read $^4T_2$ (G) instead of $^4T_2$ (C))

from defect centres to Mn^{2+} (cf. Section 4). The intensity ratio between the band at ≈ 260 nm and the most prominent d → d transition at ≈ 410 nm goes up rapidly with decreasing Mn content. After irradiation of $Cd_3Al_2Ge_3O_{12}$:Mn with $\lambda \approx 260$ nm (e.g. for 5 min) an additional broad absorption band peaks at ≈ 380 nm in the excitation spectra of the Mn^{2+} emission. For higher Mn contents ($y \geqq 0.05$) a further broad excitation band develops, visible as shoulder at ≈ 280 nm (Fig. 4).

In complete accordance with the diffuse reflectance spectra the substitution of Ca for Cd results in a shift of the UV band to higher energies (Fig. 4). Simultaneously a reduction of its intensity relative to the d → d transitions is observed; for $x \geqq 1.45$ the UV band has practically disappeared. The 77 K spectra show a uniform decrease of the UV maximum and a better resolution of the d → d transitions. Especially $^6A_1(S) \rightarrow {}^1A_4$, $^4E(G)$ exhibits a threefold splitting, consistent with the site symmetry D_2 of the dodecahedrally coordinated Mn^{2+} ion.

Excitation in the Mn^{2+} transition $^6A_1 \rightarrow {}^4A_1$, $^4E(G)$ results at RT for all Cd/Ca ratios in a moderate yellow Mn^{2+} emission at 570 nm ($^4T_1(G) \rightarrow {}^6A_1(S)$; Table 1). At 77 K the Mn^{2+} emission gains some intensity. In addition a weak Mn^{4+} emission is visible at about 690 nm ($^2E_1$, $^2T_1 \rightarrow {}^4A_2$), indicating that some manganese enters the lattice in the tetravalent state. The prominent Mn^{4+} excitation $^4A_2 \rightarrow {}^4T_1(F)$ at about 375 nm is situated in the neighbourhood of the Mn^{2+} transition $^6A_1(S) \rightarrow {}^4A_1$, $^4E(G)$ at 410 nm. At RT the Mn^{4+} luminescence is thermally quenched.

Excitation with 260 nm yields at RT for $Cd_3Al_2Ge_3O_{12}$:Mn after a slow build-up an intense yellow emission with a strong afterglow. With increasing Ca content UV excitation results in a more and more reduced Mn^{2+} luminescence (Table 1).

The phosphorescence is easily quenched by cooling the sample to 77 K. Elevating the temperature causes again luminescence. These processes are repeatable until all traps (cf. Section 4) are emptied. Correspondingly, excitation at 77 K within the UV band generates a much weaker Mn^{2+} emission compared to RT.

A crystal field analysis of the Mn^{2+} excitation and emission transitions according to [9] yields the following values for the parameters: $B = 530\ cm^{-1}$, $C = 3800\ cm^{-1}$, and $10Dq = 9400\ cm^{-1}$, which are close to the values for other garnets with Mn^{2+} [10].

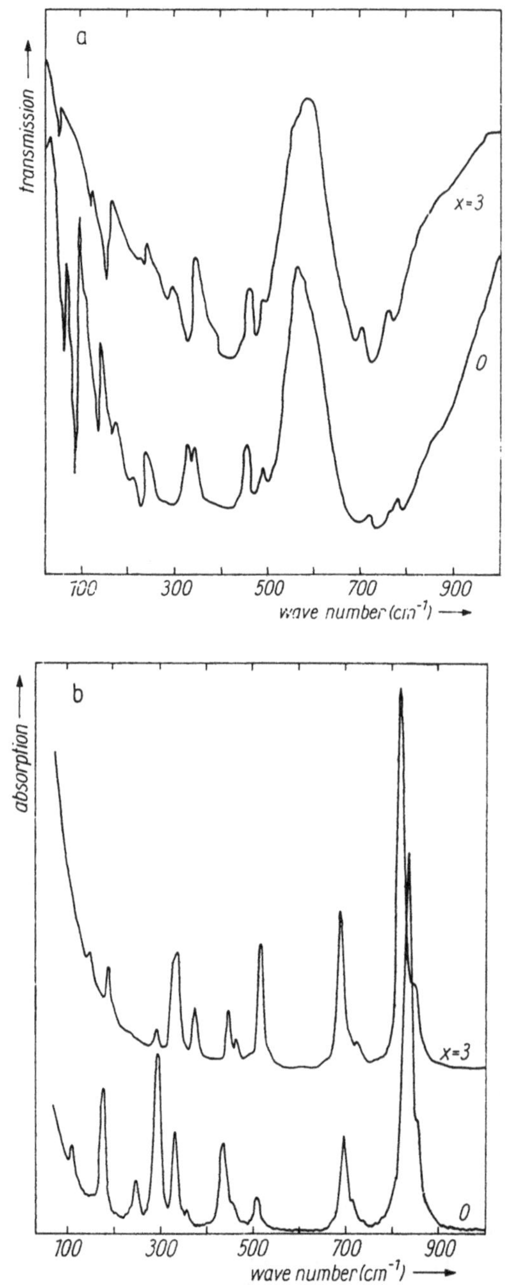

Fig. 5. a) IR and b) Raman spectra of $Cd_{3-x}Ca_xAl_2O_{12}$

3.4 Raman and IR spectra

The Raman and IR spectra of the pure Cd and Ca host show essentially the same features (Fig. 5 a and b). Only minor changes in the band positions are observed, mainly in the low energetic region of the lattice vibrations. These are mostly influenced by the mass difference between Cd and Ca. The higher energetic vibrations are mainly due to the stretching modes of the GeO_4 tetrahedra.

4. Discussion

In the pure Ca host $Ca_3Al_2Ge_3O_{12}$ the band gap is situated at $\approx$210 nm ($\approx$ 6.0 eV). By replacing divalent Cd for divalent Ca a whole series of solid solutions $Cd_{3-x}Ca_x \cdot$ $\cdot$ $Al_2Ge_3O_{12}$ is formed. Both divalent ions have about the same ionic radii and we would expect little lattice distortion and about the same phonon spectra (in agreement with the experimental results). An inspection of the CL and PL properties demonstrates that the incorporation of Cd has an immense influence on the optical properties by introducing defect levels within the band gap. In CL the defect centres can trap the lattice excitation energy. The energy can then be radiatively or thermally released from the trapping state and subsequently trapped at a deeper defect centre. In dependence on the Cd/Ca ratio (series $Cd_{3-x}Ca_xAl_2Ge_3O_{12}$) the emission of at least two defect centres is detected. At smaller Cd content an emission at $\approx$ 380 nm is prominent; for higher Cd concentrations an emission at $\approx$ 430 nm is more abundant (Fig. 3). The different defect centres are believed to have localized character because the emission shows no uniform shift with varying Ca/Cd ratio. The detailed structure of the emission centres has yet to be clarified. Most surely they are associated with the incorporation of Cd in the host, probably by the formation of lattice defects

$$Cd_{Cd}^{\times} + O_O^{\times} \rightarrow CdO(g) + V_{Cd}'' + V_O^{\cdot\cdot}$$

due to a small volatilization of CdO.

In the manganese activated Cd host $Cd_3Al_2Ge_3O_{12}$:Mn the defect centres as well as the Mn^{2+} activators can act as traps for the lattice excitation energy. At low Mn^{2+} concentration the cross section of the defect centre is large compared with the Mn^{2+} activator, so that at low Mn^{2+} content it can dominate the recombination process. With increasing Mn^{2+} concentration the intensity of the defect emission decreases rapidly

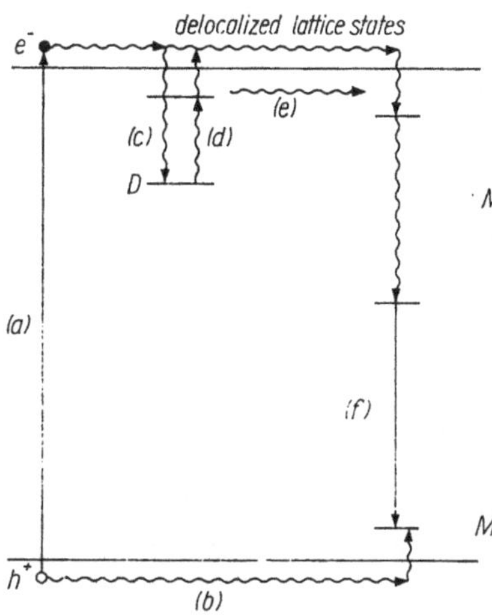

Fig. 6. Schematic diagram of the proposed mechanism for Mn^{2+} luminescence in $Cd_3Al_2Ge_3O_{12}$:Mn

This decrease is strongly correlated with an increase in activator emission (cf. Section 3.2), suggesting that excitation energy is partitioned between defect and activator centres, the latter dominating at concentrations $y \geqq 0.01$. A comparison between the CL properties of $Cd_3Al_2Ge_3O_{12}$:Mn and the corresponding Ca compound $Ca_3Al_2 \cdot Ge_3O_{12}$:Mn clearly demonstrates that defect centres can strongly influence the lattice–activator energy transfer process. The enhancement of the Mn^{2+} emission in the Cd containing garnets is due to a transfer of excitation between the defect centres and Mn^{2+}, already recognized in $Y_3Al_5O_{12}$:Ce^{3+} [11, 12].

In complete correspondance with the CL properties the PL properties of $Cd_3Al_2 \cdot Ge_3O_{12}$:Mn are again dominated by an energy transfer between defect centres and the Mn^{2+} activator. In the excitation spectra for the Mn^{2+} emission of $Cd_3Al_2 \cdot Ge_3O_{12}$:Mn the most prominent defect excitation is situated exactly in the region of the absorption edge in the diffuse reflectance spectra. A uniform band shift to higher energies with increasing x points to the delocalized character of the defect states. Additional broad excitation bands (cf. Section 3.3) implies the presence of further defect states. The presence of deeper and shallower trap levels in $Cd_3Al_2Ge_3O_{12}$:Mn was confirmed by photostimulated luminescence measurements after X-ray irradiation. Photostimulated luminescent emission was observed with e.g. $\lambda_{ex} \approx 400$ or 750 nm, respectively.

The assumed mechanism leading to Mn luminescence is schematically pictured in Fig. 6; competitive processes, e.g. thermal or optical release of the captured charge carriers and retrapping at deeper trapping levels (e.g. Mn^{4+} impurities (cf. Section 3.1 and 3.3)), are neglected but are still present.

In a first step electrons are excited by UV irradiation into delocalized lattice states, thus leaving mobile holes in the valence band (a). These photoexited holes are captured at Mn^{2+} ions which in situ change to "Mn^{3+}" (b). The excited electrons are believed to be trapped (c) in defect centres D. After thermal or optical stimulation the trapped electrons may be released either to the delocalized lattice states or to shallow tunneling states (d) and be captured at the manganese ions forming Mn^{2+} in excited state (Mn^{2+*}; (e)). The luminescence occurs when this excited state decays radiatively to the ground state (f). This mechanism is supported by the observed decrease of host lattice excitation with increasing Mn concentration (cf. Section 3.3). According to [13] for constant exciting light intensity, the absorption by the host lattice is nearly constant and, therefore, the density of excited states which can transfer electrons to Mn^{2+} is constant. Accordingly, if the density of these states is not large enough compared with that of Mn^{2+}, the overall emission intensity of Mn^{2+} for the UV irradiation increases more slowly than the Mn^{2+} concentration does, while, in the case of d $\rightarrow$ d excitation, the overall emission intensity increases proportionally to the Mn^{2+} concentration because of the large exciting photon density until quenching effects become appreciable (Table 1). A further support of the proposed model is given by the observed temperature dependence. Increase in temperature will promote the thermal release from the states D, reducing the net capture rate at the defect state and increasing the capture probability for the Mn^{2+} activator.

Acknowledgements

We want to thank the Deutsche Forschungsgemeinschaft for the support of the investigation. J. Wiehl thanks the Landesgraduiertenförderung of Baden-Württemberg for granting a scholarship. We are indebted to Prof. Dr. E. Lindner and G. Farag for the measurement of the Raman and far IR spectra. We want to thank Mrs. A. Ehmann and Mrs. R. Hüpper for their helpful assistance as well as the Verband der Chemischen Industrie for valuable support.

References

[1] W. Tolksdorf and U. Wolfmeier in: Landolt-Börnstein, New Series Group B, Vol. 12, Part a, Springer-Verlag Berlin/Heidelberg/New York 1978.
[2] A. Tauber and C. G. Whinfrey, J. Phys. Chem. Solids **21**, 25 (1961).
[3] R. Braun, R. Otto, W. Wischert, and S. Kemmler-Sack, Z. Naturf. **40a**, 503 (1985).
[4] R. Braun and S. Kemmler-Sack, J. Solid State Chem. **64**, 76 (1986).
[5] H.-D. Autenrieth, S. Kemmler-Sack, and W. Wischert, phys. stat. sol. (a) **83**, 631 (1984).
[6] D. Wolf and S. Kemmler-Sack, phys. stat. sol. (a) **93**, 271 (1986).
[7] M. J. Lammers, A. Saakes, and G. Blasse, Mater. Res. Bull. **21**, 295 (1986).
[8] A. Bergstein and W. B. White, J. Electrochem. Soc. **118**, 1166 (1971).
[9] A. B. P. Lever, Inorganic Electronic Spectroscopy, Elsevier Publ. Co., Amsterdam/London/New York 1968.
[10] D. Reinen, G. Schwab, and V. Günzler, Z. anorg. allg. Chem. **516**, 140 (1984).
[11] D. J. Robbins, B. Cockayne, B. Lent, C. N. Duckworth, and F. Mansfield, in: Proc. 12th Rare Earth Research Conference 1976, Denver Research Institute, Denver 1976 (p. 1004).
[12] D. J. Robbins, B. Cockayne, B. Lent, C. N. Duckworth, and J. L. Glasper, Phys. Rev. B **19**, 1254 (1979).
[13] T. Hoshina, J. chem. Phys. **50**, 5158 (1969).

(Received August 26, 1988)

phys. stat. sol. (a) **111**, 325 (1989)

Subject classification: 77.30

Department of Physics, National Technical University of Athens[1])

The Determination of Distributions of the Parameters of Thermally Stimulated Depolarization Current Peaks: Theory

By

C. CHRISTODOULIDES, L. APEKIS, P. PISSIS, and D. DAOUKAKI-DIAMANTI

A method is described for determining the distributions of the values of activation energy or of the preexponential time constant for a peak of thermally stimulated depolarization current which is known not to be due to a single relaxation mechanism, but to a continuous distribution of either one of these two parameters. The peak is first transformed into a density function of the peak temperature. This may then be used to obtain the distributions of activation energy or pre-exponential time constant, or both, provided a relation connecting these two parameters can be found. The resolution and applicability of the method is discussed and theoretical examples are used as illustrations.

Es wird eine Methode zur Bestimmung der Verteilung von Werten der Aktivierungsenergie oder des Zeitvorfaktors für einen Peak des thermisch stimulierten Depolarisationsstroms beschrieben, wenn bekannt ist, daß er nicht von einem einfachen Relaxationsprozeß herrührt, sondern von einer kontinuierlichen Verteilung einer dieser beiden Parameter oder beider Parameter verursacht wird. Der Peak wird zunächst in eine Verteilungsfunktion der Peaktemperatur transformiert. Diese kann dazu verwendet werden, um die Verteilungen der Aktivierungsenergie oder des Zeitvorfaktors oder beider zu bekommen, unter der Voraussetzung, daß eine Beziehung zwischen diesen beiden Parametern gefunden werden kann. Das Auflösungsvermögen und die Anwendbarkeit der Methode werden diskutiert und theoretische Beispiele zur Erläuterung verwendet.

1. Introduction

In the phenomenon of thermally stimulated depolarization, it is known in many cases that peaks observed in the thermograms of depolarization current (TSDC) as a function of temperature cannot be attributed to a single relaxation mechanism or to a finite number of such mechanisms, but rather to a continous distribution of these [1, 2]. Assuming that only one of either the activation energy or the pre-exponential time constant is distributed, or that they are both distributed but not independently, it is possible to deconvolute the TSDC peak and derive the density function (or functions) of the distribution.

One such method will be described in this work. It will be shown that, under certain assumptions, the TSDC thermogram may be transformed into a distribution (density function) in terms of peak temperature and provided a relation exists between the peak parameters, this may then be transformed into the distributions of either the activation energy or the pre-exponential time constant, or both. The basic assumption made will be the same as that of Pfister and Abkowitz [3], namely that the broad TSDC peak resulting from a distribution in the peak parameters may be considered

---

[1]) Zografou Campus, GR-157 73 Athens, Greece.

to be the sum of a continuous distribution of much narrower peaks each representing a single relaxation mechanism. The mathematical analysis and the results will, however, be different, as demonstrated in Sections 4 and 5, and discussed in Section 7. The method will be demonstrated by applying it to two theoretically derived TSDC peaks assumed to originate from a distribution in activation energy in one case and in the pre-exponential factor in the other.

2. The Polarization Process

The differential equation governing the polarization, in an electric field F, of a sample with one kind of dipoles is [1, 2]

$$\frac{\mathrm{d}P}{\mathrm{d}t} = \frac{P_e - P}{\tau}. \tag{1}$$

$P(t)$ is the polarization of the sample. The relaxation time $\tau(T)$ of the dipoles involved is usually assumed to be of the Arrhenius type,

$$\tau(T) = \tau_0 \exp\left(\frac{E}{kT}\right), \tag{2}$$

with $\tau_0 = \tau(\infty)$ the relaxation time constant, T the absolute temperature, E the activation energy, and k Boltzmann's constant. The other parameter in (1) is the equilibrium polarization which is, according to Langevin,

$$P_e = \frac{\gamma \varkappa F p^2 N}{kT}, \tag{3}$$

where γ is a geometrical factor ($\gamma = 1/3$ for free rotating dipoles), $\varkappa F$ is the local electric field at the dipoles, p their dipole moment, and N their concentration. P_e is the equilibrium polarization reached after a long time at steady state conditions.

Assuming zero initial polarization, the solution of (1) is

$$P(t) = \exp\left(-\int_0^t \frac{\mathrm{d}t}{\tau}\right) \int_0^t \frac{P_e}{\tau} \exp\left(\int_0^t \frac{\mathrm{d}t}{\tau}\right) \mathrm{d}t, \tag{4}$$

with P_e being a function of temperature as shown in (3), and of time, in the sense that F may be turned on and off as well as vary in magnitude, when different from zero.

The polarization reached at the end of a particular polarizing procedure followed is P_0 and is usually expressed in terns of the effective temperrature of polarization T_{eff} as

$$P_0 = \frac{\gamma \varkappa F p^2 N}{k T_{eff}}. \tag{5}$$

T_{eff} is determined from (4) and (5) if T and F are known as functions of time.

The whole process may be characterized by the polarization efficiency η, defined here as the fraction of available dipoles which have been polarized,

$$\eta = \frac{P_0}{pN} = \frac{\gamma p \varkappa F}{k T_{eff}}. \tag{6}$$

3. Analysis for a Single Relaxation Mechanism

For a single relaxation mechanism characterized by unique values of E and τ_0 ,the thermally stimulated depolarization current density during heating at a rate $b = \mathrm{d}T/\mathrm{d}t$ is

$$J(T) = \frac{P_0}{\tau(T)} \exp\left(-\int\limits_{T_0}^{T} \frac{\mathrm{d}T}{b\tau(T)}\right). \tag{7}$$

For an Arrhenius $\tau(T)$ and constant b, the maximum of $J(T)$ occurs at a temperature $T_{\max}$ satisfying

$$\frac{kT_{\max}^2}{bE} = \tau_0 \exp\left(\frac{E}{kT_{\max}}\right). \tag{8}$$

In terms of the variable $\varepsilon = E/kT$, its value $\varepsilon_{\max} = E/kT_{\max}$ at $T_{\max}$ and assuming $T_0 = 0$, (7) may be written as [4]

$$J(T) = \frac{P_0}{\tau_0} \exp\left(-\varepsilon - \varepsilon_{\max}^2 \exp(\varepsilon_{\max}) \frac{E_2(\varepsilon)}{\varepsilon}\right), \tag{9}$$

$E_2(\varepsilon)$being the transcendental function [5]

$$E_2(\varepsilon) = \int\limits_1^\infty \frac{\mathrm{e}^{-\varepsilon t}}{t^2}\,\mathrm{d}t = \varepsilon \int\limits_\varepsilon^\infty \frac{\mathrm{e}^{-z}}{z^2}\,\mathrm{d}z\,. \tag{10}$$

For sample cross-sectional area A, a total charge $Q_0 = AP_0$ is available for liberation during the TSDC measurement and the current is $I(T) = AJ(T)$. In a time interval $\mathrm{d}t$, or temperature interval $\mathrm{d}T = b\mathrm{d}t$, the number $\mathrm{d}n$ of dipoles per unit volume which are disorientated is

$$\mathrm{d}n = \frac{J(T)\,\mathrm{d}t}{p} = \frac{J(T)\,\mathrm{d}T}{bp} \tag{11}$$

and therefore

$$\frac{\mathrm{d}n}{\mathrm{d}T} = \frac{J(T)}{bp} = \frac{I(T)}{bpA}. \tag{12}$$

4. The TSDC Thermogram Interpreted as a Distribution

The term *density function* will be used in its usual probability-theory context as being a function having the following properties [6, 7]: it is a non-negative function $f(x)$, whose integral over all values of x is equal to unity (i.e. it is normalized), and has the property that $f(x_1)\,\mathrm{d}x$ is the probability that the variable x has a value between x_1 and $x_1 + \mathrm{d}x$. If $z(x)$ is a single-valued function of x, the density function $f(x)$ may be transformed into another, $g(z)$, in terms of the variable z, using

$$g(z) = f\big(x(z)\big)\left|\frac{\mathrm{d}x}{\mathrm{d}z}\right|. \tag{13}$$

The density function is also called *differential distribution function* or *frequency function*. The name *distribution function*, which is in fact defined as the integral of the density function between $-\infty$ and x, has sometimes been wrongly used in place of the term density function. When presenting a density function $f(x)$ in graphical form, it is important, in order to convey the information contained in the function without distortion, to plot $f(x)$ as a function of x and not some other variable derived from x (e.g. ln x etc.).

A simple TSDC peak with one discrete value of E and one discrete value of τ_0, would result in a density function $h_p(T_{max})$ of T_{max}, which consists of a delta function at the value T_{max}. Using (8) and additional information to enable the determination of E and τ_0 for the peak, $h_p(T_{max})$ could then be transformed into the density functions $g_p(E)$ of E and $f_p(\ln \tau_0)$ of $\ln \tau_0$. All three density functions will be delta functions, since the peak is characterized by single discrete values of E and τ_0 and therefore of T_{max}. The subscript p signifies the fact that these are the density functions of the polarized dipoles. To obtain the density function $h(T_{max})$ of the parent population, i.e. of the dipoles initially available for polarization, one has to divide $h_p(T_{max})$ by the polarization efficiency η given by (6), and normalize $h(T_{max})$. Naturally, for a single peak of discrete E and τ_0, this procedure will still lead to the same delta functions for the density functions.

If now a TSDC peak is considered, which is known to be due to a distribution of either E or τ_0, the density functions of these may be approximately determined, provided certain conditions are satisfied. The plot of $I(T)$ as a function of T is regarded to be the sum of a continuous distribution of much narrower peaks, each with a distinct pair of E and τ_0 values. Next, it is considered that a narrow band between T and $T + \mathrm{d}T$, at which the current is $I(T)$, is a peak at peak temperature $T_{max} = T$ and having a height of $I(T)$. This peak is considered to have such a pair of E and τ_0 values as to appear at T_{max} for the heating rate used. Using (12) it is seen that $\mathrm{d}n = (I(T)/bpA)\,\mathrm{d}T$ is the volume density of polarized dipoles with E and τ_0 values leading to peak temperatures between T and $T + \mathrm{d}T$. In other words,

$$h_p(T_{max}) = c' \left.\frac{\mathrm{d}n}{\mathrm{d}T}\right|_{T=T_{max}} = c' \left.\frac{I(T)}{bpA}\right|_{T=T_{max}}, \tag{14}$$

where c' is a normalization constant. Dividing by the polarization efficiency η and renormalizing (constant c), we obtain

$$h(T_{max}) = c \left.\frac{kT_{eff}I(T)}{\gamma p^2 \varkappa F A b}\right|_{T=T_{max}}, \tag{15}$$

the density function of T_{max} for the dipoles present in the sample. Obviously, the possibility of transforming the $I(T)$ versus T curve into the density function $h(T_{max})$, using (15), depends on p being known (since it may be a distributed parameter, too), and on being able to evaluate T_{eff}, which may require the knowledge of E. When the linear cooling rate during polarization is not very different from the heating rate during measurement of the TSDC, and for large ε_{max}, a good approximation is $T_{eff} \approx T_{max}$ [3, 8].

To be able to proceed, some further assumptions have to be made. The first is that p must either be known as a function of E and τ_0 or be assumed constant. The second is that, besides (8), there exists another relation involving one, two, or all of the parameters E, τ_0, and T_{max}, so that any one of them may be expressed in terms of either one of the other two. Special cases of this relation may be that either E = const or τ_0 = const. Under these conditions, (15) may be used to express $h(T_{max})$ in terms of T_{max} alone and $h(T_{max})$ may also be transformed into $g(E)$ and $f(\ln \tau_0)$ using (13).

5. Density Functions in Some Special Cases

Equations (13) and (15) will be used in some special cases in order to derive the density functions $g(E)$ and $f(\ln \tau_0)$ from $h(T_{max})$. The assumption made is that $\gamma p^2 \varkappa F$ is constant for a given distribution. Defining

$$C = \frac{ck}{\gamma p^2 \varkappa F A} \,. \tag{16}$$

(15) becomes

$$h(T_{max}) = \frac{C}{b} T_{eff} I(T)|_{T=T_{max}} \,. \tag{17}$$

b was not included in C, to enable normalization in cases where TSDC thermograms recorded with different heating rates are used to obtain $h(T_{max})$.

The density functions for some special cases will be presented below. T_{max} will be replaced by T in $g(E)$ and $f(\ln \tau_0)$. Even though $g(E)$ and $f(\ln \tau_0)$ will be expressed in terms of both T and E (due to the transcendental nature of (8)), they should be evaluated and plotted as functions of E and $\ln \tau_0$ respectively. This can be done using (8) together with the additional relation connecting these parameters in each particular case. Also, in all expressions below, T_{eff} should be estimated from the knowledge of E and $T_{max} = T$ for each case, when the conditions making $T_{eff} \approx T_{max}$ are not satisfied.

5.1 $E = const$

In this case, (8), (13), and (17) give (with $T_{max} = T$)

$$\frac{dT_{max}}{d(\ln \tau_0)} = \frac{T_{max}}{\varepsilon_{max} + 2} , \tag{18}$$

$$f(\ln \tau_0) = \frac{Ck}{bE} T_{eff} T^2 I(T) \frac{\varepsilon}{\varepsilon + 2} . \tag{19}$$

5.2 $\tau_0 = const$

Again, (8), (13), and (17) give

$$\frac{dT_{max}}{dE} = \frac{T_{max}}{E} \cdot \frac{\varepsilon_{max} + 1}{\varepsilon_{max} + 2} , \tag{20}$$

$$g(E) = \frac{C}{b} \frac{T_{eff} T I(T)}{E} \frac{\varepsilon + 1}{\varepsilon + 2} . \tag{21}$$

5.3 $\tau(T_{max}) = const$

This empirical relation has been found to hold in some cases, such as the broad distributions met in the TSDC of polymers [9]. It implies $E = aT_{max}^2$, with $a =$ const, as can be seen from (8). Both $g(E)$ and $f(\ln \tau_0)$ may be found in this case, the two being of course interdependent. It is found that

$$\frac{dT_{max}}{dE} = \frac{1}{2aT_{max}} , \tag{22}$$

$$g(E) = \frac{C}{2ab} \frac{T_{eff}}{T} I(T) \,. \tag{23}$$

Also, in conjunction with (8), $E = aT_{max}^2$ gives

$$\frac{dT_{max}}{d(\ln \tau_0)} = -\frac{k}{a}, \tag{24}$$

$$f(\ln \tau_0) = \frac{Ck}{ab} T_{eff} I(T). \tag{25}$$

5.4 $\tau_0 = \tau_c \exp(-E/kT_c)$

Such a relation, with τ_c and T_c constant, represents the so-called "compensation effect" observed in many phenomena, including the TSDC of some materials [10 to 15]. Then, if $\varepsilon_c = E/kT_c$,

$$\frac{dT_{max}}{dE} = \frac{T_{max}}{E} \frac{\varepsilon_{max} - \varepsilon_c + 1}{\varepsilon_{max} + 2}, \tag{26}$$

$$g(E) = \frac{C}{b} \frac{T_{eff} T I(T)}{E} \left| \frac{\varepsilon - \varepsilon_c + 1}{\varepsilon + 2} \right|. \tag{27}$$

Also,

$$\frac{dT_{max}}{d(\ln \tau_0)} = -\frac{k}{E} T_c T_{max} \frac{\varepsilon_{max} - \varepsilon_c + 1}{\varepsilon_{max} + 2}, \tag{28}$$

$$f(\ln \tau_0) = \frac{CkT_c}{b} \frac{T_{eff} T I(T)}{E} \left| \frac{\varepsilon - \varepsilon_c + 1}{\varepsilon + 2} \right|. \tag{29}$$

In both (27) and (29), given T, E and $\ln \tau_0$ can be found using (8) and the compensation law relation given at the subheading of this section.

6. Criterion of Applicability of the Method and Examples

A depolarization peak with single E and τ_0, such as given by (9), would, if deconvoluted using a method of infinite resolution, yield density functions which are delta functions at E and $\ln \tau_0$. The resolution of the method described is limited by the assumptions made. The main one is that a broad peak consists of a continuous distribution of narrow peaks of unique E and τ_0 values. These peaks do in fact have nonzero widths and this sets a limit to the resolution of the method. It is known that if T_1 and T_2 are the temperatures at which the current is half the maximum, then [4]

$$r = \frac{T_2 - T_1}{T_{max}} \approx \frac{2.44}{\varepsilon_{max} + 1.25}, \tag{30}$$

which may be called the reduced full temperature width at half maximum. The criterion of applicability of the method is therefore that the broad peak whose distributions are to be determined should have an r value which is as much higher as possible than those of the simple peaks of which it is assumed to be composed.

To illustrate the loss in detail due to the limited resolution of the method, two examples are presented below. In both cases, the density functions of the dipoles which have been polarized are considered. In the first example the energy is assumed to be constant $E = 1.2$ eV and $f_p(\ln \tau_0)$ constant between $\tau_0 = 10^{-13}$ and 10^{-9} s and zero outside this range. The resulting TSDC peak for $b = 0.1$ K/s is shown in Fig. 1a. The result of Section 5.1 will be used, after deleting T_{eff} from (19) since the required density function if $f_p(\ln \tau_0)$ and not $f(\ln \tau_0)$, and the polarization efficiency η is inversely proportional to T_{eff} as seen in (6). The recovered density function $f_p(\ln \tau_0)$ is shown in

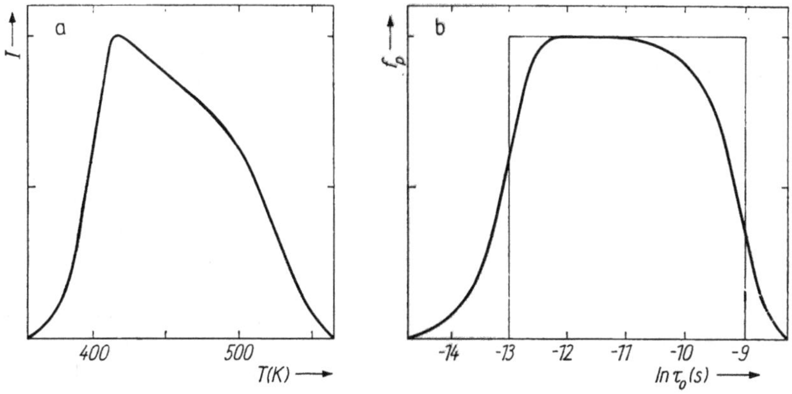

Fig. 1. Application of the method described in the text to a theoretical example of a distribution with unique activation energy $E = 1.2$ eV but distributed relaxation time constant, such that $f_p(\ln \tau_0)$ for the polarized dipoles is nonzero and constant between $\tau_1 = 10^{-13}$ s and $\tau_2 = 10^{-9}$ s and zero outside this range. The heating rate assumed is 0.1 K/s. The resulting TSDC peak is shown in a). In b), the thin line shows the original square density function of $\ln \tau_0$, while the thick line shows the d.f. $f_p(\ln \tau_0)$, recovered as described in the text

Fig. 1b (thick line). Also shown, for comparison, is the actual density function (square distribution, in thin line). The second example is for $\tau_0 =$ const and $b\tau_0 = 10^{-11}$ K, while $g_p(E)$ is constant between $E = 1$ and 1.3 eV and zero outside this range. The TSDC peak is shown in Fig. 2a and the density function $g_p(E)$ recovered using the results of Section 5.2 (with T_{eff} again deleted from (21)) is shown in thick line in Fig. 2b, together with the actual square density function of E in thin line.

The wide peak of Fig. 1a has full width at half maximum equal to 120 K and for $T_{max} = 425$ K its r value is 0.28. For a 1.2 eV peak at 400 K, (30) gives $r = 0.07$ and at 550 K, $r = 0.09$. For the peak of Fig. 2a, $r = 0.27$. Also, for $b\tau_0 = 10^{-11}$ K, a 1 eV peak would have maximum at 414 K and $r = 0.0834$, while a 1.3 eV peak would have

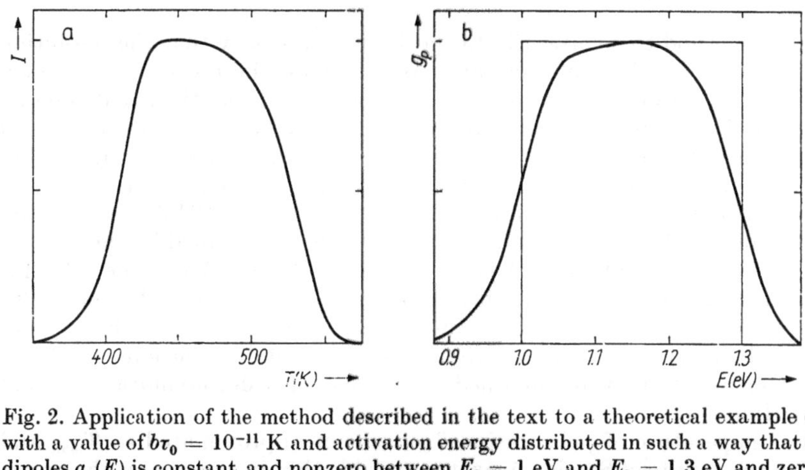

Fig. 2. Application of the method described in the text to a theoretical example of a distribution with a value of $b\tau_0 = 10^{-11}$ K and activation energy distributed in such a way that for the polarized dipoles $g_p(E)$ is constant and nonzero between $E_1 = 1$ eV and $E_2 = 1.3$ eV and zero elsewhere. The resulting TSDC peak is shown in a). In b), the thin line shows the original square density function of E, while the thick line shows the density function recovered as described in the text

maximum at 534 K and $r = 0.0827$. The effect of a ratio of r values about 3, which is true for both examples, is seen by comparing the recovered and actual density functions of Fig. 1b and 2b. Higher r ratios would improve the quality of recovery.

7. Discussion and Conclusions

The basis of the method described above is that a TSDC peak originating from a distribution of the two parameters E and τ_0, may be considered to consist of a continuous distribution of much narrower TSDC peaks with discrete parameters. This is the same assumption made by Pfister and Abkowitz [3]. They, however, proceed to convert the temperature axis into one of relaxation time $\tau(T)$ (or its inverse, frequency $\varkappa$). Our mathematical analysis, presented in Section 4, has shown, on the other hand, that the density function directly obtainable from the TSDC thermogram is that of peak temperatures T_{max} (15). The results derived from (15) depend on the additional relation assumed to exist between E and τ_0, as shown in Section 5. The conversion of the temperature axis into one of relaxation time is not unique and depends on the above-mentioned additional relation between the peak parameters. This is the main reason why our results differ from those of Pfister and Abkowitz [3]. The main result of our analysis is given by (15), which together with (13) may be used, when any additional relation between E and τ_0 is given, to derive the corresponding density function. Within the limitations set by the original assumptions, our results are therefore quite general.

Apart from cases when it is arbitrarily assumed that either E or τ_0 is constant, an additional relation connecting these two parameters may be found experimentally. One such method is that of thermal sampling, by which narrower bands of a continuous distribution may be isolated and if treated as single peaks with discrete E and τ_0 values, they may yield the required relation between these two parameters [11, 14 to 17]. This or other techniques may also be used to establish the existence of a relation of the form of the compensation effect [10 to 15], or such as in the examples of materials for which $\tau(T_{max})$ is found to be constant for all the components of the distribution [9, 17]. It may not however be possible to find such a relation, and this is the case, for example, when both E and τ_0 are distributed either without correlation or with a correlation too complex to be determined by elementary experimental techniques.

In applying the method to a certain TSDC peak, it is necessary that the resolution criterion based on (30) is taken into account. As examples, the results of Kovarskii et al. [18] on the distributions of five polymers may be cited. From the results reported, the observed reduced full width at half maximum may be found for each broad TSDC peak, and also the ratio of this to the r value of each peak according to (30), using for E the dominant energy value in the density function. These ratios are: 1.0 for PVA, 1.1 for N6, 1.5 for PVC, 2.1 for PTFCE, and 2.8 for PMMA. It appears that the distributions for PVA and N6 are too narrow to be determined by the method described, while the best results should be expected for PMMA. The same ratio for polyvinylidene fluoride, as found from data reported by Pfister and Abkowitz [3], is about 7, which signifies that the distribution is broad enough to justify the application of the method. It may be mentioned that in the cases of distributions which are too narrow, better results may be obtained by the use of a deconvolution method such as that of Hino [19].

The effective temperature of polarization T_{eff} was not approximated to T_{max} and therefore to T in the results of Section 5, as it is possible that in some cases the procedure of polarization may be such, and the distribution so wide, as to have different values of T_{eff} from one end of the spectrum to the other, or from T at each point.

Our experimental results which will be published elsewhere [17], will demonstrate further the use of the method described above, especially in conjunction with the method of thermal sampling as a means of obtaining the relation between E and τ_0 which is required, but also as another related but not identical method for determining the density functions for TSDC thermograms with distributed parameters.

References

[1] J. Vanderschueren and J. Gasiot, in: Thermally Stimulated Relaxation in Solids, Ed. P. Bräunlich, Springer-Verlag, 1979 (p. 135).
[2] J. van Turnhout, in: Electrets, Ed. G. M. Sessler, Springer-Verlag, 1980 (p. 81).
[3] G. Pfister and M. A. Abkowitz, J. appl. Phys. **45**, 1001 (1974).
[4] C. Christodoulides, J. Phys. D **18**, 1501 (1985).
[5] W. Gautschi and W. F. Cahil, in: Handbook of Mathematical Functions, Eds. M. Abramowitz and I. A. Stegun, Dover Publ., Inc., New York 1965 (p. 227).
[6] W. Feller, An Introduction to Probability Theory and its Applications, Vol. I, 3rd ed., Wiley 1965 (p. 179).
[7] A. Papoulis, Probability, Random Variables and Stochastic Processes, McGraw-Hill, 1965 (p. 94).
[8] C. Christodoulides, phys. stat. sol. (a) **109**, 611 (1988).
[9] A. L. Kovarskii and V. N. Saprygin, Polymer **23**, 974 (1982).
[10] C. Lacabanne, D. Chatain, J. C. Monpagens, A. Hiltner, and E. Baer, Solid State Commun. **27**, 1055 (1978).
[11] M. Zieliński, T. Swiderski, and M. Kryszewski, Polymer **19**, 883 (1978).
[12] E. Peacock-López and H. Suhl, Phys. Rev. B **26**, 3774 (1982).
[13] J. J. Del Val, A. Alegria, J. Colmenero, and J. M. Barandiarán, Polymer **27**, 1771 (1986).
[14] P. Pissis, L. Apekis, and C. Christodoulides, Nuovo Cimento D **9**, 195 (1987).
[15] P. Pissis, L. Apekis, and C. Christodoulides, Nuovo Cimento D **9**, 815 (1987).
[16] M. Zieliński and M. Kryszewski, J. Electrostat. **3**, 69 (1977).
[17] C. Christodoulides, P. Pissis, L. Apekis, and D. Daoukaki-Diamanti, to be published.
[18] A. L. Kovarskii, S. A. Mansimov, and A. L. Buchachenko, Polymer **27**, 1014 (1986).
[19] T. Hino, J. appl. Phys. **46**, 1956 (1975).

(Received July 4, 1988)

phys. stat. sol. (a) 111, 335 (1989)

Subject classification: 78.60; S10.1

Institute of Solid State Physics, Latvian State University, Riga[1]) *(a)*
and Institut für Experimentalphysik der Karl-Franzens-Universität, Graz[2]) *(b)*

On the Nature of Afterglow of the X-Ray Induced Luminescence in Crystalline and Glassy SiO_2

By

I. GODMANIS (a) and W. HOHENAU (b)

The X-ray induced luminescence and its afterglow is investigated in natural and synthetic quartz crystals as well as in different types of glassy SiO_2. In all types of SiO_2 the luminescence intensity cannot be correlated directly with the impurity content of the investigated samples. A comparison of luminescence afterglow measurements of crystalline samples containing different impurities indicates that at low temperatures the luminescence is caused by forbidden intracenter transitions (constant afterglow below 150 K) of radiation induced defects (oxygen vacancy–peroxy linkage). Above 180 K the decrease of the afterglow decay time is explained by additional thermally activated ((370 ± 50) meV) radiationless (Frank-Codon) transitions, competing the luminous transitions. At high temperatures an impurity dependent luminescence occurs, detectable till 800 K, whose afterglow behavior is to be connected with the well known thermoluminescence. As in the case of silver or copper doped quartz, because of its afterglow behavior, the luminescence of quartz glasses is attributed to intracenter excitations at isolated not yet identified luminescence centers.

Sowohl in natürlichen und synthetischen Quarzkristallen als auch in verschiedenen Quarzglastypen wird die Röntgenlumineszenz beschrieben. Die Intensität der registrierten Lumineszenz kann in allen untersuchten SiO_2-Typen nicht direkt mit dem Fremstoffgehalt korreliert werden. Ein Vergleich von Lumineszenz-Nachleuchtuntersuchungen bei kristallinen Proben mit unterschiedlichem Verunreinigungsgrad zeigt, daß bei tiefen Temperaturen die Lumineszenz durch an sich verbotene innere Übergänge (konstantes Nachleuchten unterhalb 150 K) strahlungsinduzierter Defekte (Sauerstoffleerstelle–Peroxydbindung) verursacht wird. Die Abnahme der oberhalb 180 K registrierten Abklingzeit des Nachleuchtens wird durch thermisch aktivierte ((370 ± 50) meV) strahlungslose Übergänge (Frank-Condon) erklärt, die in Konkurrenz zu den lumineszierenden Übergängen stehen. Bei hohen Temperaturen wird eine verunreinigungsabhängige Lumineszenz bis etwa 800 K beobachtbar, deren Nachleuchtverhalten mit der bekannten Thermolumineszenz zusammenhängt. Ebenso wie im Falle silber- oder kupferdotierter Quarze wird die Lumineszenz von Quarzgläsern aufgrund ihres Nachleuchtverhaltens inneren Übergängen, ansonsten isolierter Lumineszenzzentren, zugeschrieben, welche allerdings bis jetzt unidentifiziert sind.

1. Introduction

1.1 Crystalline SiO_2

Griscom and Sigel [1 to 3] have reported for the first time about the exponential decay of the "blue" (2.8 eV) luminescence, excited by electron pulses (500 keV, 3 ns). Because of nearly the same decay time ($\tau_{1/2} \approx 700\,\mu s$ at 77 K) as for the transient E′-center absorption at 5.8 eV, the luminescence was attributed to radiation-induced transient oxygen defects (oxygen vacancy–oxygen peroxy linkage) [4].

[1]) Kengaraga 8, SU-226063 Riga, USSR.
[2]) Universitätsplatz 5, A-8010 Graz, Austria.

Luminescence at 2.7eV with the same decay (600 to 700 μs at 77 K) was also excited by X-ray flashes (0.8 to 1.5 μs) [5]. Because of its dependence on silver electrolysis, the luminescence was interpreted by a recombination phenomenon in impurity-related Al–Li^+/Na^+/K^+ centers. The activation energy of (300 ± 50) meV, obtained from the temperature-dependent decay time, was attributed to the energetic depth of the traps, taking part in recombination [5].

Contrary to this recombination mechanism, Trukhin and Plaudis [6] have proposed an intracenter transition character for the "blue" luminescence, because of fulfilling the Mott rule for quantum yield, temperature-dependent decay time, and the degree of polarization. The luminescence centers should be created by self-trapped excitons as excited Si–O bonds, where the decay of luminescence ($\tau = 1$ ms at 77 K) is given by the lifetime of electrons in Si–O antibonding orbitals, corresponding also to the creation of transient E′-centers [6].

Gritsenko et al. [7] have obtained two components of the decay i.e. $\tau_1 \leqq 20$ ns and $\tau_2 = 1.02$ ms at 80 K for the blue (2.6 eV) luminescence, excited by electrons (200 keV, 10 ns). The activation energy $E_a = 270$ meV, obtained from the temperature-dependent decay of the luminescence as well as from the transient absorption band at 4.1 eV, was interpreted in terms of self-trapped excitons [7].

Tanimura et al. [8] measured an identical thermal quenching for the decay of the blue luminescence at 2.8 eV, transient absorption at 5.4 eV, and volume change. The mentioned phenomena were interpreted, similar as in [4], to be connected with radiation-induced Frenkel defect pairs (E′ oxygen–peroxy linkages) created by exciton self-trapping processes. According to the ODMR results [9], the decay of luminescence at 80 K was explained by intracenter recombination of excited states (triplet configuration of self-trapped excitons) [8].

Two different stages in the luminescence afterglow, giving $E_{a1} \approx 8$ meV in the temperature range 77 to 120 K and $E_{a2} \approx 260$ meV at $T > 130$ K were registered in [10]. The experimentally observed shortening of luminescence afterglow with increasing temperature, but without any increase of the peak intensity was explained by thermally activated ($E \approx 260$ meV) radiationless transitions, competing the forbidden intramolecular luminous recombinations ($\tau \approx 1$ ms, $T < 120$ K) in impurity-related centers [11].

Finally Trukhin [12] investigated the temperature dependence of the blue luminescence (2.6 eV) and found two stages both in quantum yield and in afterglow with corresponding energies $E_{a1} = (200 \pm 50)$ meV at $T < 145$ K and $E_{a2} = (300 \pm \pm 50)$ meV at $T > 145$ K connected with different degrees of polarization of the luminescence. The interpretation includes the creation of transient defect pairs of E′-centers (three coordinated silicon and non-bonding oxygen), which can bind two spatially different surrounding neighbor oxygens [12].

1.2 Crystalline SiO_2, doped with impurities

Crystalline quartz, doped with silver and copper impurities after using an electrodiffusion method [13], shows fundamentally different patterns of luminescence:

a) New UV and X-ray induced luminescence bands occur at 2.3, 2.8, 4.75 eV in SiO_2–Ag^+ [13, 14] and 2.4, 3.4 eV in SiO_2–Cu^+ [15].

b) Contrary to undoped quartz the afterglow of luminescence becomes independent of temperature between 80 and 300 K [10, 14, 15] (Table 2).

c) Spectra and glow peaks of thermoluminescence are also completely altered.

d) The absorption spectra as well as their dependence on temperature and preirradiation, change in a typical manner also [16].

e) The well-known blue luminescence of natural and synthetic quartz is replaced by a new, cation-dependent luminescence.

The proposed models for these luminescence centers are

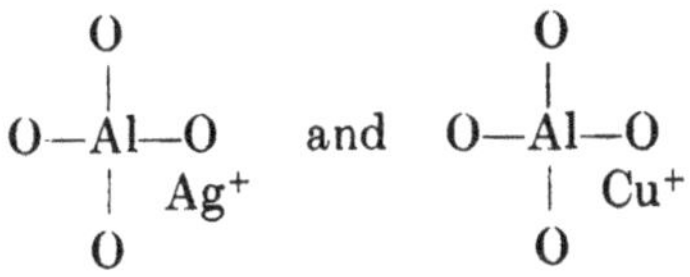

for the 4.75 and 3.4 eV luminescence, respectively [14, 15]. The luminescence in the visibile may be emitted from excited Ag^0, Ag_2^+ centers [13, 14, 17] at unknown places in the lattice. These centers can be generated only by X- or VUV-irradiation above 200 K and are destroyed at temperatures above 500 K.

The luminescence in Ge-doped crystalline quartz induced by X-ray flashes or VUV light pulses [18] has practically the same decay of luminescence afterglow as natural quartz ($\tau = 1.1$ ms at 77 K). But the spectral peak intensity at 2.3 eV differs clearly from the blue luminescence of natural quartz. Therefore the luminescence was interpreted close to natural quartz by two differently oriented self-trapped excitons at substitutional Ge defects [12].

1.3 Glassy SiO_2

The decay of the blue luminescence at 2.5 eV in glassy SiO_2 (Infrasil, Suprasil, . . .), contrary to crystalline quartz, is non-exponential [4] and the relative quantum yield of that luminescence is more than two orders lower. Similar as in the case of natural quartz, the luminescence is attributed to the radiation-induced creation of intimate oxygen vacancy–peroxy linkage pairs. Tanaka et al. [19] interpreted the multi-exponential decay of the 2.5 eV luminescence as annihilation of Frenkel-type defects (E′ and O_2^-) whose configuration is dependent on the widely distributed local structure, specific for glasses. Contrary to [4, 19], Stathis and Kastner [20], using pulse excimer laser excitation (6.4 and 7.5 eV) have measured an exponential decay ($\tau \approx 10$ ms) of luminescence afterglow, independent of temperature.

Our paper is devoted to the study of the influence of usual and specially added impurities on the X-ray flash induced luminescence and especially on its decay in crystalline and glassy SiO_2.

2. Experimental

Our experimental set-up comprises an X-ray flash (0.4 to 1 μs, 80 kV) apparatus [10]. Passing through a quartz prism monochromator (Zeiss M4Q3) the X-ray induced luminescence becomes registered by a photomultiplier (EMI 6256). The electrical signal was stored in a digital oscilloscope (Iwatsu TS 8123) and afterwards transferred to an Apple-II computer. Using interactive software and high resolution graphics the decay time of luminescence was obtained. The values of decay times, obtained at different temperatures, were presented on an Arrhenius plot.

The measured SiO_2 samples, listed in Table 1, had a size between $8 \times 8 \times 6$ and $10 \times 10 \times 1$ mm³.

We have investigated samples of other quartz glasses (Herasil, Plasmennij) and crystals with unknown impurity content, as well as samples of crystalline and glassy SiO_2, doped with silver and copper (10 to 100 ppm) by electro-diffusion. Finally luminescence of crystalline GeO_2 was measured for comparison.

Table 1
Manufacture's data about the impurity content of the investigated samples

		monovalent cations	c_{Al}	c_{Ge}	c_{OH^-}
	origin	Li^+, Na^+, K^+ (ppm)	(ppm)	(ppm)	(ppm)
natural quartz	Brazil	$\approx$ 80	$\approx$ 100	—	$\approx$ 20
	smoky	$\approx$ 200	$\approx$ 200	—	$\approx$ 12
	Morion	?	$\approx$ 1000	—	
synthetic quartz	C 5-I/900	170	170	—	23
	734/5	62	130	—	20
	21-81*)	< 10	< 10	—	
Ge-doped quartz	8 a-I / 5-5	10	10	850	
	8a-3/1	60	100	7500	
fused SiO_2	Corning	< 20	< 5	—	1200
	Suprasil	< 10	< 10	—	5

*) The sample 21-81 was grown under special conditions (from a solution containing NH–F), causing the extremely small (< 10 ppm) concentration of Al, Li^+, Na^+ impurities.

3. Experimental Results

3.1 Luminescence of crystalline quartz

The X-ray flash induced luminescence was registered in different synthetic and natural quartz crystals. The luminescence spectrum consists of broad, temperature-dependent bands lying between 2.5 and 3.3 eV. The relative intensities of these bands change slightly in different quartz crystals and are dependent on temperature [11]. However, the blue luminescence (2.6 eV) in all measured samples of crystalline quartz has the same exponential decay time ($\approx$ 1 ms at 80 K), depending on temperature only. The temperature dependence of the decay time corresponds to Mott's rule, $\tau = \tau_0/(1 + A\, e^{-E_a/kT})$.

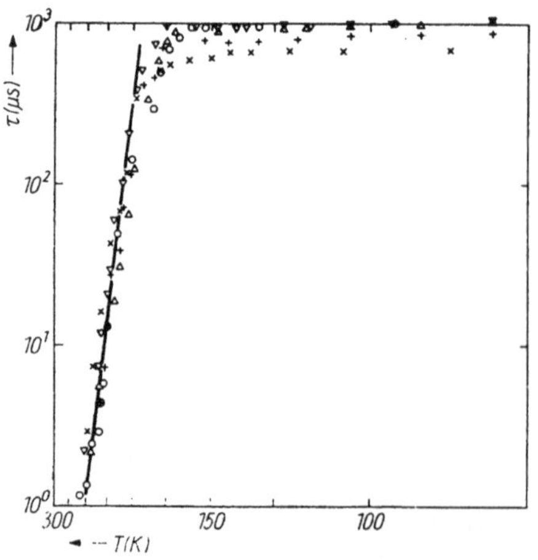

Fig. 1. Temperature-dependent decay time of the X-ray induced luminescence afterglow presented in an Arrhenius plot. ○ Morion quartz, △ Brasil quartz, ▽ synth. quartz, + Ge-doped quartz, × GeO_2; $E_a = (370 \pm 50)$ meV

The obtained activation energy $E_a = (360 \pm 50)$ meV[3]) has the same value in both synthetic and various natural quartz crystals (Fig. 1).

Above 250 K other, impurity-dependent, luminescence bands occur (at 2.5, 3.3, 2.7 eV). These bands are also sensible to temperature. But the decay time of the luminescence afterglow now is dependent on the position of the glow peaks of the thermoluminescence as well as on temperature and has values of the order of minutes (0.1 to 120 min) [11].

3.2 Luminescence of germanium-doped quartz and crystalline GeO_2

The X-ray flash induced luminescence in Ge-doped crystalline quartz spectrally differs ($E_{max} = 2.3$ eV) from the blue luminescence (2.6 eV) in Ge-free quartz. However, the decay time (1 ms at 80 K) and obtained value of the activation energy $E_a = (370 \pm 50)$ meV) remain the same in Ge-doped, as well as in Ge-free quartz (Fig. 1, Table 2).

The same activation energy was obtained also in crystalline GeO_2. But the decay time of the 2.3 eV luminescence in GeO_2 is shorter (660 μs at 77 K) than in Ge-doped quartz (1 ms) (Fig. 1, Table 2).

3.3 Luminescence of copper and silver containing quartz

In doped crystalline quartz the luminescence is more intense than in natural samples. In crystalline SiO_2–Cu^+, X-ray flashes induce a broad luminescence at 3.4 eV. The decay time of the afterglow ((41 ± 3) μs) does not change with temperature between 80 and 450 K. Above 500 K the decay time of the copper-related luminescence decreases exponentially with increasing temperature, characterized by an activation energy of (700 ± 50) meV. In silver containing crystalline quartz the well-known luminescence bands have different decay times (see Fig. 2). Similar to the copper luminescence no temperature dependence of the afterglow decay is detectable till 500 K in silver containing crystalline quartz.

Above 500 K, initially an irreversible thermal annihilation of the 2.8 eV luminescence centers occurs. Beside that, also a decrease of the afterglow decay time becomes detectable. But because of its overlapping with the more intense 2.2 eV luminescence,

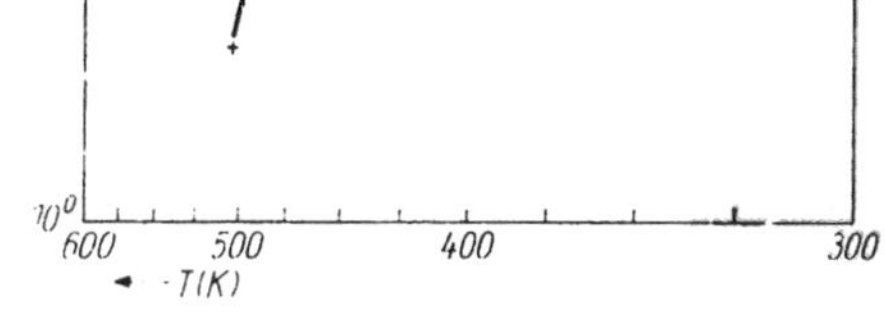

Fig. 2. Temperature-dependent decay of the X-ray induced luminescence afterglow presented in an Arrhenius plot. △ Ag-quartz 4.75, □ 2.2, ◇ 2.8, and + Cu-quartz 3.4 eV luminescence

3) The deviations are caused by the three errors:

a) In measurement of the decay time: in the case of only one single exponential decay of luminescence an error of 5 to 10% occurs, caused by noise, time base error, and excitation pulse variation.

b) Measurement of the sample temperature includes a maximum error of 25 K (measured at the α–β transition of quartz at 573 °C).

c) Sample-dependent errors caused mainly by the thermoluminescence induced afterglow.

Table 2

Decay times of luminescence afterglow dependent on temperature

type of SiO_2	luminescence band at	temperature 80	150	200	300	400	500	600	700 K
crystalline quartz	2.6 eV	1 ms**)	850 µs	65 µs	—				
(Brazil,	2.9 eV	1 ms	850 µs	65 µs	< 1 µs				
Madagascar,	3.15 eV	1 ms	850 µs	65 µs	< 1 µs				
smoky,	3.3 eV				≈ 1200 s	*)	*)	—	—
Morion,	2.5 eV				≈ 1200 s	*)	—	—	—
synthetic)	2.7 eV				—	*)	*)	*)	*)
Ag-doped quartz	4.75 eV	26 µs	26 µs	26 µs	26 µs	26 µs	26 µs	20 µs	12 µs
	2.8 eV	70 µs	70 µs	70 µs	70 µs	70 µs	≈ 60 µs	< 20 µs	?
	2.2 eV	140 µs	140 µs	140 µs	140 µs	140 µs	≈ 100 µs	≈ 40 µs	?
Cu-doped quartz	3.4 eV	41 µs	41 µs	41 µs	41 µs	41 µs	45 µs	10 µs	<1 µs
Ge-doped quartz	2.3 eV	1 ms	1 ms	850 µs	65 µs	21 µs	?	?	?
quartz glass		50/450 µs			50/450 µs		50/450 µs	?	?
GeO_2	2.3 eV	600 µs	500 µs	60 µs	<1 µs				

*) The decay of the luminescence is dependent on the relevant glow peak of thermoluminescence and varies between seconds and hous [11].
**) The underlined values of the decay time correspond to the temperature-dependent spectral maximum of the luminescence.

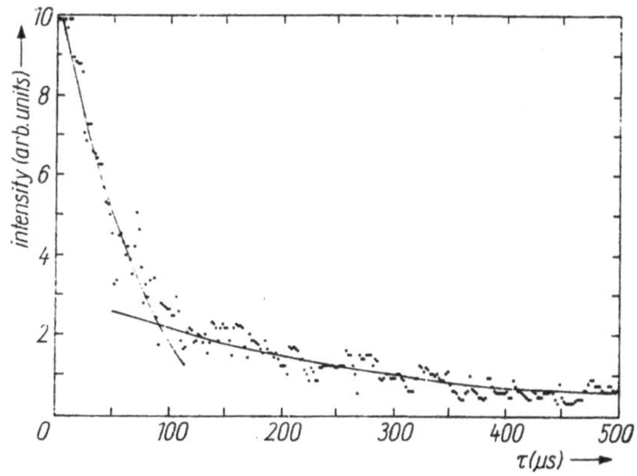

Fig. 3. Oscillogram of the luminescence decay of quartz glass. The afterglow of the X-ray induced luminescence, consisting of two decays (50 and 450 μs), is independent of temperature till 600 K. Above 250 K the slow afterglow component decreases with increasing temperature

the measured values of the 2.8 eV luminescence decay time are too inaccurate for calculating an activation energy. At 550 K an equivalent behavior can be detected for the 2.2 eV luminescence; the intensity, and later also the decay time, decreases with increasing temperature. Finally above 600 K the 4.75 eV luminescence of silver containing quartz decreases in intensity as well as in afterglow decay. The registered weak decrease of the afterglow decay time with increasing temperature did not permit a calculation of activation energies but it indicates low frequency factors connected with isolated luminescence centers.

3.4 *Luminescence of glassy SiO_2 (quartz glasses)*

The decay of X-ray flash induced luminescence in all measured types of quartz glasses (Suprasil, Infrasil, Herasil, etc.) is non-exponential. The quantum yield of luminescence is more than two orders of magnitude lower than in crystalline quartz and therefore a spectral analysis of the emitted light was not possible with our experimental set-up. The total (integrated) luminescence in quartz glasses could be characterized approximative by two decay states ($\approx$ 50 and $\approx$ 450 μs) Fig. 3.

In contrast to crystalline quartz, the spectral maximum of X-ray flash induced luminescence both in silver- and copper-doped quartz glasses is the same (2.4 eV). However, there is a clear difference in the decay.

In copper containing glass the decay time is 150 μs and does not change in the temperature range 77 to 300 K. In silver containing glass the decay time is shorter than 1 μs, thus was not detectable with our experimental set-up. Similarly as in silver containing crystalline quartz, the intensity of the silver-dependent luminescence in quartz glasses drastically decreases above 500 K.

4. Discussion

The obtained results clearly indicate that the presence of the most usual impurities in crystalline quartz, like aluminium, monovalent alkali metal ions (Li^+, Na^+, K^+), and OH groups in different concentrations, does not change the decay parameters of the blue luminescence ($\approx$ 1 ms, E_a = 370 meV) at low temperatures. Especially the same luminescence decay in natural quartz (Morion), which contains $\approx$1000 ppm Al and in synthetic quartz (sample 21-81) with practically no Al ($<$ 10 ppm), experimentally proves that the broad blue luminescence at 80 K is not connected with aluminium-related defects, in contrast to earlier proposals [5, 10]. Thus, this blue luminescence with its characteristic decay parameters ($\tau_{80K} \approx 1$ ms, $E_a = 370$ meV) can be attributed only to radiation-induced, (transient) fundamental defects in the quartz lattice. The absence of any saturation effects in the luminescence behavior of quartz, because of the long lifetime of those excited states, especially expected below 150 K, indicates also the existence of radiation-induced luminescence centers.

An argument for radiation-induced transient luminescence centers consists also in the registered weak quantum yield (≈ 1) of that luminescence, indicating usual intracenter excitations (in contrast to the motion of excited electrons to pre-existing recombination centers). An additional fact for the existence of radiation-induced luminescence centers at low temperatures consists in the experimental result that the intensity of that luminescence, excited by X-ray flashes, is independent of the afterglow decay time: The luminescence intensity in the pulse peak region becomes not more intense with afterglow shortening during the increase of temperature. This behavior is in contrast to a recombination mechanism of luminescence. The applied X-ray flashes generate the same number of electrons and holes at each temperature. At low temperatures a certain part of electrons (holes) is trapped and could be released only thermally, thus producing a temperature-dependent afterglow of luminescence. Therefore these electrons are lost for the recombination process during excitation and the intensity of luminescence, detected in this time range, should be lower. With increase of temperature, more electrons (holes) can be released, producing shorter decay times, and therefore the luminescence in its pulse peak region should be more intense. But we have not detected such a trend. On the other hand, in terms of a recombination mechanism the temperature independence of the luminescence peak intensity is to be explained only by competing processes: The thermal quenching of luminescence should be characterized by the same activation energy as needed for thermal release of trapped electrons. But it seems improbable that two different centers (trap and recombination center) have exactly the same characteristics. (Such a quenching could be thought of as originating from thermally activated trapping of electrons from the valence band by holes in recombination centers and thus preventing the recombination of electrons from the conduction band.)

In the opposite case, for an intracenter transition mechanism, the luminescence intensity in the pulse peak region depends not on the number of electrons and holes but on the concentration of luminescence centers and on the exciting pulse intensity only (which are the same at each temperature). Therefore the luminescence intensity in the pulse peak region is expected to be independent of temperature, which corresponds to our experimental results.

As mentioned in [11, 12] the approximately constant afterglow of luminescence below 150 K indicates forbidden intramolecular transitions. The independence of that afterglow decay of cations like Si, Ge, Al as well as of the usually existing monovalent impurities (H, Li, Na, K) indicates that radiation-induced defects in the oxygen sublattice are responsible for the luminescence afterglow (oxygen vacancy–peroxy linkage pair). In the well established nomenclature of a self-trapped exciton, forbidden transitions (triplet–singlet) are responsible for the afterglow. The observed decrease of decay time of the registered luminescence with increasing temperature can be explained by thermally activated radiationless transitions in the well-known Frank-Condon model.

In opposition to natural and Ge-doped synthetic quartz the registered new X-ray flash induced luminescence bands in crystalline SiO_2–Cu and SiO_2–Ag are dependent on copper and silver impurities only and can be erased by reverse electrolysis.

Cu^+ and Ag^+ ions replace during electrolysis the usual monovalent interstitials like Li^+, Na^+, K^+ and take their original places in the quartz lattice. Therefore, the proposed models for copper and silver luminescence centers include the aluminum places in quartz, where the charge compensators Li^+, Na^+ are replaced by Ag^+, Cu^+, producing $AlO_{4/2}$–Ag or $AlO_{4/2}$–Cu. The temperature independence of the decay times of silver and copper luminescence between 80 and 500 K indicates also the existence of partially forbidden intramolecular-like optical transitions at these centers.

At temperatures above 500 K, a thermally induced ionization connected with an annihilation of radiation-induced 2.2 and 2.8 eV luminescence centers occurs in silver-containing quartz. But one part of the 2.2 eV luminescence centers cannot be destroyed by thermal treatment till 1000 K. The structure of these luminescence centers is still unknown.

Especially the possibility of exciting silver- and copper-related luminescence by UV irradiation with energies below 6 eV demonstrates the intracenter character of that luminescence (up to 6 eV there does not exist absorption in natural clear quartz).

The luminescence and its afterglow in quartz glasses cannot be interpreted at present because of the lack of spectral data and their connection with the afterglow behavior. Looking at the luminescence data of silver or copper containing quartz glasses it becomes obvious that also at low temperatures the luminescence centers are expected to be at pre-existing defects. As well as in crystalline SiO_2 the luminescence in quartz glass is most probably also an intracenter excitation process, causing low quantum yield and constant afterglow up to high temperatures. The difference in luminescence behavior of glasses compared with that of crystalline SiO_2 can only be explained by a special luminescence center neighborhood, specific for glasses.

References

[1] D. L. Griscom and C. H. Sigel, Bull. Amer. Phys. Soc. **13**, 1474 (1968).
[2] G. H. Sigel and D. L. Griscom, Bull. Amer. Ceram. Soc. **48**, 447 (1969).
[3] G. H. Sigel, J. non-crystall. Solids **13**, 372 (1973/74).
[4] D. L. Griscom, Proc. 3rd Frequency Control Symp., Atlantic City (N.J.) 1979 (p. 98).
[5] W. Hohenau, W. Meyer, and F. Gross, Acta phys. Austriaca **44**, 337 (1976).
[6] A. W. Trukhin and A. E. Plaudis, Fiz. tverd. Tela **21**, 1109 (1979).
[7] B. P. Gritsenko V. M. Lisitsyn, and V. N. Stepanchuk, Fiz. tverd. Tela **23**, 393 (1981).
[8] K. Tanimura, T. Tanaka, and N. Itoh, Phys. Rev. Letters **51**, 423 (1983).
[9] W. Hayes M. J. Kane, O. Salminen, R. L. Wood, and S. P. Doherty, J. Phys. C **17**, 294 (1984).
[10] W. Hohenau, K. Schwingenschuh, and F. Gross, J. Phys. Chem. Solids **45**, 61 (1984).
[11] W. Hohenau, phys. stat. sol. (a) **88**, 267 (1985).
[12] A. N. Trukhin, phys. stat. sol. (b) **143**, K83 (1987).
[13] W. Hohenau and F. Gross, Acta phys. Austriaca **37**, 385 (1973).
[14] A. N. Trukhin, S. S. Etsin, and A. V. Shendrik, Izv. Akad. Nauk SSSR, Ser. fiz. **40**, 2329 (1976).
[15] A. N. Trukhin, Optika i Spektroskopiya **40**, 756 (1976).
[16] W. Hohenau, phys. stat. sol. (a) **82**, 119 (1984).
[17] A. N. Trukhin and A. V. Shendrik, phys. stat. sol. (b) **98**, K13 (1980).
[18] A. E. Plaudis and M. E. Springis, Izv. Akad. Nauk Latv. SSR, Ser. fiz. tekh. Nauk No. 1, 107 (1986).
[19] T. Tanaka, T. Eshita, T. Tanimura, and N. Itoh Crystal Lattice Defects amorph. Mater. **11**, 221 (1985).
[20] Y. H. Stathis and M. A. Kastner, Phys. Rev. B **35**, 2972 (1987).

(Received July 17, 1987; in revised form June 17, 1988)

phys. stat. sol. (a) **111**, 345 (1989)

Subject classification: 78.55; 71.55; 78.30; S7.12

Institute of Physics, Czechoslovak Academy of Sciences, Prague[1]) *(a) and ZSNP n.p., Žiar n. Hronom (b)*

Lateral Distribution of Si-Related Defects and Carrier Density in GaAs Crystals

By

J. PASTRŇÁK (a), J. OSWALD (a), I. GREGORA (a), V. VORLÍČEK (a), and M. BABINSKÝ (b)

The spatial distribution of luminescence intensity for edge emission and extrinsic Si related bands is studied for Si-doped LEC grown GaAs crystals. Free carrier distribution is determined from the halfwidth and peak position of intrinsic luminescence and from the position of coupled LO-phonon–plasmon modes in Raman spectra. An increase of the luminescence intensity with carrier concentration is observed up to $n \approx 10^{18}$ cm^{-3}, the sharp decrease for higher concentrations is ascribed to Auger recombination. The temperature dependence of three Si-related luminescence bands at 1.35 eV (A), 1.2 eV (B), and 1.0 eV (C) is measured. The similarity of the temperature behaviour of the A band with EL2 centres supports its assignment to ($Si–As_{Ga}$) donors.

Пространственное распределение интенсивности краевой люминесценции и примесной полосы кремния изучалось для кремнием легированнух кристаллов GaAs выращиваемых методом Чохральского. Распределение свободных носителей было определено из полуширины и положения максимума краевой люминесценции и из положения связанной ЛО-фонон–плазмонной моды в спектрах комбинационного рассеяния. Возрастание интенсивности люминесценции с ростом плотности свободных носителей наблюдалось вплоть до $n \approx 10^{18}$ cm^{-3}; последующее резкое падение для более высоких плотностей было присуждено рекомбинации Оже. Измерялась температурная зависимость трех люминисцентных полос связанных с кремнием у 1,35 eV (A), 1,2 eV (B) и 1,0 eV (C). Похожее температурное поведение A полосы и полосы ЕЛ2 центров говорит в пользу интерпретации A полосы при помощи ($Si–As_{Ga}$) доноров.

1. Introduction

Silicon is the most commonly used doping element for n-type GaAs. Numerous studies have been devoted to the problem of incorporation of Si atoms into GaAs lattice under different growth conditions, and to properties of Si related defects and their electrical activity (see e.g. [1 to 23]). Contradictions remain as to the origin of Si-related complexes, their relation to the growth conditions, spatial distribution of different Si-related defects in crystals as a function of growth direction, stoichiometry, and Si concentration. Also their microphysical models are still a subject of discussion.

In our studies we concentrated our attention on the spatial distribution of Si-related centres in LEC grown Si-doped GaAs crystals, and on relation between their concentration and that of free carriers.

2. Experimental

2.1 Samples

Standard chemico-mechanically polished GaAs wafers with (100) orientation cut from ingots pulled in [100] or [111] directions were used for measurements. The average

[1]) Na Slovance 2, CS-180 40 Prague 8, Czechoslovakia.

concentration of free carriers was determined by galvanomagnetic measurements. The used doping levels varied from 10^{16} to 5×10^{18} cm^{-3}, the main attention was paid, however, to the heavily doped LEC grown samples. The dislocation density determined by molten KOH etching was typically ≈ EPD-40-80 mm^{-2} for crystals grown in [111] direction with free carrier concentration n ranging from (2.35 to 4.05) $\times 10^{18}$ cm^{-3} (samples No. 1850 to 1854) and EPD equal to 590—1000 mm^{-2} for crystals grown in [100] direction with n from (1 to 2) $\times 10^{18}$ cm^{-3} (samples No. 1855 to 1864). The highest value of n above 4×10^{18} cm^{-3} was found in sample No. 1865.

2.2 Measurements

The homogeneity of samples was determined by luminescence topography. The luminescence was excited by the 632.8 nm line of a He–Ne 30 mW laser focused on a spot of 20 μm in diameter and its local intensity was plotted as a function of the spot position for two different wavelengths — one corresponding to the band edge emission and the second to the emission of the complex Si related band. Also the complete spectra were measured at a set of individual spots along one line chosen across a given sample.

Low-temperature luminescence spectra and their temperature dependence were measured using a SDL1 double grating monochromator with cooled R1767 photomultiplier and He–Ne laser as an excitation source. The obtained spectra were decomposed into individual maxima using standard minimization program [24]. The long-wavelength luminescence was measured using a cooled PbS detector.

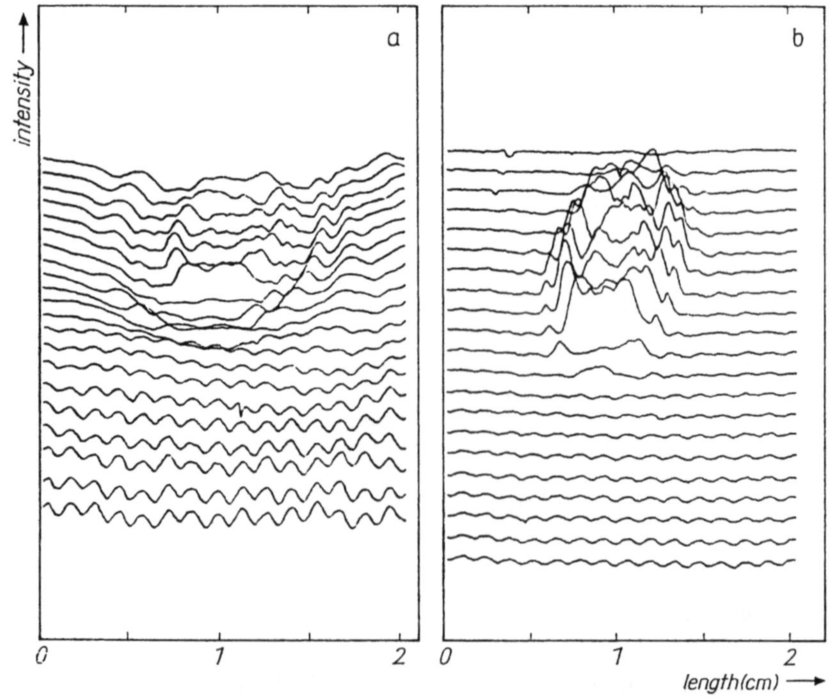

Fig. 1. Luminescence topogram of the 1.5 × 1.5 cm² central area of the LEC GaAs:Si sample of (100) orientation No. 1865 taken at 300 K in the band edge emission at a) 860 nm and b) 960 nm

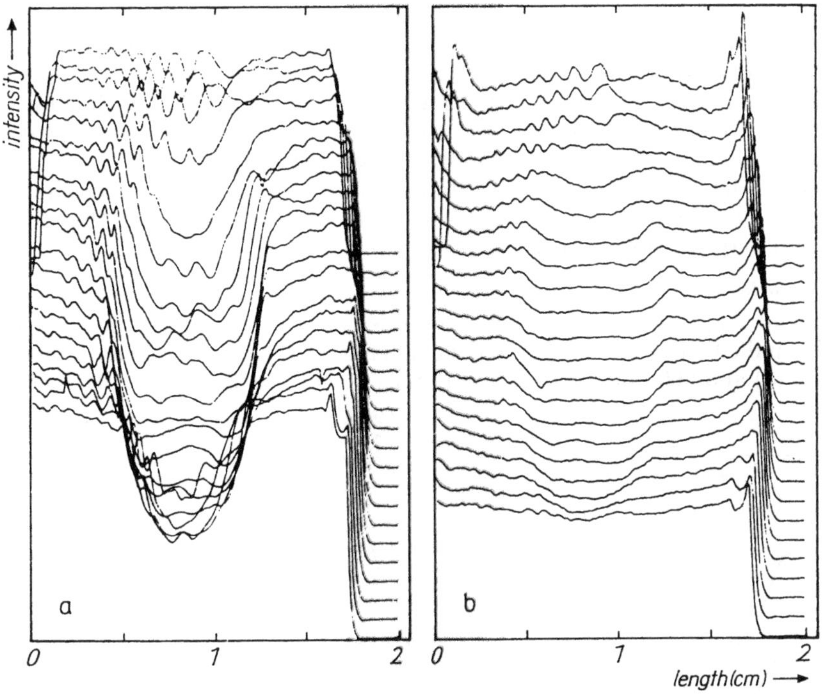

Fig. 2. Luminescence topogram of the 2.0 × 1.5 cm^2 central area of the LEC GaAs sample No. 185 grown in [111] direction taken at 300 K at a) 860 nm and b) 960 nm

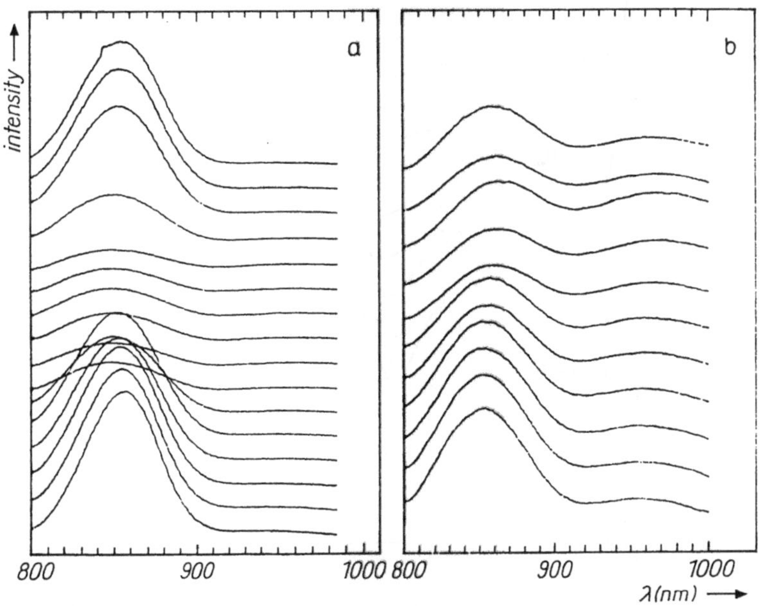

Fig. 3. Emission spectra taken at 300 K at different spots along the line crossing samples a) No. 1853 and b) 1865

Raman spectra were measured in backscattering geometry, using near Brewster angle incidence on (100) surface of the sample. The laser beam (Ar laser line: 200 mW at 514.5 nm) was focused on the sample surface by means of a cylindrical lens giving a line focus ≈1.5 mm long, the scattered light was collected by imaging the line focus with a fast camera lens onto the entrance slit of a double spectrometer Spex 14018. A cooled photomultiplier RCA C31034 and standard photon counting electronics were employed for detection. Both incident and scattered light was polarized parallel to the [011] axis. The sample could be shifted laterally without disturbing the alignment in order to map the sample in 1 mm steps.

2.3 Results

In Fig. 1 a, b and 2 a, b we can see typical luminescence topograms of heavily Si-doped samples pulled in [100] (Fig. 1) and [111] (Fig. 2) directions taken at wavelengths 860 nm (a) and 960 nm (b). In both samples we can see distinct oscillations in

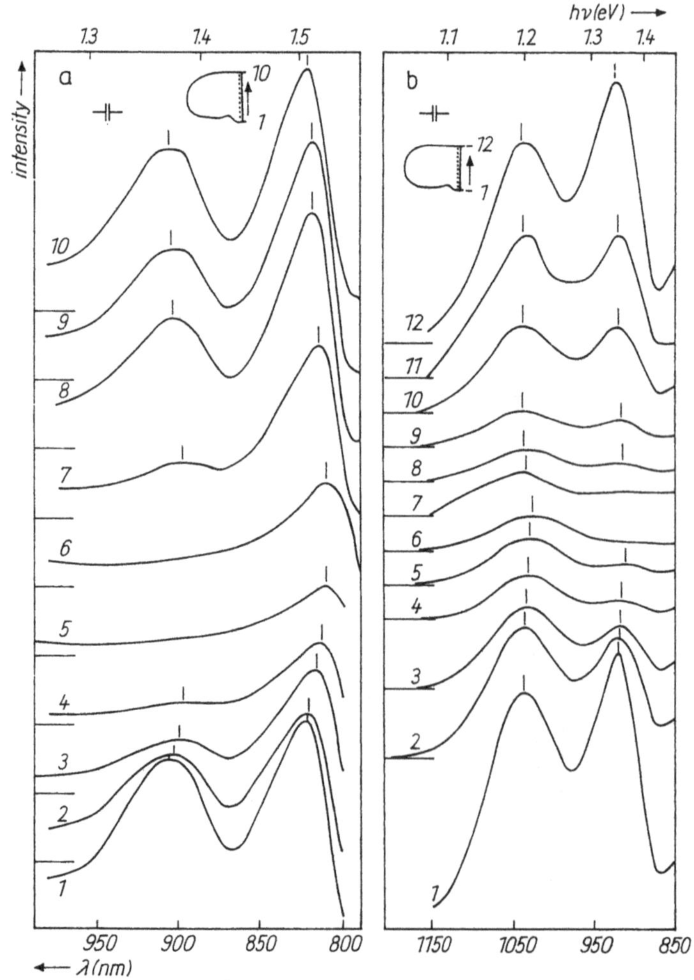

Fig. 4. Luminescence spectra taken at 4.2 K at spots along the line crossing sample No. 1853 (see inset) in the spectral regions a) 800 to 970 nm and b) 850 to 1150 nm. Uncorrected for the spectral sensitivity of the detection system

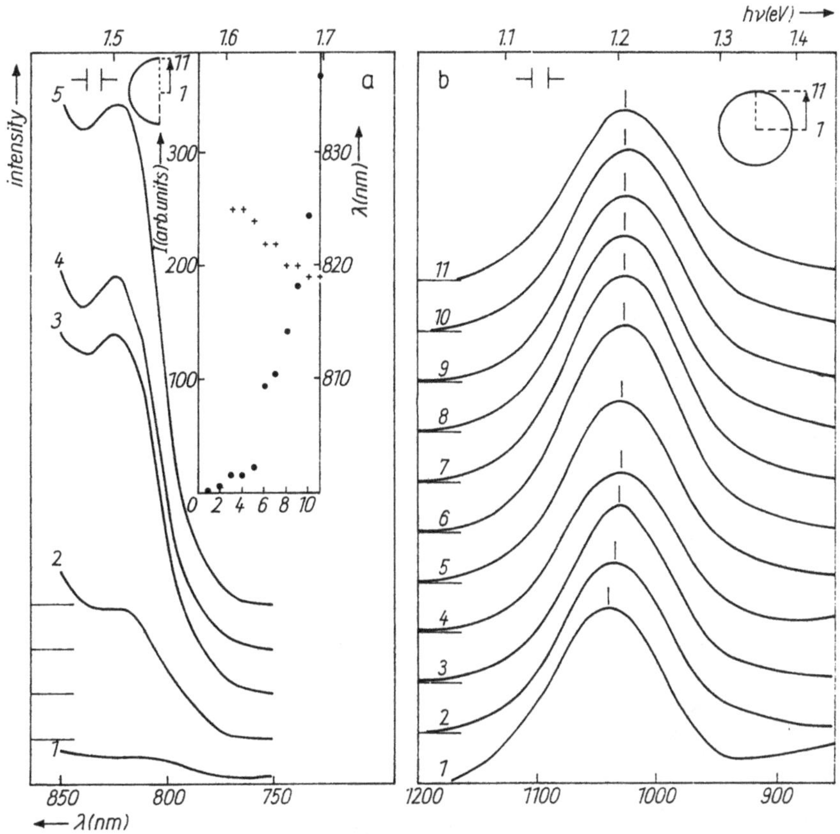

Fig. 5. Luminescence spectra taken at 4.2 K in eleven spots along the line from the central part to the margin of sample No. 1865 for the spectral regions a) 750 to 850 nm and b) 850 to 1200 nm. In the inset of Fig. 5a the intensities (●) and the wavelengths (+) of the band edge emission maxima are given. Uncorrected for the spectral sensitivity of the detection system

the band edge luminescence intensity in marginal parts of samples and its prominent decrease in the middle part of topograms. In crystals with lower Si content these growth striations manifesting themselves by variations in luminescence intensity were usually less distinct.

Oscillations of the luminescence intensity for both the edge emission as well as of complex band are in phase in the marginal region of the crystal, also the decrease in the luminescence intensity when approaching to the central part of the sample is a common effect. For the crystal No. 1865 grown in [100] direction, however, the intensity of the complex band sharply increases in the central part of the sample, while in crystals No. 1850 to 1854 it remains of low intensity.

The changes of the complex band intensity are apparent also from Fig. 3a and b, where room temperature emission spectra taken at points along the line crossing the sample are shown. The intensity of band edge emission is lower in the central part of both samples, for sample No. 1865 the intensity of the complex band increases in this area. These spectra were measured using FU 62 photomultiplier with S1 photocathode and were not corrected for spectral sensitivity.

Low-temperature luminescence spectra measured at several different points across these samples in spectral regions covering edge emission and complex band maxima are shown in Fig. 4 a, b and 5 a, b. While in sample No. 1865 only band edge emission and one emission band due to complexes was observed, in other samples an additional, more narrow band at 920 nm appeared. The temperature dependence of these complex spectra is given in Fig. 6 a and b.

An additional comparatively weak broad emission band has been found at 1203 ± ± 3 nm. The temperature dependence of the emission in this spectral region is given in Fig. 7 a, b for the central (a) and marginal parts (b) of sample No. 1865 together with decomposition into individual bands.

The spectra measured at individual points across the samples (Fig. 3 a, b) differ not only in the intensity of bands. Also the position of the maximum and the half-width of the band edge emission change. In Fig. 8 these values are plotted against the position of the spot on the sample. Abrupt decrease in the luminescence intensity is accompanied by the increase of the halfwidth and by the shift of the maximum towards shorter wavelengths.

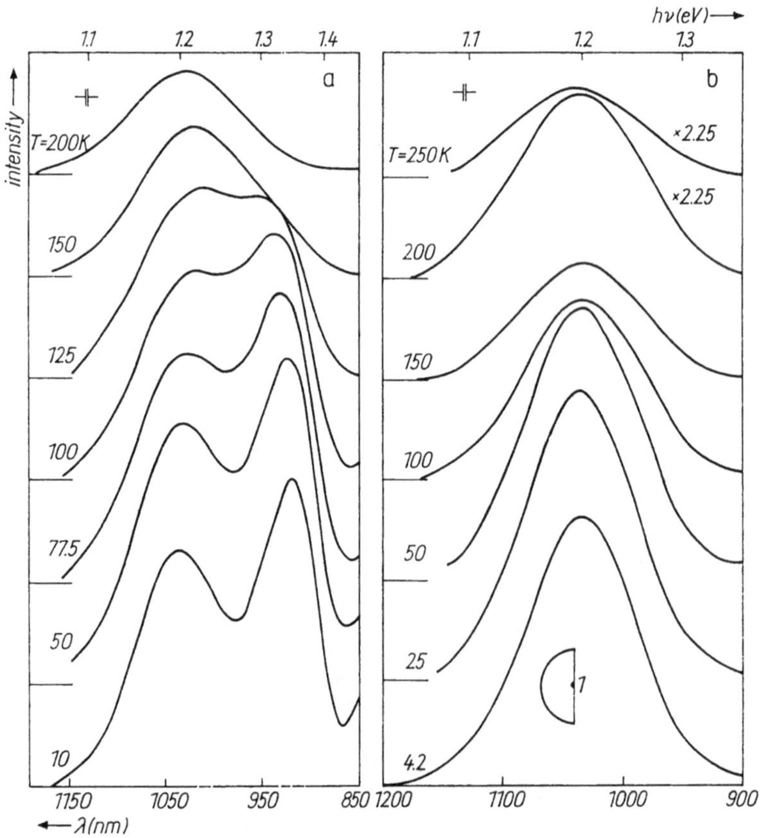

Fig. 6. Temperature dependence of the emission bands of Si-related complexes for samples a) No. 1853 and b) No. 1865. Spectral curves were not corrected for the spectral sensitivity of the detection system.

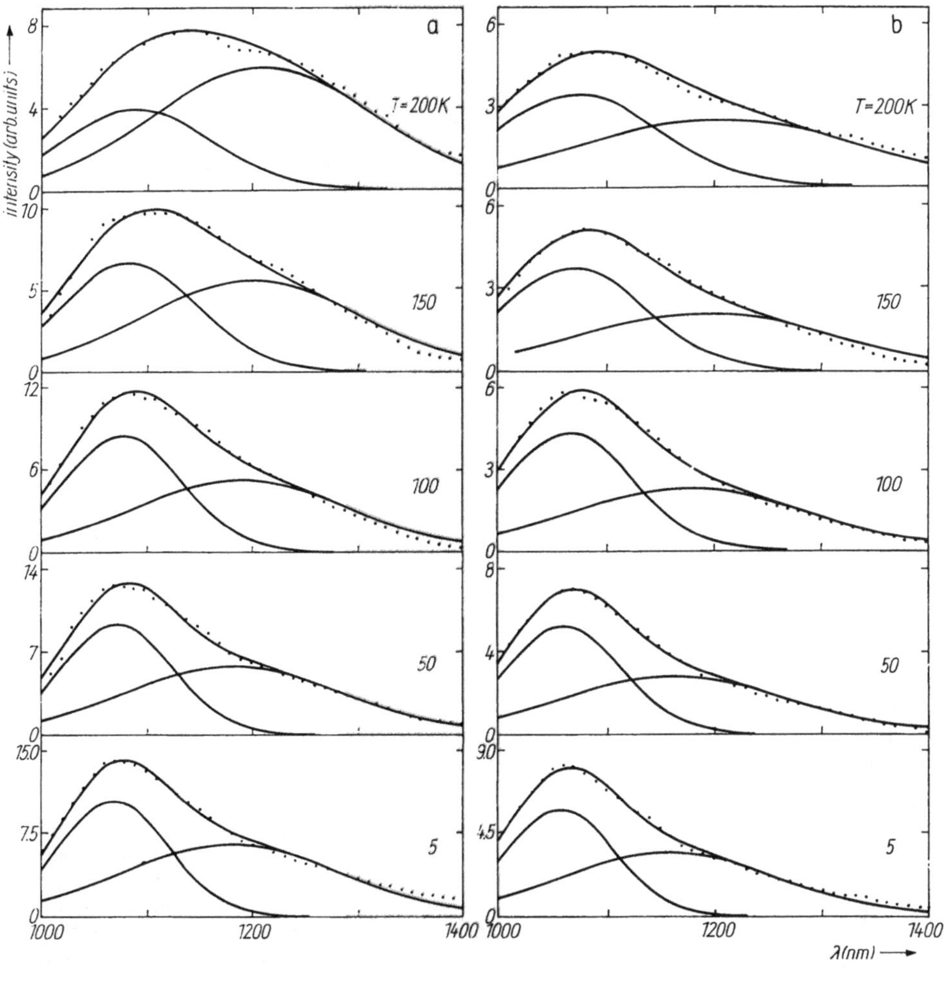

Fig. 7. Temperature dependence of the long-wavelength emission bands in a) central and b) marginal parts of sample No. 1865, corrected for spectral sensitivity of the detection system

As we assumed that the observed changes in the halfwidth and peak position are due to the local changes in the free electron concentration we measured this value independently using spatially resolved Raman spectra. Typical examples are shown in Fig. 9. The most prominent features are the two peaks due to coupled LO-phonon–plasmon modes: the lower L^- at about 267 cm^{-1} and the upper, broader L^+ mode observed between 500 to 800 cm^{-1}. Also seen in the spectra is a sharp peak due to scattering by an unscreened "pure" LO phonon ($\omega_{LO} = 292$ cm^{-1}) from the depleted surface layer, as well as some structure of two-phonon origin between 500 and 550 cm^{-1}, and at about 350 cm^{-1}. The profile of the electron concentration determined from (L^+) mode is shown in Fig. 10.

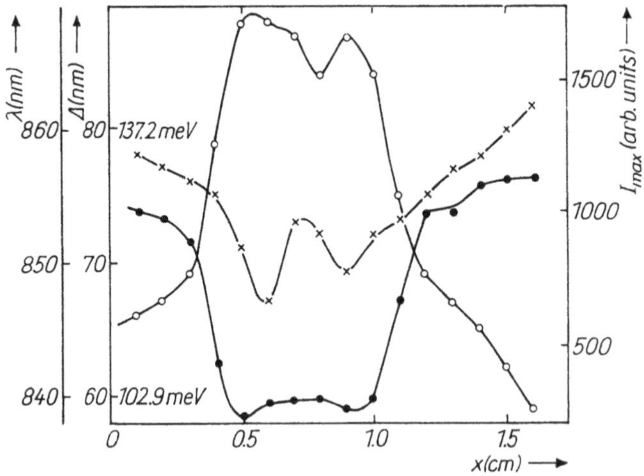

Fig. 8. Lateral dependence of the band edge luminescence intensity (crosses), position of the maxima (full circles), and halfwidth (open circles) for the central part of sample No. 1853. $T = 300$ K

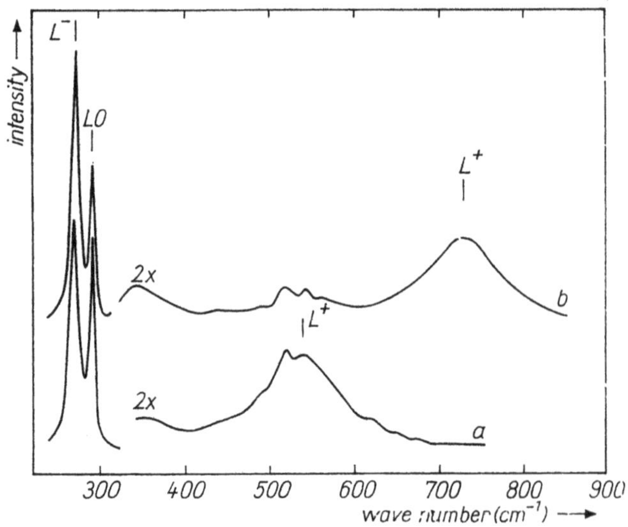

Fig. 9. Raman spectra at two points on the sample surface No. 18 (a) near the margin ($n = 2.1 \times 10^{18}$ cm^{-3}), (b) at the centre ($n = 4.5 \times 10^{18}$ cm^{-3}), $T = 300$ K; $\lambda_L = 514.5$ nm

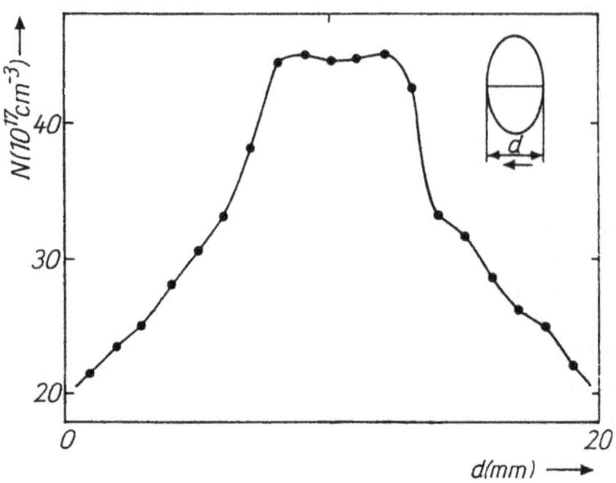

Fig. 10. Profile of the electron concentration across the sample surface determined from (L⁺). $T = 300$ K

3. Discussion

3.1 Si-related defects and their properties

The major part of Si atoms is incorporated into GaAs lattice in the form of substitutional point defects, forming shallow donor (Si_{Ga}) or acceptor (Si_{As}) levels. In luminescence spectra they manifest themselves as bound exciton lines, (e, A), (D, h), and (D, A) transitions [41]. While in LPE layers the preferential incorporation of either Si_{Ga} donors or Si_{As} acceptors is given by growth temperature, in MBE layers the substrate orientation plays important role [10, 13, 14, 34 to 37]. The n-type layers grown onto (100) and (311) B planes exhibited, according to [13], increasing concentration of free carriers with increasing Si content up to $n \approx 3 \times 10^{18}$ cm^{-3}. (D–A) transitions were dominating in luminescence spectra for low concentrations, for higher ones a peak at 1.27 eV appeared and prevailed at concentrations above 3×10^{18} cm^{-3}. This peak was attributed to self-activated centres, namely to (Si_{Ga}–V_{Ga}) complexes [13]. When studying the mechanism of compensation in heavily Si-doped GaAs MBE grown layers Maguire et al. found by local vibrational mode spectroscopy besides (Si_{Ga}–Si_{As}) complexes also (Si–X) centres of acceptor type, where X was assigned as V_{Ga} [10].

Another Si-related peak in photoluminescence spectra at 1.4 eV was found by Queisser [26] and others [27 to 32] in Si-implanted GaAs (so-called Q peak) and ascribed to (V_{As}–Si_{As}) [28, 29] or (V_{As}–Si_{As}) and (Ga_i–Si_{As}) [31] complexes. Their characteristic feature is a large shift with excitation intensity about 10 meV per decade. In implanted and annealed layers also 1.32 and 1.28 eV maxima were observed and related to the negative charge state of the 77 meV double acceptor (presumably Ga_{As}) [12], its neutral state is connected with the 1.44 eV band, not necessarily connected with the presence of Si [11, 12, 33].

Orientation dependent incorporation of Si into GaAs lattice has been observed also in HB grown crystals [38]. The influence of solid–liquid interface orientation on the way of Si incorporation can be also expected in melt-grown crystals. As far as we know the quantitative study of orientation dependence of segregation coefficient, however, has not yet been performed.

In LEC grown crystals the maximum at 1.2 eV has been reported already in 1967 by Williams et al. [1, 2, 40] and identified as (Si_{Ga}–V_{Ga}) complex. A complex band located at 1.35 eV has been found also in Si-doped GaAs and AlGaAs layers grown by metalorganic chemical vapour deposition [39, 5] and also ascribed to (V_{Ga}–Si_{Ga}) complex.

The incorporation of Si into GaAs lattice in the form of substitutional point defects, (Si_{Ga}–Si_{As}) pairs, and (Si–X) complexes, has been also confirmed by local vibrational mode spectroscopy [7, 8, 10, 18, 20, 42]. The unambiguous interpretation of the nature of these complexes is still lacking.

In our samples we observed three complex Si-related luminescence bands, A at 1.35 eV, B at 1.15 eV, and C at 1.00 eV. They differed in halfwidth and in the temperature dependence of peak position, halfwidth, and intensity. For all three maxima the halfwidth could be described with sufficient accuracy by the usual expression

$$FWHM(T) = FWHM(0) \sqrt{\coth \frac{T_0}{T}} . \quad (1)$$

The values of $FMHW(0)$ and T_0 for these maxima are given in Table 1. For the description of the temperature shift of the peak position we used the formula describing the temperature dependence of the gap [43, 44],

$$E(T) = E(0) - \frac{\alpha T^2}{T + \theta} , \quad (2)$$

Table 1

Parameters of complex maxima

	band A	band B	band C
E_M at 4.2 K	1.35 ± 0.01	1.17 ± 0.01	1.03 ± 0.03
α (meV/K)	0.61 ± 0.05	0.08 ± 0.03	0.22 ± 0.07
θ (K)	134.5 ± 6.0	221 ± 174	269 ± 227
FMHW (meV) at 4.2 K	89 ± 3	126 ± 5	206 ± 7
T_0 (K)	71 ± 30	116 ± 4	267 ± 206
C	0.000054	0.42	1.77
T_{eff} (K)	1240 ± 90	130 ± 15	

where constants α and θ were determined by least-squares fitting. They are also given in Table 1. We can see that the temperature shift of band A is much higher than the corresponding shift of the band gap energy (0.58 meV/K), while the shift of band B is negligible. The properties of band B can be compared to those found for self-activated complex in GaAs [1, 2, 9], the observed thermal quenching is in our samples, however, much smaller. The relevant data are presented in Table 2 for both bands. If we suppose the thermal quenching in the form

$$\tau_{NR} = C \exp \frac{T_{eff}}{T}, \tag{3}$$

then we obtain by least-squares fitting the values of T_{eff} and C_0 as given in Table 1. They were obtained from the normalized data summarized in Table 2. The observed sharp decrease in the luminescence intensity above 130 K was also found in [45] for the comparatively weak, narrow A2 band at 1.34 eV ascribed to transitions involving the deep acceptor at $E_v + 0.174$ eV due to a Ga_{As}–V_{Ga} pair defect. The temperature shift of the A2 band differed from that of the A band in our work. The band at 1.35 eV was also observed in [5] and identical assignment was given.

The temperature dependence of the total intensities of A, B, and C bands in our samples is given in Table 2 and could be described by the formula for temperature quenching

$$I_{PL} = \frac{A}{1 + 1/\tau_{NR}} \tag{4}$$

for temperatures above ≈ 30 K.

Table 2

Temperature dependence of the normalized luminescence intensity of complex maxima

T (K)	intensity A	intensity B	intensity C
5	1.00	1.00	1.00
50	1.02 ± 0.03	0.85 ± 0.09	0.75 ± 0.12
100	0.92 ± 0.04	0.66 ± 0.13	0.65 ± 0.19
150	0.24 ± 0.03	0.52 ± 0.11	0.65 ± 0.21
200	0.02 ± 0.03	0.41 ± 0.08	0.71 ± 0.33

Having in mind calculations of Jansen et al. [52] we can conclude that the highest equilibrium concentrations of native defects in n-type GaAs for positive stoichiometry (excess of As) have V_{Ga} and As_{Ga} or As_i defects, closer to stoichiometry also the Ga_{As} defects may contribute to the total defect atmosphere in crystal [55]. These defects are also decisive for complex formation with the Si doping element. For high doping levels complexes (Si_{Ga}–Si_{As}) may be of importance as well.

We have not found any fact contradicting the original assignment of the self-activated band B to (Si_{Ga}–V_{Ga}) complexes [1, 2, 40]. For band A at 1.35 eV large thermal shift and pronounced thermal quenching starting between 100 and 150 K are typical. In this respect it behaves in a similar way as the maximum A2 reported by Yu et al. [45] and ascribed by them to D–A transitions involving the acceptor level $E_v + 0.174$ eV, tentatively assigned to the (V_{Ga}–Ga_{As}) pair.

The temperature behaviour of 1.35 eV band found by Glinchuk et al. in Te-doped GaAs samples and ascribed to (V_{Ga}–Te_{As}) V_{As} complexes also strongly resembles that of band A [63]. However, these models are not Si related. We assign this band to a complex centre containing an As_{Ga} defect, say the (Si_{Ga}–As_{Ga}) complex. The metastable behaviour of the As_{Ga} defect may be responsible for the temperature dependence of the luminescence intensity like in the case of EL2 centres [53].

One of possible candidates for the C band is the Si_{Ga}–Si_{As} complex. The energy levels of an impurity pair have been calculated by a tight-binding Koster-Slater model and a deep level should be expected [54]. Formation of such defects for high doping levels was predicted [22, 23], unambiguous evidence, however, is still lacking.

It should be reminded that luminescence maxima at 1.0 eV, 1.2 eV [60, 61], and 1.35 eV [62, 63] were also observed in GaAs without Si doping. In our samples the intensity of bands A and B is comparable to that of band edge emission and cannot be ascribed to any background impurity. The relation of the weaker C band to Si content is less obvious. However, it has been also confirmed in [65 to 68].

3.2 Spatial distribution of carrier concentration and luminescence

The spatial variations in luminescence intensity of band edge emission seen in Fig. 1 and 2 can be understood if we accept the concentration dependence of luminescence intensity found in [6, 18], with the intensity growing monotonously with increasing concentration, followed by a sharp decrease for concentrations above 3×10^{18} cm^{-3}. This dependence (see [18], Fig. 1) can be fitted by the formula

$$I_{PL}(n) = \frac{K_0 n}{n + K_1 n^2 + K_2}, \tag{5}$$

where the proportionality coefficient for the linearly concentration-dependent term in the radiative transition probability is taken as unity. The quenching processes which are concentration independent and proportional to the squared concentration are taken into account by coefficients K_2 and K_1, respectively. The least-squares fitting gives $K_0 = 45.8$, $K_1 = 7.46 \times 10^{-18}$, and $K_2 = 67.1 \times 10^{18}$ for n in cm^{-3}. The quenching proportional to the squared carrier concentration could be ascribed to an Auger processes. Using formulae of [46] satisfactory quantitative agreement for the value of K_1 can be obtained.

Another possibility of quenching processes at high concentration is the formation of some new efficient quenching centres starting at definite concentration. The abrupt change in the Si diffusion constant at 3×10^{18} cm^{-3} with formation of (Si_{Ga}–Si_{As}) pairs has been reported [22, 23]. For an explanation of the concentration dependence of luminescence intensity this process should have been more general, not Si dependent, as the concentration quenching was observed also in Te-doped samples [18].

The (Si_{As}–Si_{Ga}) complexes are electrically neutral so that variations in their concentration could not explain the observed spatial distribution in electron density.

To check the concentration dependence of the luminescence intensity on the free carrier concentration we determined its value directly from the luminescence spectra. We applied the approach of De-Sheng et al. [25] and determined the concentrations directly from the halfwidth of the band edge luminescence using the formula

$$FMHW = Dn^{2/3}\ (\mathrm{eV})\,, \tag{6}$$

where for the constant D we used the value found in [25] for GaAs:Te samples and measurements at 1.8 K, $D = 3.84 \times 10^{-14}$. The carrier concentrations obtained in this way from known halfwidths obtained in our measurements at room temperature were approximately twice as large as the average values obtained by Hall effect measurements. Simultaneously with changing concentration also the maximum shifts towards higher energies were observed as in [25] so that the agreement of the model describing the connection between concentration of free carriers, halfwidth, and peak position was excellent.

For an independent determination of the carrier concentration we used the Raman spectra measurements. Unless the damping is not excessively high, the frequencies of the LO-phonon-plasmon coupled modes L^-, L^+ depend on the electron concentration in the sample according to the formula

$$\omega^2(L^\pm) = \tfrac{1}{2}\left\{\omega_p^2 + \omega_{LO}^2 \pm \sqrt{(\omega_p^2 - \omega_{LO}^2)^2 + 4\omega_p^2(\omega_{LO}^2 - \omega_{TO}^2)}\right\}, \tag{7}$$

where ω_{TO} and ω_{LO} stand for transverse and longitudinal phonon frequency, respectively, and ω_p is the conduction electron plasma frequency, given by

$$\omega_p = \left(\frac{ne^2}{\varepsilon_0\varepsilon_\infty m^*}\right)^{1/2}. \tag{8}$$

Our results indicate that the electron concentration in the investigated sample exceeds 10^{18} cm^{-3}. Since in this concentration range the position of the lower peak L^- closely approaches the value of ω_{TO}, we can get a reasonable estimate of the electron concentration in the sample from the position of the L^+ peak, using e.g. the $\omega(L^+)$ versus n dependence presented in [47, 48].

The values obtained at different points across the diameter of the sample are shown in Fig. 10. A remarkable increase of the electron concentration is observed in the central part of the sample (more than a factor of two higher as compared to the paramarginal region). This result corresponds to the photoluminescence mapping of the same sample.

One should note that the data for the curves presented in [47, 48] were obviously taken at liquid nitrogen temperature. However, they may be safely taken in the analysis of our room temperature spectra, since for $n \geqq 10^{17}$ cm^{-3} the electron gas becomes degenerate in GaAs and the phonon frequency shift with temperature (4 cm^{-1}) is not larger than the uncertainty in the position of the broad L^+ peak (see e.g. [49]).

Raman spectra of opaque materials give information on the sample from a layer extending from the surface down to a depth comparable with the penetration depth of the incident laser light, i.e. $(2\alpha)^{-1}$, where α is the absorption coefficient. In our case (GaAs, $\lambda_L = 514.5$ nm) $\alpha^{-1} = 100$ nm [50]. The signal at the LO-phonon frequency, however, originates from the surface depletion layer and is proportional to $(1 - \exp(-2\alpha\delta))$, where δ is the width of this layer (see e.g. [47, 54]),

$$\delta = \left(\frac{2\varepsilon_0\varepsilon_s\Phi_B}{ne^2}\right)^{1/2}. \tag{9}$$

Here Φ_B is the surface barrier height and ε_s the static dielectric constant. At electron concentrations higher than 1×10^{18} cm^{-3}, $\delta < \alpha^{-1}$ in our case of GaAs (taking $\Phi_B = 0.7$ eV and $n = 1 \times 10^{18}$ cm^{-3}, we obtain $\delta = 30$ nm [47]).

Although no attempt to measure the absolute value of the scattered intensity was made in our experiments, all the spectra were taken under identical conditions so that we can compare the intensities at ω_{LO} frequency from different points on the sample surface. It follows then from (8) and (9) that we should expect

$$I_a/I_b = [(1 - \exp(-2\alpha\delta_a))/(1 - \exp(-2\alpha\delta_b))] \approx (n_a/n_b)^{1/2}, \qquad (10)$$

provided n_a and n_b are not very different. The experimental value of the ratio of the LO-peak intensities for the points between the margin and the centre of the sample is approximately 1.4 ± 0.1. This value is consistent with the results derived from the position of L^+ peaks, which indicate that the electron concentration at the central part of the sample is by about a factor of two higher than that near the margins.

Using the values for the free electron concentration determined from Raman spectra we can calculate the proportionality constant D in (6). We obtained for Si-doped GaAs samples and spectra measured at 4.2 K, $D = (4.8 \pm 0.2) \times 10^{-14}$ cm^2, in good agreement with [25].

3.3 Mutual relations among growth conditions, carrier concentration, and total amount of Si in the samples

The crucial problem for the explanation of prominent differences in free electron concentrations between central and marginal parts of the samples was the determination of total Si concentration along the diameter of the sample. By careful calibrated measurements using the microprobe JEOL 733 the total amount of Si was determined at three points across the sample No. 1850 of (100) orientation, cut from the ingot grown in [111] direction. The total concentrations of Si atoms found in the left, central, and right parts of the sample were $(4.55 \pm 0.91) \times 10^{18}$, $(5.33 \pm 1.07) \times 10^{18}$, and $(6.48 \pm 1.10) \times 10^{18}$ cm^{-3}, respectively. The Si concentration is monotonously increasing in the direction towards the lower part of the ingot as expected from its segregation coefficient lower than unity. No higher amount of Si was found in the central part of the sample. That means the carrier concentration is not determined by the total amount of Si, but by the electrical activity of the centres it forms. Besides Si_{As} centres also V_{Ga} centres and their complexes are the most probable candidates of centres compensating the main donor impurity Si_{Ga}. Indeed, in some of our samples microprecipitates composed of As and Si were found. This fact demonstrates the stoichiometry deviation in samples towards As excess.

The reason why in the central part of the sample almost all Si is incorporated into the GaAs lattice as donors while at the marginal regions partly as acceptor centres, can be understood on the basis of above-discussed results of experiments on MBE grown GaAs:Si layers [10, 13, 14, 34 to 37]. The exact form of the solid-melt interface plays the decisive role. As discussed e.g. by Nishinaga [51] the interface tilted at a small angle to the low-index plane is unstable against the appearance of wave-like oscillations in the growth step density and the originally flat interface becomes faceted. The wavelength of these oscillations decreases with increasing supercooling. Such simple waves can be observed in the marginal parts of the sample (see Fig. 1 to 4). These oscillations should be distinguished from the commonly observed growth striations connected with the temperature fluctuations at the crystal-melt interface, which are directly related to the crystal rotation and pulling rate. The more abrupt change in the slope of the solid-liquid interface may cause the abrupt change in probability for the formation of different Si-related centres. The shape of the solid-melt interface similar to that needed for the observed distribution of Si_{Ga} defects has been mentioned for the case of InP LEC grown crystals in [54] and an analogous distribution of free electron concentration in this material has been reported [64]. To our knowledge the consistent study of the relation between the growth conditions and the distribution of dopant controlled by the shape of the interface has not been performed as yet.

Independent information on the Si centres in GaAs lattice was obtained from local vibrational mode (LVM) measurements. Absorption spectra revealed unambiguously that in the central part of the samples corresponding to high electron concentration also the concentration of Si_{Ga} centres is more than twice as large as that in the marginal part [42].

4. Conclusions

The luminescence topography of Si-doped LEC GaAs crystals revealed inhomogeneities in the band edge as well as extrinsic luminescence intensity. Three complex bands at 1.35 eV (A), 1.20 eV (B), and 1.01 to 1.06 eV (C) were found in low-temperature luminescence spectra. The position and halfwidth of the edge emission band and coupled LO-phonon–plasmon modes were used for the determination of free carrier concentrations.

For high carrier concentrations the luminescence quenching was observed for all bands, but with different efficiency. Auger recombination was suggested as the most probable nonradiative recombination channel with sufficient universality to explain all observed phenomena. The apparent quenching starts at the concentration 3.0×10^{18} cm^{-3}.

The tilting of the solid–liquid interface with respect to the low index planes during the growth process is responsible for the way of incorporation of Si dopant into GaAs lattice and oscillations in impurity concentration. The detailed correspondence between the exact form of the oscillations and the corresponding theory is still lacking.

New features on the temperature dependence of the Si-related band at 1.35 eV has been established and the similarity to the EL2 centres pointed out. The origin of this band has been ascribed to the presence of (Si–As_{Ga}) complexes. The interpretation of the band at 1.2 eV by (Si–V_{Ga}) complexes was confirmed.

Acknowledgements

The authors are indebted to K. Jurek for the microprobe measurements and to V. Pourová and M. Šimečková for technical assistance.

References

[1] E. W. Williams and D. M. Blacknall, Trans. TMS-AIME **239**, 387 (1967).
[2] E. W. Williams and H. B. Bebb, Semiconductors and Semimetals, Vol. 8, Chap. 5, Academic Press, 1972.
[3] L. Samuelson, P. Omling, H. Titze, and H. G. Grimmeiss, J. Crystal Growth **55**, 164 (1981).
[4] V. F. Kovalenko and A. V. Prokhorovich, Fiz. Tekh. Poluprov. **16**, 2064 (1982).
[5] K. Okamoto, S. Onozawa, and T. Imai, J. appl. Phys. **56**, 2993 (1984).
[6] Y. M. Chu, D. B. Darby, and G. R. Booker, Proc. 2nd Oxford Conf. Microscopy of Semiconducting Materials, Inst. Phys. Conf. Ser. **60**, 331 (1981).
[7] W. M. Theis and W. G. Spitzer, J. appl. Phys. **56**, 890 (1984).
[8] J. Woodhead, P. C. Newman, A. K. Tipping, J. B. Clegg, J. A. Roberts, and I. Gale, J. Phys. D **18**, 1575 (1985).
[9] W. I. Wang, E. E. Mendez, Y. Iye, and G. E. Stillman, J. appl. Phys. **60**, 1834 (1986).

[10] J. Maguire, R. Murray, and R. C. Newman, Appl. Phys. Letters 50, 516 (1987).
[11] T. Hiramoto, Y. Mochizuki, T. Saito, and T. Ikoma, Japan. J. appl. Phys. 24, L921 (1985).
[12] T. Hiramoto, Y. Mochizuki, and T. Ikoma, Japan. J. appl. Phys. 25, L830 (1986).
[13] T. Takamori, T. Fukunaga, J. Kobayashi, K. Ishida, and H. Nakashima, Japan. J. appl. Phys. 26, 1097 (1987).
[14] T. Takamori, T. Fukunaga, and H. Nakashima, Japan. J. appl. Phys. 26, L920 (1987).
[15] N. Pau, B. Lee, S. S. Bose, M. H. Kim, J. S. Hughes, G. E. Stillman, K. Arai, and Y. Nashimoto, Appl. Phys. Letters 50, 1832 (1987).
[16] R. A. Morrow, J. appl. Phys. 62, 3671 (1987).
[17] J. van de Ven, W. J. A. M. Hartman, and L. J. Giling, J. appl. Phys. 60, 3735 (1986).
[18] M. Cocito, P. Franzosi, G. Salviati, and F. Taiariol, Scanning Microscopy, Part IV, SEM Inc./AMF O'Hare, Chicago (Ill.) 1986 (pp. 1299 to 1310).
[19] B. J. Skromme, S. S. Bose, B. Lee, T. S. Low, T. R. Lepkowski, R. Y. DeJule, G. E. Stillman, and J. C. M. Hwang, J. appl. Phys. 58, 4685 (1985).
[20] T. Nakamura and T. Katoda, J. appl. Phys. 57, 1084 (1985).
[21] T. Hyuga, K. Watanabe, J. Osaka, and K. Hoshikawa, Appl. Phys. Letters 48, 1742 (1986).
[22] J. Kobayashi, M. Nakajima, Y. Bamaba, T. Fukunaga, K. Matsui, K. Ishida, and H. Nakashima, Japan. J. appl. Phys. 25, L385 (1986).
[23] T. Narusawa, Y. Uchida, K. L. I. Kobayashi, T. Ohta, M. Nakajima, and H. Nakashima, Proc. 11th. Internat. Symp. GaAs and Related Compounds, Inst. Phys., Bristol/London 1984 (p. 127).
[24] J. Pastrňák, Czech. J. Phys. B 37, 933 (1987).
[25] J. De-Sheng, Y. Makito, K. Ploog, and H. J. Queisser, J. appl. Phys. 53, 999 (1982).
[26] H. J. Queisser, J. appl. Phys. 37, 2909 (1966).
[27] H. Kressel, J. U. Dunse, H. Nelson, and F. Z. Hawrylo, J. appl. Phys. 39, 2006 (1968).
[28] M. Jeong, J. Shirafuji, and Y. Inuichi, Japan. J. appl. Phys. 12, 109 (1973).
[29] T. Itoh and M. Takeuchi, Japan. J. appl. Phys. 16, 227 (1977).
[30] V. Svaminathan, N. E. Shumaker, and J. L. Zilko, J. Lum. 22, 153 (1981).
[31] G. S. Pomrenke, Y. S. Park, R. L. Hengehold, and M. T. Machin, Physica B 116, 414 (1983).
[32] N. Matsunaga, S. Takahashi, and N. Hashimoto, GaAs IC Symp., Tech. Digest, Boston 1984 (p. 159).
[33] D. C. Reynolds, J. appl. Phys. 53, 1263 (1982).
[34] W. I. Wang, R. F. Mark, and L. Vina, J. appl. Phys. 59, 937 (1986).
[35] S. Subbana, H. Kroemer, and J. L. Merz, J. appl. Phys. 59, 488 (1986).
[36] T. Fukunaga, T. Takamori, and H. Nakashima, Proc. 4th Internat. Conf. Molecular Beam Epitaxy, York, 1986, J. Crystal. Growth 81, 85 (1987).
[37] T. Fukunaga and H. Nakashima, Japan. J. appl. Phys. 25, L856 (1986).
[38] J. A. Mareš, J. Oswald, and J. Pastrňák, Mater. Sci. Forum 10/12, 1235 (1986).
[39] T. Oh-hori, H. Itoh, H. Tanaka, K. Kasai, M. Takikawa, and J. Komen, J. appl. Phys. 61, 4603 (1987).
[40] E. M. Williams, Phys. Rev. 168, 922 (1968).
[41] see e.g. H. Künzel and K. Ploog, Appl. Phys. Letters 37, 416 (1980).
[42] M. Zavětová, J. Pastrňák, and J. Oswald, to be published.
[43] Y. P. Varshni, Physica 34, 149 (1967).
[44] M. B. Panish and H. C. Casey, J. appl. Phys. 40, 163 (1969).
[45] P. W. Yu, D. C. Look, and W. Ford, J. appl. Phys. 62, 2960 (1987).
[46] P. T. Landsberg and M. J. Adams, J. Lum. 7, 3 (1973).
[47] G. Abstreiter, in Light Scattering in Solids IV, Ed. M. Cardona and G. Güntherodt, Topics in Applied Phys., Vol. 54, Springer-Verlag 1984.
[48] G. Irmer, J. Monecke, and W. Siegel, Cryst. Res. Technol. 20, 1125 (1985).
[49] J. S. Blakemore, J. appl. Phys. 53, R123 (1982).
[50] D. E. Aspnes and A. A. Studna, Phys. Rev. B 27, 985 (1983).
[51] T. Nishinaga, Proc. 2nd Internat. School, Cetnievo, May 6 to 13, 1978, Ed. M. A. Herman, PWN-Polish Sci. Publ. Warszawa 1980 (p. 131).
[52] R. W. Jansen and O. F. Sankey, Solid State Commun. 64, 1 (1987).

[53] See e.g. G. A. BARAFF and M. SCHLÜTER, Mater. Sci. Forum **10/12**, 293 (1986).
[54] O. F. SANKEY and J. D. DOW, Appl. Phys. Letters **38**, 685 (1981).
[55] A. N. MOROZOV, V. T. BUBLIK, and O. YU. MOROZOVA, Cryst. Res. Technol. **21**, 859 (1986).
[56] T. NAKAMURA and T. KATODA, J. appl. Phys. **57**, 1084 (1985).
[57] See e.g. A. G. MILNES, Adv. Electronics and Electron Phys. **61**, 63 (1983).
[58] W. RICHTER, in: Springer Tracts in Modern Physics, Vol. 78, Springer-Verlag 1976 (p. 121).
[59] F. ORITO, H. OKADA, M. NAKAJIMA, T. FUKUDA, and T. KAJIMURA, J. Electr. Mater. **15**, 87 (1985).
[60] K. D. GLINCHUK, V. I. GUROSHEV, and A. V. PROKHOROVICH, Cryst. Res. Technol. **22**, 969 (1987).
[61] K. D. GLINCHUK and A. V. PROKHOROVICH, phys. stat. sol. (a) **44**, 777 (1977).
[62] K. D. GLINCHUK, A. V. PROKHOROVICH, and N. S. ZAYATS, phys. stat. sol. (a) **84**, 567 (1984).
[63] K. D. GLINCHUK, A. V. PROKHOROVICH, and N. S. ZAYATS, phys. stat. sol. (a) **82**, 503 (1984).
[64] D. WRUCK and A. KNAUER, phys. stat. sol. (a) **107**, 321 (1988).
[65] J. KUNG and W. G. SPITZER, J. appl. Phys. **45**, 4477 (1974).
[66] A. LIDOW and J. W. GIBBONS, J. appl. Phys. **49**, 5213 (1978).
[67] R. T. CHEN and W. G. SPITZER, J. Electron. Mater. **10**, 1085 (1981).
[68] H. SCIBIOR, I. BRYLOWSKA, P. MAZUREK, and M. SUBOTOWICZ, Acta phys. polonica **A73**, 307 (1988).

(Received September 19, 1988)

phys. stat. sol. (a) **111**, 361 (1989)

Subject classification: 78.70; 71.25; S10

Technikum Suhl der Technischen Hochschule Ilmenau, Suhl[1]) *(a) and*
Sektion Physik der Martin-Luther-Universität Halle-Wittenberg, Halle (Saale)[2]) *(b)*

X-Ray Spectroscopic Investigations of Titanium Oxides and Basic Materials for Ferroelectric Ceramics

By

E. Schulz (a), G. Dräger (b), W. Czolbe (b), and O. Brümmer (b)

The Ti K valence band emission spectra (VBXES) and the Ti K near edge absorption spectra (XANES) of some polycrystalline titanium oxides and perovskites, including also the polytitanate $Ba_2Ti_9O_{20}$, are investigated. Characteristic relations between the spectra and the structure and composition of the materials are found. They can be used for fundamental research of the electronic structure but also for analytical applications at working with these compounds which are basic materials for ferroelectric and semiconducting ceramics.

Die Ti K-Valenzbandemission (VBXES) und die kantennahen Ti K-Absorptionsspektren (XANES) von einigen polykristallinen Titanoxiden und -perowskiten, einschließlich auch des Polytitanats $Ba_2Ti_9O_{20}$, werden untersucht. Charakteristische Zusammenhänge zwischen den Spektren und der Struktur und Zusammensetzung der Materialien werden gefunden. Sie können für Grundlagenuntersuchungen zur elektronischen Struktur genutzt werden, aber auch für analytische Zwecke bei der Arbeit mit diesen Verbindungen, welche Basismaterialien für ferroelektrische und halbleitende Keramiken darstellen.

1. Introduction

Ceramic materials enter permanently new fields of practical application. For a better understanding of their fundamental properties basic research on these materials is necessary. The objects of our investigations are some polycrystalline titanium oxides and perovskites serving as basic materials for ferroelectric and semiconducting ceramics. In particular, these compounds are TiO, Ti_2O_3, TiO_2 (rutile), $BaTiO_3$, $SrTiO_3$, $Ba_{0.7}Sr_{0.2}Ca_{0.1}TiO_3$, $BaTi_{1.04}O_3$, and also the polytitanate $Ba_2Ti_9O_{20}$.

In all of these compounds the Ti atoms are coordinated by six nearest oxygen atoms in an ideal or distorted octahedron. Greater differences exist in the arrangement and in the species of the next nearest and further neighbour atoms. These differences cause also a different electronic behaviour.

Some remarkable properties of these compounds are the existence of a ferroelectric phase and the so-called PTCR-effect (positive temperature coefficient of electric resistance [1]) in some of these materials. It is of theoretical and practical interest to explain such phenomena and to make proposals for realizing these and other effects in practice. An important step in this direction is the investigation of the crystal structure and of the electronic properties. Different methods are used for this purpose which complement each other. In this paper some results will be obtained by means of the high resolving X-ray emission and absorption spectroscopy.

---

[1]) PSF 142, DDR-6000 Suhl, GDR.

[2]) PSF, DDR-4010 Halle (Saale), GDR.

2. Method

The method is based on the measurement of the valence band X-ray emission spectra (VBXES) and of the X-ray absorption near edge structure (XANES) of polycrystalline materials. In our paper we will concentrate on the Ti K spectra.

For the VBXES the intensity $I(\omega)$ of the emitted radiation with frequency ω can be written in the form

$$I(\omega) \sim \omega^3 M_{s,p}^2(E)\, N_p(E)\, \delta(E - E_K - \hbar\omega)\,, \tag{1}$$

where E denotes the binding energy of the valence states and E_K that one of the K-core level, respectively. The formula means that the intensity $I(\omega)$ provides a reproduction of the local partial density of states $N_p(E)$ of the valence band states with p-symmetry modulated by the radial transition probability $M_{s,p}^2(E)$. The latter changes slowly for emission spectra in the energy range of interest and produces no additional structures of the spectra. For the XANES the absorption coefficient $\mu(\omega)$ can be written in a form similar to (1),

$$\mu(\omega) \sim \omega M_{s,p}^2(E)\, N_p(E)\, \delta(E - E_K - \hbar\omega)\,. \tag{2}$$

But there exist some differences between the emission and absorption spectra. First of all the latter ones reflect the local partial density of unoccupied states in the environment of the absorbing atom. Furthermore, because of the missing K-electron, this density of states need not agree with that of the ground state band structure. There exists no simple rule to estimate the influence of the remaining core hole on the density of empty states. In some cases also the so-called core excitons are discussed which can be found experimentally below the conduction band.

Secondly in the absorption spectra the energy dependence of the transition probability cannot be neglected in general. Only for the first 10 to 15 eV above the absorption threshold the fine structure can be interpreted unambiguously in terms of the local density of states. At higher energies the influence of the transition probability increases and the interpretation of the experimental spectra on the basis of (2) is no longer correct. On the other hand, the absorption spectra depend on the final state wave functions and, therefore, they contain information on the arrangement of the atoms in the sample material. So the extended X-ray absorption fine structure (EXAFS) enables to investigate the local environments of the absorbing atoms by using special mathematical methods. From the XANES region further information can be obtained such as bond angles and the symmetry of the arrangement of the adjoining atoms. To get this information it is necessary either to compare the spectra with those of well-known substances or to interpret them by means of theoretical spectra calculated for model substances.

3. Experimental

The spectra were measured with a Johann-type spectrometer [2] suitable now for recording VBXES and XANES. An X-ray tube with a chromium anode (40 kV, 45 mA) and a tungsten tube (7.4 keV, 35 mA) served as radiation sources for the excitation of the emission spectra and for the absorption spectra, respectively. The radiation was analyzed with a quartz crystal $(10\bar{1}1)$ in the second order of reflection. The spectrometer operated in a vacuum vessel. The absorption spectra were taken in an air atmosphere of about 13 kPa causing a sufficient discrimination of the continuous radiation obtained from the first-order reflection. The samples for the emission experiments were pellets of polycrystalline material, pressed and partly sintered. For the XANES measurements thin absorber foils were prepared by spreading fine powder materials onto adhesive tape. The spectra recorded on X-ray film were transformed into digital intensities and further treated by means of a photometer coupled with a control and computer unit. For better comparability all emission and absorption spectra were normalized in the same way. This means for the VBXES that all spectra were fitted on the high-energy tail of the normalized Ti K$\beta_{1,3}$ peak which appears at the low energy side of the Ti K$\beta_{2,5}$ valence band spectra. On the other hand, all XANES data were normalized to the same absorption step which is the difference in the average absorption over a wide range of energies below and above the absorption edge.

4. Results and Discussion

In Fig. 1 the measured Ti K VBXES and XANES data are presented in a common energy scale.

The low-energy satellite Ti Kβ'' represents a tightly bound subband with mainly O 2s character and some content of Ti 4p states. In terms of the MO theory these levels have t_{1u} symmetry [3]. For the compounds with Ti^{4+} ions and having a Ti–O distance of 0.2 nm and smaller the mixing of Ti 4p states into this band is considerable (for example in $BaTiO_3$, TiO_2, and $Ba_2Ti_9O_{20}$). In the spectra of the other compounds with a lower oxidation state of Ti the Ti Kβ'' satellite is not clearly visible indicating that the O 2s–Ti 4p mixing is only weak.

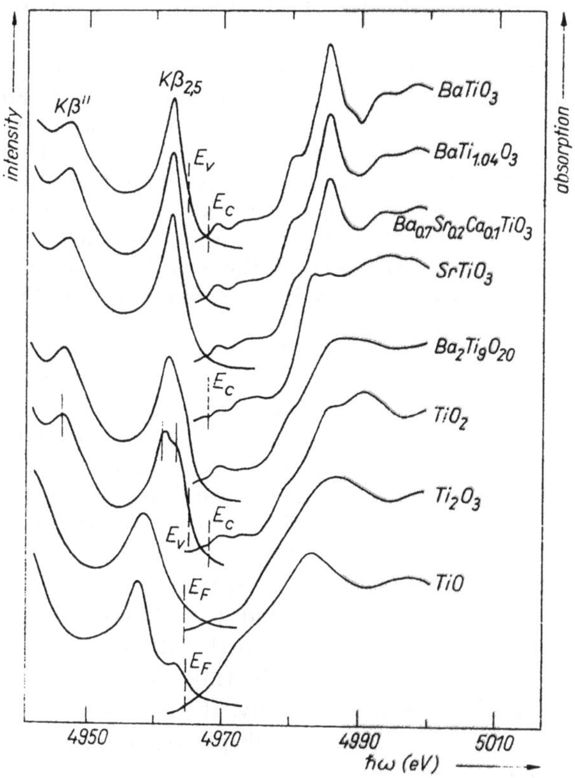

Fig. 1. The Ti K valence band emission (VBXES) and the X-ray absorption near edge structure (XANES) of some titanium oxides and titanates

The Ti $K\beta_{2,5}$ band represents the states with most Ti p character within the valence band. The peak shifts to higher energies when the ionic charge increases from Ti^{2+} to Ti^{4+}. This chemical shift is caused by the reduced screening charge around the Ti core orbitals. The small peak and the shoulder at the top of the Ti $K\beta_{2,5}$ emission of TiO indicates a strong p–d mixing as a result of destroying the inversion symmetry of the Ti lattice sites at higher oxygen defect concentration. For some compounds the experimental VBXES data were fitted to theoretical band calculations [4 to 7]. The positions of the valence and conduction band edges E_v and E_c, respectively, and of the Fermi level E_F found by this procedure are marked in the spectra of Fig. 1. They are in realistic positions confirming by this the results of the band calculations.

Within the first conduction band states extending over a few eV the Ti 3d character dominates. Therefore, the absorption near E_c or E_F is weak. Under the influence of the crystal field the Ti 3d bands are split. In the case of TiO_2 and the perovskites the resulting higher t_{2g} and lower e_g bands corresponding to a nearly octahedral crystal field can be clearly distinguished. They are reproduced in the spectra by quadrupole transitions and/or dipole transitions where the latter ones can result from a p–d mixing by hybridization. Additionally, for these compounds a weakly pronounced shoulder nearly 1 eV below E_c is visible. According to Balzarotti et al. [8] and Grunes [9] it may be a result of the so-called core excitons.

The less ionic compounds Ti_2O_3 and TiO have no clearly resolved fine structures at the beginning of absorption. This fact can be qualitatively explained by the smaller band splitting due to the lower bonding ionicity and the greater atomic distances. Moreover, the Ti d-like bands are filled with one or two additional electrons, which will reduce the absorption strength because of the decreased number of free electron states in this region.

The absorption structure at higher energies is very different even for substances with similar electronic structure and with similar first neighbour coordination. For example, the pronounced absorption peak at 4985.5 eV exists only for the Ba perovskites and may result from the existence of Ba 4f states at this conduction band energy [10]. It must be concluded that the structures in the energy region from 10 to 40 eV above the absorption threshold are strongly affected by the species, number, and arrangement of atoms in the second and possibly further coordination shells. The same result was obtained from theoretical calculations: only the complete consideration of the second shell atoms makes it possible to reproduce the experimental Ti K absorption spectra of TiO theoretically (Kutzler and Ellis [11]). Another question is, how do little variations of the composition influence the X-ray spectra. Because of the wealth of resolved structures in the XANES region this energy range seems to be suitable for giving an answer to the question. An additional Ti content of 4% in $BaTiO_3$, for example, causes a significant decrease of the peak at 4985.5 eV and a smearing of the absorption valleys at 4981 and 4990 eV. There is a similar effect of the substitution of Ba atoms by Sr and Ca atoms (Fig. 1). A quantitative analysis has shown that the normalized Ti K XANES of $Ba_{0.7}Sr_{0.2}Ca_{0.1}TiO_3$ can be reproduced in a good approximation by a weighted superposition of the normalized spectra of $BaTiO_3$, $SrTiO_3$, and $CaTiO_3$. If we suppose an ideal random distribution of the Ba, Ca, and Si atoms on their lattice sites in this mixed compound, the additivity of the spectra confirms the following two assumptions: Firstly, the little differences in the crystal structure ($BaTiO_3$, $CaTiO_3$, and $SrTiO_3$ crystallize in three different space groups) have no significant influence on the spectra. Secondly, the final state wave function of the excited electron is not strongly affected by multiple scattering processes between two atoms of the group Ba, Ca, Sr.

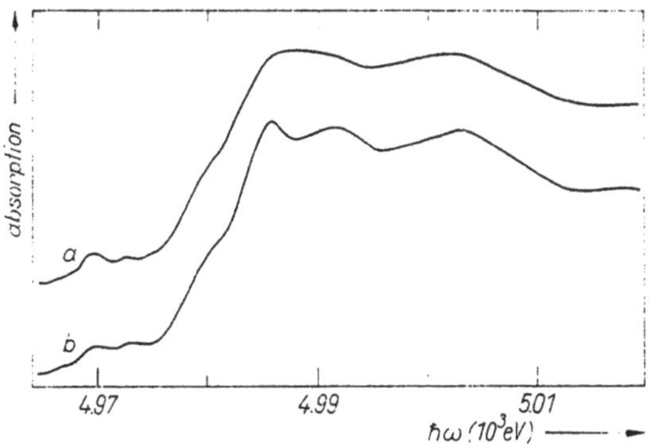

Fig. 2. The measured Ti K XANES of $Ba_2Ti_9O_{20}$ (a) in comparison to the weighted average of the experimental $BaTiO_3$ and TiO_2 (rutile) XANES (b)

In Fig. 1 there are also seen the spectra of the polytitanate $Ba_2Ti_9O_{20}$. The similarities of the Ti VBXES and of the t_{2g} and e_g pre-edge absorption structures to those of TiO_2 (rutile) and the perovskites point to corresponding similarities of the electronic structure and therefore also of the Ti^{4+} coordination. The conclusion is obvious that the first neighbours of the Ti atoms are also oxygen atoms in more or less distorted octahedra. But what can be concluded about the arrangement of the more distant neighbour atoms?

In Fig. 2 the XANES of the polytitanate $Ba_2Ti_9O_{20}$ is compared with that of a weighted average of the TiO_2 and $BaTiO_3$ XANES in the ratio 7 : 2. This corresponds to a formal composition of $Ba_2Ti_9O_{20}$ from seven parts of TiO_2 and two parts of $BaTiO_3$. Especially in the energy range from 4985 to 4995 eV there are some pronounced differences in the absorption structures, lying beyond the error bars. They should be the result of a modified configuration of the next nearest neighbours of Ti, which are the Ba and other Ti atoms [12]. In this way from the experimental Ti K XANES follows that the considered polytitanate will not be a simple mixing of $BaTiO_3$ and TiO_2 phases but represents a new phase. For getting more information about this it should be successful to compare the experimental spectra with the results of XANES model calculations at clusters containing the first and several different second coordination shells around the Ti atoms. By this not only EXAFS but also XANES can contribute to investigate complicated crystal structures or other atomic arrangements.

References

[1] J. Daniels, K. H. Härdtl, and R. Wernicke, Philips' tech. Rdsch. **38**, 1 (1979).
[2] G. Dräger and O. Brümmer, phys. stat. sol. (b) **98**, K33 (1980).
[3] D. W. Fischer, J. appl. Phys. **41**, 3561 (1970).
[4] S. Munnix and M. Schmeits, Phys. Rev. B **30**, 2202 (1984).
[5] J. Ashkenazi and T. Chuchem, Phil. Mag. **32**, 763 (1975).
[6] K. Schwarz and A. Neckel, Phys. Rev. B **19**, 5439 (1979).
[7] P. Pertosa and F. M. Michel-Calendini, Phys. Rev. B **17**, 2011 (1978).
[8] A. Balzarotti, F. Comin, L. Incoccia, M. Piacentini, S. Mobilio, and A. Savoia, Solid State Commun. **35**, 145 (1980).
[9] L. A. Grunes, Phys. Rev. B **27**, 2111 (1983).
[10] V. V. Nemoshkalenko and A. N. Timoshevskii, phys. stat. sol. (b) **127**, 163 (1985).
[11] F. W. Kutzler and D. E. Ellis, Phys. Rev. B **29**, 6890 (1984).
[12] E. Tillmanns and W. Hofmeister, J. Amer. Ceram. Soc. **66**, 268 (1983).

(Received September 13, 1988)

phys. stat. sol. (a) **111**, 367 (1989)

Subject classification: 72.70 and 73.40; 78.60; S7.15

Technical University, Brno[1]) *(a)*
and Radio Research Institute, Prague (b)

The Noise of the GaAlAs Super-LED Radiation

By

J. ŠIKULA (a), P. VAŠINA (a), and I. HÜTTEL (b)

The noise is studied of the radiation emitted by GaAlAs super-LEDs. At frequencies higher than 1 kHz only the shot noise is observed in the super-LED radiation, its spectral density being close to the theoretical value. At lower frequencies (about 100 Hz) the radiation noise spectral density S_u is of the $1/f$ type and varies as the square of the diode current, while at still lower frequencies $S_u \sim I^m$ holds, where $m > 2$. As a source of the observed noise parasitic laser oscillations are proposed which is confirmed by the shape of the spectral distribution of radiation.

Предметом работы является изучение шума излучения супер-светодиодов из GaAlAs. При частотах выше 1 kHz излучение содержит только шум Шоттки, спектральная плотность которого близка теоретическому значению. При частотах ниже 100 Hz излучение супер-светодиода содержит шум типа $1/f$ и его спектральная плотность пропорциональна квадрату тока. При очень низких частотах $S_u \sim I^m$, где $m > 2$. Мы предлагаем, что источником этого шума являются паразитные осцилляции лазерного типа, что соответствует экспериментальным данным.

1. Introduction

Among the radiation sources for optical communications the LED (light emitting diode), the LD (laser diode), and the SLD (super-LED) are the most widely used. They all make use of the GaAlAs/GaAs or GaInAsP/InP double hetero-structure, which features relatively low threshold current, i.e., a current value at which the population inversion in the active region takes place.

The optical gain is in this case higher than unity. In contrast to the laser diode, an SLD has no optical feedback. Its structure does not contain any optical resonator. The SLD generates incoherent radiation, which is further amplified by the stimulated emission in the active region. The output radiation power of the SLD is fairly high and can be compared to that of an LD, its spectral line and angular distributions being relatively narrow. The response to a current impulse is near that of the LD. Compared to the LDs there are less problems with the temperature dependence of the operational characteristics, lifetime and reliability, as well as the parasitic oscillations. In the present paper we suggest a noise diagnostics as an approach to a further technology improvement.

2. The Sample Technology

The SLD structure is produced in the same way as that of the LD. However, the optical feedback is rejected, for instance, by coating the Fabry-Perot mirrors by a suitable layer (see ref. [1] to [3]).

[1]) Barvičova 85, CS-662 37 Brno, Czechoslovakia.

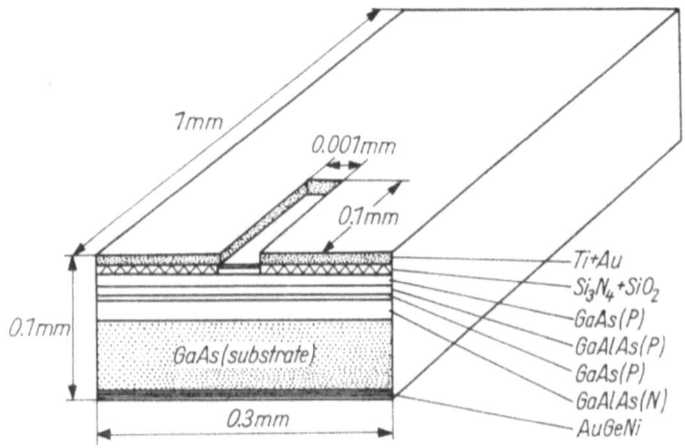

Fig. 1. Cross-section of a GaAlAs super-LED

A simplified cross-sectional view of a GaAlAs super-LED is shown in Fig. 1.

3. The Photoluminescent Characteristics

As photoluminescent characteristic we take the P–I curve, which represents the emitted radiation power P_0 versus the diode current I. A typical P_0–I curve for one of the samples is illustrated in Fig. 2. It is seen that that the curve is not linear, approximately following a square law. This is a consequence of the growth of the optical gain in the active region when the current is increasing.

Fig. 3 shows the radiation power versus wavelength plot for various values of the

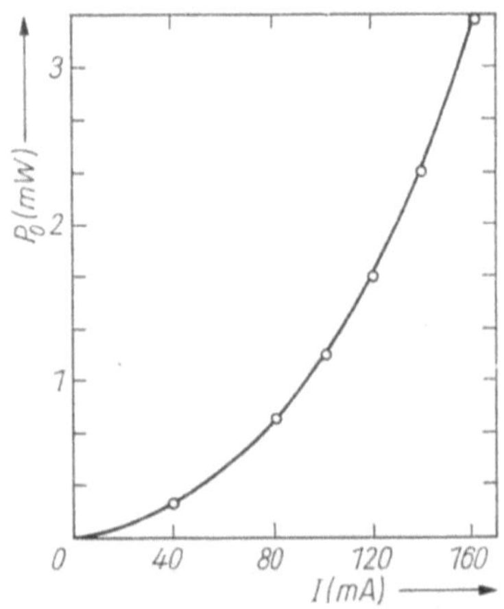

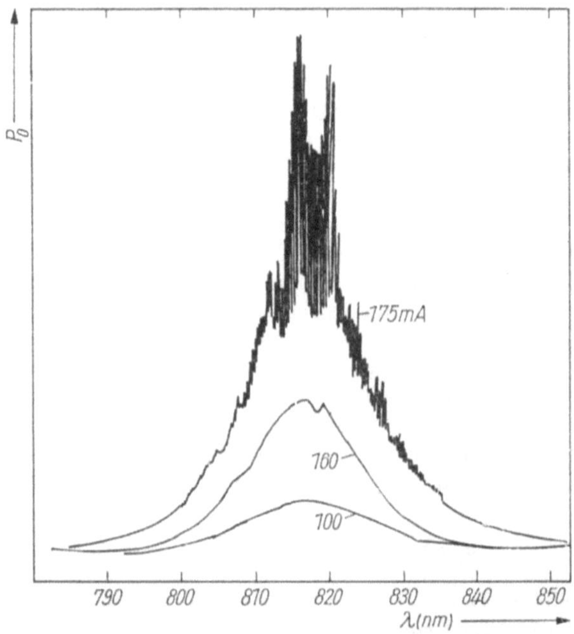

Fig. 2 Fig. 3

Fig. 2. Radiation power vs. current plot for a 1156 SLD

Fig. 3. Optical output vs. wavelength for various currents of the SLD diode current. As the current grows the influence of the back side reflection becomes more pronounced. This is evidenced by parasitic laser oscillations at higher currents.

4. The Noise Spectra

In Fig. 4 the spectral density S_I of the 4B PIN photodiode output current fluctuations as a function of frequency is drawn. In the calculations of S_I the noise background has not been allowed for. From the figure it is seen that above a frequency of about 1 kHz the spectral density S_I corresponds to that of the shot noise, $S_I = 9.6 \times \times 10^{-25}$ A^2 s for $I = 3$ µA. At frequencies lower than 1 kHz the diagram splits into two curves. Lower values (curve a) of S_I are due to an incandescent lamp illumination, the $1/f$ noise appearing below 10 Hz. When the PIN photodiode is illuminated by the SLD radiation the spectral density S_I is higher (curve b) and the $1/f$ noise appears below about 100 Hz. In both cases the photodiode dc current was equal to 3 µA.

Fig. 5 shows the plots of the 4B PIN photodiode output current spectral density versus the SLD current at various band centre frequencies; i.e. 1 kHz, 80 Hz. The noise background was not taken into account when calculating the spectral density S_I; the relative noise bandwith was 2%.

For extremely low values of the photodiode current the spectral density S_I is the same for all the frequencies and equal to that of the shot noise. When increasing the current the individual curves corresponding to various frequencies branch out and the highest slope pertains to the lowest frequency.

The spectral density of the SLD radiation noise consists of two components, the shot noise and the $1/f$ noise. The former thas the spectral density proportional to the current, while the spectral density of the latter varies as a power of the current higher than two. Consequently, the cut-off frequency, that is, the frequency at which the $1/f$ noise spectral density is just equal to that of the shot noise depends on the current.

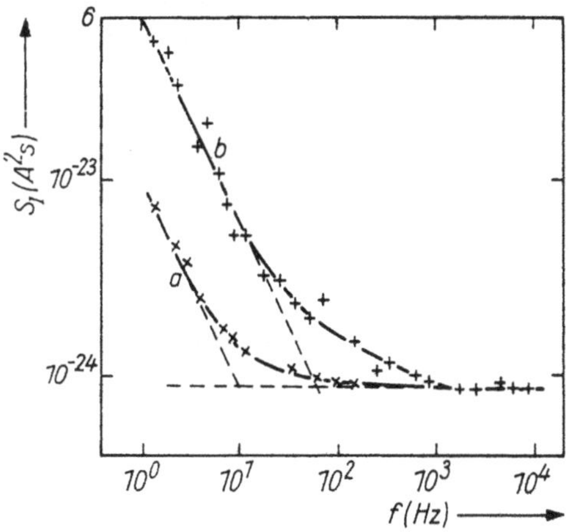

Fig. 4

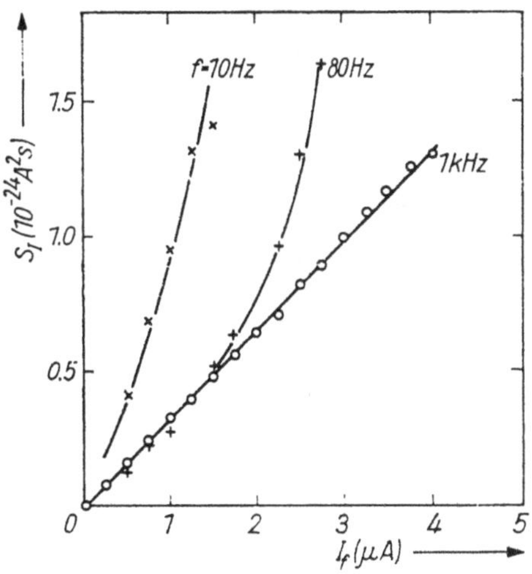

Fig. 5

Fig. 4. Spectral density of the 4B PIN photodiode output current fluctuation vs. frequency (see text)

Fig. 5. Spectral density of the 4B PIN diode output current fluctuations as a function of the 1156 SLD current

For the shot noise it holds

$$S_{Is} = 2eI\,,$$

where e is the elementary charge, I the current, while for the $1/f$ noise we assume

$$S_{If} = \frac{kI^a}{f}\,,$$

with exponent $a = 5/2$ and the coefficient k depends on the particular diode. The cut-off frequency of the composite curve (Fig. 4) is

$$f_c = \frac{k}{2e}\,I^{1.5}\,.$$

From the experimental results for the SLD 1156 we obtain $k = 10^{-8}\ A^{-1/2}$, so that

$$f_c = 30I^{1.5}$$

or

$$I = \left(\frac{f}{30}\right)^{2/3},$$

where I is in μA.

These equations bind together the current and the cut-off frequency. To give evidence for this approach, we take $f_c = 10$ and 80 Hz. From the above equation we obtain $I = 0.480$ and 1.923 μA, respectively, which is in a good agreement with Fig. 5.

5. Conclusions

A super-LED features higher efficiency than standard incoherent light emitting diodes. On the other hand, in the low-frequency region its radiation contains an excess noise of the $1/f$ type. We propose that this is due to random laser oscillations or mode interference. The noise analysis proved to be an efficient tool to get information about the process of light generation.

References

[1] J. Boeck and M. Amman, Frequenz 33, 278 (1979).
[2] T. Paoli, R. Burnham, R. Thornton, and D. L. Smith, Appl. Phys. Letters 47, 450 (1985).
[3] G. Eisenstein and L. W. Stulz, Appl. Optics 23, 161 (1984).

(Received August 5, 1988)

phys. stat. sol. (a) 111, 371 (1989)

Subject classification: 73.40 and 85

Chair of Electron Devices, Technical University of Budapest[1])

Approximate Calculation of the Switching Voltage of MISS

By

I. Zólomy

The dependence of the threshold voltage of MISS upon the doping concentration of the epilayer is discussed based upon a thyristor analogy. The threshold voltage reaches a maximum value at medium doping concentrations. At low doping concentrations the threshold voltage is limited by the punch-through voltage, at high doping concentrations by the breakdown voltage. At low doping concentrations the threshold voltage is very sensitive to the current amplification of the tunnel MIS part of the device and to two-dimensional effects.

Es wird die Abhängigkeit der Kippspannung von der Dotierungskonzentration mit Hilfe der Thyristoranalogie diskutiert. Die Kippspannung erreicht einen maximalen Wert bei mittleren Konzentrationen. Bei niedrigen Konzentrationen wird die Kippspannung von der Punch-Through-Spannung, bei höheren Konzentrationen von der Durchbruchspannung begrenzt. Bei niedrigen Konzentrationen ist die Kippspannung sehr empfindlich gegen die Stromverstärkung der Tunnel-MIS Diode und zweidimensionale Effekte.

1. Introduction

The MISS switch is a metal-(tunnel) oxide-n-layer-p-layer structure having a current controlled negative resistance, as shown in Fig. 1. The pn junction is forward biased and the p^+-layer injects many holes into the n-layer. Thus at the oxide-semiconductor interface these holes are collected forming an inversion layer. This leads to an increased field in the oxide, thus the hole and electron tunnel currents across the oxide will also increase. The electrons travel across the n-layer to the pn junction biasing it more into forward direction, thus a positive feedback loop is created.

A more detailed theory of this device can be found in [1] based upon earlier papers [2, 3]. The aim of this paper is to show the dependence of the threshold voltage upon the doping concentration of the n-layer and also to discuss the possibility of punch-through, taking into account the electron tunnel current and two-dimensional effects.

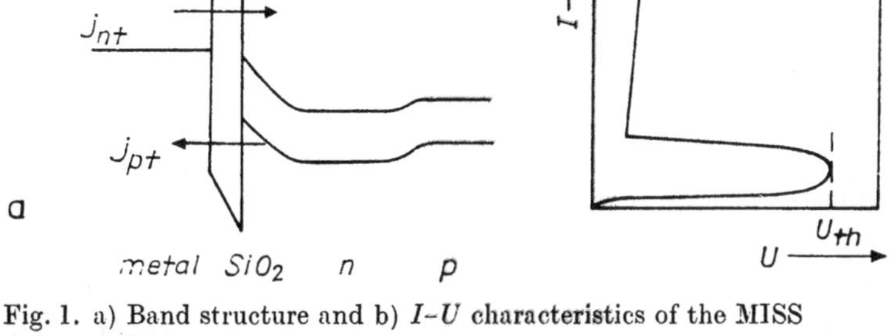

Fig. 1. a) Band structure and b) I–U characteristics of the MISS

[1]) H-1521 Budapest, Hungary.

2. Approximate Theory

The calculation is based upon the two-active device model of the MISS [4], consisting of a bipolar transistor and an amplifying tunnel MIS diode. The emitter of the transistor is the p$^+$-layer, the base is the n-layer, and the depletion layer at the semiconductor–oxide interface serves as collector. The common base current amplification of this transistor is α_{pnp}. The amplification factor of the MIS (metal–tunnel oxide–n-layer) diode, α_{MIS}, can be defined as the ratio of the electron (tunnel) current to the total current flowing through the MIS diode,

$$\alpha_{MIS} = \frac{i_{nt}}{i} . \tag{1}$$

The threshold voltage U_{th} can similarly be calculated as in the case of the thyristor [5, 6],

$$U_{th} = U_B(1 - \alpha_{pnp} - \alpha_{MIS})^{1/m} , \tag{2}$$

where U_B is the breakdown voltage of the MIS diode determined by the n-layer doping concentration and $m = 4$ for silicon. α_{pnp} is given as

$$\alpha_{pnp} = \gamma\beta , \tag{3}$$

where γ is the emitter efficiency and β the transport factor [7],

$$\beta = \frac{1}{\cosh\left(\dfrac{W_n - W_j - W_s}{L_p}\right)} \approx 1 - \frac{1}{2}\left(\frac{W_n - W_j - W_s}{L_p}\right)^2 . \tag{4}$$

W_n, W_s, and W_j are the widths of the n-layer, the surface, and the junction depletion layers, respectively. L_p is the diffusion length of holes.

At low doping concentrations the device is often in or near to the punch-through condition at the threshold voltage. In this case the neutral part of the n-layer is zero or very small, and the injected hole current is limited not by diffusion, but by the thermionic emission over the junction barrier. This can be taken into account with an additional term in the injected hole current, resulting [2] in

$$i_{ep} = \frac{qD_p n_i^2 \exp(U_j/U_T)}{N_d(W_n - W_j - W_s + 2\pi^{1/2}L_D)} , \tag{5}$$

where D_p, n_i, U_j, q, and N_d have the usual meanings, $U_T = kT/q$, and L_D is the Debye length in the n-layer.

The injected electron current is

$$i_{en} = \frac{qD_n n_i^2 \exp(U_j/U_T)}{L_n N_a} . \tag{6}$$

Using (5) and (6) the emitter efficiency can be expressed as

$$\gamma = \frac{i_{ep}}{i_{ep} + i_{en}} \approx 1 - \frac{D_n N_d}{D_p N_a} \frac{W_n - W_j - W_s + 2\pi^{1/2}L_D}{L_n} . \tag{7}$$

With (3), (4), and (7) α_{pnp} can be determined. Considering that both γ and β are near to unity, after some approximations, one gets

$$\alpha_{pnp} = 1 - \frac{D_n N_d}{D_p N_a} \frac{W_n - W_j - W_s + 2\pi^{1/2}L_D}{L_n} - \frac{1}{2}\left(\frac{W_n - W_j - W_s}{L_p}\right)^2 . \tag{8}$$

At threshold the external voltage differs very little from the voltage in the depletion layer of the oxide–semiconductor interface. Thus at homogeneous doping, with good approximation,

$$W_n - W_j - W_s = (W_n - W_j)\left[1 - \left(\frac{U}{U_{pt}}\right)^{1/2}\right], \tag{9}$$

where U_{pt} is the punch-through voltage, given by (17). Making use of (2), (8), and (9), the threshold voltage can be determined (with $U = U_{th}$) from the resulting equations

$$\left(\frac{U_{th}}{U_B}\right)^m = K_1\left[1 - \left(\frac{U_{th}}{U_{pt}}\right)^{1/2}\right]^2 + K_2\left[1 - \left(\frac{U_{th}}{U_{pt}}\right)^{1/2}\right] + K_3 - \alpha_{MIS}, \tag{10}$$

where

$$K_1 = \frac{1}{2}\left(\frac{W_n - W_j}{L_p}\right)^2, \tag{11}$$

$$K_2 = \frac{D_n N_d}{D_p N_a} \frac{W_n - W_j}{L_n}, \tag{12}$$

$$K_3 = \frac{2\pi^{1/2} D_n N_d L_D}{D_p N_a L_n}. \tag{13}$$

To determine U_{th} from (10) α_{MIS} must be known. From the more rigorous theory the value of α_{MIS} can be determined, its typical value at the threshold is very low, usually less than 10^{-4}. However, if α_{pnp} is reduced by, for example, two-dimensional effects, α_{MIS} is greater, because a higher value is necessary for switching. The value of α_{MIS} is higher at higher currents, thus the threshold current also increases.

At high doping concentrations $U_{pt} \gg U_B$ and $U_{th} < U_B$, thus $U_{th} \ll U_{pt}$ and (10) can be approximated as

$$U_{th} = U_B(K_1 + K_2 + K_3 - \alpha_{MIS})^{1/m}. \tag{14}$$

As $K_1 + K_2 + K_3 < 1$, U_{th} is lower than U_B, but proportional to it.

At low doping concentrations $U_B \gg U_{pt}$ and $U_{th} \leqq U_{pt}$, thus if the thermionic emission is neglected ($K_3 = 0$) and α_{MIS} is also neglected, then from (10) follows that U_{th} is always lower than U_{pt}, but approaches it, as N_d decreases. This was the result of Choi [6]. If $\alpha_{MIS} > K_3$, U_{th} is always smaller than U_{pt} and will not approach it. However, if $K_3 > \alpha_{MIS}$, punch-through will take place at a certain doping concentration; with $U_{th}/U_{pt} = 1$ from (10),

$$U_{th} = U_{pt} = U_B(K_3 - \alpha_{MIS})^{1/m}. \tag{15}$$

U_{pt} and U_B are functions of N_d and W_n, given in (16) and (17), thus the doping concentration can be determined, where the punch-through regime begins,

$$U_B[\mathrm{V}] = 60\left(\frac{N_d\,[\mathrm{cm}^{-3}]}{10^{16}}\right)^{-3/4}, \tag{16}$$

$$U_{pt} = qN_d\,\frac{(W_n - W_j)^2}{2\varepsilon_0\varepsilon_{Si}}. \tag{17}$$

The incorporation of two-dimensional effects strongly influences the threshold voltage. This was shown experimentally [8 to 10] and theoretically [3]. If the area of the tunnel oxide is smaller than the area of the pn junction, a certain part of the injected holes will not be collected by the surface depletion layer, thus the transport factor and therefore α_{pnp} will decrease by a factor

$$K_4 = \frac{A_{ox}}{A_{j,eff}} . \tag{18}$$

A_{ox} is the area of the tunnel oxide, and $A_{j,eff}$ is the effective forward biased area of the pn junction. Using the same procedure as the ones for obtaining (10), the following equation results for determination of U_{th}:

$$\left(\frac{U_{th}}{U_B}\right)^m = 1 - K_4\left\{1 - K_1\left[1 - \left(\frac{U_{th}}{U_{pt}}\right)^{1/2}\right]^2 - K_2\left[1 - \left(\frac{U_{th}}{U_{pt}}\right)^{1/2}\right] - K_3\right\} - \alpha_{MIS} . \tag{19}$$

3. Results

Fig. 2a shows the threshold voltage versus the donor concentration of the n-(epi)-layer, with α_{MIS} as parameter. The other parameters are: $W_n = 5\ \mu m$, $\tau_n = 10^{-6}$ s,

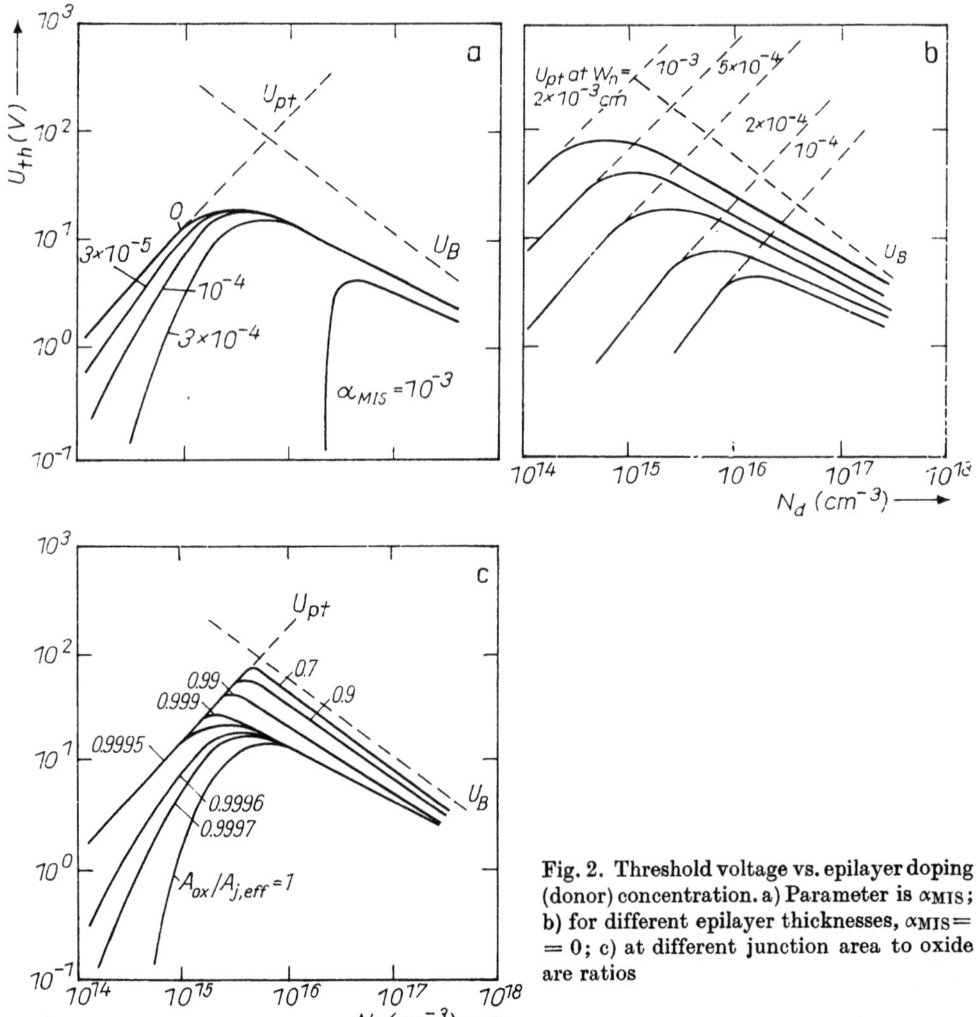

Fig. 2. Threshold voltage vs. epilayer doping (donor) concentration. a) Parameter is α_{MIS}; b) for different epilayer thicknesses, $\alpha_{MIS} = 0$; c) at different junction area to oxide are ratios

Table 1

Hole diffusion coefficient and lifetimes for calculating the threshold voltage

N_d (cm^{-3})	10^{11}	3×10^{14}	10^{15}	3×10^{15}	10^{16}	3×10^{16}
D_p (cm^2/s)	15.6	15.6	14.5	14	12.5	11
τ_p (s)	3×10^{-5}	2×10^{-5}	10^{-5}	7×10^{-6}	5×10^{-6}	3×10^{-6}
N_d (cm^{-3})	10^{17}	3×10^{17}				
D_p (cm^2/s)	9.1	7.8				
τ_p (s)	1.8×10^{-6}	10^{-6}				

$N_a = 10^{18}$ cm^{-3}, $D_n = 9$ cm^2/s. The hole lifetimes and diffusion coefficients which were used to calculate U_{th} are given in Table 1.

The highest threshold voltage is at $\alpha_{MIS} = 0$. As α_{MIS} increases, U_{th} decreases. U_{th} is much more sensitive to the increase of α_{MIS} at low doping concentrations than at high ones.

Fig. 2b shows U_{th} versus N_d at different epilayer thicknesses, if $\alpha_{MIS} = 0$. U_{th} reaches a maximum value at medium doping concentrations, and this value increases with increasing epilayer thickness.

Fig. 2c shows the effect of different oxide and effective junction areas, with K_4 as parameter, at $\alpha_{MIS} = 5 \times 10^{-4}$. As the two-dimensional effect starts (K_4 is less than unity), the threshold voltage increases. At low doping concentrations this increase is very rapid, and the device very soon reaches the punch-through. At high doping concentrations an appreciable increase in U_{th} takes place only if K_4 is decreased more drastically. That is, in the punch-through region the device is much more sensitive to two-dimensional effects than in the avalanche region.

References

[1] I. Zólomy, phys. stat. sol. (a) **100, 693 (1987)**.

[2] S. E. D. Habib and J. G. Simmons, Solid State Electronics **22**, 181 (1979).

[3] I. Zólomy, Solid State Electronics **26, 643** (1983).

[4] A. Adán and I. Zólomy, Solid State Electronics **23**, 449 (1980).

[5] I. Zólomy and A. Adán, Periodica Polytechnica Electron. Engng. **23**, 267 (1979).

[6] W. K. Choi, A. E. Owen, J. J. Delima, and S. M. Gage, Electronics Letters **23**, 1 (1987).

[7] S. M. Sze, Physics of Semiconductor Devices, Wiley, New York 1981.

[8] K. A. Duncan, P. D. Tonner, J. G. Simmons, and L. Faraone, Solid State Electronics **24**, 941 (1981).

[9] L. Faraone, J. G. Simmons, F-L. Hsueh, and U. K. Mishra, Solid State Electronics **25**, 335 (1982).

[10] I. Zólomy, phys. stat. sol. (a) **73**, K249 (1982).

(Received June 29, 1988)

phys. stat. sol. (a) **111**, 377 (1989)

Subject classification: 73.40 and 85; S7.11; S7.15

Sektion Physik der Humboldt-Universität zu Berlin[1])

Influence of Surface Recombination on the Rate of Electron Leakage in Ridge-Waveguide Injection Lasers

By

H.-J. Wünsche and H. Wenzel

Simple formulae for the rate of electron leakage in injection lasers are derived, which take into account the surface recombination and an arbitrary number of heterojunctions. Applied to RW-lasers, where the surface beside the ridge is close to the active layer, a considerable enhancement of the leakage rate by the surface recombination is obtained. The influences of internal recombination, of drift fields, and of some device parameters on this effect are quantitatively investigated.

Es werden einfache Formeln für die Elektronenleckrate in Injektionslasern hergeleitet, welche die Oberflächenrekombination und beliebig viele Hetero-Übergänge berücksichtigen. Angewendet auf RW-Laser, bei denen die Oberfläche seitlich des Stegs nahe an der aktiven Schicht liegt, ergibt sich ein spürbares Anheben der Leckrate durch die Oberflächenrekombination. Der Einfluß der inneren Rekombination, von Driftfeldern und einiger Bauelementeparameter auf diesen Effekt wird quantitativ untersucht.

1. Introduction

The leakage current in DH lasers is due to those injected carriers, which leave the active zone again over the heterobarriers into the cladding layers. In this paper, we study the influence of surface recombination on the electron leakage, which has not been regarded in the literature until now. Surface recombination should be of large influence, whenever the distance between the surface and the active zone (AZ) is comparable with or even smaller than the diffusion length of electrons. This is extremely the case in RW lasers (cf. Fig. 1) beside the ridge, where the former distance is only some hundred nm, whereas the diffusion length measures some μm. Therefore, we concentrate on RW lasers. The theoretical results, however, are applicable to other types of laser diodes, too.

A typical example of a RW laser, which we shall use as our standard case, is sketched in Fig. 1. According to [7], the ridge in the middle is produced after planar epitaxy and photolithography by etching with an etchant that selectively removes InP but not the quaternary material of layer 3. Therefore, this layer is called etch-stop layer (ESL). The ridge plays an important role in the device. It causes an effective lateral wave guiding, supposed the ESL is sufficiently thin. The ESL is cladded with an insulating oxide at the sides. At the interface ESL/oxide, which we simply call surface, a considerable amount of recombination centers may occur. Since the ESL is much smaller than the diffusion length of electrons, we expect this surface recombination to be of great importance for the electron leakage at the sides of the ridge.

Our calculation of the electron leakage bases on the stationary semiconductor device equations, which were originally proposed by van Roosbroeck [1] and have been successfully used for modeling the current flow in a variety of semiconductor devices (see, e.g. [2, 3]), among them injection lasers (see, e.g., [4]). In Section 2, we present

[1]) PSF 1297, DDR-1086 Berlin, GDR.

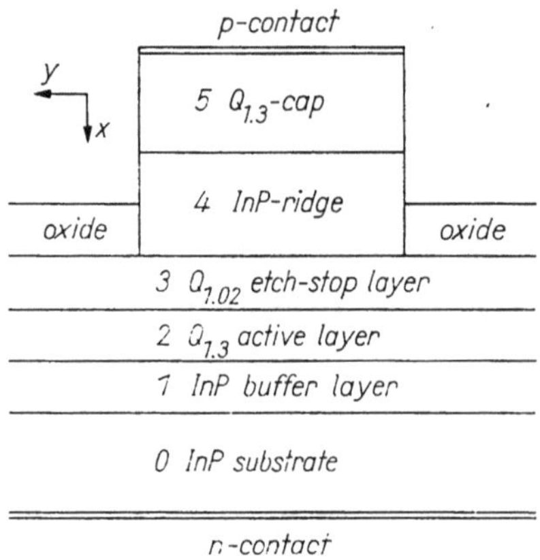

Fig. 1. Schematic cross section of a 1.3 μm ridge-waveguide laser composed of 5 epitaxial layers on InP substrate (simplified). Q_w indicates quaternary InGaAsP matched to InP, with the index w being the photoluminescence peak wavelength in μm. The structure and the data have been taken from [7]. Thickness and concentration of substrate and layers: 0: $d = 100$ μm, $n = 4 \times 10^{18}$ cm^{-3}; 1: $d = 8.0$ μm, $p = 5 \times 10^{17}$ cm^{-3}; 2: $d = 0.14$ μm, undoped; 3: $d = 0.2$ μm, $p = 1 \times 10^{17}$ cm^{-3}; 4: $d = 1.2$ μm, $p = 5 \times 10^{17}$ cm^{-2}; 5: $d = 0.3$ μm, $p = 4 \times 10^{18}$ cm^{-3}

these equations for the stationary case in the zero space charge approximation. Furthermore, in confining on the vertical dimension, a differential equation for the excess concentration N in every layer is derived. This equation is solved in Section 3 by adopting two different levels of approximation. The first kind of approximation (Section 3.1) consists in completely neglecting drift fields and recombination in the cladding layers. It yields a simple analytic formula for the leakage current density, which is also applicable for the cases of more than one cladding layer and with the pn-junction distant from the AZ. By the second type of approximation (Section 3.2), drift fields and recombination are effectively taken into account. A transcendent equation instead of an analytic formula is obtained in this case. Quantitative results are presented in Section 4 together with a discussion of the influence of different device parameters on the leakage rate.

2. Basic Assumptions and Equations

Our main assumption is charge neutrality in all layers,

$$n - p - C = 0\,, \tag{1}$$

where n, p, and C denote the densities of electrons, holes, and charged impurities, respectively. Mathematically, this represents the zero-space-charge approximation for the Poisson equation (cf. chapter 4 of [2]). From the physical point of view, the finite space charge layers at the pn-junction, at interfaces, and surfaces are modeled by infinitely thin dipole layers, when this approximation is adopted. This is reasonable at forward bias, when these space charge layers become sufficiently thin.

The particle densities in (1) are given by

$$n = N_c F_{1/2}\left(\frac{F_n + U - U_c}{U_T}\right) \quad \text{and} \quad p = N_v F_{1/2}\left(\frac{-(F_p + U - U_v)}{U_T}\right) \tag{2}$$

with

$$N_c = 2\left(\frac{2\pi m_c k T}{h^2}\right)^{3/2} \quad \text{and} \quad N_v = 2\left(\frac{2\pi m_v k T}{h^2}\right)^{3/2}. \tag{3}$$

Here, U denotes the electrostatic potential, k the Boltzmann constant, h the Planck constant, T the temperature, $U_T = kT/q$ the thermic voltage with q being the elementary charge, m_c and m_v the masses, $U_c = E_c/q$ and $U_v = E_v/q$ the edges of conduction and valence bands, and F_n and F_p the electron and hole quasi-Fermi potentials, respectively.

Laser operation requires carrier degeneracy at least in the active zone. This is accounted for in (2) by the Fermi-Dirac integral

$$F_{1/2}(x) = \frac{2}{\sqrt{\pi}} \int_0^\infty \mathrm{d}z \, \frac{z^{1/2}}{1 + \exp(z - x)} \,. \tag{4}$$

This integral cannot be expressed by elementary functions, but several approximations with more or less accuracy are available (see e.g. [5]). For our calculations, the rather simple expression [6]

$$F_{1/2}(x) = z + 0.15z^2 \quad \text{with} \quad z = \ln\left(1 + \exp(x)\right) \tag{5}$$

is used, because it allows an analytic inversion. It yields the correct asymptotics in the Boltzmann case and has an accuracy of better than 1% in the range of intermediate degeneracy ($x < 8$). For strong degeneracy, (5) becomes proportional to x^2 instead of $x^{3/2}$. In this case, however, the Fermi potential lies some hundred meV above the band edge, which is only of academic interest and not correctly described even by (4), which bases on parabolic bands.

The current densities of electrons and holes are given by

$$j_\mathrm{n} = qn\mu_\mathrm{n} \operatorname{grad}(F_\mathrm{n}) \quad \text{and} \quad j_\mathrm{p} = qp\mu_\mathrm{p} \operatorname{grad}(F_\mathrm{p}) \,, \tag{6}$$

with μ_n and μ_p being the mobilities, respectively. In the stationary case, they must obey the continuity equations

$$\operatorname{div}(j_\mathrm{n}) = qR \quad \text{and} \quad \operatorname{div}(j_\mathrm{p}) = -qR \,, \tag{7}$$

with the net recombination rate R. In InGaAsP/InP diodes, recombination via centres, radiative as well as Auger recombination contribute significantly. In the non-degenerate limit, these contributions can be described as

$$R = pn\left\{1 - \exp\left(\frac{F_\mathrm{p} - F_\mathrm{n}}{U_\mathrm{T}}\right)\right\}\left\{\frac{1}{(n + n_\mathrm{i})\, t_\mathrm{n} + (p + n_\mathrm{i})\, t_\mathrm{p}} + B_\mathrm{r} + C_\mathrm{n}n + C_\mathrm{p}p\right\} \tag{8}$$

with

$$n_\mathrm{i} = (N_\mathrm{c}N_\mathrm{v})^{1/2} \exp\left(\frac{-U_\mathrm{gap}}{2U_\mathrm{T}}\right) \tag{9}$$

being the intrinsic concentration, t_n and t_p the minority life-times of the Shockley-Read-Hall model, B_r, C_n, and C_p the coefficients of spontaneous radiative and Auger recombination, respectively, and $U_\mathrm{gap} = U_\mathrm{c} - U_\mathrm{v}$.

For the case of degenerate carriers, due to the Fermi statistics, the dependence on the densities of every contribution to R is much more complicated. Practical application, however, requires simple expressions. Therefore, polynomial fits of the form $An + Bn^2 + Cn^3$ are usually employed for describing R in the active layer under sufficiently high injection, where $p = n \gg n_\mathrm{i}$. This form is also obtained by (8), since the exponential term is much smaller than unity in this case. So we shall use (8) for degenerate carriers, too. In the active zone, (8) has to be accomplished by the stimulated recombination. We neglect this contribution, because it does not directly influence the electron leakage.

The above equations use U, F_n, and F_p as basic quantities. From this point of view, the neutrality condition (1) is a transcendent equation. Its solution becomes much simpler, however, when n and p are regarded as basic. This is achieved by the inversion of (2),

$$F_n = U_c - U + U_T F_i\left(\frac{n}{N_c}\right) \quad \text{and} \quad F_p = U_v - U - U_T F_i\left(\frac{p}{N_v}\right), \tag{10}$$

where F_i denotes the inverse of $F_{1/2}$.

Let n_0 and p_0 be the equilibrium concentrations. They are fixed quantities, easily obtained by solving (1) in the case $F_n = F_p$. Under bias, the neutrality condition is solved then by

$$n = n_0 + N \quad \text{and} \quad p = p_0 + N\,, \tag{11}$$

where N is the excess concentration. The latter quantity is uniquely connected with the so-called Fermi voltage

$$U_F := F_n - F_p = U_{gap} + U_T F_i\left(\frac{n}{N_c}\right) + U_T F_i\left(\frac{p}{N_v}\right) \tag{12}$$

and is a measure for the local deviation from equilibrium. Using now N and U as basic, the continuity equations read as

$$\operatorname{div}\left(D_n \operatorname{grad}(n_0 + N) - \mu_n(n_0 + N)\operatorname{grad}(U)\right) = R\,, \tag{13a}$$

$$\operatorname{div}\left(D_p \operatorname{grad}(p_0 + N) + \mu_p(p_0 + N)\operatorname{grad}(U)\right) = R\,. \tag{13b}$$

The electron diffusion coefficient is given by

$$D_n = \mu_n U_T F_i'\left(\frac{n}{N_c}\right)\frac{n}{N_c}, \tag{14}$$

where F_i' denotes the derivative of the inverse Fermi integral. An analogous expression holds for D_p.

Equations (13) are nonlinear second-order differential equations for N and U, which we now complete by boundary conditions. At the interfaces we assume continuity of the quasi-Fermi potentials and of the normal components of both current densities. The top of the ridge and the bottom of the substrate are assumed to be ideal ohmic contacts, i.e., there holds $F_n = F_p$ and the top and bottom Fermi potentials differ by the external voltage applied. The semiconducter–oxide interface outside the ridge is modeled by a surface recombination of Shockley-Read-Hall type. This results in the boundary condition

$$j_{nx} = -j_{px} = qnp\,\frac{1 - \exp(-U_F/U_T)}{(n + n_i)/s_p + (p + n_i)/s_n}, \tag{15}$$

where the parameters s_n and s_p denote the velocities of surface recombination. The first equality in (15) expresses that no current can flow through the interface into the insulator. With respect to the approximations to be presented in Section 3, the conditions at the vertical boundaries need not be discussed.

3. Approximate Solutions

In order to obtain a direct insight into the interplay between surface recombination and electron leakage, we shall look now for further simplifications of the basic equations (13).

First, we assume the lateral distribution $N(y)$ of excess carriers in the active zone as known and consider only the layers above. Second, we neglect all derivatives of U and N with respect to the lateral y and longitudinal z coordinates in these layers and obtain ordinary differential equations. This is reasonable, because the strongest variations of those quantities can be expected to occur in the vertical x-direction, due to the close distance between the AZ and the surface.

As a quantitative measure for the leakage rates, we shall calculate within the frame of these approximations two functions $j_{LR}(N)$ and $j_{LS}(N)$. These quantities are the current densities of electron leakage under the ridge and beside it, respectively, in dependence on the value of N in the AZ at the local position y.

3.1 Diffusion model without recombination

An analytic formula can be derived by restricting to the most simple model that still contains the main effects, which are the diffusion of minorities and their recombination at the surface. The simplifications leading to this model are

(i) Neglect of recombination, $R = 0$, except at the surfaces.

(ii) Neglect of drift fields, grad $(U) = 0$, in the minority equation (13).

(iii) Constant doping, $C =$ const, within every layer.

(iv) $N \ll |C|$ and Boltzmann statistics for the minorities in the layers above the active zone.

Under these conditions, the minority equation in (13) is easily solved, yielding

$$N(x) = N(0) + Ax\,, \tag{16}$$

with

$$A = \frac{j_n}{kT\mu_n} \quad \text{or} \quad A = \frac{-j_p}{kT\mu_p} \tag{17}$$

for n- or p-layers, respectively, where x is measured from one of the interfaces. Due to (i), the current densities j_n and j_p are constant throughout all layers and are treated as parameters, for the time being. The values $N(0)$ can be obtained from the boundary conditions. For this purpose, we use the identity

$$N = \frac{(\exp(U_F/U_T) - 1)\, n_i^2}{|C|}\,, \tag{18}$$

which follows from (10) and (iv).

Let us consider the p-layer below the oxide beside the ridge. Be $x = 0$ the oxide/semiconductor interface. The boundary condition (15) together with (18) yields

$$N(0) = \frac{j_n}{s_n}. \tag{19}$$

The corresponding boundary condition at the contact on the ridge is $N(0) = 0$. This is (19) in the limit of infinite s_n. Therefore, we can work with (19) in both cases.

Now, we turn to the interface at the upper border of the AZ. There, the Fermi voltage must be continuous. Within the AZ, we have to apply (12), because the eventual degeneration does not allow the use of Boltzmann statistics. In the layer above, (iv) is valid, and N at the boundary can be expressed by U_F with the help of (18).

At all other interfaces between layers above the AZ, assumption (iv) holds at both sides. Hence, from the necessary continuity of U_F, one deduces by (18) that the quantity $N|C|/n_i^2$ must be continuous.

Using this, (19) and (12), we obtain the relation

$$j_n = j_{ns}\left(\exp\left(\frac{U_F}{U_T}\right) + K_n j_p - 1\right) \tag{20}$$

with U_F being the Fermi voltage in the AZ,

$$j_{ns} = \left\{ \frac{N_a}{s_n n_i^2}\bigg|_S + \sum_{\text{p-layers}} \frac{N_a d}{\mu_n k T n_i^2} \right\}^{-1} \tag{21}$$

and

$$K_n = \sum_{\text{n-layers}} \frac{N_d d}{\mu_p k T n_i^2} . \tag{22}$$

For clarity, we have put here $C = N_d$ in n-layers and $C = -N_a$ in p-layers. The summations in (21) and (22) run over all p and n layers above the AZ, respectively. The first term in (21) must be taken at the surface, which is indicated by the "S" behind it. In the standard case, as sketched in Fig. 1, no n-layer exists there. Then, the sum in (22) is empty, yields $K_n = 0$ and j_n becomes independent of j_p. The situation changes, however, when the pn-junction lies in the layers above the AZ. In this case, we obtain $K_n > 0$, and the hole current density j_p is needed. For this quantity, the following assumptions seem to be reasonable. Under the stripe, j_p is estimated by the hole injection necessary for compensating the recombination in the AZ,

$$j_{rec} = q d_{AZ} R_{AZ}(N) . \tag{23}$$

So, the leakage current density in this region becomes finally

$$j_{LR} = j_{ns} \left(\exp\left(\frac{U_F}{U_T}\right) + K_n j_{rec} - 1 \right). \tag{24}$$

Since j_{rec} is positive, the electron leakage under the ridge increases, when shifting the pn-junction away from the AZ into the upper layers. This can be understood as follows.

In the n-layers above the AZ, the holes are minorities. Their density p is rather small and, hence, a comparatively large increase of F_p is required for driving the injection (cf. (6)). The change of F_n, on the other hand, is small due to the large electron conductivity in these layers. As a consequence, the Fermi voltage increases from the AZ up to the pn-junction. Above, its decrease towards zero at the contact must be stronger than in the absence of the n-layers, resulting in an enhanced electron current.

Now, we turn to the region beside the ridge. Corresponding to (15), $j_p = -j_n$ holds there. Inserting this into (20) and solving for j_n yields immediately

$$j_{LS} = \frac{j_{ns} (\exp (U_F/U_T) - 1)}{1 + K_n j_{ns}} . \tag{25}$$

Here, in contrast to (24), the quantity K_n occurs in the denominator. The electron leakage is diminished by shifting the pn-junction into the etch-stop layer. This behaviour is caused by the fact that the Fermi voltage now must fall from the AZ up to the pn-junction or driving the holes in the opposite direction towards the surface.

3.2 Model with drift field and recombination

In this section, we relax the simplifications (i) and (ii) of the preceding section, but maintain all others. An exact solution of the system of coupled nonlinear ordinary differential equations (13a) and (13b) seems to be impossible in this case. The following manipulations, however, will lead us to a quasi-analytic solution, which approximately takes into account drift fields and recombination within all layers of interest.

At first, let us solve (6) and (10) for the drift field.

$$E := -\text{grad}\,(U) = \frac{D_n j_p + D_p j_n}{D_n \mu_p p + D_p \mu_n n}\,, \tag{26}$$

and combine (13a) and (13b) so that the derivative of E disappears, yielding

$$\mu_p p \,\text{div}\,(D_n \,\text{grad}\,(N)) + \mu_n n \,\text{div}\,(D_p \,\text{grad}\,(N)) + \mu_n \mu_p C E \,\text{grad}\,(N) = \\ = (\mu_p p + \mu_n n)\, R\,. \tag{27}$$

For the layers above the AZ, we simplify this by inserting the Einstein relations

$$D_n = \mu_n U_T \quad \text{and} \quad D_p = \mu_p U_T\,. \tag{28}$$

This is the nondegenerate limit of (14), which is valid for the minority but not necessarily for the majority carriers. However, that term of the first two in (27), which contains the majority diffusion coefficient, is of small influence, because of the small minority concentration in front of it. Therefore, we can insert the constant values (28) into (27). The result can be written in the form

$$\frac{d^2}{dx^2} N + 2f \frac{d}{dx} N - \frac{N}{L_{eff}^2} = 0\,, \tag{29}$$

where

$$f = \frac{EC}{2U_T(n+p)} \quad \text{and} \quad \frac{1}{L_{eff}^2} = \frac{(p/\mu_n + n/\mu_p)\, R(N)}{(n+p)\, U_T N} \tag{30}$$

represent a drift factor and an effective diffusion length. These coefficients still depend on the unknown quantities N and E. Their spatial variation within the layers above the AZ, however, is weak, as can be seen by detailed inspection when taking into account the maintained assumptions (iii) and (iv) of Section 3.1. This supports the following approximation. Within a given layer, we replace the weakly varying quantities f and L_{eff} by their values at the boundary $x = 0$. Then, (29) is easily solved. The solution can be written similiar to (16) as

$$N(x) = \left(N(0) \cosh\,(kx) + A \frac{\sinh\,(kx)}{k}\right) \exp\,(-fx) \tag{31}$$

with

$$k^2 = f^2 + \frac{1}{L_{eff}^2}\,. \tag{32}$$

This solution contains three free constants, namely $N(0)$, A, and f (k depends on f and $N(0)$), which must be determined for every layer above the AZ. For m layers, we need $3m$ conditions, which are the same as already used in Section 3.1. Continuity of U_F, j_n, and j_p at the interfaces between these layers yield $3m - 3$ conditions. A further one is the presumed value for N in the AZ, which fixes U_F at their boundary. The remaining two conditions are different under the ridge and outside. Under the ridge, we require $j_p = j_{rec}$ at the AZ and $U_F = 0$ at the contact. Beside it, the two equations (15) must be fulfilled at the oxide. In difference to Section 3.1, these conditions yield now a set of transcendent equations, which cannot be solved analytically. A simple personal computer, however, suffices for a fast numerical solution.

4. Quantitative Results and Discussion

With both models introduced in the preceding section, we have studied the influence of some selected parameters on the leakage current density. Besides the device parameters given in Fig. 1, we have used $\mu_n = 4000$ cm²/Vs, $\mu_p = 200$ cm²/Vs, $t_n = t_p =$

$= 3$ ns, $B_r = 8.3 \times 10^{-11}$ cm^3/s, $C_n = 10^{-29}$ cm$^6/s$, and $C_p = 3C_n$, for all layers. Furthermore, the formulae

$$U_{gap} = 1.2399\ \mathrm{V\mu m}/\lambda_g = (1.35 - 0.738y + 0.138y^2)\ \mathrm{V}\,, \tag{33}$$

$$m_c/m_0 = 0.08 - 0.039y\,, \quad \text{and} \quad m_v/m_0 = 0.87 - 0.25y \tag{34}$$

have been used for deriving the energy gap and the masses of quaternary InGaAsP/InP from the luminescence peak wavelength λ_g.

The results obtained with these parameters for the RW laser of Fig. 1 are collected in Fig. 2. Three levels of approximation have been applied:

(i) the diffusion model without recombination, i.e., the simple analytic formulae (24) and (25) (dotted curves);

(ii) the model of Section 3.2 with drift fields but with zero recombination in the layers above the AZ (dashed curves);

(iii) the complete model of Section 3.2 with both drift fields and recombination (solid curves).

Under the ridge, these levels of approximation differ considerably one from another. The electron leakage is underestimated by the simple approach (i). It is enhanced by including the drift fields, which drive the holes towards the AZ and the electrons away from it. The recombination gives an additional increase, which is about 20 A cm^{-2} at $N = 2 \times 10^{18}$ cm^{-3}. This recombination occurs almost completely in the etch-stop layer, which is closest to the AZ and has a comparatively small gap. Regardless of the enhancement by drift fields and recombination, however, the leakage under the ridge is small compared with the recombination in the AZ (formula (23), dash-dotted curve) and needs no further consideration.

At the sides, the situation changes completely. The electron leakage is strongly enhanced by the surface recombination at the oxide. Acceptable small leakage (less then 10% of j_{rec}) requires recombination velocities s_n at the oxide, which are smaller than 10^5 cm/s. For $N < 10^{18}$ cm^{-3}, good agreement is obtained between the three levels of approximation, i.e. neither drift fields nor recombination play any role. For larger N, the drift fields become important. They must drive the holes towards the

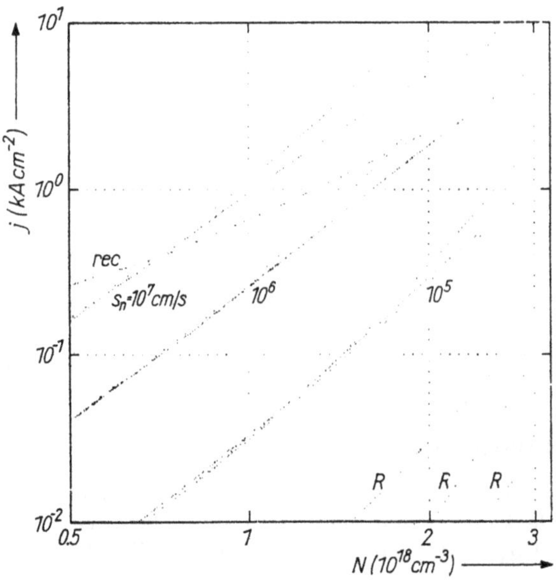

Fig. 2. Current densities of electron leakage under the ridge (marked by R) and at the sides (marked by the recombination velocities at the oxide (in cm/s) in dependence on the excess concentration N in the active layer. The dotted curves have been calculated under neglect of drift fields and recombination (model 3.1), the dashed curves with drift fields only and the solid curves with both drift fields and recombination (model 3.2). For comparison, the recombination current density is presented too (dash-dotted)

oxide and, hence, diminish the electron leakage from the dotted lines to the dashed lines. For $s_n > 10^5$, the dashed lines coincide with the full lines. This indicates that the recombination is of minor influence. According to the numerical figures, it contributes only 21, 20, 11, and 3 A cm^{-2} for $s_n = 10^4, 10^5, 10^6$, and 10^7 cm/s, respectively, and $N = 2 \times 10^{18}$ cm^{-3} in the AZ. For small s_n, these values are the same, as under the ridge. For larger s_n, the contribution of recombination decreases, which effect is again due to the drift fields. At small drift fields, the electron concentration in the etch-stop layer falls linearly towards the oxide, cf. (16). With acting drift fields, this fall becomes exponential, according to (31). As a consequence, the number of electrons in the etch-stop layer becomes smaller, what diminishes the recombination.

So far, we have regarded the structure and the parameters of Fig. 1. We have found that the electron leakage at the sides of the ridge can be large. For preventing this, one must ensure small values of the recombination velocity at the oxide when preparing the diode. However, this is only one of several possibilities. Small electron leakage — even for large surface recombination — can also be achieved by appropriately choosing other device parameters. For this purposes, one must keep the saturation current density (21) as small as possible. s_n as a parameter of influence has already been discussed. Another important quantity is the intrinsic concentration n_i. It should be as small as possible, what requires a large gap, according to (9). The reason is obvious. The gap of the ESL determines the height of the barrier to be overcome by the leaking electrons. Indeed, as can be seen from Fig. 3, the leakage current raises dramatically with diminishing the energy gap of the etch-stop layer. From this point of view, an InP ESL would be the best, because of its large gap. However, this is only a theoretical possibility. The technological aim of this layer is to stop the etching, which requires the material to be sufficiently different from InP. This difficulty can be overcome by introducing an additional InP barrier layer between the AZ and the ESL (see, e.g., [8]). For demonstrating the effectiveness of such a barrier layer, we have regarded an ESL with a relatively small gap. The results are drawn in Fig. 4. Without the barrier layer, the electron leakage is very large. A barrier layer, which is

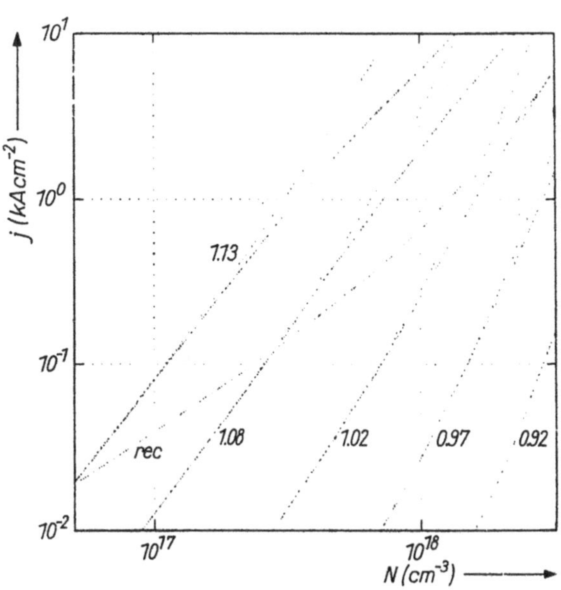

Fig. 3. Leakage current densities beside the ridge in dependence on the excess concentration N in the AZ. Different etch-stop layers have been adopted, which are quoted by the luminescence peak wavelength (in μm). The meaning of dotted, solid, and dash-dotted curves is the same as in Fig. 2 $s_n = 10^6$ cm/s

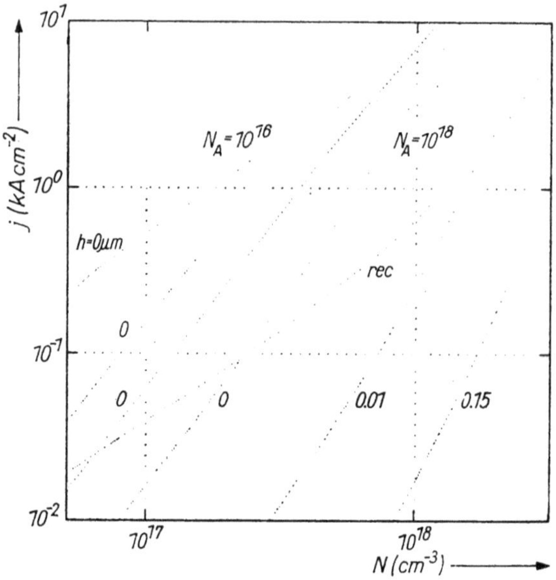

Fig. 4. Leakage current densities beside the ridge in dependence on the excess concentration N in the AZ. A $Q_{1.15}$ etch-stop layer has been adopted. The full lines show the influence of an additional InP barrier layer between the active and etch-stop layers. The thickness h of the barrier is quoted (in µm). The dotted line corresponds to $h = 0$ with the pn-junction being 0.1 µm above the AZ. The dashed lines show the influence of the etch-stop doping concentration, which is quoted in cm^{-3}. All results have been calculated with the full model of Section 3.2

only $h = 10$ nm thin, reduces the leakage already dramatically. This can again be understood by means of (21), where now the sum goes over the barrier layer and the ESL. The intrinsic concentration n_i in the barrier is more than two orders of magnitude smaller than in the ESL. Therefore, the contribution from the barrier dominates the sum in (21) already for very small thicknesses.

Except by a barrier layer, the electron leakage can also be reduced by increasing the doping or by shifting the pn-junction into the ESL. These measures, however, are not very effective, as can be seen from Fig. 4.

References

[1] V. W. VAN ROOSBROECK, Bell Syst. tech. J. 29, 560 (1950).
[2] S. SELBERHERR, Analysis and Simulation of Semiconductor Devices, Springer-Verlag, Wien/New York 1984.
[3] P. A. MARKOVICH, The Stationary Semiconductor Device Equations, Springer-Verlag, Wien/New York 1986.
[4] T. OHTOSHI, K. YAMAGUCHI, C. NAGAOKA, T. UDA, Y. MURAYAMA, and N. CHINONE, Solid State Electronics **30**, 627 (1987).
[5] F. J. F. VELICIA, Phys. Rev. A **34**, 4387 (1986).
[6] K. UNGER, private communication.
[7] M.-CH. AMANN and B. STEGMÜLLER, Appl. Phys. Letters **48**, 1027 (1986).
[8] M.-CH. AMANN and B. STEGMÜLLER, Japan. J. appl. Phys. **25**, 228 (1986).

(Received September 26, 1988)

phys. stat. sol. (a) **111**, 387 (1989)

Subject clasification: 73.40 and 85; S1.3; S6; S10

Institut für Angewandte Physik der Technischen Hochschule Darmstadt[1])

Metal Influence on Switching MIM Diodes

By

H. PAGNIA and N. SOTNIK

Post-deposition onto electroformed planar metal–insulator–metal diodes (displaying e.g. I–U characteristics with negative differential resistance due to rupturing filamentary current paths) was proved to cause an additional process, similar to basic electroforming, affecting the switching behaviour temporarily only. It cannot patch ruptured filaments.

Es wird gezeigt, daß eine nachträgliche Beschichtung elektroformierter planarer Metall–Isolator–Metalldioden (die z. B. I–U-Charakteristiken mit negativ differentiellem Widerstand infolge des Reißens fadenförmiger Strompfade zeigen) einen zusätzlichen Prozeß verursacht, ähnlich der Grundelektroformierung, der das Schaltverhalten nur zeitlich beeinflußt. Sie kann gerissene Fäden nicht ausbessern.

1. Introduction

Experiments with post-deposition of metal onto electroformed metal–insulator–metal (MIM) diodes so far have been reported in two cases. Thurstans [1] observed a regeneration-like behaviour of sandwich MIM diodes after switching to the low conductivity (OFF) state with subsequent evaporation of Au (explicitly not with Al) of small rate. Most diodes were short-circuited by this procedure, some were unaffected, but a remarkable number of them (30%) seemingly recovered stepwise towards the high conductivity (ON) state. That means more precisely: a current increase with time could be obtained reaching about half the values of the prior ON state current. This was performed with an applied voltage below the memory threshold U_{th} so that no normal regeneration was expected. These results were assumed to be a proof of the metallic nature of the current paths.

Gengenbach et al. [2] exhibited a strong increase of the whole current voltage I–U characteristic up to e.g. four times the prior current maximum after metal evaporation at $U < U_{th}$ with planar MIM diodes. This behaviour could be obtained irrespective of starting the procedure from the ON or the OFF states. Sometimes repeated voltage cycling was necessary until stable I–U traces on a higher level were established. Although these investigations were only of preliminary nature, it was already shown that various metals (Au, Ag, In, Bi, Pb, and Cr) reveal an analogous behaviour of current recoverage. A few possible explanations shall be mentioned only. They reach from simply patching ruptured filament metal contacts, or filament sections, to conductivity enhancement by doping the filament material.

The question what principal role metal plays is not yet settled for the various anomalous effects of those diodes, often described [3, 4, 5], as well as especially for the electroforming and regeneration processes. Although it is now known that diodes with all well developed anomalous features can be produced without any metal (at least at and near the active region of the device) utilizing carbon films and graphite paste

[1]) Schloßgartenstr. 7, D-6100 Darmstadt, FRG.

contacts [6], discussions on the metal influence remain rather controverse. It seemed therefore reasonable to start new investigations to clarify the effects small traces of metal induce in-situ. Experiments were performed with planar diodes, instead of the closed sandwich diodes, for they are more directly accessible to the impinging particles from the metal vapour. But even in the sandwich case one should take into consideration the holes in the top electrode [7], where the filaments intersect the upper surface, through which metal can intrude. Comprehensive conclusions of the new experiments were already published in a review article [8], but shall be discussed with more details here.

Electroformed MIM diodes can be switched to their OFF state by a high voltage (U_M) pulse and start to recover towards their ON state either after rapidly turning on a fixed regeneration voltage (U_R), or (what will be used in this investigation) starting a slow velocity linear voltage ramp (dU/dt). Then, regeneration towards the ON state current sets on at a relatively sharp threshold voltage U_{th}. Because it is helpful to understand this principle behaviour, one example is shown in Fig. 1a as an illustration. The fast return curves (after turning to the OFF state at U_M) were not actually traced in the used registration equipment. Their principal shape (without a maximum) is sketched by a dashed curve.

Another characteristic feature of these electroformed diodes is the behaviour in air (or oxygen [9]) environment. One example is demonstrated in Fig. 1b. After exposure to air at atmospheric pressure the first run remains for the present on the ON state trace up to a new threshold voltage U_{th}^* from where the current descends to the OFF state (curve 3). It is not before reestablishing high vacuum (HV) conditions that the voltage-controlled negative differential resistance (VCNR)-like characteristic recovers during further cycling the voltage.

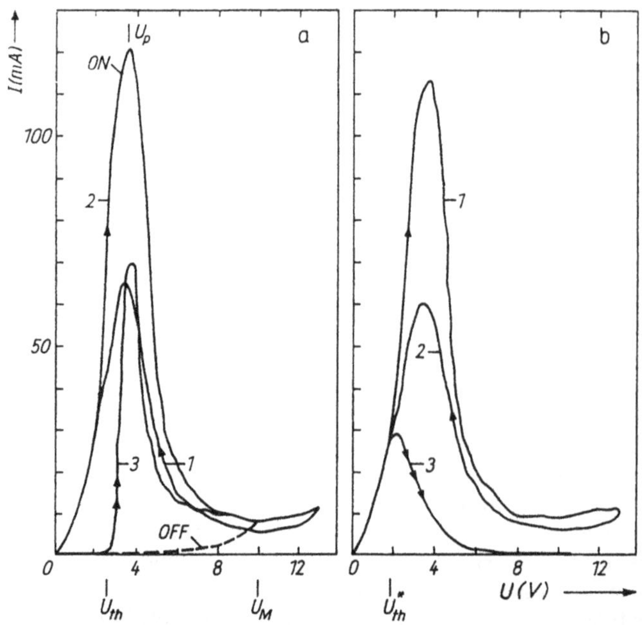

Fig. 1. Current–voltage characteristics of a planar MIM diode (Au on quartz glass). a) Numbers indicate subsequent traces with slow sweep velocity ($dU/dt \approx 1$ V/s); b) curves 1 and 2 in HV, curve 3 in air at atmospheric pressure

2. Experimental

Planar MIM diodes of usual convenience, made up of Au thin films on quartz glass slides, were mounted in a HV chamber ($p < 10^{-3}$ Pa). The necessary discontinuity was introduced by Joule heating delivering the metal–insulator–metal configuration. After this basic (B-type) electroforming process through which a high resistance device is built, subsequent sweeping a triangular voltage (0 to 12 V) produces the well-known VCNR-like I–U characteristics. Since this can advance in a vacuum contaminated with organic vapour only, it was called adsorption-assisted (A-type) electroforming [8]. Then, such diodes are ready to display I–U curves dependent upon the actual and the prior applied voltage–time shapes. Especially, they were controlled by the chosen voltage amplitude. I–U characteristics were usually plotted with an x–y recorder.

Additional metal could be evaporated from a Joule heated tungsten boat through a mask onto the device area of the planar diodes. All measurements proceeded in the same chamber without breaking the vacuum, except in the case of air inlet checks.

The amount of post-deposited metal was controlled only indirectly by the exposure times of each distinct evaporation step at a fixed power of the source. From the sum of metal observable after a given number of treatment steps (Section 3), being of the order of magnitude of 100 nm, one can estimate that not more than about 10 nm could have been deposited in each single step. That is less or at maximum the same order of magnitude for forming semicontinuous films in the case of Au and at which percolation of ohmic conduction occurs.

3. Metal Post-Deposition

Metal was post-deposited with the diodes at their OFF or ON states as well. What happens after depositing a small amount of metal onto the diodes being in the OFF state is shown in Fig. 2. Curves number 0 and 2 display the usual shape before and

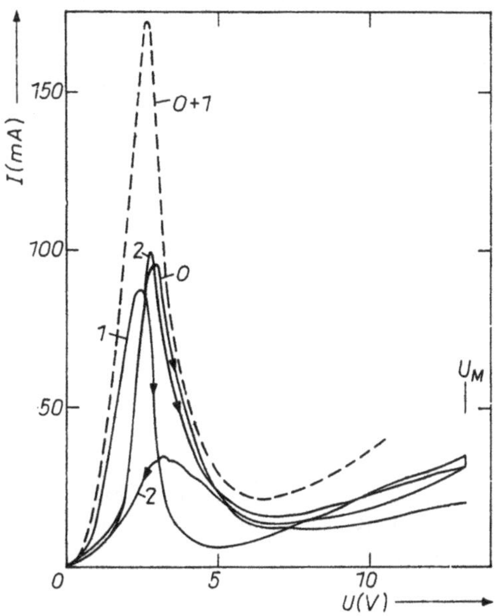

Fig. 2. I–U charactetistics of a planar MIM diode (Au on quartz glass); (0) forward trace up to U_M (fast return to zero voltage not registered), (1) after post-deposition of metal with the diode in its OFF state, (2) succeeding traces; dashed curve is the sum of (0) and (1)

after the metal evaporation and one subsequent voltage cycle. The weak difference between these curves (which increases during repeated treatments), essentially observable at high voltage, is possibly due to metal deposition beneath the actual device on the free surface (which was not satisfactorily covered by the mask), or on or beneath the filaments, building a metallic shunting resistor. Curve 1 traced immediately after the deposition is totally different from the usual regeneration threshold behaviour expected without additional metal, e.g. curve 3 of Fig. 1a. Remarkable is the abrupt fall of the current with a minimum at a voltage where regeneration should start. But after one voltage sweep the original behaviour is recovered (curve 2). The ascending branch of curve 1 does not reveal true properties of a homogeneous material at constant temperature. This could be proved by applying one single fast voltage ramp, thus preventing remarkable Joule heating. Under such circumstances, the I–U characteristic was ohmic up to even a voltage beyond the current maximum displayed with slowly increasing voltage.

Deposition of smaller amounts (measured after the one of Fig. 2 during the course of repeated treatments) has of course weaker influence. One example is presented in curve 1 of Fig. 3 which shows that usual regeneration is strongly suppressed at least. It should be mentioned that after subsequent deposition treatments with equal doses, quite identical shapes of each following first I–U run were obtained.

An explanation of the observed response to metal deposition requires the metal to
(i) temporarily establish additional conductive material which
(ii) disappears during one forward voltage sweep (up to e.g. at least about 5 V, the minimum in Fig. 2), and to
(iii) prevent regeneration to a degree dependent on the dose.

Consequently, metal post-deposition onto diodes in the ON state should deliver peak shapes due to an addition of the normal VCNR-like curves (e.g. 0 in Fig. 2) with a curve like 1 in Fig. 2. An example is shown in Fig. 4 (curve 3) (it can be compared with the sum curve in Fig. 2 (dashed)). Again common shapes are reestablished after one forward sweep (cycle 4). That means (i) and (ii) hold, and (iii) is of no relevance. Metal disappearance seems not to affect the original ON state.

Air influence was checked in both cases after metal post-deposition onto diodes in their ON or OFF states, respectively. It turned out that untreated room air (without any precaution like drying) at atmospheric pressure ceases the metal-induced conductivity during a current flow. Generally, no uniform threshold is observable, but a

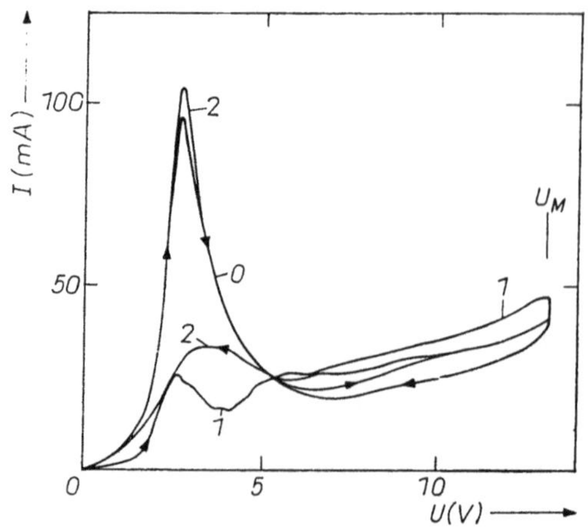

Fig. 3. I–U characteristics as in Fig. 2; same notation at the curves, smaller amount of post-deposited metal

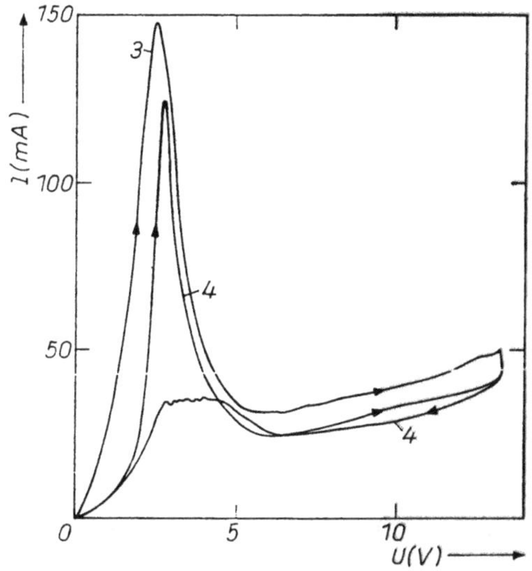

Fig. 4. I–U characteristics as in Fig. 1; (3) after post-deposition of metal with the diode in its ON state and a subsequent cycle (4)

distinct electrical power must be introduced before ceasing of the current sets on. On the other hand, in the case of a prior ON state the usual off-switching effect is superimposed. This is shown for one example in Fig. 5 (curve 1). After operation in air, evidently, also the increased current tail, due to the shunting resistor can be reduced towards an OFF state, having been obtained prior to the post-deposition treatments. In the given example the relative current decrease sets on at a low voltage ($\approx$ 1 V). The process continues up to the maximum applied voltage, so that in few cases more than one voltage cycle was necessary in order to reach the full OFF state. The general behaviour is that

(iv) in air environment at atmospheric pressure (under voltage and current stress) disappearance of the metal is faster than in HV.

Thin metal covers seem weakly bound to the underlying material and are sensitive to environmental influence.

Since after each treatment and subsequent voltage cycling the original I–U characteristics recovers, it must be expected that the microslit should be unchanged. That

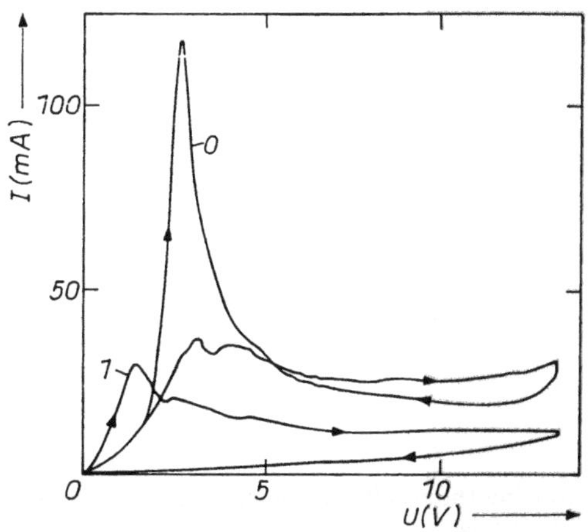

Fig. 5. I–U characteristics; (0) before and (1) after metal post-deposition in the ON state and air inlet

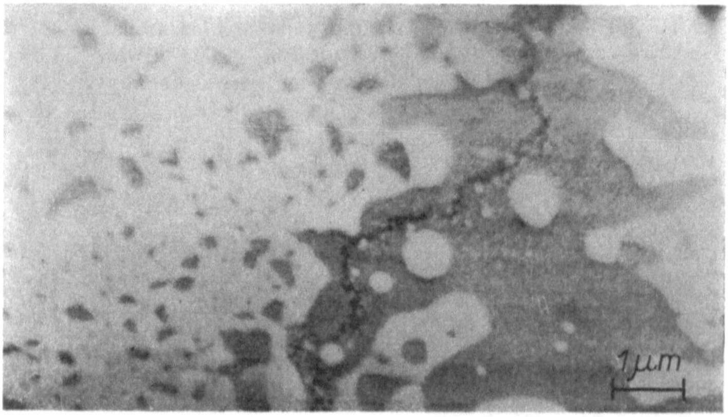

Fig. 6. Scanning electron micrograph of a planar MIM diode after usual electroforming and a metal post-deposition treatment; the original gold film and coagulated islands are bright, the post-deposited metal is grey and the narrowed microslit appears dark (see text and illustration in Fig. 7)

proved to be correct in principle but not in detail. Part of the metal covers of the discrete deposition steps sum up to continuous films covering parts of the prior microslit. By this procedure and repeated voltage cycling a new smaller microslit is built, e.g. visible in a scanning electron microscope (SEM) (Fig. 6). That means at least the active parts of the filaments are shorter than the original microslit gap. The whole situation of a diode illustrates Fig. 7 which should be obtained after usual electroforming and the subsequent post-deposition treatment as well as after metal coverage and repeated electroforming. This finding is in agreement with older ones that even small microslits (without incorporated islands) exhibit the known anomalous electrical behaviour [10].

A-type and B-type electroforming of planar diodes even proceed by cycling the voltage during metal evaporation simultaneously. But such a method proved to be worse than the usual one, often developing broader I–U shapes. On the other hand, failures may be revised by additional metal deposition, e.g. after having gained a weakly pronounced characteristic only.

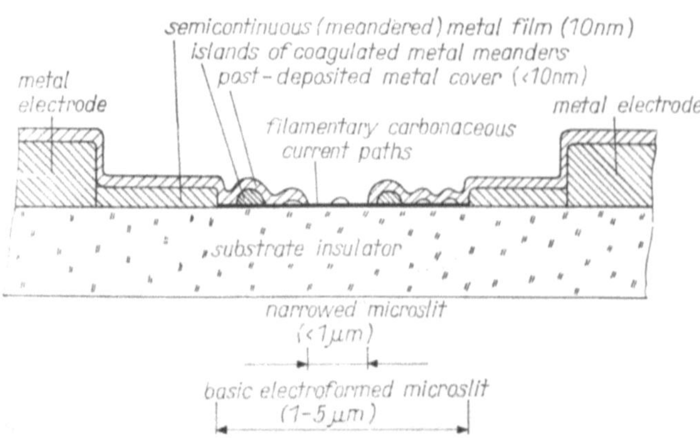

Fig. 7. Sectional view on a planar MIM diode illustrating the results of the electroforming processes, as well as of metal post-deposition and voltage cycling

4. Discussion and Conclusions

Post-deposition of metal onto electroformed planar MIM diodes temporarily causes additional conductivity bypassing the underlying OFF or ON state conductance. The resulting *I–U* characteristic can only be traced once in each case of subsequent treatments. The ascending branch of the characteristic is apparently non-ohmic at moderate voltage sweep velocities. It is at the moment an open question (which will be attempted to answer in the near future with combined in-situ electron microscopic, evaporation, and current-voltage measurements) whether clustering of the metal leads e.g. to proper configurations for intercluster tunneling, or whether simply coagulation processes of mobile metal atoms or clusters dynamically change the actual conductance. Some indication of the latter process may be deduced from the curve shapes exhibiting some similarity with those observable during B-forming with an applied ramp voltage. The falling branches are frequently steeper than the VCNR-like curves due to the common stepwise filament rupturing, even for equal peak heights. So it is clear that metal post-deposition cannot really regenerate off-switched filamentary current paths as was formerly proposed by Thurstans [1].

On the contrary, the experimental result evidences freshly deposited small metal amounts to prevent regeneration of the predominantly carbonaceous filament material. Possibly, the shunting material simply diminishes the electrical field strength locally at the ruptured filaments. But also an electrochemical process may be taken into account acting through a charge transfer between metal atoms or clusters and the constructive molecules or fragments [8].

Metal deposition of larger amount in one step will of course ultimately short-circuit the electrode gap (the microslit). Usual B-type and A-type electroforming may then proceed during voltage cycling, and that can pretend effective regeneration; this procedure may even increase the effective electrode length and thus entail the maximum possible current. Interpretations must hence be brought about with care. The latter effect assumedly caused the permanently increased peak heights mentioned in the Introduction [2].

So it turned out that none of the former interpretations correctly describes the in-situ metal post-deposition influence. It is really an additional process which proceeds essentially independently of the adsorption-assisted electroforming and its typical features.

References

[1] R. E. Thurstans, Ph. D. Thesis, Leicester Polytechnic, 1976.
[2] T. Gengenbach, H. Pagnia, and N. Sotnik, Mater. Letters **5**, 215 (1987).
[3] G. Dearnaley, A. M. Stoneham, and D. V. Morgan, Rep. Progr. Phys. **33**, 1129 (1970).
[4] D. P. Oxley, Oxide and Oxide Thin Films, Ed. A. Vigh, M. Dekker, Inc., New York 1981.
[5] A. K. Ray and C. A. Hogarth, Internat. J. Electronics **57**, 1 (1984).
[6] T. Gengenbach, H. Pagnia, and N. Sotnik, phys. stat. sol. (a) **100**, K199 (1987).
[7] H. Biederman, Vacuum **26**, 513 (1976).
[8] H. Pagnia and N. Sotnik, phys. stat. sol. (a) **108**, 11 (1988).
[9] I. Emmer, Thin Solid Films **20**, 43 (1974).
[10] H. Pagnia, K. Schlemper, and N. Sotnik, Mater. Letters **5**, 260 (1987).

(Received October 5, 1988)

Short Notes

phys. stat. sol. (a) 111, K1 (1989)

Subject classification: 61.60; S1.61

Department of Applied Physics, Faculty of Engineering, Tohoku Gakuin University, Miyagi[1)]

Note on the Crystal Structure of FeCoGe and FeNiGe

By

T. KANOMATA and T. GOTO

The crystal structure and magnetic properties of several compounds of MM'Ge type (M, M' are 3d transiton metals) were investigated by Castelliz /1/, who found that the compounds FeCoGe and FeNiGe have the hexagonal Ni_2In ($B8_2$)-type structure (Fig. 1) and are ferromagnets with Curie temperatures of 370 and 770 K, respectively. Furthermore, Buschow and van Engen /2/ reported a study of the magnetic and magneto-optical properties of FeCoGe and FeNiGe. On the other hand, X-ray studies for FeCoGe and FeNiGe revealed the existence of a cubic phase in addition to the phase with Ni_2In-type structure /3/. Similarly, we have found that the Ni_2In-type structure in the Fe_xNiGe system exists in stable form in the non-stoichiometric composition of $0.5 \leqq x \leqq 0.9$ and the ferromagnetism in FeNiGe is due to the contamination of a small amount of ferromagnetic impurity /4/. Thus, it is highly questionable that the compounds FeCoGe and FeNiGe crystallize in the single phase with Ni_2In-type structure. The purpose of the present note is to present the results of investigations of the metallurgical properties of FeCoGe and FeNiGe obtained with X-ray and Mössbauer effect methods.

Samples of FeCoGe and FeNiGe were prepared from high purity elements (Fe, Ni, Co: at least 99.9% purity, Ge: 99.999% purity). They were mixed in the desired proportion, sealed in evacuated silica tubes, and melted at 1200 °C. The reaction products were crushed, mixed, and remelted at 1200 °C. After grinding the reaction products to powder, the samples were annealed under different conditions and designated by A and B for both FeCoGe and FeNiGe: samples A were annealed at 950 °C for 2 days and then quenched in water, and samples B were annealed at 800 °C for 2 days and then quenched in water.

1) Chuo-1-13-1, Tagajo City, Miyagi 985, Japan.

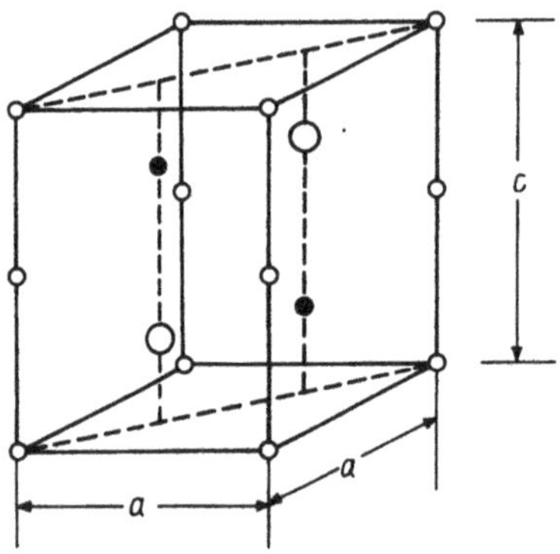

Fig. 1. Unit cell of the MM'Ge system (Ni_2In-type, space group $P6_3/mmc$)

X-ray diffraction measurements were made on a Rigaku X-ray powder diffractometer using monochromatic CuKα radiation. The lines of the X-ray pattern from FeCoGe (sample A) could not be indexed on the basis of the Ni_2In-type structure completely and another set of additional lines was found. The lattice constants of the phase with Ni_2In-type structure are found to be a = 0.4003 nm and c = 0.5045 nm. This result is in agreement with the value given by Buschow and van Engen /2/. However, it is somewhat larger than the value reported by Castelliz /1/. The similarity of X-ray scattering amplitudes of the component elements precluded a quantitative determination of the chemical order. For the Ni_2In-type compound $Fe_{0.9}Co_{0.1}Ge$, Szytula et al. /3/ reported that the Fe atoms occupy preferentially the 2a site. The results of X-ray analysis for FeCoGe (sample A) are given in Table 1. Five extra lines in Table 1 were indexed on the basis of a cubic structure. The detection of the diffraction lines of the cubic phase was troublesome, because they appear at almost the same angles as the strong lines of the hexagonal phase. The lattice parameter of this phase is found to be 0.287 nm. This value is comparable with the value of α-iron. For FeCoGe, the properties of X-ray patterns of samples A and B were essentially the same, though the intensities of five extra lines for sample B were somewhat stronger than those of sample A. The results of X-ray analysis for FeNiGe (sample A) are given in Table 2. As seen in Table 2, the compound FeNiGe (sample A) exhibits a crystallographic behaviour similar to that of the compound FeCoGe. The lattice constants derived

Table 1

X-ray diffraction pattern from FeCoGe

observed spacing d(nm)	intensity	indexing for hexagonal phase (h k l)	indexing for impurity (h k l)	d-spacing for α-Fe (nm)
0.2859	w	1 0 1		
0.2520	w	0 0 2		
0.2040	s	1 0 2		
0.2028	m		1 1 0	0.2027
0.1988	vs	1 1 0		
0.1640	vw	2 0 1		
0.1567	w	1 1 2		
0.1511	vw	1 0 3		
0.1435	w		2 0 0	0.1433
0.1429	m	2 0 2		
0.1271	vw	2 1 1		
0.1260	w	0 0 4		
0.1206	vw	2 0 3		
0.1185	vw	1 0 4		
0.1172	w		2 1 1	0.1170
0.1163	w	2 1 2		
0.1156	w	3 0 0		
0.1067	w	1 1 4		
0.1048	vw	3 0 2		
0.1034	vw	2 1 3		
0.1013	w		2 2 0	0.1013
0.1001	w	2 2 0		
0.0908	w		3 1 0	0.0906
0.0899	w	3 1 2		

vs: very strong, s: strong, m: medium, w: weak, vw: very weak

from the diffraction lines of the hexagonal phase are found to be a = 0.4024 nm and c = 0.5084 nm. This result is in agreement with the value given by Castelliz /1/. The results of X-ray analysis for sample A and sample B were essentially the same as well as in the case of the samples A and B for FeCoGe.

In order to examine the above experimental results from a different point of view, the Mössbauer effect measurements were carried out for FeCoGe (sample A) and FeNiGe (sample A). Mössbauer spectra using a radioactive source of ^{57}Co in Rh matrix were taken with a constant acceleration transmission-mode technique at room temperature. The velocity scale was calibrated with a high purity metallic iron foil. The Mössbauer spectra for FeCoGe and FeNiGe are shown in Fig. 2 and 3. In the case of FeCoGe, the

Table 2

X-ray diffraction pattern from FeNiGe

observed spacing d(nm)	intensity	indexing for hexagonal phase (h k l)	indexing for impurity (h k l)	d-spacing for α-Fe (nm)
0.2883	w	1 0 1		
0.2547	m	0 0 2		
0.2057	s	1 0 2		
0.2024	w		1 1 0	0.2027
0.2017	vs	1 1 0		
0.1580	w	1 1 2		
0.1439	m	2 0 2		
0.1434	vw		2 0 0	0.1433
0.1272	w	0 0 4		
0.1172	w	2 1 2		
0.1170	vw		2 1 1	0.1170
0.1164	w	3 0 0		
0.1057	w	1 1 4		
0.1015	vw		2 2 0	0.1013
0.1007	vw	2 2 0		
0.0936	vw	2 2 2		
0.0905	vw		3 1 0	0.0906
0.0903	w	3 1 2		
0.0858	w	4 0 1		
0.0856	vw	3 0 4		
0.0823	w	1 0 6		

vs: very strong, s: strong, m: medium, w: weak, vw: very weak

spectrum consists of three sets of lines which may be interpreted as the overlapping of a quadrupole doublet and two Zeeman sextets. This quadrupole doublet may be attributed to the existence of the paramagnetic phase with the Ni_2In-type structure. For the Zeeman sextet with large internal field, the internal field at an Fe nucleus of the impurity phase is about 340 kOe. This value is comparable with that of α-iron. So, the Zeeman sextet with larger internal field may be attributed to the ferromagnetic α-phase. On the other hand, the intensities of the Zeeman sextet with smaller internal field are weak in comparison with those of the sextet with larger internal field. In fact, we could not observe the X-ray diffraction pattern of the impurity phase with smaller internal field within experimental uncertainty as shown in Table 1. As seen in Fig. 3, the Mössbauer spectra of FeNiGe and FeCoGe exhibit a

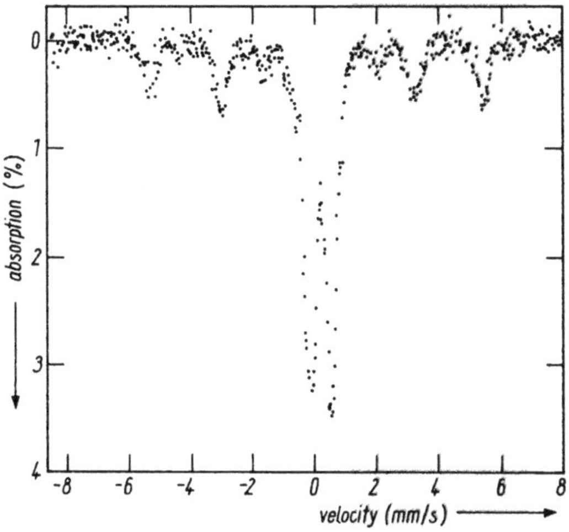

Fig. 2. Mössbauer spectrum of ^{57}Fe in FeCoGe at room temperature

similar behaviour, though the intensities of Zeeman sextets of FeNiGe are much weaker than those of FeCoGe. The internal field at the Fe nucleus of the ferromagnetic impurity phase for FeNiGe is about 310 kOe.

Taking into account the above results of X-ray and Mössbauer measurements, it becomes evident that the compounds FeCoGe and FeNiGe do not exist in a single phase: A small amount of ferromagnetic α-phase precipitates in the matrix with Ni_2In-type structure. So, the ferromagnetism which Castelliz /1/ and Buschow and von Engen /2/ reported for FeCoGe and FeNiGe is thought to be due to the existence of the ferromagnetic α-phase. As mentioned above, we observed that the intensities of Zeeman sextets for FeCoGe are much stronger than those of FeNiGe. Thus, we believe that a large amount of ferromagnetic α-phase exists in the FeCoGe matrix compared with the FeNiGe matrix. In fact, Castelliz /1/ reported that the magnetic moments for FeCoGe and FeNiGe are in the saturation state of 2.2 and 0.5 μ_B per molecule, respectively. Furthermore, Buschow and von Engen /2/ showed that the values of the saturation magnetization obtained at 4.2 K for FeCoGe and FeNiGe are found to be 2.50 and 0.70 μ_B, respectively.

On the basis of the present experimental results for FeCoGe and FeNiGe, it seems that the existence of vacancies in metal sites stabilizes the hexagonal

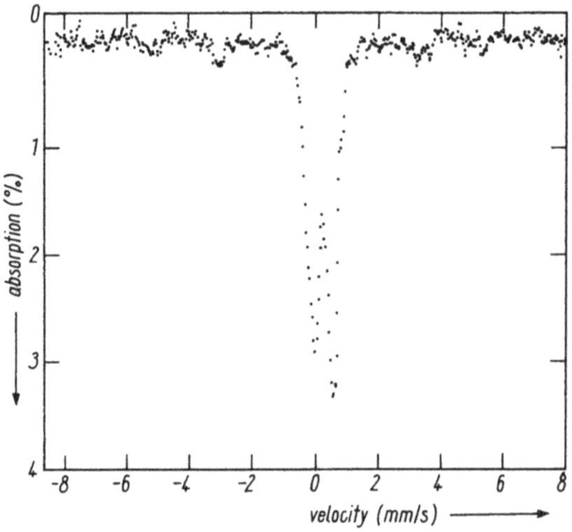

Fig. 3. Mössbauer spectrum of ^{57}Fe in FeNiGe at room temperature

Ni_2In-type structure. On the other hand, the compound PtMnGa of equiatomic composition crystallizes in the hexagonal Ni_2In-type structure /5, 6/. The stability of the Ni_2In-type structure, and the occupancy and the distributions of atoms on each site may depend on the sizes of atoms, their electronegativity, and the number of electrons in the outer shells, etc. Crystallographic and magnetic studies of the non-stoichiometric compounds $FeCo_xGe$ and $CoFe_xGe$ are in progress.

References

/1/ L. CASTELLIZ, Mh. Chem. 84, 765 (1953).

/2/ K.H.J. BUSCHOW and P.G. VAN ENGEN, phys. stat. sol. (a) 76, 615 (1983).

/3/ S. SZYTULA, A.T. PEDZIWIATR, Z. TOMKOWICZ, and W. BAZELA, J. Magnetism magnetic Mater. 25, 176 (1981).

/4/ T. GOTO and T. KANOMATA, J. appl. Phys. 57, 3450 (1985).

/5/ K.H.J. BUSCHOW and D.B. DE MOOIJ, J. less-common Metals 99, 125 (1984).

/6/ T. KANOMATA, K. SHIRAKAWA, and T. KANEKO, phys. stat. sol. (a) 97, K149 (1986).

(Received July 19, 1988)

phys. stat. sol. (a) 111, K7 (1989)
Subject classification: 66.30

Department of Physics (a) and Department of Chemistry (b), Friedrich Schiller University, Jena[1])

Diffusion from a Gaussian Half-Profile at an Absorbing Boundary in the Presence of an Electric Field

By

H. KAHNT (a) and T. POSSNER (b)

Recently Kaneko /1/ derived an analytical solution of a two-step diffusion process in an electric field. The first step was determined by diffusion from an infinite reservoir with a constant concentration and the second step by an absorbing boundary. This model was applied to the preparation of optical waveguides in glass by field assisted Ag-Na ion exchange. We here consider a similar two-step field assisted diffusion problem, which is of interest not only for optical waveguide processing. It also appears in the study of radio tracer diffusion in an electric field for the simultaneous determination of the diffusion coefficient and mobility. A detailed description of an experimental arrangement for such measurements basing on a modified Chemla experiment is given in /2/.

The first diffusion step is characterized by a finite plane source (δ-like) at the surface of the sample and the absence of the electric field. This step results in a Gaussian half-profile giving the initial condition for the second diffusion step which proceeds in an electric field with an absorbing boundary. Such an initial distribution can also be obtained by field assisted ion exchange through a small step-like mask opening, with a special combination of temperature and electric field /3/. The boundary condition is given by the contact to a salt melt which contains the respective alkali ion that was replaced in the sample at the preceeding step by radio tracer ions or by other ions for changing the refractive index, respectively. The problem is described by the diffusion equation

$$\partial c/\partial t = D \partial^2 c/\partial x^2 - V \partial c/\partial x \,, \quad V = \mu E \,; \tag{1}$$

D is the diffusion coefficient and V is the constant particle velocity given by the electric field E and the mobility μ. The transformation

1) Steiger 3, Haus 3, DDR-6900 Jena, GDR.

$$c = c^{*} \exp((2Vx - V^2 t)/4D) \tag{2}$$

leads to the simpler expression

$$\partial c/\partial t = D\,\partial^2 c^{*}/\partial x^2 . \tag{3}$$

The general solution of (2) is well known and can be obtained easily by separating the variables /4/:

$$c^{*}(x, t) = \int_0^{\infty} (A(u)\cos(ux) + B(u)\sin(ux))\exp(-u^2 Dt)du . \tag{4}$$

Next the boundary condition

$$c(0, t) = 0 \text{ for } t > 0 \tag{5}$$

and the initial condition given by the Gaussian half-profile obtained in the first diffusion step

$$c(x, 0) = \frac{2N}{(2\pi)^{1/2}\sigma} \exp(-x^2/2\sigma^2) \quad \text{for } x > 0 \tag{6}$$

must be incorporated into (4). N is the initial number of particles/cm^2 and σ is the standard deviation of the Gaussian half-profile. From (5) it is evident that $A(u) = 0$. The initial condition $c(x, 0)$ must be transformed into $c^{*}(x, 0)$ according to (2). This leads to

$$c^{*}(x, 0) = \frac{2N}{(2\pi)^{1/2}\sigma} \exp\left(-\frac{x^2}{2\sigma^2} - \frac{Vx}{2D}\right) . \tag{7}$$

For clearer presentation we rename $c^{*}(x, 0)$ into $f^{*}(x')$.

$$f^{*}(x') = \frac{2N}{(2\pi)^{1/2}\sigma} \exp\left(-\frac{x'^2}{2\sigma^2} - \frac{Vx'}{2D}\right) . \tag{7'}$$

Then (4) is written as

$$c^{*}(x, t) = \int_0^{\infty} B(u)\sin(ux)\exp(-u^2 Dt)du . \tag{8}$$

B(u) is given by the Fourier integral including the initial condition $f^{*}(x')$,

$$B(u) = \frac{2}{\pi}\int_0^{\infty} f^{*}(x')\sin(ux')dx' . \tag{9}$$

Inserting (9) into (8) and making use of

$$\frac{1}{2}\sin(ux')\sin(ux) = \cos(u(x'-x)) - \cos(u(x'+x)) \tag{10}$$

we get

$$c^* = c_1^* - c_2^* \tag{11}$$

with

$$c_1^* = \frac{1}{\pi} \int_0^\infty du \int_0^\infty f^*(x')\exp(-u^2 Dt)\cos(u(x' - x)\, dx') \tag{12}$$

and

$$c_2^* = \frac{1}{\pi} \int_0^\infty du \int_0^\infty f^*(x')\exp(-u^2 Dt)\cos(u(x' + x)\, dx'). \tag{13}$$

From

$$\int_0^\infty \exp(-a^2u^2)\cos(bu)du = \frac{\pi^{1/2}}{2a} \exp(-b^2/4a^2) \tag{14}$$

/5/, we get

$$c_1^* = \frac{1}{(2\pi Dt)^{1/2}} \int_0^\infty f^*(x')\exp(-(x' - x)^2/4Dt)\, dx' \tag{15}$$

and

$$c_2^* = \frac{1}{(2\pi Dt)^{1/2}} \int_0^\infty f^*(x')\exp(-(x' + x)^2/4Dt)\, dx'. \tag{16}$$

Inserting (7) in (15) and (16) we find

$$c_1^* = \frac{N \exp(-x^2/4Dt)}{(2\pi Dt)^{1/2}} \int_0^\infty \exp(-(ax'^2 + b_1 x')/c)dx' \tag{17}$$

and

$$c_2^* = \frac{N \exp(-x^2/4Dt)}{(2\pi Dt)^{1/2}} \int_0^\infty \exp(-(ax'^2 + b_2 x')/c)dx' \tag{18}$$

with

$$a = 2Dt + \sigma^2, \tag{19}$$

$$b_1 = 2\sigma^2(Vt - x), \tag{20}$$

$$b_2 = 2\sigma^2(Vt + x), \tag{21}$$

$$c = 4\sigma^2 Dt. \tag{22}$$

The integral

$$J = \int_0^\infty \exp(-(ax'^2 + bx')/c)dx' \tag{23}$$

is now rewritten as

$$J = \exp(b^2/4ac) \int_0^\infty \exp(-\frac{a}{c}(x' + b/2a)^2)dx' \tag{24}$$

and transformed into

$$J = \exp(b^2/4ac)(c/a)^{1/2} \int_{b/(4ac)^{1/2}}^{\infty} \exp(-\xi^2)d\xi \tag{25}$$

using the substitution

$$\xi = (a/c)^{1/2}(x' + b/2a) . \tag{26}$$

Inserting (26) into (17) and (18) and taking into consideration (11), (19) to (22) and

$$\mathrm{erfc}(x) = \frac{2}{\pi^{1/2}} \int_x^{\infty} \exp(-z^2)dz \tag{27}$$

we find

$$c^*(x, t) = \frac{N}{(2\pi(2Dt + \sigma^2))^{1/2}} \exp(-x^2/4Dt) \times$$

$$\times \left(\exp(Y^2)\mathrm{erfc}(Y) + \exp(Z^2)\mathrm{erfc}(Z)\right) \tag{28}$$

with

$$Y = \frac{Vt - x}{2(Dt(2Dt + \sigma^2))^{1/2}} \tag{29}$$

and

$$Z = \frac{Vt + x}{2(Dt(2Dt + \sigma^2))^{1/2}} \tag{30}$$

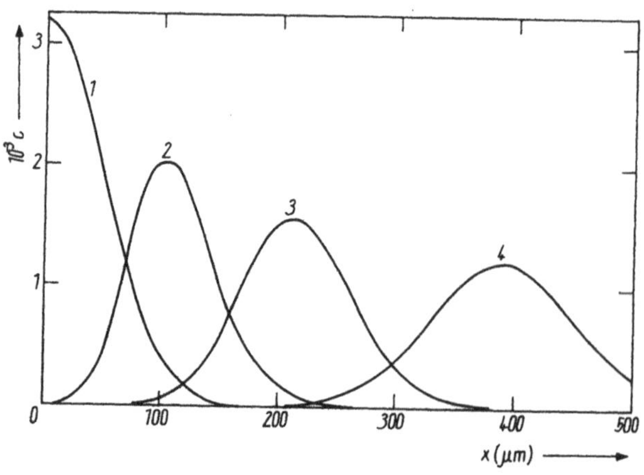

Fig. 1. Particle distribution for a set of typical experimental parameters ($D^* = 3\times10^{-10}$ cm^{2}s^{-1}, $\mu = 10^{-8}$ cm^{2}V^{-1}s^{-1}, E = 20 Vcm^{-1}, $\sigma = 5\times10^{-3}$ cm) for different times t(s): (1) 0, (2) 1×10^4, (3) 2.5×10^4, (4) 5×10^4

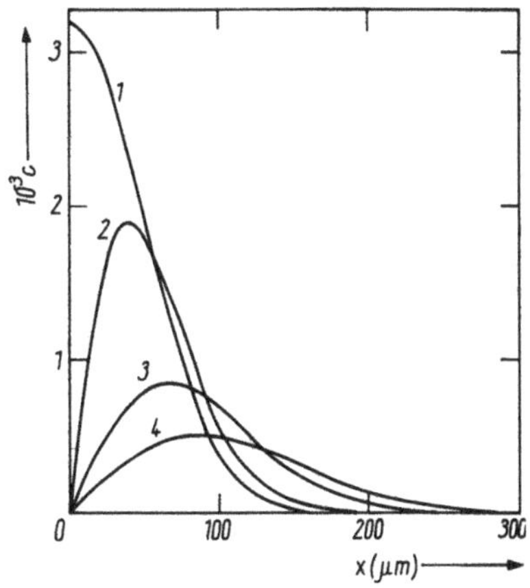

Fig. 2. Particle distribution at zero electric field with the same sample parameters as in Fig. 1 for different times t (s): (1) 0, (2) 1×10^4, (3) 5×10^4, (4) 1×10^5

Finally, using (2) $c^*(x, t)$ is re-transformed into $c(x, t)$.

$$c(x, t) = \frac{N}{(2\pi(2Dt + \sigma^2))^{1/2}} \exp\left[-\frac{(Vt - x)^2}{4Dt}\right](\exp(Y^2)\,\mathrm{erfc}(Y) - \exp(Z^2)\,\mathrm{erfc}(Z)). \tag{31}$$

Fig. 1 shows the diffusion profiles for a set of typical experimental parameters. With increasing time the particles are shifted into the sample and the distribution is broadened. For zero electric field (Fig. 2) the maximum of the particle distribution moves into the sample due to the loss of particles through the absorbing boundary as can be seen from the decreasing area of the distribution.

References

/1/ T. KANEKO, J. Phys. D 18, 1597 (1985).

/2/ H. KAHNT, CH. KAPS, and H. OPFERMANN, submitted to Solid State Ionics.

/3/ T. POSSNER and G. SCHREITER, to be published.

/4/ W. JOST and K. HAUFFE, Fortschritte der Physik. Chemie, Vol. 1, Dr. Dietrich Steinkopff Verl., Darmstadt 1972.

/5/ I.N. BRONSTEIN and K.A. SEMENDJAJEV, Taschenbuch der Mathematik, Izd. Nauka, Moskau and B.G. Teubner, Leipzig 1979.

(Received September 1, 1988)

phys. stat. sol. (a) 111, K13 (1989)

Subject classification: 61.80 and 68.48; S1.2

Belorussian State University, Minsk[1] (a) and
Friedrich-Schiller-University, Jena[2] (b)

Study of Ion Bombardment of Double-Layer Thin Zr/Ni Films

By

V.P. GOLTSEV (a), V.V. KHODASEVICH (a), G. GÖTZ (b),
I.I. PRIKHODKO (a), V.M. DRAKO (a), and V.V. UGLOV (a)

Atomic mixing of two solid phases using ion beams can result in the formation of stable, metastable, amorphous, and crystalline structures /1 to 3/.

The exact knowledge of the elemental composition and distribution of impurities in the depth of the modified layer will promote an understanding of the mechanism of their formation.

The present note gives results on the elemental and phase composition of the thin-layer system Zr/Ni mixed by argon ions. The largest mixing effect was achieved by choosing the argon ion energy equal to 150 keV so that the released energy maximum was at the interface of both films. The radiation doses were 10^{16} and $5x10^{16}$ ions/cm^2. The elemental composition and distribution of impurities in the depth were analysed using Rutherford backscattering of helium ions of 1.4 MeV energy. The depth scale of concentration profiles is given in Nt units, because the density of our samples changed with depth, especially in the mixing region. The phase composition was studied by electron microscopy.

The analysis of the elemental composition has shown that argon ion irradiation up to doses of 10^{16} and $5x10^{16}$ ions/cm^2 causes a considerable concentration of oxygen, carbon, and argon in the surface region. The concentration of these elements increased with irradiation dose (see Table 1).

The presence of oxygen and carbon impurities in the films and the increase of their concentration with radiation dose can be attributed to several causes. The first reason is the ability of thin films to absorb residual gases, oxygen, and carbon from the atmosphere. The second one is the intensive saturation of the upper film layer by these elements during irradiation due to the inverse

---

1) Leninskii Prospekt 4, SU-22080 Minsk, USSR.

2) Max-Wien-Platz 1, DDR-6900 Jena, GDR.

Table 1

Relative elemental concentration (%) in the layer calculated with respect to nickel

dose (ions/cm^2)	C	O	Ar	Zr
$1x10^{16}$	10	24	4	38
$5x10^{16}$	11	39	13	37

effect. The high concentration of oxygen and carbon in our experiments is determined by the high absorption capability of the upper zirconium film.

Along with changes in the elemental composition, irradiation-induced redistribution of impurities in the sample depth occurs. It should be noted that the oxygen concentration is highest in the upper zirconium film layers (Fig. 1).

Atomic mixing promotes the redistribution of zirconium in nickel and nickel in zirconium deep into the sample and at the surface. Thus, a mixing region is formed whose dimensions depend on the radiation dose and amount

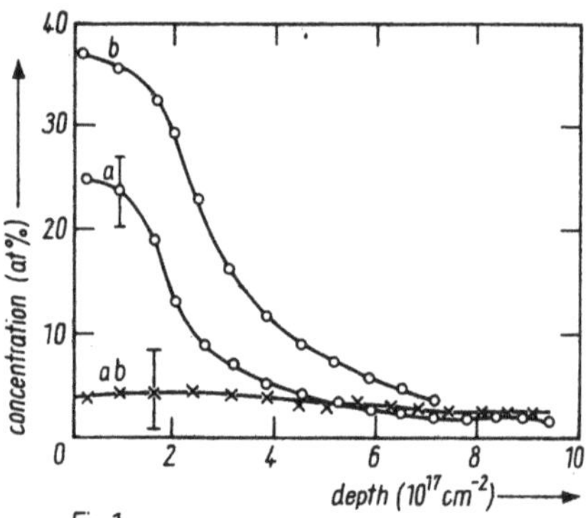

Fig.1

Fig. 1. Oxygen (O) and carbon (×) concentration profiles for films mixed by argon ions with doses of (a) 10^{16} and (b) $5x10^{16}$ ions/cm^2

Fig. 2. Nickel and zirconium concentration profiles for films mixed by argon ions with doses of (a) 10^{16} and (b) $5x10^{16}$ ions/cm^2

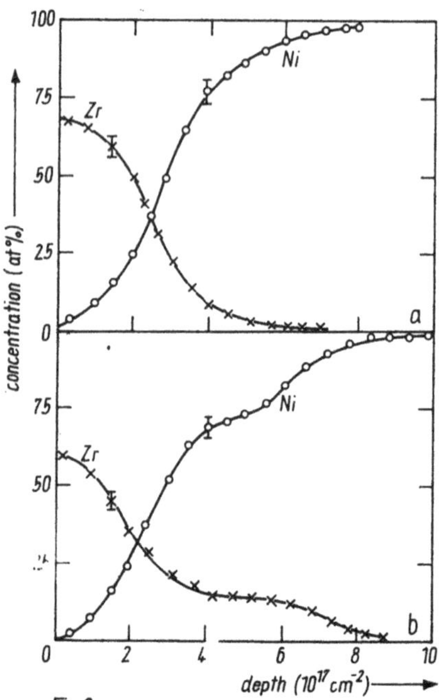

Fig.2

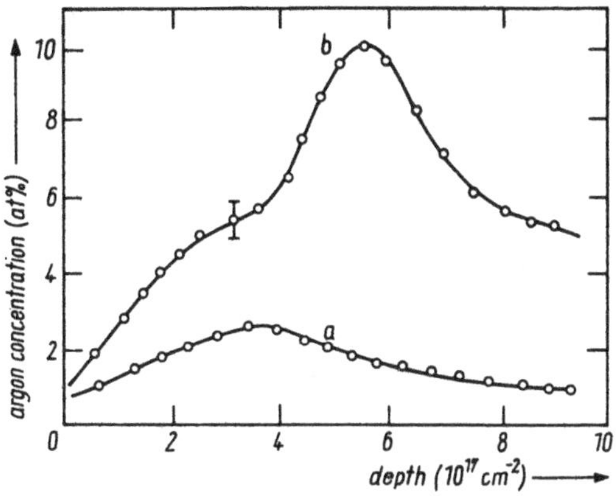

Fig. 3. Argon concentration profiles in films mixed by argon ions with doses of 10^{16} (a) and $5x10^{16}$ (b) ions/cm^2

to $6x10^{17}$ and $8x10^{17}$ cm^{-2} in Nt units (Fig. 2).

The high argon ion irradiation doses used for an effective mixing of thin zirconium and nickel films cause an excessive argon concentration in the mixing region. Its distribution profile is characterized by a maximum concentration at a depth of $6x10^{17}$ cm^{-2} (Fig. 3). For irradiation doses of 10^{16} ions/cm^2, this maximum is less pronounced and the argon concentration amounts to about 2 at%. An increase of the dose up to $5x10^{17}$ ions/cm^2 leads to the formation of a more pronounced maximum with concentration of about 10 at%.

The high concentration of elements observed in our experiment (see Table 1) may provide the formation of oxygen and carbon compounds with zirconium and nickel as well as zirconium and nickel compounds. Fig. 4 shows the photometric results for electron patterns obtained for samples irradiated by 10^{16} and $5x10^{16}$ ions/cm^2. This figure shows that ZrO_2 and $NiZr_2$ compounds are generated. ZrO_2 formation is due to the high oxygen concentration in a narrow surface region achieving 25 and 40 at% for doses of 10^{16} and $5x10^{16}$ ions/cm^2, respectively (Fig. 1). Evidently the entire zirconium which is not redistributed into the mixing region chemically reacts with oxygen.

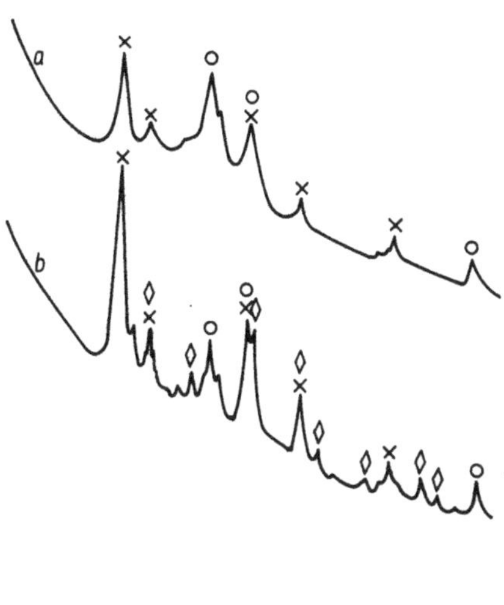

Fig. 4. Densitometric curves of electron patterns of films mixed by argon ions with doses of 10^{16} (a) and $5x10^{16}$ (b) ions/cm^1; o Ni, x ZrO_2, ◇ $NiZr_2$

This is confirmed by the absence of pure zirconium reflexes for both irradiation doses. No compound formation between zirconium and carbon is found which is perhaps due to its moderate, in comparison with oxygen, concentration in the film being 5 at% (Fig. 1).

The formation of a mixing region and its increase with irradiation dose results in the phase $NiZr_2$ for a mixing dose of $5x10^{16}$ ions/cm^2. The appearance of this phase is associated with a step in the nickel and zirconium distribution profiles which allows to estimate the depth of this phase being equal to $6x10^{17}$ cm^{-2} (in Nt units).

References

/1/ L.S. HUNG and J.W. MAYER, Nuclear Instrum. and Methods B7/8, 676 (1985).

/2/ V.V. KHODASEVICH, I.I. PRIKHODKO, and V.V. UGLOV, Coll. Abstr. VIII. All. Union Conf. Interaction of Atomic Particles with Solids, RTI, Minsk 1987 (p. 283).

/3/ V.V. KHODASEVICH, I.I. PRIKHODKO, and V.V. UGLOV, Coll. Abstr. XVII. All. Union Conf. Interaction of Charged Particles with Crystals, MGU, Moscow 1987 (p. 172).

(Received March 15, 1988; in revised form October 24, 1988)

phys. stat. sol. (a) 111, K17 (1989)

Subject classification: 61.70; 72.80; 75.50; S8

Polytechnical Institute, Academy of Sciences of the Azerbaidzhan SSR, Baku[1])

Magnetic Susceptibility and Electrophysical Properties of SiTe Single Crystals

By

F.M. KAMARLY, N.G. GUSEINOV, and A.YU. YANGIROV

It is known from published literature that the A^{IV}-group monotellurides exhibit several kinds of defects /1 to 3/, in particular those due to A^{IV}-group elements and tellurium, as well as singly charged and antisite (a.s.) defects, all of them being dependent on the relations between the components of the compounds under discussion.

A deviation from the stoichiometry with excessive tellurium is typical of the A^{IV}Te compounds having the following compositions: 50.012, 50.40, and 50.61 at% Te in PbTe, SnTe, and GeTe, respectively. It may be assumed that SiTe also belongs to this series, and hence the SiTe-based homogeneity region will be in the range of excessive tellurium content with respect to the stoichiometric composition.

The alloys prepared for the investigation were close to SiTe compounds with 49.02 to 50.20 at% Te contents. All of the investigated crystals with excessive tellurium had p-type conductivities.

In order to reveal the nature of defects in SiTe, the magnetic susceptibility χ was studied and its components, i.e. the magnetic susceptibilities of the crystal lattice $\chi_{latt.}$, the free carriers $\chi_{f.c.}$, and the defects $\chi_{def.}$ were calculated with the use of appropriate parameters.

Besides the magnetic susceptibility, we studied the concentration dependences of thermo-e.m.f. and electric conductivity of the Si-Te system close to the stoichiometric composition of SiTe.

A magnetic susceptibility-composition microdiagram for Si-Te at room temperature shows the maximum diamagnetic susceptibility at a chemical composition with 50.08 at% Te (Fig. 1).

An increase in the concentration of tellurium or silicon in the alloy brings about a decrease in the diamagnetism of the system. The isotherms of the com-

1)

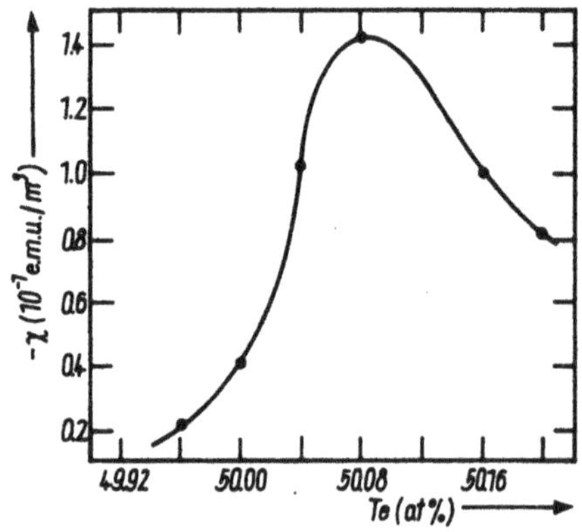

Fig. 1. Concentration dependence of the magnetic susceptibility of SiTe single crystals

position dependence of electrical conductivity and thermo-e.m.f. exhibit sharply defined extrema for chemical compounds corresponding to the composition deviating from stoichiometry (Fig. 2). The existence of extrema in the isotherms showing the relationship between the composition and the properties (χ, α, σ) bears witness to the conservation of a strong chemical interaction. Thus, an analysis of (χ, α, σ) ~ f(c) has shown that SiTe can exist not only at an exact stoichiometric ratio between the Si and Te components but also near this ratio.

A shift of the χ, α, and σ extrema on the tellurium side may be associated with the formation of cationic (Si) vacancies.

At minor deviations (of the order of 1 at%) of the SiTe components the parameters involved in the expression for the lattice share of the total susceptibility /1/ ($\chi_{latt.}$ = -1×10^{-6} ϱZ/M e.m.u./m^3, where ϱ is the density, Z is the total number of electrons, and M is the molecular weight) vary only slightly and that is why its value is almost constant and amounts to $\chi_{latt.}$ = = -1.660 e.m.u./m^3.

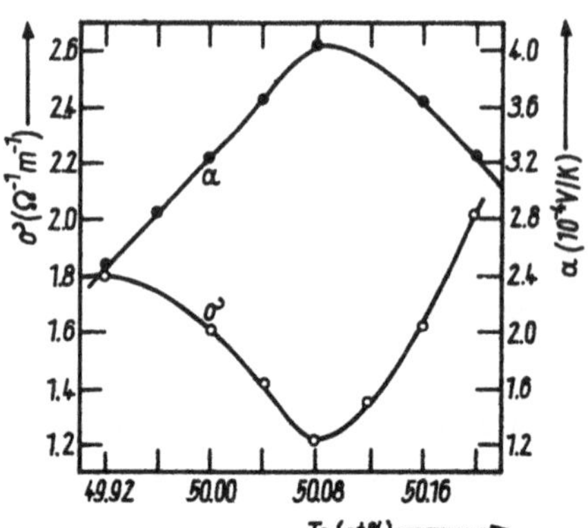

A calculation of the effective mass of the density of states for stoichiometric SiTe at a concentration of current carriers P = 1×10^{23} m^{-3}, performed with the formula /1/

$$m^*/m_o = \frac{1}{T}\left(\frac{P}{5.44\times10^{15}\; F_{1/2}(\mu^*)}\right)$$

Fig. 2. Concentration dependences of the electrical conductivity and thermo-e.m.f. of SiTe single crystals has resulted in a value of $m^*/m_o = 0.14$, and hence $\chi_{f.c.} = -0.007$ e.m.u./m^3. The value is beyond the accuracy of our measurements and $\chi_{f.c.} \ll \chi_{latt.}$, and that is why we shall not take it into account henceforth.

The magnetic susceptibility of the defects was determined by subtracting the lattice share of susceptibility from the total (measured) susceptibility.

In this case the magnetic susceptibility must be positive, if the main type of defects consists of Si vacancies presented as doubly plus-charged centres (similar to Ge vacancies in GeTe and Sn vacancies in SnTe). If the main type of defects is represented by singly plus-charged defects (p-type conductivity of SiTe) brought about according to the pattern $Si^{2-}_{a.s.} \rightleftarrows Si^{1-}_{a.s.} + \ominus$, i.e. each vacancy gives one carrier (hole) to the conduction band, or by antisite defects produced according to the pattern:

$$Si^{2-}_{a.s.} \rightleftarrows Si^{1-}_{a.s.} + \ominus \quad ,$$

$$Te^{2+}_{a.s.} \rightleftarrows Te^{1+}_{a.s.} + \oplus \quad ,$$

the magnetic susceptibility is negative.

From $\chi_{def.} = \chi_{total} - \chi_{latt.}$ we obtain the susceptibility of defects at room temperature $\chi_{def.} = 0.297$ e.m.u./m^3.

With the expression $\chi_{def.} = n_d \left(\frac{\mu_B}{kT} - \frac{e^2 m_o^2 \varepsilon^2 d^2}{a_B\, m^{*3}} \right)$ /1to 3/the concentration of defects n_d can be determined if one knows the values of the other terms in the expression, where μ_B is the Bohr magneton, k is the Boltzmann constant, e is the electronic charge, m_o is the electron mass, ε is the dielectric constant of the medium (in our case it is air), a_B is the radius of the first Bohr orbit, and m^* is the effective mass of the density of states.

The concentration of defects, determined by us, is $n_d = 1.70 \times 10^{23}$ m^3. With the concentration of current carriers being known, one can find the total value for singly plus-charged and antisite defects:

$$n_{+1} + n_{a.s.} = n_d \, ,$$

$$n_{+1} - n_{a.s.} = P \, .$$

The concentration of singly plus-charged defects, determined in this way is $n_{+1} \approx 1.33 \times 10^{23}$ m^{-3}, whereas that of antisite defects is $n_{a.s.} \approx 0.37 \times 10^{23}$ m^{-3}.

It is evident that the antisite defects of Te as a carrier donating to the valence band (according to the above pattern) are added to the singly charged Si vacancies. Hence, the main defects in SiTe are the singly plus-charged Si vacancies and the antisite Te defects that can give a hole concentration in the valence band amounting to $P = 1.33 \times 10^{23}\ m^{-3}$, and a concentration of antisite $Si_{a.s.}$ defects of the order of $n_{a.s.} \approx 0.37 \times 10^{23}\ m^{-3}$.

References

/1/ I.M. TSIDILKOVSKII, Thermomagnetic Phenomena in Semiconductors, Fizmatgiz, 1960 (in Russian).

/2/ S.V. VONSOVSKII, Magnetism, Nauka, Moscow 1971 (in Russian).

/3/ V.F. ORMONT, Introduction to the Physical and Crystals Chemistry of Semiconductors, Vysshaya Shkola, Moscow 1973 (in Russian).

(Received June 19, 1987)

phys. stat. sol. (a) 111, K21 (1989)
Subject classification: 64.70 and 64.75; S8.15

National Research Institute for Metals,
Tsukuba-Laboratories, Ibaraki[1])

Solid Solubility Range and Lattice Parameter of Semiconducting $Pb_{1-x}Cd_xS_{1-y}Se_y$ for Mid-Infra-Red Lasers

By
T. KIYOSAWA, S. TAKAHASHI, and N. KOGUCHI

Introduction Extensive research has been done on the physico-chemical and semiconducting properties of PbS, PbSe, CdS, CdSe, and their pseudo-binary alloys. However, there has been no report regarding the quarternary system with PbS, PbSe, CdS, and CdSe as terminal compounds.

PbS /1/ and $Pb_{1-x}Cd_xS$ /2, 3/ have been fabricated into laser diodes that emit light at short wavelengths of the mid-infra-red region. Application of lattice-matched double-hetero and/or quantum well structures of laser diodes to the materials is expected to greatly improve the performance of PbS-based laser diodes. This makes the $Pb_{1-x}Cd_xS_{1-y}Se_y$ quarternary alloy a promising material and the knowledge on its lattice constant, band gap energy, and refractive index is required.

As a first step to better understand the quarternary system, this note investigated the solid phase equilibria in the PbS-PbSe-CdSe-CdS system, as well as the solid solubility range and lattice constant of the $Pb_{1-x}Cd_xS_{1-y}Se_y$ alloy.

Preparation of materials and measurement techniques Various compositions in the PbS-PbSe-CdSe-CdS system were prepared from 6 N grade elemental materials (i.e. Pb ingot, Cd and Se shot, and S powder) by the standard method of melting and water-quenching. To achieve this, several grams of the elemental mixture were sealed in a cleaned quartz ampoule under a vacuum of 10^{-6} Torr. After the ampoule was first heated carefully with a hand torch to react the elements, it was introduced into a furnace and kept at about 1130 $^{\circ}$C with the sample molten for 5 h and subsequently water-quenched.

To examine the solid phase equilibria within the PbS-PbSe-CdSe-CdS system samples with x = 0.5, and y = 0.25 and 0.75 were prepared. The bulk

---

1) 1-2-1 Sengen, Tsukuba-Shi, Ibaraki 305, Japan.

materials obtained by water-quenching from the melt were heat treated at 818 and 963 °C for eight days and water-quenched. Care was taken to keep a uniform temperature within the ampoule. The chemical composition of the phases in the samples was measured by X-ray micro-analysis using Cd, PbS, and PbSe as standards. The phases were also identified by an X-ray powder diffraction method. Samples with identical composition to those above were annealed for only four days to check attainment of an equilibrium state. Comparison of their X-ray diffraction results with those of samples annealed for eight days showed no difference in the existing phases and their lattice parameters.

The lattice constant of the $Pb_{1-x}Cd_xS_{1-y}Se_y$ phase was measured by an X-ray powder diffraction method for samples with y = 0, 0.5, and 1.0 and various x values to determine the variation of the lattice constant in the $Pb_{1-x}Cd_xS_{1-y}Se_y$ phase and the solubility limit of Cd. Samples for X-ray diffraction measurements were prepared by grinding the entire water-quenched bulk material to a powder and annealing it in an evacuated quartz ampoule with subsequent water-quenching. The annealing time was changed to assure obtainment of an equilibrium state. Typically, four days at a temperature between 400 and 700 °C or one day above 800 °C was sufficient. Powder diffraction patterns were obtained by a conventional X-ray diffractometer using CuK_α radiation. Lattice constants were calculated from extrapolation of the least-squares equation for the spacing versus $\cos^2\Theta$ plot for reflections higher than (622).

Results and discussion From the results of X-ray micro-analysis and X-ray powder diffraction, the isothermal phase diagram at 963 °C of the PbS-PbSe-CdSe-CdS system was constructed as shown in Fig. 1. There are two solid phases in equilibrium. One is the $Pb_{1-x}Cd_xS_{1-y}Se_y$ solid solution with the cubic NaCl structure (denoted α phase hereafter) and the other is the $CdS_{1-z}Se_z$ solid solution with the wurtzite structure in which Pb was not detected by X-ray micro-analysis (denoted β phase hereafter). The hatched area in the figure represents the α phase range. The heavy line denotes the phase boundary and the thin lines denote tie lines. The solubility limit of Cd in the α phase for y = 0.5 and 1.0 was obtained from the variation of the lattice constant with Cd content as described below, while that for y = 0 was cal-

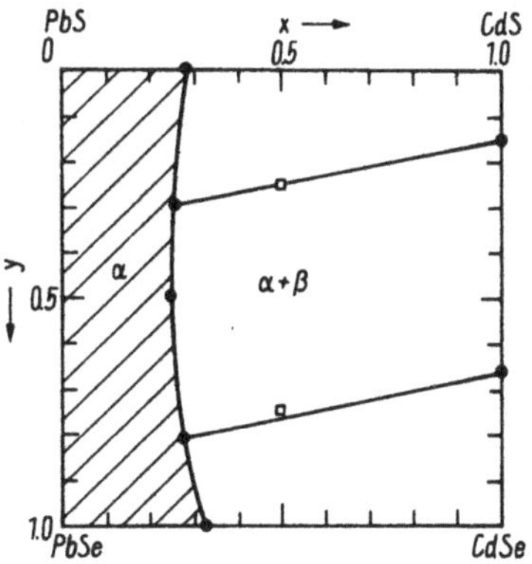

Fig. 1. Isothermal phase diagram of the PbS-PbSe-CdSe-CdS system at 963 °C. α : $Pb_{1-x}Cd_xS_{1-y}Se_y$ solid solution, β : $CdS_{1-z}Se_z$ solid solution, thin line: tie line, □ sample composition, ● after /4/

culated from an equation given by /4/. The isothermal phase diagram at 818 °C was found to be very similar to that at 963 °C except that the α phase range was reduced, as shown in Fig. 4.

The α phase lattice constant is shown in Fig. 2 for y = 0 ($Pb_{1-x}Cd_xS$), y = 0.5 ($Pb_{1-x}Cd_xS_{0.5}Se_{0.5}$), and y = 1.0 ($Pb_{1-x}Cd_xSe$). The samples were annealed at 952 °C. The variation of the lattice constant with Cd content can be parameterized for x less than the solubility limit as

$$a(x) = 0.5936-0.0417x\,(\mathrm{nm}) \quad \text{for } y = 0\,, \tag{1}$$

$$a(x) = 0.6032-0.0419x\,(\mathrm{nm}) \quad \text{for } y = 0.5\,, \tag{2}$$

$$a(x) = 0.6127-0.0418x\,(\mathrm{nm}) \quad \text{for } y = 1.0\,. \tag{3}$$

Equation (1) agrees well with the result reported by /4/ but not with those by /5 to 7/. Equation (3) agrees favourably with the result of /8/ but not with that of /9/. It should be noted that all the equations are linear with similar gradients, and that the lattice constant of $PbS_{0.5}Se_{0.5}$ ((2) with x = 0) is close to the lattice constant average of PbS and PbSe. As a result, the variation of the lattice constant with x and y in the α phase is given by

$$a(x,\ y) = 0.5936 - 0.0418\ x + 0.0191\ y\ (\mathrm{nm})\ . \tag{4}$$

The solubility limit of Cd in the α phase was determined at y = 0.5 and 1.0 by observing the deviation from linearity of its lattice constant versus x relation at various temperatures, as shown in Fig. 2 for 952 °C. The lattice constant of samples with fixed value of y is expected to vary even in the two phase (α+ β) region except for y = 0 or 1.0, since the α phase composition moves along the phase boundary such that increase of the sample composition

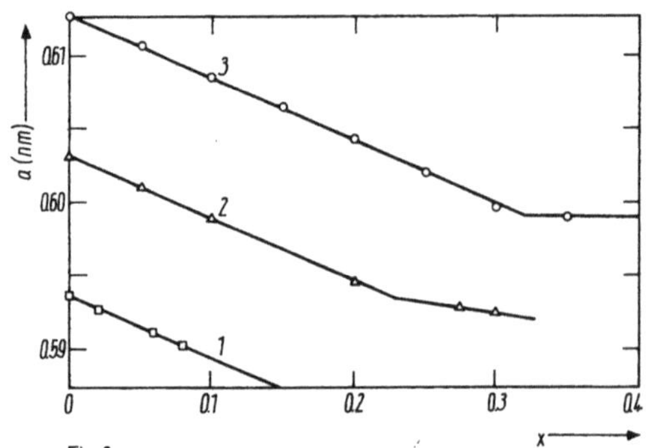

Fig.2

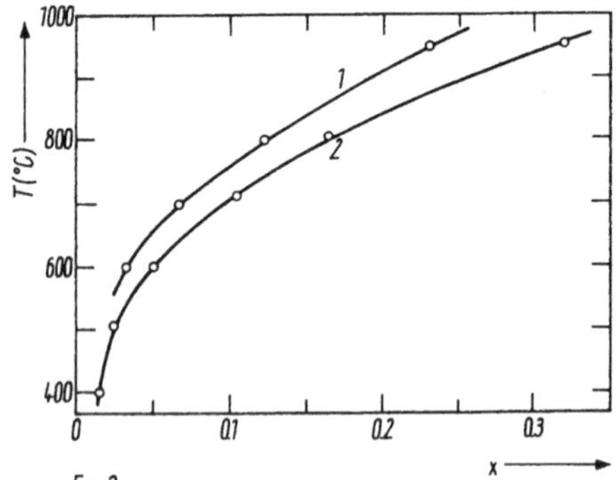

Fig.3

Fig. 2. Change of the α phase lattice constant with Cd content, x, for: y = (1) 0, (2) 0.5, (3) 1.0

Fig. 3. Temperature dependence of the solubility limit of Cd in the α phase: y = (1) 0.5, (2) 1.0

x causes increase of y in the α phase, as suggested by location of the tie lines shown in Fig. 1. The temperature dependence of the solubility limit thus obtained is plotted in Fig. 3. Our result for the solubility limit of Cd for y = 1.0 is larger than that given by /8, 10/ at 800 and 940 °C. Concerning the discrepancy, it may be reasonable to point out that the solubility limit of Cd in PbSe strongly depends on its deviation from stoichiometry /9/.

Contours of constant lattice parameter calculated from (4) are shown in Fig. 4 together with the range of α phase at 818 and 963 °C, which serves to show the compositional condition for the lattice-matched

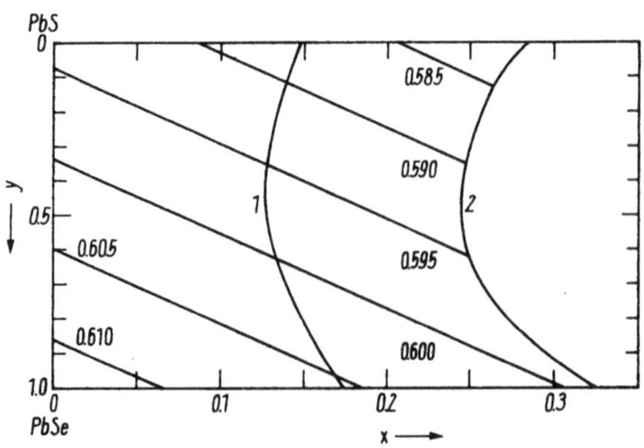

Fig. 4. Contours of constant lattice parameter and range of α phase at (1) 818 °C and (2) 963 °C

hetero-structure of PbS-based laser diodes. Our preliminary result was reported in /11/ on fabrication of lattice-matched double hetero-structure laser diodes using the $Pb_{1-x}Cd_xS_{1-y}Se_y$ alloy and their lasing characteristics. Variation of band gap energy and refractive index in the α phase will be reported elsewhere.

The authors are indebted to T. Kimura for X-ray micro-analysis. They wish to thank S. Tsukamoto and Y. Usui for their assistance in the experiments, and Dr. K. Nii of their Institute and Prof. K. Masumoto of Tohoku University for their encouragement. The first author appreciates a discussion with Dr. D.R. Dietderich.

References

/1/ H. PREIER, Appl. Phys. 20, 189 (1979).

/2/ A.R. CALAWA, J.A. MROCZKOWSKI, and T.C. HARMAN, J. electronic Mater. 1, 191 (1972).

/3/ K.W. NILL, A.J. STRAUSS, and F.A. BLUM, Appl. Phys. Letters 22, 677 (1973).

/4/ P.M. BETHKE and P.B. BARTON, Amer. Mineral. 56, 2034 (1971).

/5/ G.S. OLEINIK, P.A. MIZETSKII, A.I. NIZKUVA, L.A. POLIVTSEV, and I.A. RYADNINA, Izv. Akad. Nauk SSSR, Ser. neorg. Mater. 19, 1587 (1983).

/6/ P. STETIU, phys. stat. sol. (a) 15, K19 (1973).

/7/ A.K. SOOD, K. WUD, and J.N. ZEMEL, Thin Solid Films 48, 73 (1978).

/8/ F. WALD and A.J. ROSENBERG, J. Phys. Chem. Solids 26, 1087 (1965).

/9/ B.J. SEALY and A.J. CROCKER, J. Mater. Sci. 8, 1247 (1973).

/10/ Z.F. TOMASHIK, G.S. OLEINIK, and V.N. TOMASHIK, Izv. Akad. Nauk, SSSR, Ser. neorg. Mater. 16, 168 (1980).

/11/ K. KOGUCHI, T. KIYOSAWA, and S. TAKAHASHI, J. Crystal Growth 81, 400 (1987).

(Received September 26, 1988)

phys. stat. sol. (a) 111, K27 (1989)

Subject classification: 64.70; 77.20; S12

Institute of Chemistry, University of Wrocław[1] (a) and
Institute for Low-Temperature and Structure Research,
Polish Academy of Sciences, Wrocław[2] (b)

Phase Transition in $(CH_3NH_3)_3Bi_2Cl_9$

By

R. JAKUBAS (a), P.E. TOMASZEWSKI (b), and L. SOBCZYK (a)

Alkylammonium halogenoantimonates appeared to be interesting crystals showing in many cases phase transitions to the polar state /1 to 5/. On the other hand, the exchange of antimony by bismuth creates further possibilities of modification of the properties and stimulates searches for new materials of practical importance. The studies performed to date prove that such an exchange can lead to substantial changes of properties in spite of similarities in the crystal structure /6/. In the present communication we would like to report on the growth of $(CH_3NH_3)_3Bi_2Cl_9$ crystal (abbreviated as MACB) and compare its properties with the antimony analogue (MACA). The MACA crystal is built up of infinite chloroantimonate polyanions of double-chain-stack structure directed along the a-axis with three crystallographically inequivalent $CH_3NH_3^+$ cations /1/. The structural phase transition at 208 K observed by using various techniques is related to a freezing of the orientational motion of those cations. The behaviour of dielectric properties and second moment of proton magnetic resonance showed that above 208 K a continuous release of orientational motion takes place at least for one of the three types of cations.

The single crystals of MACB were grown similarly as previously reported for MACA /1/. The Weissenberg photographs allow to classify the crystal to be orthorhombic, space group Pmcn, with the lattice parameters measured by the Bond method:

$$a = 0.7697\ \text{nm},\quad b = 1.3248\ \text{nm},\quad c = 2.0422\ \text{nm},$$
$$D_M = 2.67\ \text{Mg/m}^3, D_C = 2.66\ \text{Mg/m}^3, Z = 4.$$

The crystal was found to be isomorphous with MACA and $Cs_3Sb_2Cl_9$ /7/. Prismatic-shaped rodes elongated along the a-axis with very small unitary (010) and

---

1) Joliot-Curie 14, PL-50-383 Wrocław, Poland.

2) Pl. Katedralny 1, PL-50-950 Wrocław, Poland.

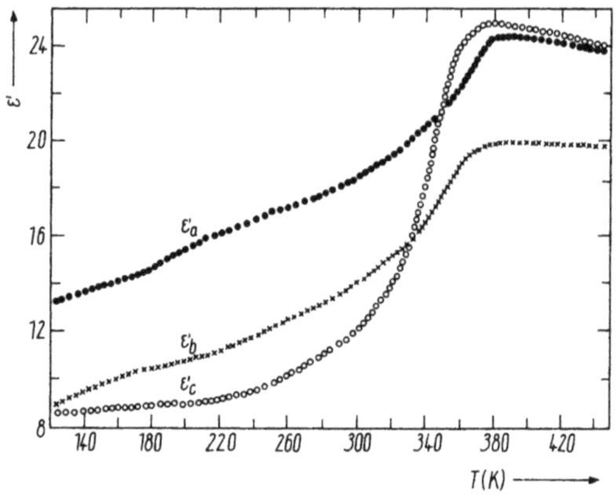

Fig. 1. ε' plotted against temperature for ● a, × b, and ○ c axes (on heating)

and (001) faces were usually obtained, contrary to MACA. Thus, the orientation of samples for dielectric studies should be carefully verified by X-rays. The correct orientation of the sample may also be made by using a polarizing microscope (the c-axis coincides with the direction of extinction).

The measurements of the static electric permittivity at 1 kHz have been carried out by using an automatic capacitance bridge E 315 A. The samples of 0.5x0.5x0.1 cm^3 dimensions were covered with silver paste.

The preliminarily powder diffraction studies show that the crystal structure motif is the same as for the antimony analogue MACA. The isomorphism of MACA and MACB implies a similar ordering of methylammonium cations in the polyanion sublattice and, therefore, we should expect considerable similarities of both crystals. As far as dielectric properties are concerned, we find a distinct resemblance but some essential differences have also to be noticed. In Fig. 1 the temperature dependence of the static electric permittivity is shown along three crystallographic axes when heating. The temperature dependences of ε' and dielectric anisotropy are similar to those obtained for MACA. The a direction is distinguished in accordance of the orientation of polyanions. However, contrary to MACA crystal, MACB does not show the low temperature first-order phase transition which is characterized by a stepwise change of permittivity. On the other hand, for MACB another phase transition at (385 $\pm$ 2)K is observed. The change of the symmetry at this temperature was confirmed by a preliminary powder high-temperature X-ray examination. The temperature dependence of ε'_c shows a weakly marked maximum ($\varepsilon'_{c, max} \approx 25$) as shown in Fig. 2 for both heating and cooling cycles. For the remaining axes the maximum is much less pronounced. The lack of temperature hysteresis of maxima and the character of the ε'(T) plot speak in favour of a continuous structural phase transition. Somewhat lower values of ε' when cooling are most

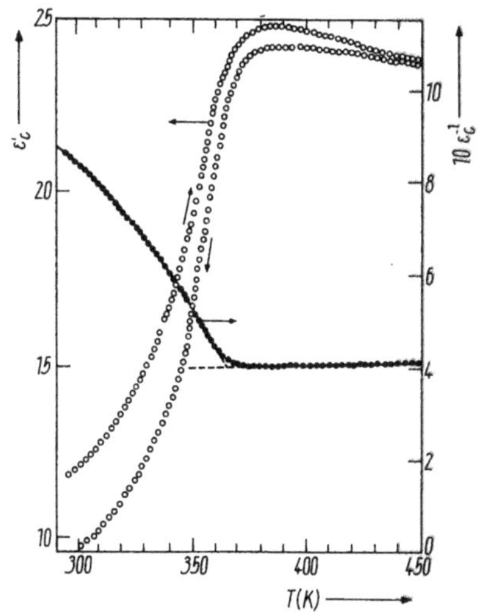

Fig. 2. ε_c' plotted against temperature in the vicinity of the phase transition point (open circles) and Curie-Weiss law (on heating, full circles)

probably due to an increase of defects, particularly at the surfaces of the crystal. The dielectric losses do not show any anomaly. They increase continuously, the slope being enlarged above 400 K. It is worth pointing out that the maxima on the plot of $\varepsilon'(T)$ are observed only for crystals of high quality which are distinguished by low values of tan δ. The Curie-Weiss law is nearly obeyed in the low temperature phase as shown in Fig. 2. The high temperature phase of MACB above 385 K which has no equivalence in MACA, resembles the plastic crystals, i.e. all methylammonium cations possess presumably a freedom of isotropic rotation. Some difference in permittivity can be due to the anisotropy of atomic and ionic polarizability /1/. In the low temperature phase a continuous freezing of orientational motion takes place similarly to MACA crystals.

References

/1/ R. JAKUBAS, Z. CZAPLA, Z. GALEWSKI, L. SOBZZYK, O.J. ŻOGAŁ, and T. LIS, phys. stat. sol. (a) 93, 449 (1986).

/2/ R. JAKUBAS, Z. GALEWSKI, L. SOBCZYK, and J. MATUSZEWSKI, J. Phys. C 18, L857 (1985).

/3/ R. JAKUBAS, Solid State Commun. 60, 389 (1986).

/4/ R. JAKUBAS, Z. CZAPLA, Z. GALEWSKI, and L. SOBCZYK, Ferroelectrics Letters 5, 143 (1986).

/5/ R. JAKUBAS, L. SOBCZYK, and J. MATUSZEWSKI, Ferroelectrics 74, 339 (1987).

/6/ R. JAKUBAS, U. KRZEWSKA, G. BATOR, and L. SOBCZYK, Ferroelectrics 77, 129 (1988).

/7/ K. KIHARA and T. SUDO, Acta cryst. B30, 1088 (1974).

(Received July 25, 1988

phys. stat. sol. (a) 111, K31 (1989)

Subject classification: 73.30; S5.11

CNR - Istituto LAMEL, Bologna[1)]

Schottky Barrier Height Modification on n-Type Silicon by Wet Chemical Etching

By

G.A. ADEGBOYEGA[2)]

In a number of semiconductors (Si, GaAs, InP), it is now known that donor-like defects are generated by the damage induced by a variety of methods including ion implantation /1 to 5/, sputtering /6 to 10/, reactive ion etching /11, 12/, ion beam etching /12 to 14/, and polishing /15/. These donor states close to the semiconductor surface alter the electrical properties of subsequently formed Schottky barriers by lowering the barrier height on n-type materials thereby making the junctions more leaky and increasing it on p-type, thus making the junctions less leaky. The dependences of these effects on the sputter power and time /6, 7/ and also on ion energy and dose /11, 12/ have been well documented.

In this note, we present experimental data to show that similar defects are also generated by wet chemical etching. The results of dark current-voltage (J-U) and capacitance-voltage (C-U) measurements on Au/n-Si contacts which were made on substrates subjected to wet chemical etch for different lengths of time are presented. A gradual monotonic change of the J-U and C-U characteristics with etching time has been observed.

The starting material is n-type phosphorus-doped Si(100) with resistivity of about 10 Ωcm. A solution of $HF:HNO_3:CH_3COOH$ (1:1:2) is used to etch the substrates prior to the regular cleaning process. The standard cleaning procedure contained the following steps:

(i) decreasing step in trichlorethylene and acetone,

(ii) 15 to 60 s etch in $HF:HNO_3:CH_3COOH$ (1:1:2),

(iii) 10 min boil in $NH_4OH:H_2O_2:H_2O$ (1:1:10),

(iv) 10 min boil in $HCl:H_2O_2:H_2O$ (1:1:6),

(v) 2 min dip in $HF:H_2O$ (1:5).

---

1) Via Castagnoli 1, I-40126 Bologna, Italy.

2) Permanent address: Department of Electronic and Electrical Engineering, Obafemi Awolowo University, Ile-Ife, Nigeria.

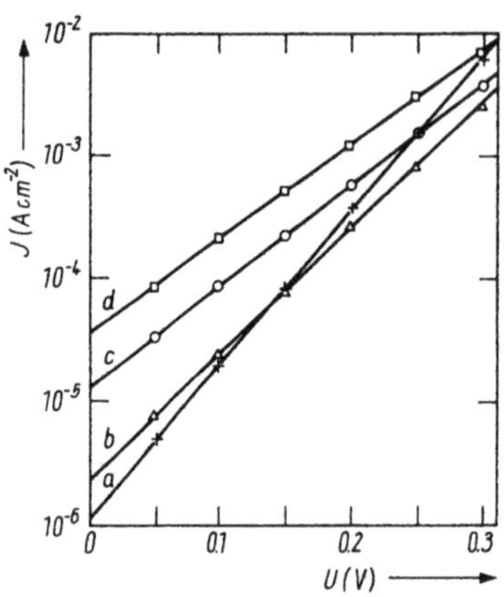

Fig. 1. Dark forward J-U characteristics of Au/n-Si diodes fabricated on substrates with different etching times; (a) 0 s, (b) 15 s, (c) 30 s, (d) 60 s

Specimens prepared with the omission of step (ii) are used as control and are referred to as unetched in this report. Gold was thermally evaporated through a 1.2 mm diameter mask to a thickness of about 50 nm under a vacuum pressure of about 10^{-4} Pa. Apart from the cleaning process, there was no intentional heating of the substrate before or during metallization. The entire back surface of the substrate was covered with silver paste for ohmic contact. Electrical connection to the barrier was made via a gold wire attached to it using a small quantity of silver paste. Current-voltage plots were obtained by means of a combination of a HP6963A power supply, a HP3438A digital multimeter, and a Keithley 172 autoranging digital multimeter. Dark C-U measurements were carried out by means of a computer controlled C-U programme of a Polaron-Biorad DL4600 DLTS system based on a Booton capacitance meter.

Fig. 1 shows the room temperature dark forward J-U characteristics for the Au/n-Si diodes fabricated on substrates with different etching times. In all cases an exponential J-U relationship from which the ideality factor, n, was calculated, was observed. The barrier height, Φ_B, was obtained by using the standard procedure of extrapolating the value of J towards zero voltage. The value used for the effective Richardson constant A^* was 112 A $cm^{-2}K^{-2}$. The Φ_B and n values as well as the values of the saturation current density, J_s, obtained from Fig. 1 are shown in Table 1. The n-factors are significantly higher than unity and hence the current flow is not likely to be dominated by thermionic emission. In such cases, one can only deduce effective barrier heights from the J_s values. Table 1 shows that the unetched control specimen has an effective Φ_B of 0.78 eV, while the Φ_B values for the etched samples ranged between 0.76 and 0.68 eV depending on the etching time employed. The reduction in Φ_B values for the etched specimens could be attributed to the

Table 1

Data obtained from J-U and C-U measurements on Au/n-Si fabricated on substrates with different wet etching times

etching time (s)	J_s ($A\,cm^{-2}$)	n	Φ_B(J-U) (eV)	Φ_B(C-U) (eV)	N_d (10^{14} cm^{-3})
0	$1.1x10^{-6}$	1.2	0.78	1.02	4.39
15	$2.0x10^{-6}$	1.6	0.76	0.99	4.55
30	$1.3x10^{-5}$	2.0	0.71	0.94	4.97
60	$4.0x10^{-5}$	2.2	0.68	0.90	4.92

effect of etching. The tendency of the junction to become more leaky with increasing etching time is manifested in the increase in J_s values as observed in Table 1.

Further evidence of the barrier height modification is seen in the C^{-2}-U plots of Fig. 2. The curves do not display any significant change in slope, but only changes in the intercept voltage consistent with the barrier height change inferred from the J-U data of Fig. 1. All relevant data from Fig. 2 are also shown in Table 1. While the slope of the C^{-2}-U plots gave donor concentration values consistent with the known resistivity of the n-Si bulk, we observe a slight increase of the donor concentration in the etched substrates when compared with the control sample. This increase might not be unconnected with the surface etch-induced defects. There is also virtually no difference between the N_d values obtained for etching periods of 30 and 60 s. This might be an indication of a tendency towards saturation in the number of the etch-induced defects. The barrier height values (see Table 1) obtained from the C-U measurements are larger than the effective barrier height values determined from the saturation current. This discrepancy has also been observed by other authors

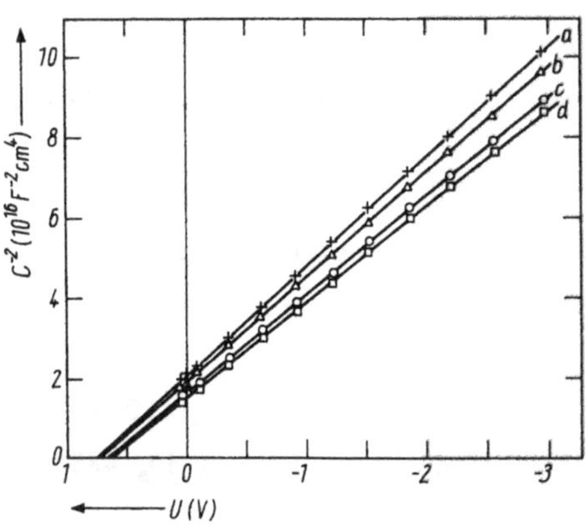

Fig. 2. Dark C^{-2}-U characteristics of Au/n-Si diodes of Fig. 1; (a) 0 s, (b) 15 s, (c) 30 s, (d) 60 s

/5, 12/ and has been attributed to the possible presence of a thin interfacial layer.

Like earlier authors, we do not understand the reason for the lower barrier height obtained in this work as a result of the etching process. However, we are of the opinion that the etching process might create an activated surface layer probably through the creation of dangling bonds. Such a layer would be more conducting than the bulk material and can lead to a lowering of the barrier height as observed.

The author acknowledges the support of the ICTP Programme for Training and Research in Italian Laboratory, Trieste, Italy, for his stay at the Istituto LAMEL, Bologna.

References

/1/ S. ASHOK and A. MAGRO-CAMPERO, IEEE Electron Device Letters **5**, 48 (1984).

/2/ S. ASHOK, H. KRAUTLE, and H. BENEKING, Appl. Phys. Letters **45**, 431 (1984).

/3/ O. AINA and K.P. PANDE, J. appl. Phys. **56**, 1717 (1984).

/4/ S. ASHOK and K. GIEWONT, Japan. J. appl. Phys. **24**, L533 (1985).

/5/ S. ASHOK and K. GIEWONT, IEEE Electron Device Letters **6**, 462 (1985).

/6/ F.H. MULLINS and A. BRUNSHWEILER, Solid State Electronics **19**, 47 (1976).

/7/ E. GRUSELL, S. BERG, and L.P. ANDERSON, J. Electrochem. Soc. **127**, 1573 (1980).

/8/ M. FINETTI, I. SUNI, M. BARTUR, T. BANWELL, and M-A. NICOLET, Solid State Electronics **27**, 617 (1984).

/9/ G.J.A. HELLINGS, A. STRAYER, and A.H.M. KIPPERMAN, J. appl. Phys. **57**, 2067 (1985).

/10/ W. DE BOSSCHER, R.L. VAN MEIRHAEGHE, A. DE LAERE, W.H. LAFLERE, and F. CARDON, Solid State Electronics **31**, 945 (1988).

/11/ S. ASHOK, T.P. CHOW, and B.J. BALIGA, Appl. Phys. Letters **42**, 687 (1983).

/12/ X.C. MU and S.J. FONASH, IEEE Electron Device Letters **6**, 410 (1985).

/13/ S.J. FONASH, A. SHOK, and R. SINGH, Appl. Phys. Letters **39**, 423 (1981).

/14/ Y-X. WANG and P.H. HOLLOWAY, J. Vacuum Sci. Technol. B 2, 613 (1984).

/15/ L.M.O. VAN DEN BERGHE, R.L. VAN MEIRHAEGHE, W.H. LAFLERE, and F. CARDON, Solid State Electronics 29, 1109 (1986).

(Received July 19, 1988)

phys. stat. sol. (a) 111, K37 (1989)

Subject classification: 68.55; 73.60; S1.3; S4

Thin Film Laboratory, Department of Physics,
Indian Institute of Technology, Madras[1)]

Large Scale Coalescence in Terms of a New Agglomeration Rate

By

M.S. MURALI SASTRY[2)]

The phenomenon of large scale coalescence (LSC) in thin films and cermet films with the formation of a metallic network has been discussed by many researchers /1 to 4/. In an earlier study /5/, we had investigated the LSC by the repeated deposition technique in Cu, Ag, and Al films. An agglomeration rate was defined on the basis of the functional dependence of the post deposition resistance increase on time and a large fall in the agglomeration rate beyond a particular deposition cycle was observed and attributed to LSC. In this communication, we shall evaluate the change in the tunneling length (interisland spacing) during aging and define a new agglomeration rate called the "coalescence rate" in terms of the rate of change of the tunneling length with time. For full experimental details, the reader is referred to /5/. This communication makes no attempt to give a physical interpretation of the process of LSC which has been dealt with in the earlier work /5/. The accent here is on the application of an approximation to Andersson's /6/ approach as elucidated below and the use of the new approach to the results on the aging of Cu, Ag, and Al films studied by the repeated deposition technique.

The time variation of the resistance of a discontinuous film based on the quantum mechanical tunneling model (after Hill /7/) is given by

$$R(t) = (Dd^2(t)/\Phi)\exp(1.025(m^*\Phi)^{1/2}d(t))\exp(\delta E/kt)\,, \quad (1)$$

where d(t) is the time dependent tunneling length; D is a constant; Φ means tunneling barrier for the electron; m* effective mass of the tunneling electron; δE activation energy for charge carrier creation and the other symbols have the usual meaning.

---

1) Madras 600 036, India.

2) Present address: Department of Physics, University of Rome II (Tor Vergata), Via O. Raimondo, I-00173 Rome, Italy.

For the Cu, Ag, and Al films of initial resistances R_0 = 82, 20, and 65 M Ω/□, respectively, a good fit is got for the R(t) versus time data if the tunneling length is assumed to vary logarithmically with time. Following Andersson /6/, with suitable modifications,

$$d(t) = d_I + d_S (A \ln(t + 1)) , \tag{2}$$

where

$$d_S = d_F - d_I ; \tag{3}$$

d_I is the tunneling length at the stoppage of deposition and d_F is the tunneling length at the end of the period of aging. The constant A is such that $A \ln(t + 1) = 1$ for $t = t_{max}$ where t_{max} is the time interval of aging. This condition must be satisfied as can be seen from (1) to (3). For large time intervals, $\ln(t+1) \approx \ln(t)$. Substituting (2) in (1) and neglecting time dependent changes in $d^2(t)$ and $\exp(\delta E/kT)$ terms compared to the $\exp(1.025(m^*\Phi)^{1/2} d(t))$ term, we get

$$\ln R(t) = \ln R_I + \ln(R_F/R_I) (A \ln t) , \tag{4}$$

where R_I the initial resistance of the film at the start of aging and R_F the final resistance of the aging as determined from a least-squares fit of the data. The constant of proportionality in the ln R(t) versus ln t curves, $A \ln (R_F/R_I)$, has been termed the agglomeration rate in the earlier study /5/. From (1) and (3), the change in the tunneling length is evaluated to be

$$d_S = 1.54 (\ln R_F - \ln R_I) , \tag{5}$$

assuming $m^* = 0.4$ and $\Phi = 1$ eV. Since these values are not very different for different metals, they have been assumed the same for Cu, Ag, and Al for simplicity. Knowing R_F and R_I, the change in the tunneling length may be obtained. For commonly obtained values of the activation energy and tunneling length in island films, a reasonable approximation would be

$$\ln R_F/\ln R_I \approx d_F/d_I . \tag{6}$$

This is also readily seen from (1) through substitution of typical values for d for a particular resistance film and evaluation of $(D d^2/\Phi) \exp(\delta E/kT)$. For the Ag film of R_0 = 20 M Ω/□, assuming d = 2.5 nm, the above expression is calculated to be 1.7528 which is negligible compared to $\exp(1.025(m^*\Phi)^{1/2} d)$ which has the value 7.6401×10^7. A lower value has been deliberately assumed for d. The approximation would have been even better for the calculated value

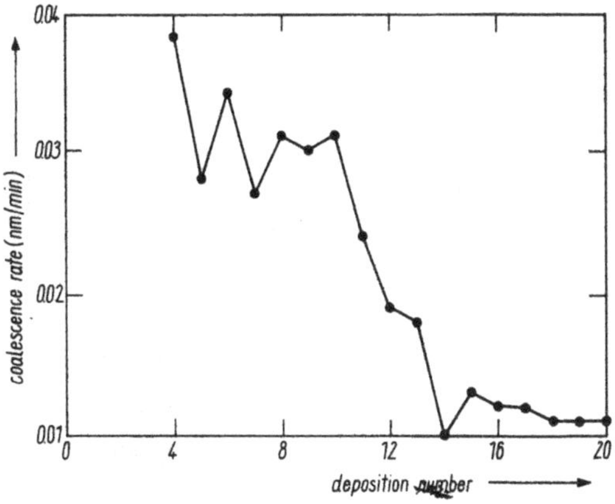

Fig. 1. Coalescence rate versus deposition number for an Al film of $R_o = 65\ M\ \Omega/\square$

of d = 2.8 nm. Coupling (5) and (6), the values of the tunneling length before and after aging may be calculated. From the change in the tunneling length d_S and the time interval of study t_{max} is defined the "coalescence rate" and is given by d_S/t_{max} nm/min. The coalescence rates for the Cu and Ag films are given in Tables 1 and 2, respectively. As an example, the variation of the coalescence rate with deposition number is given in Fig. 1 for the Al film with a view to show graphically the change. The tunneling length calculated using the above procedure is the tunneling length corrected for image force effects. From Hill /7/, the reduced tunneling length (in nm) due to image force effects is $d(t) = d'(t) - 1.2/(\varepsilon \Phi_o)$ where $d'(t)$ is the actual tunneling length in nm, ε is the relative dielectric constant of the medium through which the electron tunnels, and Φ_o is the tunneling barrier in eV. If one assumes a value of 2.25 for ε and $\Phi_o = 1$ eV, then the real tunneling length $d'(t)$ is obtained by adding 0.53 nm to the calculated values of the tunneling lengths given in the tables.

As mentioned earlier, Tables 1 and 2 give the initial (d_I) and final (d_F) values of the tunneling length, the coalescence rates evaluated over the period of aging, and the agglomeration rate data for the Cu and Ag films of the earlier study /5/ for easy comparison. In order to show clearly the occurrence of large scale coalescence, the difference in the successive values of the coalescence rates are given in the last column. The point at which there is the greatest change in the coalescence rates is taken to mirror the onset of large scale coalescence and is underlined in the tables. From Fig. 1, a large fall in the coalescence rate for the Al film can be seen beyond the tenth deposition which continues till the forteenth deposition. The steady tail is due to secondary nucleation and the incorporation of the secondary nuclei with the metal backbone formed during LSC. As can be seen from the tables, the onset of LSC occurs beyond the first cycle for the Cu film and beyond the second cycle for the Ag

Table 1

Coalescence rate data for the Cu film of R_0 = 82 M Ω/□ during successive depositions, t_{max} = 60 min

deposition number	agglomeration rate	d_I (nm)	d_F (nm)	coalescence rate(nm/min)	$\lvert m'_n - m'_{n-1}\rvert$ (nm/min)
1	0.0998	2.85	2.91	0.0010	
2	0.0455	2.82	2.85	0.0005	0.0005
3	0.0293	2.82	2.84	0.0003	0.0002
4	0.0182	2.82	2.83	0.0002	0.0001

Table 2

Coalescence rate data for an Ag film of R_0 = 20 M Ω/□ during successive depositions, t_{max} = 60 min

deposition number	agglomeration rate	d_I (nm)	d_F (nm)	coalescence rate (nm/min)	$\lvert m'_n - m'_{n-1}\rvert$ (nm/min)
1	0.5101	2.81	3.22	0.0069	
2	0.4837	2.66	3.00	0.0055	0.0014
3	0.1837	2.61	2.72	0.0019	0.0036
4	0.1029	2.59	2.65	0.0011	0.0008
5	0.0728	2.60	2.64	0.0008	0.0003

film. Hence, it is seen that the coalescence rate approach gives a similar picture of the process of LSC but with certain distinct advantages. A detailed explanation of the process of LSC and the metal-insulator transition in metal films is given in /5/.

Perhaps the greatest advantage of this technique (evaluation of the "coalescence rate") is that it puts the quantification of the aging process on a concrete footing. In the study of aging in Ag island films /8/, it was found that depending on the condition of study, the agglomeration rate had to be defined either as $d(\ln R)/d(t)$ or $d(\ln R)/d(\ln t)$ with difficulty arising when the agglomeration rates had to be compared. This problem is removed in the case of evaluation of the coalescence rate which just gives the rate of change of the tunneling length. There is no cause for ambiguity. Another advantage is that the aging rates of different metals deposited on different substrates may be compared by choosing appropriate values of m* and Φ(1). In a way, the "coalescence rate" is a universal aging rate with obvious advantages.

The author would like to thank the Council for Scientific and Industrial Research (CSIR), Government of India, for the Senior Research Fellowship. The assistance of Dr. Manjanatha Pattabi and Mr. P.J. Sebastian is gratefully acknowledged.

References

/1/ T. ANDERSSON, J. Phys. D 9, 973 (1976).

/2/ J.D. BENJAMIN, C.J. ADKINS, and J.E. VAN CLEVE, J. Phys.C 17, 559 (1984).

/3/ B. ABELES, PING SHENG, M.D. COUTTS, and Y. ARIE, Adv. Phys. 24, 407 (1975).

/4/ T.J. COUTTS, Electrical Conduction in Thin Metal Films, Elsevier, Amsterdam 1974 (p. 220).

/5/ V. DAMODARA DAS and M.S. MURALI SASTRY, Phys. Rev. B 34, 8528 (1986).

/6/ T. ANDERSSON, J. appl. Phys. 47, 1752 (1976).

/7/ R.M. HILL, Proc. Roy. Soc. A309, 377 (1969).

/8/ V. DAMODARA DAS and M.S. MURALI SASTRY, J. appl. Phys. 59, 3184 (1986).

(Received April 29, 1988; in revised form September 29, 1988)

phys. stat. sol. (a) 111, K43 (1989)
Subject classification: 73.30; S7.12

V. Kapsukas State University, Vilnius[1] (a)
and Science Research Institute of Radio Measuring Devices, Kaunas (b)

Unified Inverted Defect Model for GaAs Schottky Barrier Formation

By
A. SAKALAS (a) and S. ZHUKAUSKAS (b)

Experiments have demonstrated that the Fermi level position is not affected by the metal work function after deposition of a metal coverage on the GaAs surface /1/. A unified defect model of Spicer et al. /2/ explains the Fermi level pinning on the semiconductor surface in the limit of metal or dielectric coverages. The model predicts two types of defects on the semiconductor surface - acceptors and donors. The Fermi level is, then, claimed to be pinned at the surface acceptor level $E_V + 0.75$ eV for n-GaAs and surface donor level $E_V + 0.5$ eV for p-GaAs. However, there appear some data to be inconsistent with the proposed model. First, for contacts of Au, Cu, Ag on n-GaAs the Fermi level is reported to be pinned at $E_V + 0.5$ eV instead of $E_V + 0.75$ eV, while for Gd, Ti, Ni on p-GaAs the pinning position has been registered at $E_V + 0.75$ eV /1/. Second, current evidence has been provided for the Fermi location being dependent on the free charge carrier concentration in the bulk as well as on the density of the surface state distribution /3 to 5/.

Furthermore, significant increase of arsenic depletion at the GaAs surface has been found to lead to the disappearance of defect states at $E_V + 0.75$ eV and emergence of new states in the vicinity of the conduction band. Simultaneously, Fermi level delocalization has been observed /4, 5/. In this way, the above stated experimental observations are not in contrast to the unified defect model, but, instead, evidence the presence of donor states in the upper half of the band gap, acceptor states being in the lower half of the band gap. Virtually, the derived model resembles the unified defect model, though it is inverted.

In this study the proposed unified inverted defect model is supported by theoretical simulation of the Fermi level position as a function of the surface state

---

1) SU-232734 Vilnius, Lithuanian SSR, USSR.

density and carrier concentration for n- and p-type GaAs in consistency to calculations of Zur and McGill /6/. Alongside, comparison with the existing experimental data is presented.

Theoretical simulation of the Fermi pinning position on the surface According to Zur and McGill /6/, the total charge in the bulk is related to the Fermi level position on the surface,

$$Q(\eta_0) = \sqrt{\frac{2\,\varepsilon_s}{q^2} \int_{\eta_\infty}^{\eta_0} (N_D^+ + p - N_A^- - n)\,d\eta}\,; \tag{1}$$

$$\eta = E_c - F\,, \tag{2}$$

here, $N_D^+(N_A^-)$ is the concentration of ionized donors (acceptor), n(p) the electron concentration in the conduction band (hole concentration in the valency band), ε_s dielectric constant of the semiconductor, $\eta_0(\eta_\infty)$ the Fermi level position on the surface (bulk) of the semiconductor.

The charge of the defects is written as a total number of charges on donor and acceptor defects:

$$Q_{ss} = \sum_{i=1}^{m} Q_{iD}^+ - \sum_{j=1}^{n} Q_{jA}^- \;; \tag{3}$$

$$Q_{iD}^+ = \frac{\sigma_{iD}}{1 + g_{iD} \exp\left(\frac{F - E_{iD}}{kT}\right)}\;; \tag{4}$$

$$Q_{jA}^- = \frac{\sigma_{jA}}{1 + g_{jA} \exp\left(\frac{E_{jA} - F}{kT}\right)}\;; \tag{5}$$

where $E_{iD}(E_{jA})$, $\sigma_{iD}(\sigma_{jA})$, $g_{iD}(g_{jA})$ ionization energies and defect densities of the surface acceptor (donor) defects and degeneracy factor, respectively, k Boltzmann constant, T temperature.

The Fermi level position on the surface is now, determined by the overall charge neutrality condition

$$Q_{ss} = Q(\eta_0)\;. \tag{6}$$

Equations (1) to (6) were solved numerically.

As revealed by calculations for GaAs, in case of both types of defects - donors and acceptors (E_v + 0.75 eV) and (E_v + 0.5 eV), respectively, being

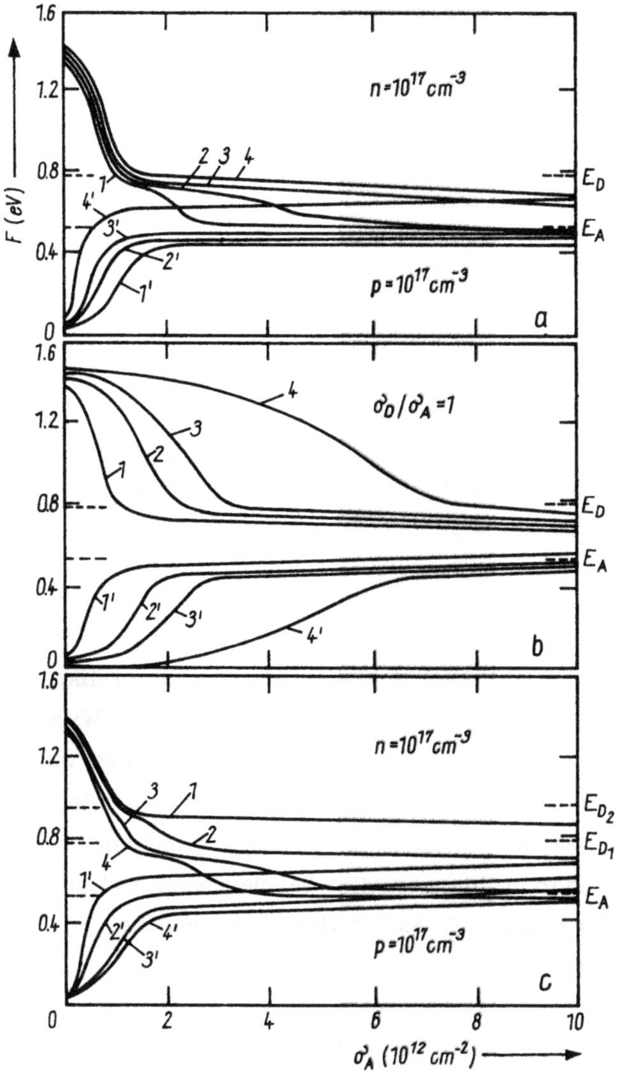

Fig. 1. The Fermi level position at the GaAs surface versus defect density for different donor and acceptor density ratio a) σ_D/σ_A = 0.5 (1); 0.75 (2); 0.1 (3); 2.0 (4), c) σ_{D1}/σ_A = 1.0 (1); 0.5 (2); 0.25 (3): 0.1 (4); and different main charge carrier concentration b) n = 1×10^{17} cm^{-3} (1); 5×10^{17} cm^{-3} (2); 1×10^{18} cm^{-3} (3); 5×10^{18} cm^{-3} (4). Dashes mark the positions of the defect levels

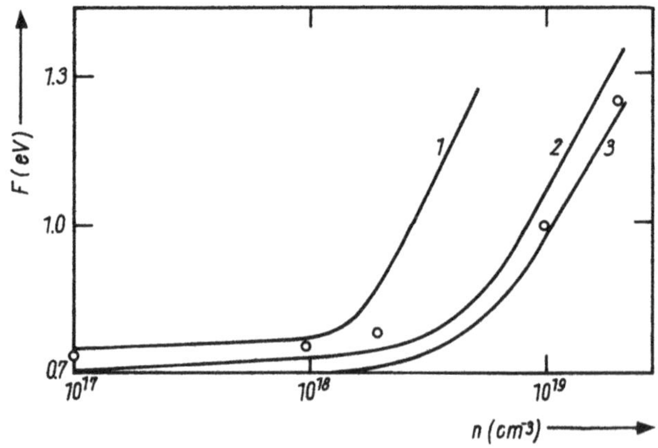

Fig. 2. The position of the Fermi level at the surface as a function of electron concentration for n-GaAs at different defect densities $\sigma_D = \sigma_A$ = (1) 4, (2) 6, (3) $8x10^{12}$ cm^{-2}. Points indicate experimental data by the Kelvin probe /3/

present on the surface - the Fermi level may be equally pinned at donor-like and acceptor-like levels depending on the defect density ratio (Fig. 1a). With increasing main charge carrier concentration in the bulk, the Fermi level was observed to be pinned by high defect densities (Fig. 1b). The decrease of the donor ionization energy as well as the emergence of new donor defects of lower ionization energy appeared to lead tc a Schottky barrier height reduction. Hence, it is evident that barrier height of GaAs Schottky diodes may be governed by the defect density distribution.

Correlation between theoretical and experimental results, conclusion
Fig. 2 illustrates the Fermi level location on the surface of n-GaAs in dependence on the electron concentration in the conduction band. The calculation has been performed according to the afore-cited approach. The theoretical calculation of the Fermi level location at $\sigma_D = \sigma_A = 6x10^{12}$ cm^{-2} seems to correlate well with the Kelvin probe experimental data for heavily doped n-GaAs /3/. It is apparent that the developed unified inverted defect model allows the interpretation of the Fermi level pinning positions for contacts of Cu, Ag, Au, on n- as well as p-GaAs.

By assuming that Cu, Ag, Au atoms form additional acceptor-like centres in the lower half of the band gap and, provided condition $\sigma_D < \sigma_A$ is satisfied, the Fermi level is pinned at $E_v - 0.5$ eV (Fig. 1a), which is in agreement with experimental observations. On the other hand, contacts of Gd, Ni, Ti on p-GaAs can be described by the inequality $\sigma_D > \sigma_A$. The Fermi level is then pinned at $E_v + 0.75$ eV.[2)]

As commonly known, deposition of both thin insulator and metal films on the GaAs surface produce defect levels at approximately the same energy. Studies of the defect density distribution for low-dose inert gas ion pre-treated Si_3N_4-GaAs structures indicate surface depletion by As atoms. The missing As atoms were identified to cause defect levels at $E_v + 0.75$ eV, the replacement by new defect levels is followed by a Fermi level shift by 0.15 eV towards the conduction band. This is in excellent agreement with theoretical simulations displayed in Fig. 1c.

To summarize, the established experimentally verified unified inverted defect model is able to reliably explain a wide scope of Schottky barrier formation problems, the interpretation of which has been complicated due to prediction of additional mechanisms /2/.

References

/1/ N. NEWMAN, W.E. SPICER, T. KENDELEWICZ, and I. LINDAU, J. Vacuum Sci. Technol. B4, 931 (1986).

/2/ W.E. SPICER, P.W. CHYE, P.R. SHEATH, C.Y. SU, and I. LINDAU, J. Vacuum Sci. Technol. 16, 1422 (1979).

/3/ V. FILIPAVIČIUS, R. GAIDYS, V.A. MATULAITIS, G. PETRAUSKAS, A. SAKALAS, and S. SAKALAUSKAS, phys. stat. sol. (a) 99, 543 (1987).

/4/ A.C. WARREN, S.D. OFFSEY, J.M. WOODOLL, P.D. KIRCHER, and T.I. CHOPPEL, J. Vacuum Sci. Technol. B4, 1115 (1986).

/5/ N.A. IVES, G.W. STUPIAN, and M.S. LENNG, Appl. Phys. Letters 50, 256 (1987).

/6/ A. ZUR and T.C. MCGILL, J. Vacuum Sci. Technol. B2, 440 (1984).

/7/ K.K. CHIN, S.H. PAN, D. MO, P. MAHOVALD, N. NEWMAN, I. LINDAU, and W.E. SPICER, Phys. Rev. B 32, 918 (1985).

---

2) Exchange reactions are known to occur between Ag, Au /7, 8/ as well as Ti, Ni /9, 10/, and GaAs, which result in gallium replacement. Subsequently, additional acceptor-like defects originate in case of Au, Ag and donor-like defects for Ni, Ti.

/8/ A.G. MILNES, Adv. Electronics and Electron Phys. 61, 63 (1983).
/9/ R LUDEKE and G. LANDGREN, Phys. Rev. B 33, 5526 (1986).
/10/ A. LAHOV and M. EIZENBERG, Appl. Phys. Letters 46, 430 (1985).

(Received July 4, 1988)

phys. stat. sol. (a) 111, K49 (1989)
Subject classification: 71.55; S5.11

Department of Physics, Sichuan University, Chengdu[1)]

The Deep Levels in Nickel-Doped Silicon

By
MIN GONG and ZHI-PU YOU

Deep level transient spectroscopy (DLTS) is a powerful technique to investigate the deep level impurities in semiconductors /1/. Many DLTS studies on nickel-doped silicon have been reported, but the number and the energy values of nickel-related deep levels are quite divergent in the publications /2 to 9/. In Ni-doped n-type silicon, levels of E(0.19), E(0.36) /2/; E(0.38) /3/; and E(0.46) /4/ were observed, but not any peak appeared in the DLTS scan of Indusekhar and Kumar's work /5/. In Ni-doped p-type silicon, the investigators have reported levels of H(0.15) /6/; H(0.33) /7/; H(0.15), H(0.21), H(0.33) /8/; H(0.28), H(0.32) /5/; and H(0.17) /3, 4/, however, Rohatgi et al. did not find any level in Ni-doped p-type silicon /9/. Thus the nickel levels in silicon are not yet reliably established.

The divergence of the Ni-related levels in silicon could be caused by additional undesired impurities which came from the contaminations of the nickel sources or the diffusion processes at high temperatures. As Indusekhar and Kumar pointed out, the level at H(0.15) measured in Ni-doped silicon must be due to some other chemical present in the plating solution /5/. On the other hand, there was another possibility that the nickel atoms might not diffuse into the samples in experiments in which no Ni-related level was observed.

The purpose of the present note is to examine the deep levels in Ni-doped n- and p-type silicon by atomic absorption spectroscopy (AAS) and DLTS. It was verified by means of AAS that the nickel atoms had diffused into the samples. The contaminations were studied using various silicon substrates and various nickel sources.

The starting materials were 3, 5, and 20 Ωcm p-type; and 1, 5, and 30 Ωcm n-type Cz-Si. The n^+-p diode structures were fabricated on p-type materials by phosphorus diffusion, and the p^+-n structures on n-Si by boron diffusion.

---

1) Chengdu, Sichuan, People's Republic of China.

Nickel was deposited on the backside of the n^+-p and the p^+-n wafers by three methods: $NiCl_2$ solution spreading, nickel wire scraping, and evaporating in vacuum. The nickel diffusion was carried out in a quartz tube for 1 h at 1100 ^oC under flowing nitrogen gas. Two cooling modes were employed, rapidly quenching to room temperature, and slowly cooling in the switched-off furnace. Control samples with the same heat treatment but without nickel doping were also prepared in each case. Every piece of wafers was divided into two parts, one of which, dissolved in a mixture of highly pure nitric acid and hydrofluoric acid, was for the AAS measurements, the other part was for the DLTS.

A strong nickel atom absorbing signal was observed in all Ni-doped n^+-p and p^+-n samples. The concentrations of nickel in these samples were about $3x10^{17}$ atoms/cm^3, this value agrees with that of Weber /10/. The signals in the control samples, however, were on the background level. This result confirmed that nickel had diffused into all Ni-diffused samples in our experiments.

None of the DLTS peaks appears in most of the Ni-doped quenched n^+-p samples, and in all the Ni-doped slowly cooled ones as shown in Fig. 1, curve a. A level at H(0.34) was observed only in a few of the $NiCl_2$ solution spreaded samples as shown in Fig. 1, curve b. But if nickel was deposited using a new $NiCl_2$ solution, this level disappeared. It means that the level at H(0.34) was introduced by some contaminations in the nickel source. Many Ni-related levels in p-Si have been reported by several authors /3 to 8/. These levels existed only in Ni-doped silicon, and did not exist in the undoped control ones, so they were considered to be associated with nickel. But it must be noted that the levels in the various papers were quite discrepant in number, location, or capture cross-section /3 to 8/.

Our experiments in which no deep level appeared in almost all Ni-contained p-Si samples support Rohatgi and co-workers' /9/ conclusion that nickel does not introduce any deep level in the lower half of the band gap of silicon.

Fig. 2 shows a DLTS signal located at E(0.40) in all Ni-doped quenched p^+-n samples. The control samples did not show any peak in the DLTS scans. Several investigators have also observed a level at E(0.36)-E(0.46) in Ni-doped

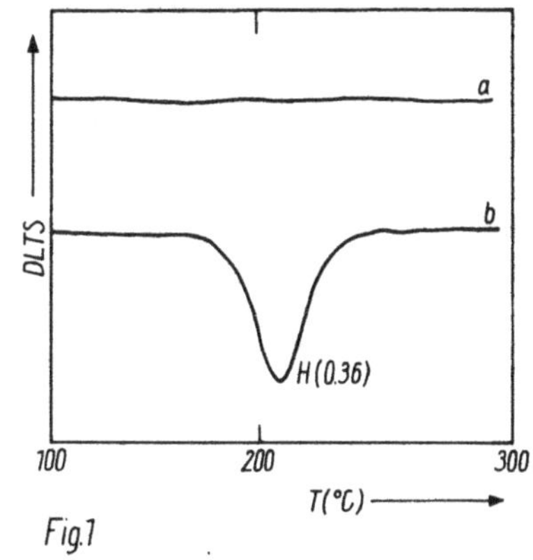

Fig.1

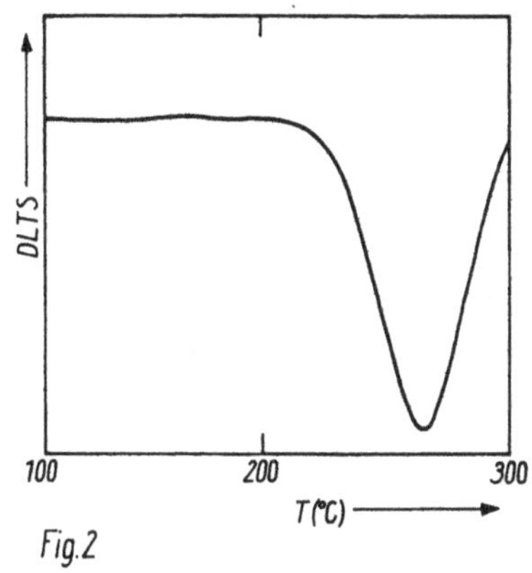

Fig.2

Fig. 1. (a) Typical DLTS spectrum of Ni-doped n-p structure; (b) DLTS spectrum of a few of the Ni-doped n-p structures

Fig. 2. Typical DLTS spectrum of Ni-doped structure

n-Si /2 to 5/. Yoshida and Saito /11/ reported a level at E(0.41) whose concentration in quenched samples was nearly the same as that in the slowly cooled ones. In our works, however, no DLTS signal was observed in most of the slowly cooled samples. Fig. 3 shows the discrepancy of the 20 min isothermal annealing behaviours of the level at E(0.40) in the different samples. In some of the $NiCl_2$ spreaded quenched and slowly cooled samples, it did not anneal out at temperatures up to 550 ^{o}C as shown in Fig. 3, curve a. But in the other quenched samples, the level E(0.40) annealed out at lower temperatures. Fig. 3, curves b, c shows the annealing

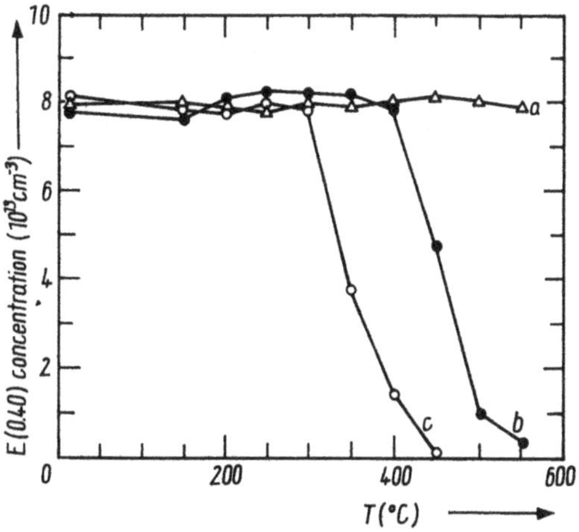

Fig. 3. 20 min isothermal annealing behaviours of E(0.40) in the different batches of samples

temperatures of two batches of samples, it is about 500 °C in one batch, and about 400 °C in the others. The discrepancy of the annealing behaviours is very large and beyond the experimental errors. This result suggests that the levels at E(0.40) in various samples are not due to the same kind of defect.

The investigation shows that the level E(0.40) in Ni-doped n-Si might be associated with nickel. Because of the large differences in the annealing temperature, this level also might come from some contaminations in our experiments. So it is worth noting that no DLTS signal appeared in Ni-doped n-Si in Indusekhar and Rohatgi's work /5, 9/.

This work is supported by the Science Fund of Sichuan Province and the National Science Fund of China.

References

/1/ D.V. LANG, J. appl. Phys. 45, 3023 (1974).

/2/ H.H. BUSTA and H.A. WAGGNER, J. Electrochem. Soc. 124, 1424 (1977).

/3/ H. LEMKE, phys. stat. sol. (a) 99, 205 (1987).

/4/ H. KITAGAWA and H. NAKASHIMA, phys. stat. sol. (a) 99, K49 (1987).

/5/ H. INDUSEKHAR and V. KUMAR, J. appl. Phys. 61, 1449 (1987).

/6/ K. GRAFF and H. PIEPER, in: Semiconductor Silicon, 1981 (p. 331).

/7/ S.J. PEARTON, Solid State Electronics 25, 499 (1982).

/8/ S.J. PEARTON and A.J. TAVENDALE, J. appl. Phys. 54, 1376 (1983).

/9/ A. ROHATGI, J.R. DAVIS, R.H. HOPKINS, and P.G. MC MULLIN, Solid State Electronics 26, 1039 (1983).

/10/ E.R. WEBER, Appl. Phys. A30, 1 (1983).

/11/ M. YOSHIDA and K. SAITO, Japan. J. appl. Phys. 6, 573 (1967).

(Received April 15, 1988; in revised form August 29, 1988)

phys. stat. sol. (a) 111, K53 (1989)

Subject classification: 61.70 and 72.70; S7.12

Department of Electrical Engineering, National Tsing Hua University, Hsinchu[1)]

Annealing Temperature Dependence of the 1/f Noise in Si-Implanted GaAs

By

J. GONG and S.H. YANG

Ion implantation has been widely used in semiconductor industry, because it offers good control in geometry and doping concentration. These advantages are even more important for GaAs for which a diffusion doping has not been fully developed. However, due to the collisions of incident ions with lattice sites, cascades of displacements and zones of amorphization may occur. It has been found that initially the radiation damage increases with dose, but for sufficiently large beam current and dose the energy dissipated in the substrate may be large enough to avoid the formation of an amorphous layer /1/. Therefore, we will concentrate, mainly, on the high temperature annealing stage.

Cr-O doped semi-insulating GaAs substrate with resistivity greater than 10^6 Ωcm was used. After standard cleaning procedures, the substrate was implanted with Si ions with a dose of 10^{13} cm^{-2}, then an SiON layer was grown on GaAs with plasma enhanced CVD as the passivation layer. Samples were then annealed at different temperatures, from 600 to 900 ^oC, each for 30 min. Following annealing, the SiON layer was removed and the implanted layer was formed into resistors.

Noise measurements were performed on implanted resistors. The noise was first amplified by an Ithaca 1201 low noise preamplifier and then measured with an HP 3562A dynamic signal analyzer, whose frequency span was set from 10 mHz to 10 Hz. Typical 1/f noise spectra were observed for all samples. According to Hooge's formula the 1/f noise power spectral density is given as $S_V = \alpha V^2/Nf$, if the voltage across sample, U, is normalized to 1 V then the normalized power spectral density should be given as $S_{vn} = S_V/U^2 = \alpha/Nf$. Fig. 1 shows S_{vn} measured at 3 Hz as a function of annealing temperature, T_a.

1) Hsinchu, Taiwan.

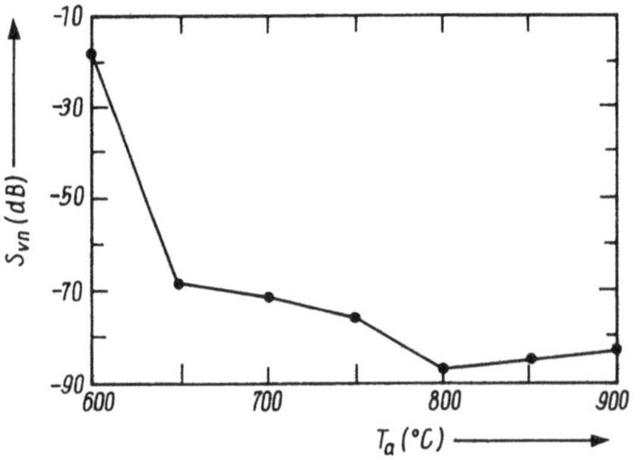

Fig. 1. Normalized noise power spectral density measured at 3 Hz versus annealing temperature

Before the implanted atoms were activated, they may exist in some interstitial sites or in the form of some defect compound. Once implanted atoms are activated, the defects in crystalline structure are also annealed out. Therefore, noise due to crystal defects will be decreased with increasing annealing temperature. Meanwhile, the activation percentage also increases with T_a, which increases the total number of carriers in the sample and then decreases the 1/f noise. For an implantation dose of 10^{13} cm^{-2} as we have in this work the electrical activity reaches almost 100% for T_a around 800 °C /2/. Experimental results show that S_{vn} decreases with annealing temperature for $T_a < 800$ °C and reaches a minimum at $T_a = 800$ °C.

In our work, the sample resistance was first decreased with increasing annealing temperature in the range from 600 to 800 °C, which implies the increase in the activation percentage, see Table 1. For $T_a > 800$ °C, the resistance increases with annealing temperature, we believe that this is due to the growth of dislocations and some dislocations may even form networks. So, the conducting cross section area was reduced with increasing annealing temperature.

Table 1

Sample resistance at different annealing temperature

T_a (°C)	sample resistance (10^3 Ω)
600	200 000
650	44.8
700	24.5
750	6.98
800	0.935
850	191
900	1.566

The above statement was supported by the report that dislocations may propagate in GaAs and their movement is associated with enhanced EL2 centers in the immediate vicinity of the dislocations /3/. Also, the EL2 concentration will increase with annealing temperature and finally will become uniform throughout the whole wafer /4/. The growth of dislocations will also increase the 1/f noise, the experimental results showed that noise was indeed increased with annealing temperature for $T_a > 800\ ^oC$.

The 1/f noise is important for GaAs field-effect transistors. From this note, we find that dislocations are one of the noise origins. This kind of noise can be minimized by choosing suitable annealing temperatures after ion implantation, which makes high impurity activation percentage and low dislocation density.

References

/1/ W.J. ANDERSON and Y.S. PARK, Japan. appl. Phys. 8, 4568 (1978).

/2/ W.J. HOWES and D.V. MORGAN (Ed.), GaAs Materials, Devices, and Circuit, John Wiley and Sons, New York 1985.

/3/ S. CLARK, M.R. BROZEL, and D.J. STIRLAND, in: Defect Recognition and Image Processing in III-V Compounds, Ed. J.P. FILLARD, 1985 (p. 201).

/4/ S. MARTIN and M. DUSEAUX, ibid. (p. 287).

(Received August 18, 1988)

phys. stat. sol. (a) 111, K57 (1989)
Subject classification: 72.80; S7.11

Institute of Physics, Technical University of Wrocław[1)]

Electrical Properties of Zn_3P_2 at Room Temperature

By
K. SIERAŃSKI and J. SZATKOWSKI

Introduction Zn_3P_2 appears to be a promising semiconductor for low-cost solar cells (see, e.g. /1/). Electrical and photoelectrical properties of metal-Zn_3P_2 diodes are widely investigated /1 to 5/. To date only limited research has been carried out on the electrical properties of Zn_3P_2 /6 to 8/. In this note, we report an investigation of the electrical properties of Zn_3P_2 at room temperature.

Experimental Large-grain polycrystalline samples of Zn_3P_2 were obtained in our laboratory by means of the close tube vapour transport method /9/. The standard six-probe dc configuration method was applied for the Hall effect and conductivity measurements. For all of the measurements Au vacuum deposited contacts were used. Three types of samples (denoted hereafter A, B, and C), with different hole concentration at room temperature were investigated. The room temperature hole concentrations were equal to $2x10^{14}$, $2x10^{15}$, and $4.5x10^{17}$ cm^{-3}, for types A, B, and C.

Results Typical representative temperature dependences of the electrical conductivity and Hall coefficient for the samples of A and B types are shown in Fig. 1 and 2, respectively. For these samples the Hall coefficient obeys the exponential dependence

$$R_H = R_o \exp(-E/kT) \qquad (1)$$

with activation energy E close to 140 meV. This exponential dependence indicates the existence of an acceptor level with ionization energy about 140 meV.

For the samples with the highest hole concentration (C type) the measurements of the Hall coefficient were carried out within the temperature range 200 to 300 K. The Hall coefficient remained constant within the whole temperature range measured.

For this class of samples the measurements of the electrical conductivity were carried out in the temperature range from 77 to 300 K. A representative

---

1) Wybrzeże Wyspiańskiego 27, PL 50-370 Wrocław, Poland.

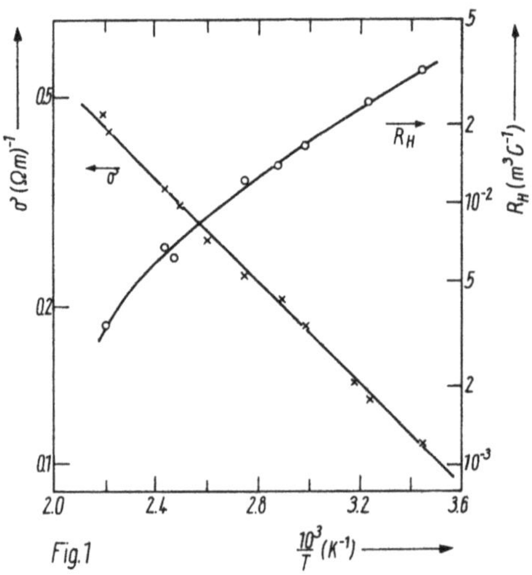

Fig.1

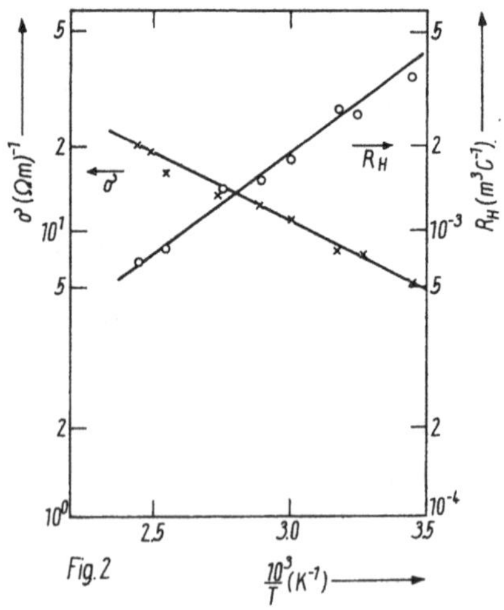

Fig.2

Fig. 1. Temperature dependences of conductivity and Hall coefficient for A-type samples

Fig. 2. Temperature dependences of conductivity and Hall coefficient for B-type samples

temperature dependence of the electrical conductivity is shown in Fig. 3. Between 130 and 180 K the electrical conductivity varies approximately as

$$\sigma = \sigma_0 \exp(-E/kT) \tag{2}$$

with an activation energy E = 40 meV. This indicates the presence of a shallow acceptor level with ionization energy about 40 meV. Between 200 and 300 K the electrical conductivity decreases with increasing temperature. Within this temperature range the temperature dependence of the conductivity appears to be completely controlled by the temperature dependence of the mobility, while the hole concentration remains constant.

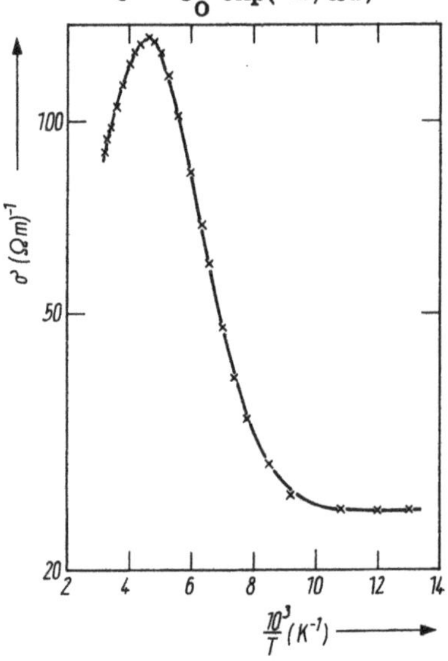

Fig. 3. Temperature dependence of conductivity for C-type samples

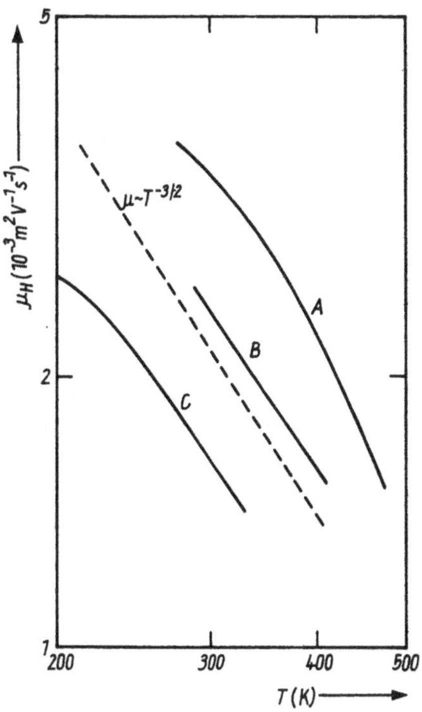

Fig. 4. Temperature dependence of Hall mobility for all three types of samples (solid lines). The dashed line illustrates the $\mu \sim T^{-1.5}$ relation

To obtain the temperature dependence of the hole mobility we multiply the Hall coefficient by the conductivity obtained from the fitting of curves presented in Fig. 1 and 2. The results are shown in Fig. 4.

The hole mobility obtained varies with temperature as

$$\mu = \mu_0 T^{-\alpha}, \qquad (3)$$

where the value of the coefficient α changes from 1.35 for A-type samples to 1.7 for C-type samples. This indicates the acoustic phonon scattering mechanism for hole scattering in Zn_3P_2 at temperatures close to 300 K.

Conclusions The electrical properties of Zn_3P_2 samples with different hole concentration have been investigated. The investigation of low hole concentration samples indicates the existence of an acceptor level with ionization energy about 140 eV. For high hole concentration samples a shallow acceptor level with ionization energy about 40 meV has been found. The latter result is the same as obtained by us for annealed samples /7, 8/.

For all types of samples the hole mobility at room temperature seems to be restricted by acoustic phonon scattering.

We would like to thank Prof. J.M. Pawlikowski for helpful discussion, and Dr. J. Misiewicz and Dr. F. Królicki for sample fabrication.

This work was supported by the Polish Academy of Sciences under contract 01.0.4.I.2.7.

References

/1/ J.M. PAWLIKOWSKI, Infrared Phys. 28 (1988).

/2/ J. SZATKOWSKI and K. SIERAŃSKI, phys. stat. sol. (a) 106, 473 (1988).

/3/ J. SZATKOWSKI, K. SIERAŃSKI, and J. MISIEWICZ, phys. stat. sol. (a) 106, 551 (1988).

/4/ P.S. NAYAR, J. appl. Phys. 53, 1069 (1982).

/5/ M. BUSHAN, J. appl. Phys. 53, 514 (1982).

/6/ A. CATALANO and R.B. HALL, J. Phys. Chem. Solids 41, 635 (1980).

/7/ J. SZATKOWSKI and K. SIERAŃSKI, Acta phys. Polon. A71, 297 (1987).

/8/ K. SIERAŃSKI and J. SZATKOWSKI, phys. stat. sol. (a) 94, K133 (1986).

/9/ J. MISIEWICZ, F. KRÓLICKI, M. LEWICKI, and J.F. KASPRZAK, Acta phys. Polon. A69, 1127 (1986).

(Received May 19, 1988)

phys. stat. sol. (a) 111, K61 (1989)

Subject classification: 72.80; 73.25; S12

Department of Spectroscopy,
Indian Association for the Cultivation of Science, Calcutta[1)]

Electrical Conduction in Ferrocene with Adsorbed Vapours

By

B. MALLIK

There is a renewed research interest /1, 2/ in the electrical conductivity properties of ferrocene [bis(cyclopentadienyl)iron: $(C_5H_5)_2Fe$], the first synthesized organo-metallic sandwich compound. The recent studies /1, 2/ involved the effect of mechanical pressure on the electrical properties. Inconsistent results related to the pressure dependence of the electrical conductivity have been reported /2, 3/ in the literature.

Indirect evidence was found /1/ in favour of a continuous and reversible phase transition in ferrocene under pressure at room temperature. Adsorption of minute amounts of moisture on ferrocene was thought /1/ to have an additional effect on the pressure dependence of conductivity. This can be the reason of the inconsistent conductivity-pressure characteristics reported in the literature.

It appears that systematic studies on the adsorption induced changes in the pressure dependence of the conductivity of microcrystalline organics can provide new insights into both their bulk and surface properties. Therefore, a programme was set up for such studies in case of ferrocene and some of its derivatives. Before starting this investigation it was thought worthwhile to study the change in electric properties of these samples on adsorption of different vapours in a sandwich type of cell configuration. In the course of such exploratory measurements some unusual findings were encountered, which will be briefly described for ferrocene below.

High-purity microcrystalline ferrocene (in powder form) was obtained from Aldrich Chemical Co., Inc. (Milwaukee, Wisconsin, USA) and was used after further purification by repeated recrystallization. Reagent chemicals used were of spectrograde (E. Merck, FRG) or equivalent quality. The usual sandwich cell technique /4, 5/ with a conducting glass and a stainless steel electrode was

---

1) Jadavpur, Calcutta 700 032, India.

used and the cell was placed inside a suitably designed conductivity chamber made of brass and fashioned with teflon. The experimental arrangement for the measurement of semiconduction parameters with various vapours adsorbed was the same as described in our earlier communications /4, 5/. A dc voltage of 27 V from dry batteries was applied across the electrodes as a bias potential and teflon spacers, 2 mils (0.0058 cm) thick, were used for this experiment. The cell temperature was maintained at about 258 K during adsorption. When the measured current attained a saturation value, the sample cell was cooled down to about 252 K. Then the sample was heated (at a constant rate) and the value of dark current was recorded with temperature. Ferrocene is reported /6/ to undergo an order-disorder phase transition at 250 K while cooling down from room temperature (300 K). Therefore, this cooling procedure was carefully controlled within this temperature.

The effect of adsorbed ethanol on the electrical conductivity (proportional to current) of ferrocene as a function of temperature is illustrated in Fig. 1 (curve a). At temperatures below 255 K the current was lower than 10^{-14} A. On heating the sample from low temperature, the current rises gradually with temperature and attains a maximum value of 3.1×10^{10} A at about 274.5 K. Then the current starts to decrease sharply with increasing temperature, showing a hump-like structure at about 278 K. Thus a conductivity peak is observed. In contrast, the conductivity variation (curve b in Fig. 1, plotted in a magnified scale) in the same temperature range in pure ferrocene in dry nitrogen atmosphere does not show any conductivity peak and the variation is the usual one as expected for a pure organic semiconductor. The activation energy of pure ferrocene measured in the temperature range 285 to 350 K is about 1.40 eV which agrees satisfactorily with the reported value /7, 8/. The adsorption induced conductivity reaches the pure sample value near room temperature and above this temperature the variation is similar to the pure sample in nitrogen atmosphere. In the higher temperature region this sample on adsorption of ethanol vapour gives almost the same value of conductivity and activation energy as the pure sample shows in dry nitrogen atmosphere indicating the complete desorption of adsorbed vapours, i.e. a case of physical adsorption.

The influence of different vapour pressures on the peak value of current (I_{max}) as well as the temperature (T_{max}) corresponding to this peak value was

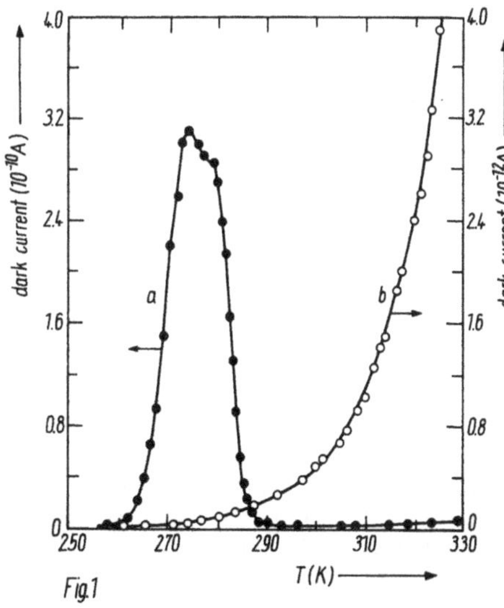

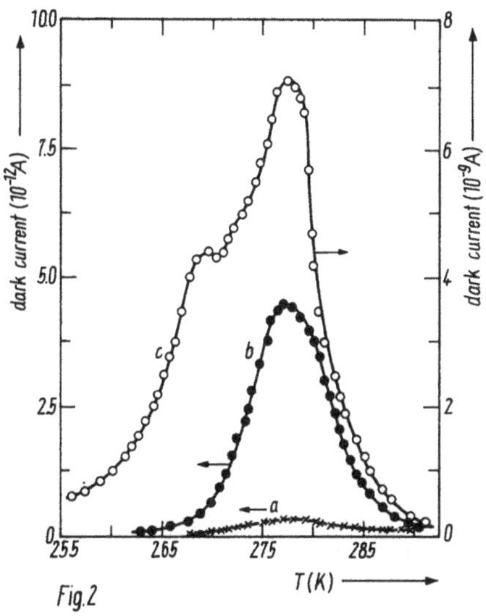

Fig. 1. Variation of dark current (conductivity) as a function of temperature in a powder cell of ferrocene: (a) sample with adsorption of ethanol vapour, (b) pure sample in dry nitrogen atmosphere

Fig. 2. Variation of dark current as a function of temperature in a powder cell of ferrocene with adsorption of ethanol vapour at different pressures; (a) 16.5 Torr, (b) 22.5 Torr, (c) 30.5 Torr. The sample was heated with a constant rate 0.01 K/s

studied. Results for three representative vapour pressures (16.5, 22.5, and 30.5 Torr) of ethanol vapour are presented in Fig. 2. The value of I_{max} and the shape of the observed dark current versus temperature curve change significantly with the increase of vapour pressure. At 30.5 Torr vapour pressure, an additional peak (at 269.5 K) overlapped with the original main peak is observed to appear on the lower temperature side of the main peak. But the position of the main peak remains almost at the same value. With further increase of vapour pressure the nature of conductivity variation with temperature changes drastically as shown in Fig. 3, curve a for ethanol vapour adsorption at 40 Torr pressure. In this case the above mentioned additional peak is more pronounced and shifts a little towards lower temperature (at 267.25 K) while the main peak is observed at a relatively higher temperature (289.5 K). For

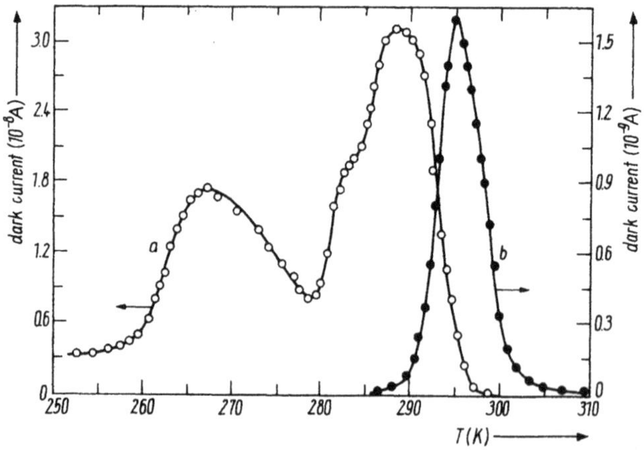

Fig. 3. Variation of dark current as a function of temperature in a powder cell of ferrocene with adsorption of (a) ethanol vapour (40 Torr), (b) carbontetrachloride vapour (40 Torr). The sample was heated with the same rate as mentioned in Fig. 2

adsorption of ethanol vapours at lower pressures as mentioned in Fig. 2 the currents above room temperature (300 K) show nearly the same value as for the pure sample in dry nitrogen atmosphere indicating a case of physical adsorption. But for adsorption of ethanol vapour at 40 Torr pressure the currents above room temperature are slightly higher than the corresponding values for pure sample in nitrogen atmosphere which indicates a case of incomplete desorption. Here a peak shift by about 10 K at 40 Torr pressure indicates the population of different molecular sites upon varying the pressure.

The appearance of conductivity peaks is not a typical property for adsorption of ethanol vapour only. With other vapours, e.g. ethyl acetate, carbontetrachloride, benzene, toluene, etc., such conductivity peaks have also been observed. In Fig. 3, curve b, the results of adsorption of carbontetrachloride vapour at 40 Torr pressure are presented. With change of vapour a change in the shape of the curve as well as in the values of I_{max} and T_{max} is evident from this figure. Usually the adsorbed vapours form charge-transfer complexes /4, 5/ with the solid surface. Ferrocene is reported to be a p-type semiconductor /7/, hence enhancement of its semiconduction current on adsorption of acceptor vapour is expected for the formation of charge-transfer type complexes of adsorbed vapours with ferrocene. However, no linear correlation of the electron affinity values of the vapours studied with the observed current on adsorption of vapours has been observed which excludes the possibility of formation of such charge-transfer complexes.

The value of activation energy of many organic semiconductors decreases /5/ or increases /9/ gradually on adsorption of vapours depending on the amount and the nature of adsorbed vapour molecules. The activation energy values evaluated by the usual procedure /5/ for adsorption of ethanol vapours at 22.5, 30.5, and 40.0 Torr pressure are 4.38, 1.51, and 0.64 eV, respectively. The conductivity variation for ethanol vapour at 16.5 Torr pressure (Fig. 2) being very low and within a narrow temperature range, the evaluation of such activation energy could not be done. In the present experiment the change in activation energy is unusual because the maximum charge is observed with adsorbed vapour at a lower vapour pressure (i.e. with less amount of adsorbed vapour). With the increase in the amount of adsorbed vapours the activation energy is observed to decrease and gradually approaches (even exceeds) the value of activation energy of the pure sample in dry nitrogen atmosphere. The activation energy of 4.38 eV observed on adsorption of ethanol vapour at 22.5 Torr pressure is much higher than the activation energy of the pure sample in nitrogen atmosphere. The reason for such a higher value is not clear. Another interesting observation in this experiment is the appearance of multiple conductivity peaks due to adsorption of vapours. In the earlier experiments with many organic semiconductors /5, 9, 10/ no such multiple peaks were observed on adsorption of vapours. These results are certainly unusual and detailed investigation on this topic may give new information about the interaction of vapours with the solid surface. Investigations into further aspects of this problem are in progress. A more complete account of these studies will be given elsewhere.

The author is thankful to Prof. G.S. Kastha and Prof. T.N. Misra for fruitful discussions.

References

/1/ B. KARVALY, B. MALLIK, and G. KEMENY, J. Mater. Sci. Letters 4, 912 (1985).

/2/ A.V. MAKSIMYCHEV, V.A. ZHORIN, A.T. PONOMARENKO, and

N.S. ENIKOLOPYAN, Dokl. Akad. Nauk SSSR, Ser. khim. 241, 414 (1978); Soviet Phys. - Doklady 241, 593 (1978).

/3/ Y. OKAMOTO, J.Y. CHANG, and M.A. KANTOR, J. chem. Phys. 41, 4010 (1964).

/4/ B. MALLIK, A. GHOSH, and T.N. MISRA, Proc. Indian Acad. Sci., Chem. Sci. 88A, 25 (1979).

/5/ B. MALLIK, A. GHOSH, and T.N. MISRA, Bull. Chem. Soc. Japan 52, 2091 (1979).

/6/ G. CALVARIN, G. CLECN, J.F. BERAR, and D. ANDRE, J. Phys. Chem. Solids 43, 785 (1982).

/7/ H. EICHHORN and T. VOIGTLÄNDER, Z. Naturf. 18, 1373 (1963).

/8/ H. WOTANABE, I. MOTOYAMA, and K. HATA, Bull. Chem. Soc. Japan 39, 850 (1965).

/9/ B. MALLIK, A. GHOSH, and T.N. MISRA, phys. stat. sol. (a) 62, 267 (1980).

/10/ A. SIRCAR, B. MALLIK, and T.N. MISRA, phys. stat. sol. (a) 69, 767 (1982).

(Received August 8, 1988)

phys. stat. sol. (a) 111, K67 (1989)

Subject classification: 72.80; 72.40; 78.40; S12.1

Institute of Chemistry, Academia Sinica, Beijing[1)]

Carrier Transport Properties of Poly(1,6-Bis-9-Carbazolyl-2,4-Hexadiyne)

By

RENYUAN QIAN, SONGQING SHI, TONGMING LIU, and DAOBEN ZHU

Since the discovery of the solid state polymerization of macro-single crystals of substituted diacetylenes by Wegner /1/, studies of the carrier transport properties have been of great interest, especially when Donovan and Wilcon /2/ discovered a very high carrier mobility at low fields and a carrier drift velocity independent of applied field strength. On the other hand, poly(1,6--bis-9-carbazolyl-2,4-hexadiyne) (PDCH) /3, 4/provides us an interesting example of Little's idea /5/ of having a conjugated backbone chain with highly polarizable pendant carbazole groups. Crystal structure studies show that the backbone chain in PDCH is a planar conjugated system of alternating double-single-triple-single bonding pattern /6/. Some indirect evidence for the high carrier mobility along the conjugated chain has also been obtained by other workers /7, 8/ to show a value in the order of 10^3 $cm^2s^{-1}V^{-1}$. In this note time of light drift mobility measurements on a sandwich cell for electrons as well as holes along directions parallel and transverse to the conjugated chain are described and clear cut non-dispersive transit times for a γ-radiation polymerized PDCH crystal have been found.

Experimental The preparation of DCH monomer has been described elsewhere /9/. Hexagonal and triangular prism-like crystals of dimensions up to 2.3 mm^2x13 mm were grown from acetone solution by evaporation at room temperature. DCH single crystals were polymerized by ^{60}Co γ-ray at 60 Mrad dose. The polymerized single crystals were found to have a flotation density of 1.313 g/cm^3 at 25.0 $^{\circ}C$ as measured in a H_3PO_4-H_2O density gradient column. For measurements of dark and photoconductivities of the single crystal along the conjugated chain direction in a sandwich configuration a crystal was put in a concentrated solution of PMMA in $CHCl_3$. After evaporating off the solvent at room temperature the crystal was fixed in PMMA. The embedded

1) Beijing 100080, People's Republic of China.

crystal was then cut perpendicular to its needle axis using a razor blade. The end surfaces were found to be the (112) crystal plane from WAXD. The cut slice of the crystal was clamped tight between two ITO conducting glass plates for measurements of volt-ampere (j-V) characteristics and steady state and transient photoconductivities. The end surfaces had an area of 1.38 mm^2 each and the thickness of the slice was 0.48 mm. A 80 W high pressure Hg lamp was used as the light source for steady state photoconductivity measurements. For the measurement of lux-ampere (j_{ph}-L) characteristics a series of calibrated neutral density filters of 7 to 90% transmission were used. For transient photoconductivity measurements a N_2 laser (λ = 337.1 nm) of 500 W peak power and a pulse width of 5 ns was used as a light source. The transient photocurrent across a signal sampling resistance of 100 kΩ was fed to a preamplifier of high input impedance. The rise time of the preamplifier was 0.1 μs, the output of which was fed to a Riken Denshi Model TCH-1000 transient recorder. The photoconductivity action spectrum was measured also with the cut crystal slice, onto the end surfaces of which a semi-transparent Au layer was vacuum evaporated. The crystal was set in a Hitachi model MPF-4 fluorescence spectrophotometer and the photocurrent was measured by a Riken Denshi Model TR-84M electrometer. For measurements on the crystal in the direction transverse to its needle axis a crystal was clamped between two ITO glass plates set parallel to the (102) crystal plane checked by WAXD. The electrode surface area was 6.1 mm^2 and the thickness of the slice was 0.58 mm.

Results and discussion Dark and photoconductivities The field (E) dependence of the current density j passing through the crystal in parallel and transverse directions to the conjugated chains in the crystal in dark and under illumination of irradiance of 28 mW cm^{-2} are shown in Fig. 1. $j_{\parallel}$ is ohmic up to a field of $1x10^3$ V/cm for both dark and photoconduction, and the photocurrent showed no difference whether the (-) or the (+) electrode was illuminated. In the SCLC region $j_{\parallel} \propto E^{1.5}$, $j_{\perp}$ is ohmic up to a field of (3 to 4)$x10^2$ V/cm and $j_{\perp} \propto E^{2.2}$ in the SCLC region for both dark and photoconduction. The photocurrent when the (+) electrode was illuminated (hole transport) was slightly higher than when the (-) electrode was illuminated (electron transport). The dark conductivity values obtained are $\sigma_{\parallel} = 4.2x10^{-12}$ S/cm and $\sigma_{\perp} = 1.5x$ $x10^{-12}$ S/cm showing rather small anisotropy.

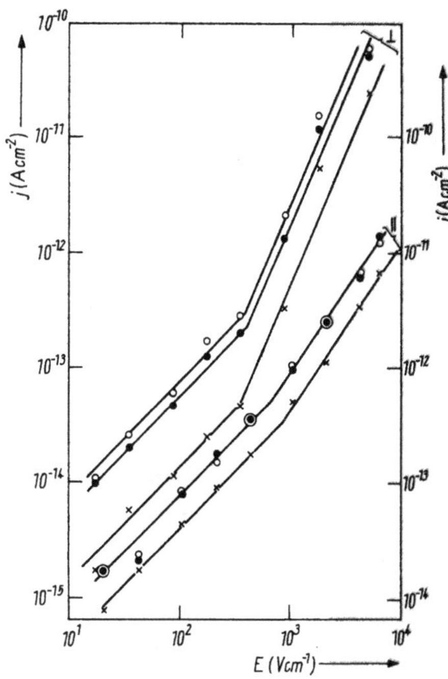

Fig. 1. Volt-ampere characteristics of dark and photoconduction in PDCH; o j^+, photo; ● j , photo; × j dark

Photoconduction action spectrum and lux-ampere characteristics The photoconduction action spectrum obtained when the applied field was parallel to the conjugated chain is roughly symbatic with its absorption spectrum (KBr compaction) as shown in Fig. 2. An analogous experiment using a surface cell gave similar results in contrast to the findings of Yee and Chance /4/ but in agreement with Lochner et al. /10/. Photoconduction peaks observed at λ= 250 and 370 nm and at λ = 460 nm could be attributed to the excitation of the carbazole chromophore and the interband transition of the conjugated chain, respectively. The quantum yield in the λ = 340 to 370 nm region was an order of magnitude higher than that in the visible region.

The lux ampere characteristics showed that $j_{ph} \propto L$ with the applied field parallel to the conjugated chain, while in the transverse direction $j_{ph} \propto L^{0.6}$

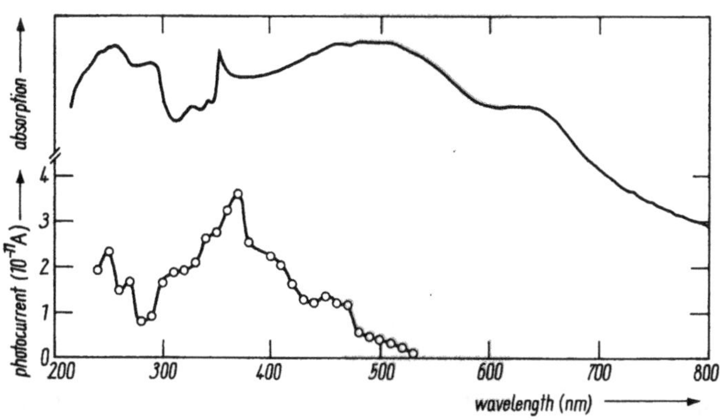

Fig. 2. Absorption and photoconduction action spectra of PDCH

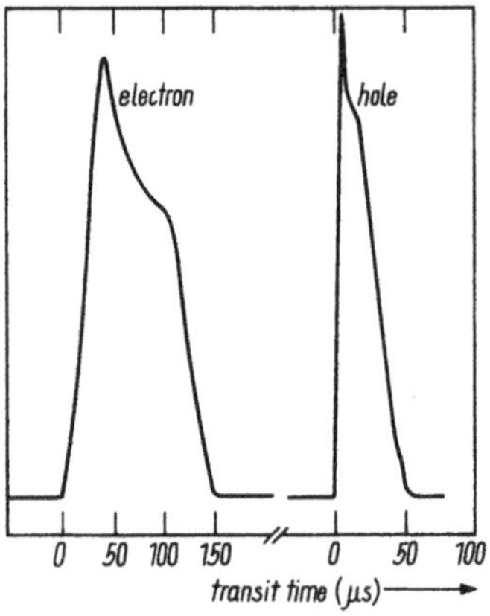

Fig. 3. Transient photocurrent in PDCH crystal after a 5 ns 337 nm pulse excitation

for both electrons and holes indicating carrier recombination or trapping. In both cases the applied field strength was in the ohmic region.

Drift mobility measurements Time of flight mobility measurements using a 5 ns N_2 laser pulse along the conjugated chain and transverse directions of the PDCH crystal showed well defined transit times for both electrons and holes as shown in Fig. 3, in contrast to trap controlled dispersive transport observed in the tosylate substituted polydiacetylene crystal /11/. This indicates that the single crystals used in this work contained much less impurities and defects so that trapping is apparently not significant. Our measurements were limited to transients in time scale of μs. The results show that the electrons have a drift velocity almost independent of the applied field of (5 to 50)x10^2 V/cm both in parallel and transverse directions to the conjugated chain. The value of the drift velocity is 3.7 m/s up to a field of 2x10^3 V/cm, declining somewhat at higher fields. The holes, on the other hand, show a decreasing drift velocity with increasing field, the value for fields parallel to the conjugated chain being 21 m/s at E < 5x10^2 V/cm and levelling off to 3 m/s at E > 3x10^3 V/cm. The

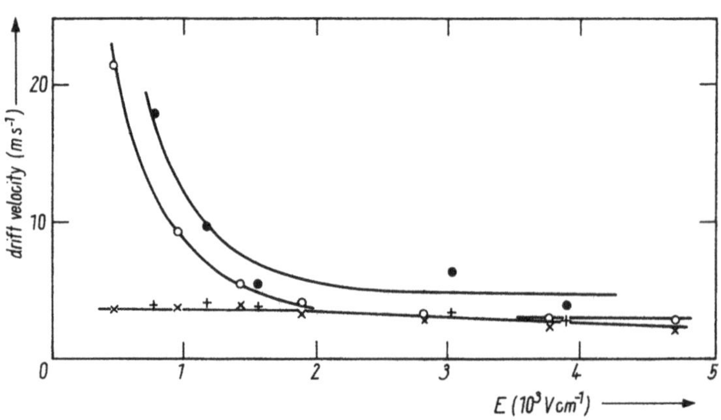

Fig. 4. Field dependence of the carrier drift velocities in PDCH crystal; o hole ∥, ● hole ⊥ , × electron ∥ , + electron ⊥

results are summarized in Fig. 4. The concept of mobility itself is no longer applicable in the present case. Wilson /12/ has suggested a solitary wave acoustic polaron theory for the ultra-high drift mobility at very low fields and a drift velocity independent of field and close to the sound velocity in the crystal. In view of these considerations it appears to us that very probably the transient photocurrent in PDCH crystal consisted of two components, one fast component for the carrier drift along the conjugated chain which was not properly resolved by the electronics used in this investigation. What we have observed is a slow component of the carrier drift with a rather low value of the drift velocity and low anisotropy. However, the peculiar field dependence or independence of the carrier drift velocity in PDCH remains to be understood.

This work was supported by the Science Fund of Academia Sinica.

References

/1/ G. WEGNER, Z. Naturf. 24b, 824 (1969).

/2/ K.J. DONOVAN and E.G. WILSON, Phil. Mag. B44, 9 (1981).

/3/ V. ENKELMANN, G. SCHLEIER, G. WEGNER, H. EICHELE, and M. SCHWOERER, Chem. Phys. Letters 52, 314 (1977).

/4/ K.C. YEE and R.R. CHANCE, J. Polymer. Sci. 16, 431 (1978).

/5/ W.A. LITTLE, Phys. Rev. A 134, 1416 (1964).

/6/ P A. APGAR and K.C. YEE, Acta cryst. B34, 957 (1978).

/7/ W. SPANNRING and H. BAESSLER, Chem. Phys. Letters 84, 54 (1981).

/8/ L. SEBASTIAN and G. WEISER, Phys. Rev. Letters 46, 1156 (1981).

/9/ D. ZHU, S. SHI, and R. QIAN, Makromol. Chem. Rapid Commun. 7, 313 (1986).

/10/ K. LOCHNER, H. BAESSLER, L. SEBASTAIN, G. WEISER, G. WEGNER, and V. ENKELMANN, Chem. Phys. Letters 78, 366 (1981).

/11/ K.J. DONOVAN, P.D. FREEMAN, and E.G. WILSON, in: Electronic Properties of Polymers and Related Compounds, Ed. H. KUZMANY, M. MEHRING, and S. ROTH, Springer-Verlag, Heidelberg 1985 (p. 256).

/12/ E.G. WILSON, J. Phys. C 16, 6739 (1983).

(Received March 21, 1988; in revised form October 3, 1988)

phys. stat. sol. (a) 111, K73 (1989)
Subject classification: 72.40 and 73.40; S8.16

Physics Department, Azerbaidzhan Institute of Oil and Chemistry, Baku[1)]

Photoelectrical Properties of $CuInSe_2$-GaAs Heterojunctions

By

D.D. ASKEROV, T.D. DZHAFAROV, A.M. AGAEV, S.G. ABDINOVA, and M.S. SADYGOV

The semiconducting $A^IB^{III}C_2^{VI}$ type compounds have recently attracted special attention due to their photovoltaic properties. Among them $CuInSe_2$ has become an important semiconductor for solar cell applications /1, 2/. However, the preparation of photoelements with large active surfaces is limited by the dimensions of the obtained $CuInSe_2$ single crystals. The preparation of photoelements on $CuInSe_2$ thin film base shows promise. The present note deals with some photoelectrical properties of $CuInSe_2$-GaAs heterojunctions which are of interest as photoelectric transformers. The preparation of CuInSe-GaAs heterojunctions has been described in /3/. X-ray research and photoconductivity measurements have shown that at evaporator and substrate temperatures of T_{ev} = 1500 K and T_{sub} = 700 K the composition of the obtained crystalline films is close to the initial $CuInSe_2$. The experimental results have shown that p-$CuInSe_2$-GaAs(111)B heterojunctions have the best photoelectrical characteristics. So this note presents results for such structures. It should be noted that thermal treatment in the air that is connected with the moulding of $CuInSe_2$ film and copper diffusion into GaAs /3 to 5/, influences considerably the properties of heterostructures. At constant temperature the photosensitivity increases first with increasing process duration, then it attains a maximum, further increase of the thermal treatment time causes a decrease of the sensitivity. The temperature of the thermal treatment influences in the same manner. This is connected probably with the degradation of $CuInSe_2$ films on increasing the temperature and duration of the treatment, i.e. the evaporation of Se and the diffusion of uncontrolled impurities take place /6/. It should be noted that the lower the thermal moulding temperature, the more time it will take to get the maximum photosensitivity. The present experimental results concern structures annealed at the chosen optimal conditions of T = 600 K and t = 10 min.

1) Lenina avenue 20, SU-370010 Baku, USSR.

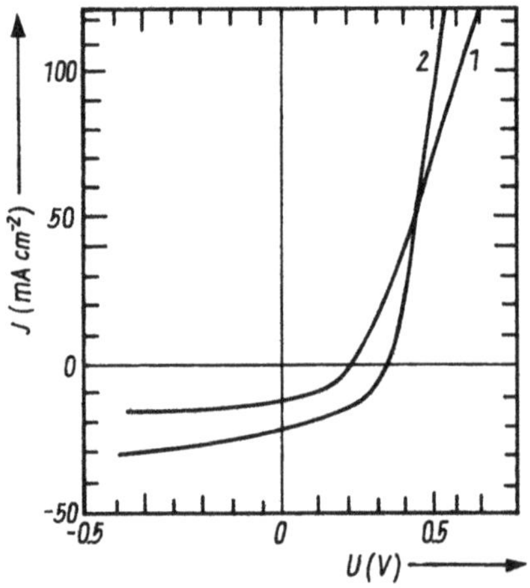

Fig. 1. Light current-voltage characteristics of a $CuInSe_2$-GaAs heterojunction (1) before and (2) after annealing (T = 300 K)

The current-voltage characteristics of p-$CuInSe_2$-n-GaAs(111)B heterojunctions taken with white light illumination and power density of 100 mW/cm^2 before (curve 1) and after (curve 2) the thermal treatment are presented in Fig. 1. The annealing produces a considerable increase of the short-circuit photocurrent J_{sc} from J_{sc} = 10 mA/cm^2 up to J_{sc} = = 22 mA/cm^2 and of the open-circuit voltage from U_{oc} = 0.23 V up to U_{oc} = 0.34 V. The efficiency of transformation is equal to 4.6%.

Fig. 2 shows the spectral dependence of the photosensitivity for one of the investigated samples. The composition displays photosensitivity in the region (0.4 to 1.2) μm. According to the energy band structure of heterojunctions, presented in Fig. 3, the photosensitivity at $\lambda < 0.9$ μm becomes apparent because of photon absorption both in $CuInSe_2$ film and GaAs substrate, but at $\lambda > 0.9$ μm carriers appear primarily in $CuInSe_2$ thin films. It should be noted that thermal treatment of the initial p-$CuInSe_2$-n-GaAs heterojunctions results in increasing photosensitivity without changing the nature of the spectral distribution. Probably the annealing of the structure produces an improvement of the characteristics without changing the film composition.

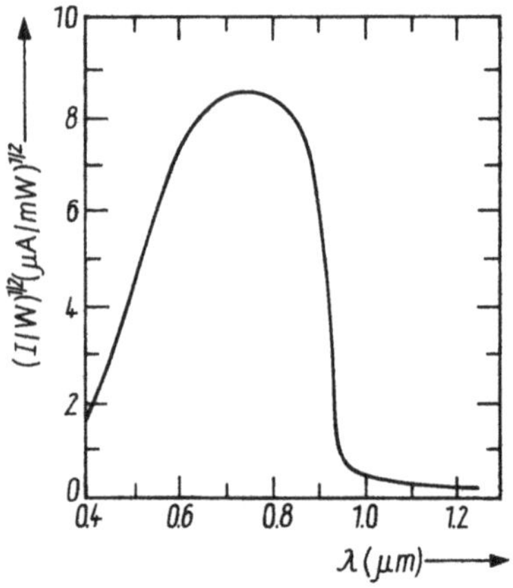

Fig. 2. Spectral distribution of the short-circuit photocurrent for a $CuInSe_2$-GaAs heterojunction (T = 300 K)

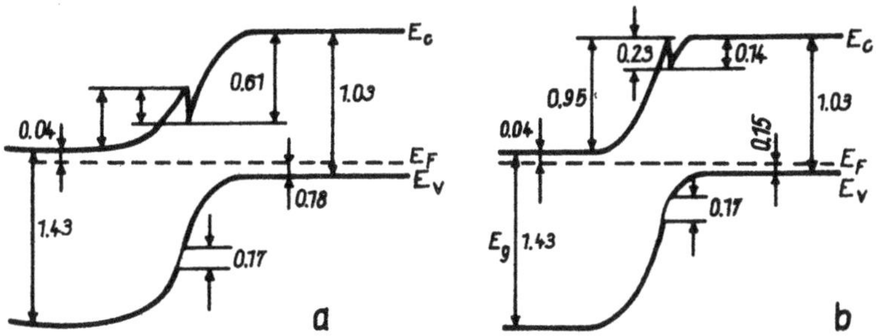

Fig. 3. Energy band diagram of p-$CuInSe_2$-p-GaAs heterojunctions a) before and b) after annealing

Thus $CuInSe_2$-GaAs heterojunctions have been obtained by evaporating $CuInSe_2$ thin films on GaAs substrate. It has been shown that the heterojunctions received on (111)B surfaces have the best photoelectrical parameters.

References

/1/ V. MIROVSKY and D. CHEN, Appl. Phys. Letters 40, 727 (1982).

/2/ L.L. KAZMERSKI, F.R. WHITE, and G.K. MORGAN, Appl. Phys. Letters 29, 269 (1973).

/3/ D.D. ASKEROV, T.D. DZHAFAROV, M.S. SADICHOV, and A.M. AGAEV, Preprint Inst. Fiz. Akad. Nauk Azerb. SSR, Baku, No. 191 (1986).

/4/ H. NEUMANN, R.D. TOMLINSON, E. NOWAK, and N. AVGERINOS, phys. stat. sol. (a) 56, K137 (1979).

/5/ H. NEUMANN and E. NOWAK, Thin Solid Films 74, 197 (1980).

/6/ L.L. KAZMERSKI and C.C. SHIEH, Thin Solid Films 91, 35 (1977).

(Received August 5, 1988)

phys. stat. sol. (a) 111, K77 (1989)

Subject classification: 74.70; S10.15

Institute of Chemical Technology, Prague[1] (a) and
Institute of Physics, Czechoslovak Academy of Sciences, Prague[2] (b)

Preparation of Pb-Doped Bi-Sr-Ca-Cu-O Superconductors

By

O. SMRČKOVÁ (a), D SÝKOROVÁ (a), L. SMRČKA (b),
and P. VAŠEK (b)

The discovery of superconductivity in the Bi-Sr-Cu-O system by Michel et al. /1/ was the first step toward high-T_c materials without the need for the rare earth elements. Subsequently the new high-T_c oxide superconducting system Bi-Sr-Ca-Cu-O was discovered by Maeda et al. /2/ which consists of the high-T_c phase with the superconducting transition onset at above 110 K, other phases with lower T_c, and non-superconducting components. Recently it was shown that the long heat treatment times and the addition of Pb /3 to 7/ promote the growth of the high-T_c phase and samples with large fractional volume of this phase were prepared with zero resistance above 100 K.

Here we report a study of the influence of preparation conditions on the superconducting properties of samples fabricated by conventional solid state reaction starting from the powder reagents.

Appropriate amounts of p.a. powders of Bi_2O_3, Pb_3O_4, $SrCO_3$, $CaCO_3$, and CuO were mixed together in the molar ratio of Bi:Pb:Sr:Ca:Cu equal to 1.4:0.6:2:2:3.6 and ground in a ball mill. The resulting mixed powder was pre-fired in air at 800 °C for 20 h, then reground and pressed into pellets of 13 mm diameter and 1 mm thickness.

One of the pellets was then sintered at 845 to 850 °C for 93 h. In spite of the experience expressed in /4 to 6/ we have found that the resulting material was almost pure 70 K superconductor without any trace of a high-T_c phase.

For this reason we decided to investigate the preparation conditions in more detail. Since the annealing experiments at various temperatures are very time-consuming we measured the thermal gradient in the furnace and placed pellets into the three regions with different temperatures. The pellet A was in the

1) Suchbátarova 5, CS-166 28 Praha 6, CSSR.

2) Na Slovance 2, CS-180 40 Praha 8, CSSR.

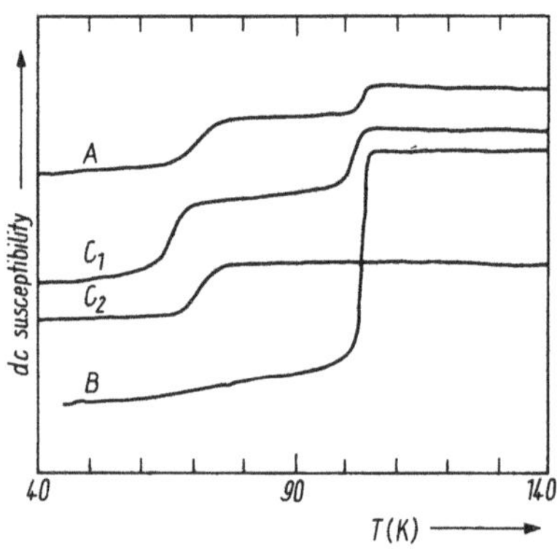

Fig. 1. Static susceptibility of samples sintered at various temperatures: (A) 875 °C, (B) 860 °C, (C_1) 855 °C, and (C_2) 830 °C

central part at 875 °C, the next pellet B in the part with temperature T close to 860 °C, and the last pellet in the region where T varied from 855 to 830 °C. All the pellets were sintered at the above given temperatures in air for 191 h and then furnace cooled.

Critical temperatures were determined by measurements of the resistivity, ac susceptibility, and Meissner effect. The resistivity was measured by the standard four-probe technique with the help of a low-dissipation ac bridge. Voltage and current leads were connected to samples with silver paste. The ac susceptibility and Meissner effect were measured using a S.H.E SQUID susceptometer. The experimental set-up allowed to carry out both measurements on a single small piece of material using the same detection coils. The Meissner effect was measured in the magnetic field of earth frozen in the superconducting shield. The S.H.E. mutual inductance bridge with the frequency 160 Hz and the excitation voltage of 5 V was used to measure the ac susceptibility $\chi' - i\chi''$.

Results of the static susceptibility measurements are shown in Fig. 1. Two samples of material were taken from the pellet C, the sample C_1 from its "hot" side and the sample C_2 from its "cool" side. The sample C_2 (sintered at temperature close to 830 °C) exhibits almost exclusively the diamagnetic shift at about 70 K. The sample C_1 from the "hot" side (sintering T approximately 855 °C) has a two-step temperature dependence of the static susceptibility. The step corresponding to the high-T_c phase and the step due to the transition at low T_c are comparable in magnitude. A similar behaviour is observed also for the sample A sintered at 875 °C. Only the sample B shows the one-step diamagnetic shift which corresponds to the transition of the high-T_c phase. It does not mean that the sample contains only the high-T_c phase. Nevertheless, the result indicates that its volume fraction

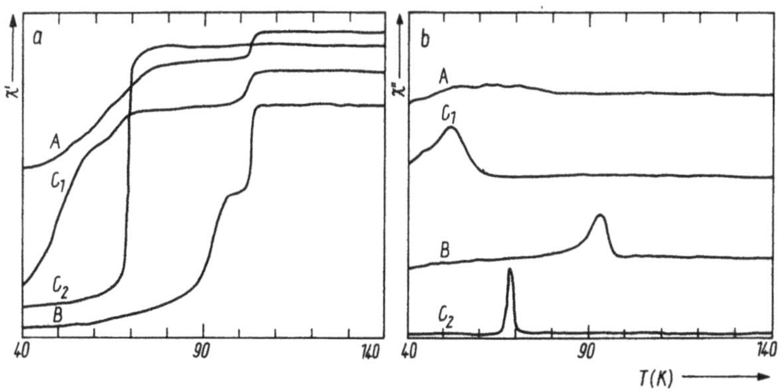

Fig. 2. The ac susceptibility of samples sintered at various temperatures: (A) 875 °C, (B) 860 °C, (C_1) 855 °C, and (C_2) 830 °C; a) real part, b) imaginary part

is large enough that this phase surrounds the low-T_c phase and a one-step transition is observed due to the shielding currents.

Fig. 2 plots the variation of the ac susceptibility with temperature. The results for both χ' and χ'' are strongly influenced when the connectivity is established in the developing superconducting phase. The superconducting loops are formed and shielding currents contribute to the formation of an additional step in χ' and a peak in χ'' which corresponds, according to /8/, to dissipation of energy by weak links in superconducting loops. In the sample C_2 the connectivity is achieved in the 70 K phase quite close to the onset temperature. As seen from static susceptibility measurements and from the χ' curve, a certain amount of the high-T_c phase is developed in C_1 but the connectivity is not reached for this phase. A similar behaviour is observed for sample A, but the peak becomes broader and is shifted slightly above the 70 K transition onset. Connectivity of the high-T_c phase is found only in sample B where both the step in χ' and the peak in χ'' are close to the onset of the high-T_c transition.

Fig. 3 plots the variation in the sample resistivity with temperature for samples A and B, the inhomogeneous sample C is omitted. Both samples demonstrate a stepped transition. The step is more pronounced for sample A

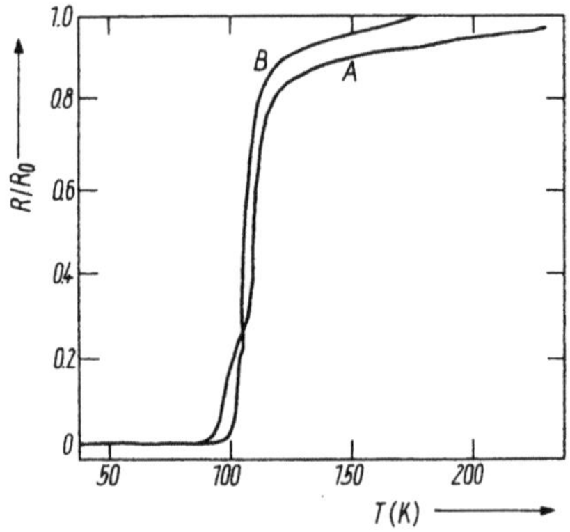

Fig. 3. Ratio of resistivity R(T) to resistivity at 0 °C, R_0, for sample A sintered at 875 °C and sample B sintered at 860 °C

and only a tiny foot is observed for sample B. If the majority transition is extrapolated, zero resistance would occur at 101 K. These results are in agreement with conclusions drawn from the susceptibility measurements.

This note confirms that adding Pb to the Bi-Sr-Ca-Cu-O system increases the volume fraction and continuity of the high-T_c phase in superconductors prepared by the standard ceramic technique. We have found that the formation of the high-T_c phase is very sensitive to the sintering temperature and that the optimum value is close to 860 °C, in agreement with /7/. We obtained a smaller fractional volume of the high-T_c phase for annealing temperatures above 870 °C and almost no high-T_c phase for temperatures below 850 °C.

References

/1/ C. MICHEL, M. HERVIEU, M.M. BOREL, A. GRANDIN, F. DESLATES, J. PROVOST, and B. RAVEAU, Z. Phys. B 68, 421 (1987).

/2/ M. MAEDA, Y. TANAKA, M. FUKUTOMI, and T. ASSANO, Japan. J. appl. Phys. Letters 27, L209 (1988).

/3/ S.A. SUNSHINE, T. SIEGRIST, L.F. SCHNEEMEYER, D.W. MURPHY, R.J. CAVA, B. BATLOGG, R.B. VAN DOVER, R.M. FLEMING, S.H. GLARUM, S. NAKAHARA, R. FARROW, J.J. KRAJEWSKI, S.M. ZAHURAK, J.V. WASZCZAK, J.H. MARSHALL, P. MARSH, L.W. RUPP, Jr., and W.F. PECK, Phys. Rev. B, submitted.

/4/ M. TAKANO, J. TAKADA, K. ODA, H. KITAGUCHI, Y. MIURA, Y. IKEDA, Y. TOMII, and H.H. MAZAKI, Japan. J. appl. Phys., submitted.

/5/ K. TOGANO, H. KUMAKURA, H. MAEDA, E. YANAGISAWA, and K. TAKAHASHI, Appl. Phys. Letters, submitted.

/6/ E. YANAGISAWA, D.R. DIETDERICH, H. KUMAKURA, K. TOGANO, H. MAEDA, and K. TAKAHASHI, Preprint.

/7/ S.M. GREEN, C. JIANG, Y. MEI, H.L. LUO, and C. POLITIS, Phys. Rev. B Letters, in the press.

/8/ T. ISHIDA and H. MAZAKI, J. appl. Phys. 52, 6798 (1981).

(Received October 3, 1988)

phys. stat. sol. (a) 111, K83 (1989)

Subject classification: 72.30

Ames Laboratory-USDOE and Department of Physics,
Iowa State University[1)]

Anomalous Skin Effect in a Cylindrical Conductor

By

R. FUCHS and K. GHOSH

A semiclassical theory of the anomalous skin effect in a cylindrical conductor has been presented recently by Dasgupta /1/. An electric field $E_z(r, t) = E_z(r)e^{-i\omega t}$ is applied parallel to the axis (z) of a cylindrical metallic conductor of radius a. Using a boundary condition analogous to "specular scattering" for a plane surface, and neglecting displacement current, expressions for the current density $j_z(r)$ and electric field $E_z(r)$ within the cylinder are derived, which involve the nonlocal bulk transverse conductivity $\sigma(k, \omega)$ of the conducting material. Solutions are obtained in which j_z and E_z depend only on the distance ϱ from the cylinder axis. In this note, we shall show results of calculations using a free-electron model for the conductor.

The expressions for the electric field and current density inside the cylinder ($\varrho < a$) are

$$E_z(\varrho) = \Delta \int_0^\infty \frac{J_0(ka)J_0(k\varrho)}{k^2 - k_1^2} k\,dk \tag{1}$$

and

$$j_z(\varrho) = \Delta \int_0^\infty \frac{J_0(ka)J_0(k\varrho)\sigma(k,\omega)}{k^2 - k_1^2} k\,dk\ , \tag{2}$$

where

$$k_1^2 = (4\pi i\omega/c^2)\sigma(k, \omega)\ , \tag{3}$$

and Δ is an unknown constant which can be related to the total current I through the conductor[2)]

$$I = 2\pi\Delta a \int_0^\infty \frac{J_0(ka)J_1(ka)\sigma(k,\omega)}{k^2 - k_1^2} dk\ . \tag{4}$$

---

1) Ames, Iowa 50011, USA.

2) Equation (4) corrects a misprint in equation (19) of /1/.

It is also of interest to calculate the surface impedance of the conductor /2/,

$$Z(\omega) = \frac{E_z(a)}{I} . \qquad (5)$$

Writing $Z(\omega) = \mathrm{Re}[Z(\omega)] + i\mathrm{Im}[Z(\omega)]$, the resistance and reactance per unit length are, respectively, $\mathrm{Re}[Z(\omega)]$ and $\mathrm{Im}[Z(\omega)]$.

We have used (1) to (5) to calculate the current distribution $j_z(\rho)$ and surface impedance Z of a free-electron wire, taking $\sigma(k,\omega) = \omega(\varepsilon(k,\omega) - 1)/(4\pi i)$, where $\varepsilon(k,\omega)$ is the Lindhard transverse dielectric function /3/. The parameters are for a free-electron gas with a density corresponding to Cu: $v_F = 1.57\mathrm{x}10^8$ cm/s, $r_s = 2.67\ a_o$, $\hbar\omega_p = 10.8$ eV. The relaxation time which appears in $\varepsilon(k,\omega)$ is $\tau = l/v_F$, where the mean free path $l = 2a$, the diameter of the wire. The local results are found by replacing the Lindhard dielectric function $\varepsilon(k,\omega)$ by the local (Drude) dielectric function $\varepsilon(\omega) = 1 - \omega_p^2/(\omega(\omega + i/\tau))$.

The absolute value of the ratio of the current density at a distance ρ from the cylinder axis to the current density at the outer surface of the cylinder at a distance a, $|j(\rho)/j(a)|$, is plotted as a function of the relative distance ρ/a at the dimensionless frequency ratio $\omega/\omega_p = 10^{-3}$ for three values of the cylinder radius $a = 2.5\mathrm{x}10^{-5}$, $2.5\mathrm{x}10^{-6}$, $2.5\mathrm{x}10^{-7}$ cm (Fig. 1 to 3). Also, the ratios $\mathrm{Re}(Z)/R_o$ and $-\mathrm{Im}(Z)/R_o$, where $R_o = \omega_p^2\tau/4\pi$ is the dc resistance per unit length of the cylinder, are plotted as functions of ω/ω_p for the three values of the radius (Fig. 1 to 3).

Several features of the calculated results can be understood qualitatively. In the local theory, the penetration depth of the current is given approximately by the classical skin depth, $\delta = c/(2\pi\sigma^{dc}\omega)^{1/2}$. Since the scattering time $\tau = 2a/v_F$, and $\sigma^{dc} = \omega_p^2\tau/4\pi \propto a$, one has $\delta \propto 1/\sqrt{a}$. At the three radii, $a = 2.5\mathrm{x}10^{-5}$, $2.5\mathrm{x}10^{-6}$, and $2.5\mathrm{x}10^{-7}$ cm, we have respectively, $\delta/a = 0.045$, 1.42, and 44.9. These values of δ/a are consistent with Fig. 1a, 2a, and 3a. For the largest cylinder (Fig. 1a), the skin depth is much smaller than the cylinder radius, so the current density decreases rapidly as one moves away from the surface. For the intermediate cylinder (Fig. 2a), the current density is more nearly constant, and for the smallest cylinder (Fig. 3a), it is almost exactly constant. It is well-known that the skin depth for a plane surface increases when a nonlocal dielectric function $\varepsilon(k,\omega)$ is introduced. This also occurs for a cylinder: the nonlocal current density penetrates more deeply into the interior

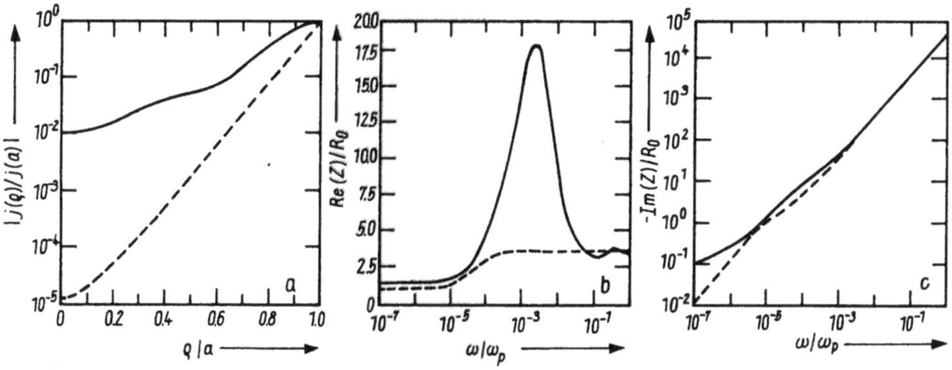

Fig. 1. a) Absolute value of the ratio of the current density at a distance ϱ from the cylinder axis to the current density at the outer surface of the cylinder at a distance a, $|j(\varrho)/j(a)|$, plotted as a function of the relative distance ϱ/a. The dimensionless frequency ratio is $\omega/\omega_p = 10^{-3}$; b) ratio of the real part of the surface impedance and the dc resistance, $Re(Z)/R_o$; c) ratio of the negative of the imaginary part of the surface impedance and the dc resistance, $-Im(Z)/R_o$, plotted as a function of the dimensionless frequency ratio ω/ω_p. Calculated values using both the local (Drude) $\varepsilon(\omega)$ and the nonlocal (Lindhard) $\varepsilon(k,\omega)$ are shown. The cylinder material is Cu (see text). The radius of the cylinder $a = 2.5 \times 10^{-5}$ cm, —— nonlocal, - - - local

than the local current density. Indeed, for the two smaller cylinders, the current density increases slightly as one moves away from the surface.

If $\delta \ll a$, the surface impedance Z, defined by (5), can be expressed in terms of the surface impedance Z_s of a plane interface. Using Ampère's law to relate the total current I to the tangential magnetic field at the surface, and neglecting the displacement current, we find

$$Z = \frac{E_z(a)}{I} = \frac{2}{ac} \frac{E_z(a)}{H_\theta(a)} = \frac{2}{ac} Z_s .$$

In the local limit, $Z_s = (4\pi i\sigma(\omega)/\omega)^{-1/2}$, which gives

$$\frac{Z}{R_o} = \frac{a\omega_p}{4c} - i\frac{a\omega_p}{2c}\omega\tau , \ \delta \ll a . \tag{6}$$

The general result for the local surface impedance, obtained by using the local conductivity $\sigma(\omega)$ in (1), (4), and (5), is

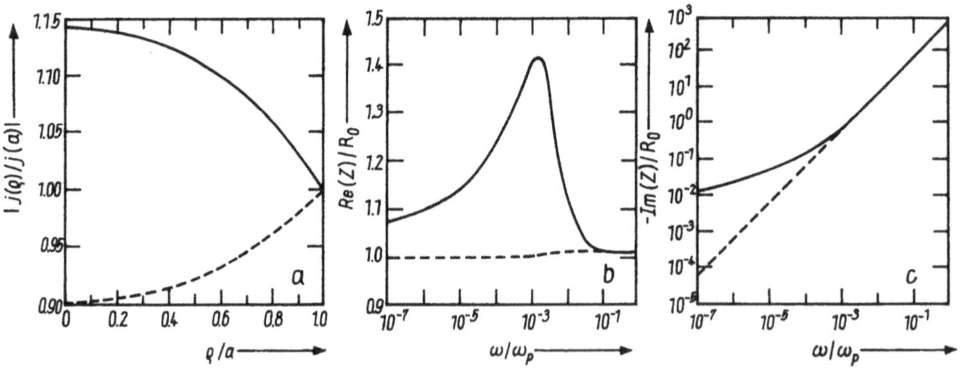

Fig. 2. Same as Fig. 1 with the radius of the cylinder $a = 2.5 \times 10^{-6}$ cm

$$Z = \frac{k_1 J_0(k_1 a)}{2\pi a \sigma(\omega) J_1(k_1 a)} \, . \tag{7}$$

If $\delta \gg a$ (or $k_1 a \ll 1$), (7) reduces to

$$\frac{Z}{R_0} = 1 - i\omega\left[\tau + \pi a^2 \sigma^{dc}/2c^2\right], \; \delta \gg a \, . \tag{8}$$

For a fixed value of a, the crossover from (6) to (8) occurs as the frequency decreases. The crossover is gradual, and is easily visible for the largest value of a (Fig. 1b, c), where it occurs for $\omega/\omega_p \approx 10^{-5}$ to 10^{-4}, is less visible in Fig. 2b, c, where it occurs for $\omega/\omega_p \approx 10^{-3}$ to 10^{-2}, and cannot be seen at all in Fig. 3b and c.

For a plane surface, the inclusion of nonlocality increases Re(Z): electrons in the metal can transfer energy from the surface region deep into the interior. If $a \lesseqqgtr \delta'$, where δ' is the anomalous skin depth, this energy transfer is restricted by the cylinder size, and Re(Z), which is proportional to the rate of energy transfer, decreases. As the frequency decreases, δ' increases, and Re(Z) increases. The fall-off in the nonlocal Re(Z) that occurs in Fig. 1b as ω/ω_p decreases from 10^{-3} to 10^{-5} occurs where δ' becomes larger than a. The condition $\delta' > a$ holds over a wider frequency range for the smaller values of a, so the nonlocal values of $\text{Re}(Z)/R_0$ never rise much above 1 in Fig. 2b and 3b.

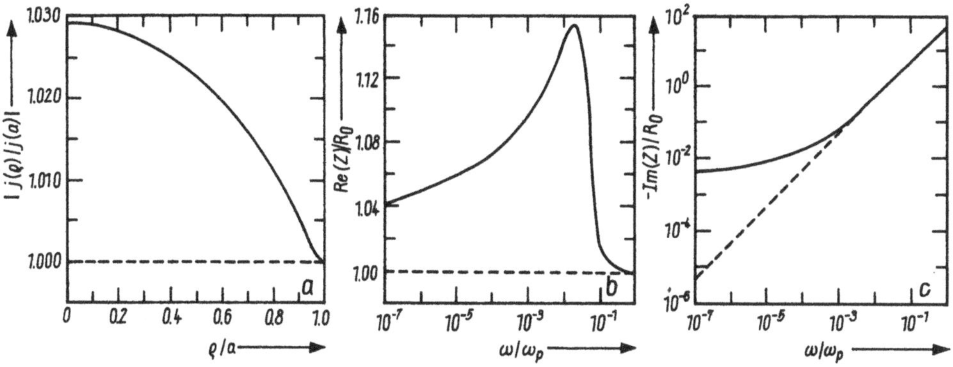

Fig. 3. Same as Fig. 1 with the radius of the cylinder $a = 2.5 \times 10^{-7}$ cm

The k-dependence of $\sigma(k,\omega)$ appears primarily in a parameter $u \equiv kv_F(1/\tau - - i\omega)^{-1}$, such that $\sigma(k,\omega) \approx \sigma(0,\omega)$, if $|u| \ll 1$. Since the most significant values of k in a cylinder of radius a are $k \approx 1/a$, we have $u_{eff} \approx (v_F/a)(1/\tau - - i\omega)^{-1}$. If $\omega\tau \gg 1$, or $\omega a/v_F \gg 1$, then $|u_{eff}| \ll 1$, and the nonlocal effects go away. Therefore, at high frequencies, nonlocal and local values of $Re(Z)/R_0$ and $Im(Z)/R_0$ become equal.

A part of this note was performed at the Institute for Theoretical Physics, Freie Universityt Berlin (West). We thank Prof. K.H. Bennemann, and other members of the Institute, for their hospitality. The Ames Laboratory is operated for the U.S. Department of Energy by Iowa State University under contract no. W-7405-Eng-82. This work was supported by the Deutsche Forschungsgemeinschaft through Sonderforschungsbereich 337 and by the Director for Energy Research, Office of Basic Energy Sciences.

References

/1/ B.B. DAGUPTA, phys. stat. sol. (a) 91, 235 (1985).

/2/ S. RAMO, J.R. WHINNERY, and T.V. DUZER, Fields and Waves in Communication Electronics, 2nd ed., Wiley, New York 1984 (Sections 3.16-3.17; 4.4-4.5).

/3/ J. LINDHARD, Kong. Danske Vid. Selsk., mat.-fys. Medd. 28, No. 8 (1954).

(Received August 22, 1988)

phys. stat. sol. (a) 111, K89 (1989)

Subject classification: 74.70; S10.15

Institute of Physics of Solids and Semiconductors, Academy of Sciences of the Belorussian SSR, Minsk[1)]

Anomalous Electrical Resistance of Y-Ba-Cu-O System Ceramics

By

B. B. BOIKO, A.I. AKIMOV, V.I. GATALSKAYA, S.E. DEMYANOV, L.A. KUROCHKIN, M.L. PETROVSKII, E.K. STRIBUK, and V.M. FINSKAYA

Up to now there has been published a large number of papers on the important discovery /1, 2/ of high-T_c superconductivity in La-Ba(Sr)-CuO and Y-Ba-Cu-O ceramic compounds. The latter system exhibits a resistance variation of the samples with temperature change from metallic, semimetallic, to semiconducting with typical values of high-T_c superconductivity transition temperatures in the range from 90 to 100 K depending on the composition and conditions of sample preparation /3/. At the same time the papers /4 to 8/ are known to report anomalies in the temperature dependences of the resistance R in the temperature range above 200 K.

In /4/, in particular, a two-step resistivity transition of multiphase $Y_{1.8}Ba_{0.2}CuO_{4-y}$ ceramics has been obtained at temperatures of ≈ 240 K and 90 K, with the superconductivity at these temperatures being confirmed by the reverse Josephson effect. The authors of /5/ have noted a rapid decline of resistivity in a $Y_{0.3}Ba_{1.0}Cu_{0.7}O_{\gamma}$ sample starting from 350 K. Below 220 K and up to $T_c = 92$ K at which the total superconducting transition is attained there occurs a very weak increase in the resistance with hysteresis-like character of the thermal dependences to room temperature. Besides for the $Y_{0.4}Ba_{0.9}Cu_{0.7}O_y$ sample the effect of thermocycling on the character of the R(T) dependence has been noted in the range from 300 to 85 K.

For one of the $Y_1Ba_2Cu_3F_2O_y$ samples containing fluorine the authors of /6/ observed a marked hysteresis in the variation of the resistance with temperature. The temperature of the total superconductivity transition was rather high (155 K) when attaining the zero-resistance state. Another sample of the

1) P. Brovkii 17, SU-220 726 Minsk, USSR.

above mentioned composition displayed an anomalous rapid decrease in the resistance according to a $T^{8.3}$ law. The zero-resistance state was achieved at 91 K. The drop of R, starting from room temperature, was associated with the presence of phases exhibiting superconducting onset above room temperature. One more study /8/ on the $Y_{0.05}Ba_{0.045}Cu_{0.5}O_y$ ceramics is worth mentioning where an anomalous behaviour of the resistance with temperature has been noted. Apart from the usual transition at T = 91 K the zero-resistance state has been observed at 230 K. On warming the sample up to 300 K the high temperature resistance minimum shifted to 275 K and on subsequent cooling no anomalies in the R(T) dependence have been found at temperatures above T_c = = 91 K.

In our studies on high-T_c superconductivity in ceramics an anomalous resistance behaviour in the range of high temperatures along with a typical superconducting transition at 90 to 100 K has been found. The ceramic Y-Ba-Cu-O samples were obtained by the following procedure. The mixtures of appropriate amounts of Y_2O_3, CuO, and $BaCO_3$ were thoroughly mixed, ground, and pressed into pellets under a pressure of 500 atm. After the calcination of pellets in air for 5 h at 950 °C the samples were reground, pressed again in pellets of 10 to 15 mm in diameter and 5 mm in thickness, and sintered for 5 h at temperatures of 950 to 1050 °C. Bulk-rectangular samples with appropriate dimensions of 2x3x10 mm^3 cut from the sintered pellets were used for the dc resistivity measurements. Besides the usual complete transition to the superconducting state at T = 90 to 100 K observed in our Y-Ba-Cu-O ceramic samples /9/ an anomalous behaviour of the resistance of several $Y_1Ba_2Cu_3O_y$ samples immediately after the preparation of the material (Fig. 1, solid line) was noted. In this sample an abnormally rapid fall of resistivity (by two orders of magnitude compared to the R value at room temperature) with a sharp minimum at T = 193 K has been observed. Below this temperature the resistance increased in a manner typical for semiconductors. After four to five fold thermal recyclings in the range from 300 to 77 K the type of R(T) dependence changed drastically (dashed line): the anomaly at 193 K disappeared, the resistivity decreased in a manner typical for "metallics" up to the value ≈ 0.5 mΩcm at the onset superconducting transition temperature T_c^{on}. The superconducting transition to the zero-resistance state (with an accuracy of 10^{-4} Ω restricted to the apparatus sensitivity) is attained at 82.7 K.

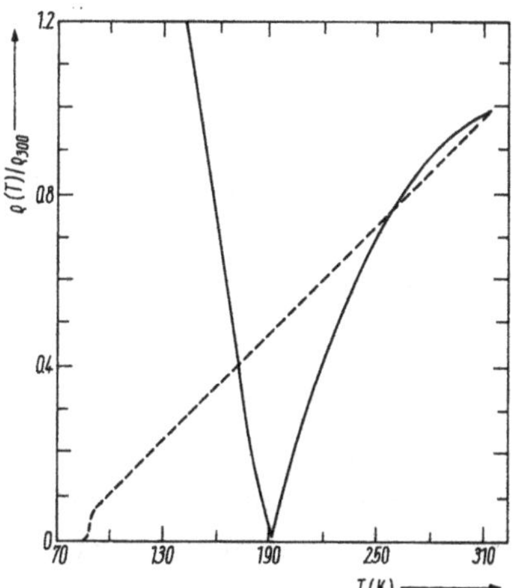

Fig. 1

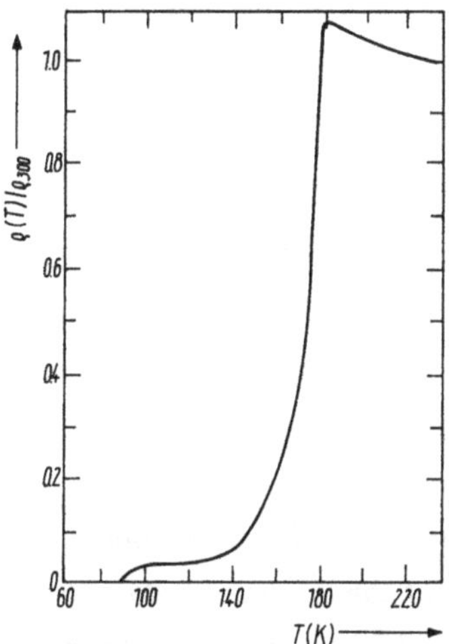

Fig. 2

Fig. 1. Resistance versus temperature in $Y_1Ba_2Cu_3O_y$ sample

Fig. 2. Resistance versus temperature in $Y_{0.9}Sc_{0.1}Ba_2Cu_3O_y$ sample

Fig. 2 gives the temperature dependence of the resistivity of $Y_{0.9}Sc_{0.1}Ba_2Cu_3O_y$ ceramics, its resistivity at 300 K is 7 mΩcm. On cooling the sample from room temperature to T = 182 K a resistivity growth typical for semiconductors is observed and at 182 K an abrupt jump is noted and the resistivity drops by more than an order of magnitude. It is possible that at this temperature the semiconductor-metal transition occurs. As follows from Fig. 2, the resistivity of yttrium-scandium ceramics varied slightly up to T = 91 K at which the superconducting transition begins and completes at 88.5 K. Such a behaviour of resistivity can be due to the presence of two phases with different transition temperatures.

Then Sm doped yttrium barium ceramic $Y_{0.9}Sm_{0.1}Ba_2Cu_3O_y$ samples were synthesized under normal conditions, the temperature dependence for these samples is illustrated in Fig. 3. The resistivity of the ceramics at 300 K was 2 mΩcm. The dc measurements were carried out at currents ranging from 1 mA to 20 μA. On cooling the sample down from room temperature the resistance fell and the zero-resistivity state was achieved at 124 K (when the cur-

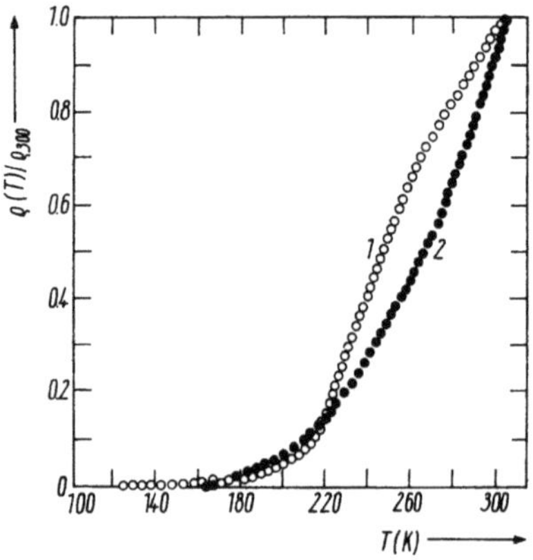

Fig. 3 Resistance versus temperature in $Y_{0.9}Sm_{0.1}Ba_2Cu_3O_y$ sample. (1) 1 mA, (2) 20 μA

rent was 20 μA the superconductivity transition was completed at T_c^o = 168 K). The phase responsible for the high superconducting transition is stable, the results are reproducible and no signs of degradation have been observed for a few months.

Thus, we have prepared ceramic samples revealing high-T_c superconductivity, the temperature dependences of the resistances of these compounds show anomalies which appear at temperatures of the order of 20 K. Further research in the properties of superconductivity and the identification of phases associated with them is nessecary in order to better understand the nature of high-T_c superconductivity in ceramic materials.

References

/1/ J.G. BEDNORZ and K.A. MÜLLER, Z. Phys. B 64, 189 (1986).

/2/ W.K. WU et al., Phys. Rev. Letters 58, 908 (1986).

/3/ C. POLITIS, J. GEERK, M. DIETRICH, R. OBST, and H.L. LUO, Z. Phys. B 66, 279 (1987).

/4/ J.T. CHEN, L.E. WENGER, C.J. MC EWAN, and E.M. LOGOTHETOS, Phys. Rev. Letters 58, 1972 (1987).

/5/ Y. ODA et al., Japan. J. appl. Phys. 26, 807 (1987).

/6/ S.R. OVCHINSKII, R.T. YAING, D.D. ALLRED, G. DE MAGGIO, and G.A. VAN DER LEEDEN, Phys. Rev. Letters 58, 2579 (1987).

/7/ K. TAKITA, T. SOPOSHI, T. UCHINO, T. GOCHAU, and MASUOLA, Japan. J. appl. Phys. 26, 306 (1987).

/8/ L.C. BOURNE, M.L. COHEN, W.N. CREAGER, M.F. CROMNIE, and and A. LETTL, Phys. Letters 123, 34 (1987).

/9/ B.B. BOIKO et al., Izv. Akad. Nauk BSSR, Ser. fiz. mat. 6, 105 (1987).

(Received February 4, 1988; in revised form June 17, 1988)

phys. stat. sol. (a) 111, K93 (1989)
Subject classification: 74.70 and 76.30; S10.15

Mineralogisch-Kristallographisches Institut der Universität Göttingen[1] (a),
Institut für Anorganische Chemie der Universität Göttingen[2] (b),
and Institut für Physikalische Chemie der Universität Freiburg[3] (c)

EPR Studies of Cu^{2+} from Impurity Phases in the High Temperature Superconducting $YBa_2Cu_3O_{7-x}$

By
K. BENTE (a), E. SCHWARZMANN (b), and D. SIEBERT (c)

Introduction EPR studies on sintered powder material of orthorhombic superconducting $YBa_2Cu_3O_{7-x}$, synthesized under dry conditions starting from Y_2O_3, $BaCO_3$, and CuO, were carried out by different authors (Kohara et al. /1/, Mehran et al. /2/, Stankowski et al. /3/, Bartucci et al. /4/). The obtained signals were typical for Cu^{2+} as an isolated paramagnetic center. Measurements of superconducting material in the Q-band at room temperature (Bente and Siebert /5/) showed typical Cu^{2+} signals with $g_1 = 2.05$, $g_2 = 2.10$, and $g_3 = 2.23$. The tetragonal non-superconducting material can show an additional EPR signal with g = 2.13 (Fig. 1). Because no EPR signals are detected for single crystals grown from the melt (Albino et al. /6/), the EPR signals of powder materials are suggested to come from impurity phases. To verify this suggestion EPR studies of different synthesized and treated powder materials have to be carried out.

Results Superconducting (standard) and non-superconducting (tetragonal) powder materials of $YBa_2Cu_3O_{7-x}$ produced by dry synthesis were hydrated and dried several times. The corresponding EPR signals were qualitatively identical with the signals of the educt materials, but show higher intensities (Fig. 2, Table 1). By these treatments an increase of an impurity phase or phases has to be suggested.

Drily synthesized superconducting materials additionally annealed at different temperatures in air and in vacuum sealed tubes showed typical EPR signals for Cu^{2+} at room temperature (Fig. 3) with g-values as described

---

1) Goldschmidtstr. 1, D-3400 Göttingen, FRG.
2) Albertstr. 21, D-7800 Freiburg i. Br., FRG.
3) Tammannstr. 4, D-3400 Göttingen, FRG.

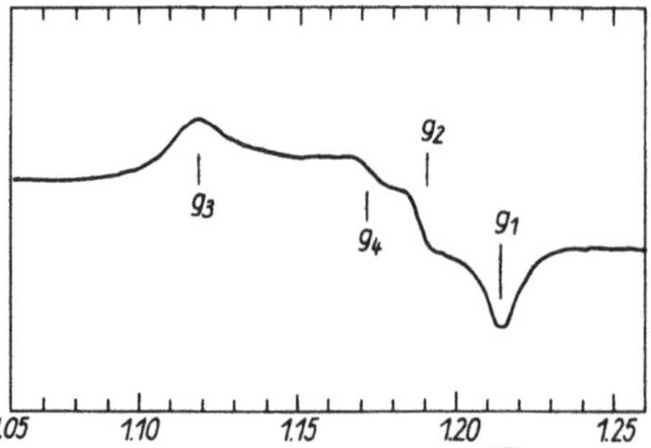

Fig. 1

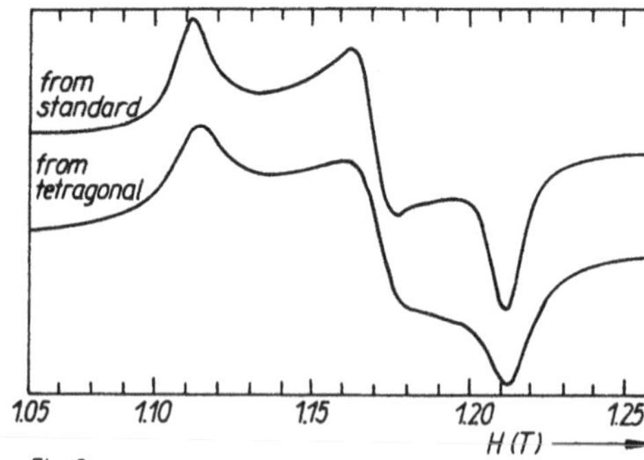

Fig. 2

Fig. 1. EPR signal of non-superconducting tetragonal powder material, 34.9 GHz at 300 K, $g_1 = 2.05 \pm 0.005$, $g_2 = 2.10 \pm 0.005$, $g_3 = 2.23 \pm 0.005$, $g_4 = 2.13 \pm 0.005$

Fig. 2. Cu^{2+} signals of non-superconducting material, hydrated and dried, 34.8 GHz

above, but contained different Cu^{2+} amounts (Table 1). Depending on the treatment the superconducting materials became non-superconducting or the volume of the superconduction portions became smaller. In the latter case the deviation of the resonator from balance was smaller (weaker) than for the superconducting material of high density. We have characterized these facts in the T_c column of Table 1 by adding "w" (= weak). The Cu^{2+} concentration is given in I_{pp}, which means the peak-to-peak intensity of the Cu^{2+} signals in arbitrary units referring to the same amount of $YBa_2Cu_3O_{7-x}$. For the sample Y41 the absolute Cu^{2+} concentration was determined by comparison with $CuSO_4(NH_4)_2SO_4{\cdot}6H_2O$ yielding (2.0 ± 0.5)at% with regard to the bulk composition. For the samples with smaller I_{pp} the conductivity at room temperature produced a higher damping of the resonator, so that these I_{pp} values underestimate the Cu^{2+} content up to a factor of 4 for the smallest I_{pp} value. Apart from the Fe samples the increase of the Cu^{2+} signal indicates a deterioration of the superconducting properties.

Materials with increasing Fe contents show a decreasing height of the Cu^{2+} signals (Fig. 4). Up to 2.3% Fe T_c only decreases slowly to about 89 K and decreases strongly above 2.3% (Bremert et al. /7/). It may be suggested that at first Cu^{2+} is replaced by Fe predominantly in the impurity phase and then Cu

Table 1

Quantitative Cu^{2+} measurements

special treatments	Cu^{2+}/I_{pp}	T_c	g_1	Fig.
from standard				
Y42, 700°C, closed, 48 h	2.9	no	broad	
Y41, 700°C, closed, 2 h	1.0	no	broad	3
Y22, 400°C, closed, 48 h	0.41	yes, w	broad	3
Y32, 700°C, open, 48 h	0.21	no	narrow	3
Y31, 700°C, open, 2 h	0.2	no	narrow	
hydrated, dried				
from standard	9.8	no	narrow	2
from tetragonal	11.1	no	narrow	2
standard, T_c = 92 K	0.15	yes	narrow	3, 4
Fe-1%, T_c = 90 K	0.06	yes, w	narrow	4
Fe-10%, T_c = 79 K	0.0	(yes)	broad	4

is substituted by Fe in the high-temperature superconducting phase.

Checking the results of Albino et al. /6/ EPR studies on twinned crystals grown from the melt with CuO excess were carried out. The crystals were mounted in identical orientation on a corundum needle. In agreement with Albino et al. /6/ no EPR signals could be detected. After annealing the single crystals at 800 °C the typical Cu^{2+} signals known from the powder materials appeared.

All these results allow the inductive suggestion that the EPR signals of sintered powder materials of $YBa_2CU_3O_{7-x}$ come from an impurity phase. This interpretation can only be verified, if powder materials without impurity phases are available. Pure powder materials of $YBa_2CU_3O_{7-x}$ without isolated Cu^{2+} centers could be expected, if very high densities relative to the theoretical density maximum can be reached. Because powders obtained by the citrate synthesis allow very high densities (Blank et al. /8/), sintered materials were produced by this method showing a density of 96%. Compared with corresponding powder materials synthesized by other methods this material showed a temperature independent paramagnetism above T_c and the highest Meissner effect of the studied samples (Wördenweber et al. /9/.)

Starting from superconducting pellets fine grained powders (< 60 μm) were studied by EPR. The corresponding runs in the Q- and X-bands did not reveal

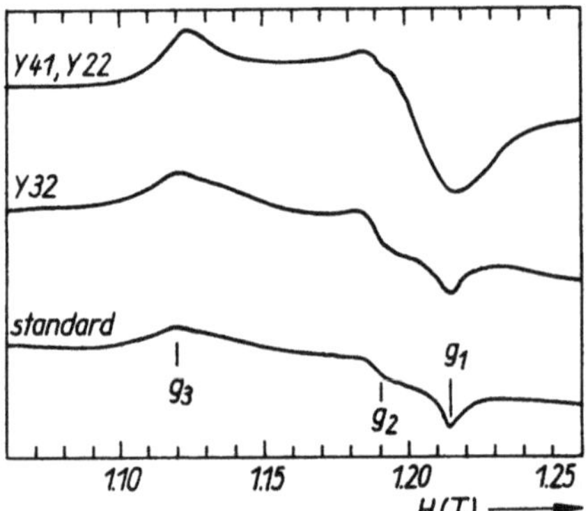

Fig. 3

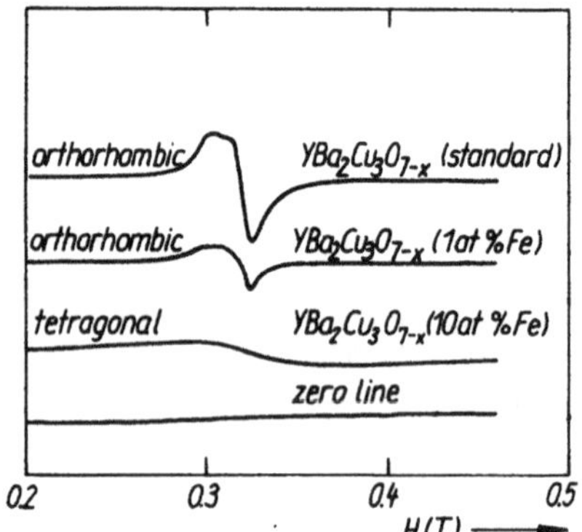

Fig. 4

Fig. 3. EPR spectra of different powders after special treatments of Table 1, 34.9 GHz at 300 K; $g_1 = 2.05 \pm 0.005$, $g_2 = 2.10 \pm 0.005$, $g_3 = 2.23 \pm 0.005$

Fig. 4. EPR spectra of superconducting powders containing Fe, 9.26 GHz at 300 K

any signal. Temperature dependent measurements in the X-band showed the expected disequilibration of the microwave resonator at T_c.

Therefore superconducting $YBa_2Cu_3O_{7-x}$ material itself is verified not to contain localized Cu^{2+} spins.

In EPR studies on powders of superconducting $Bi_2Sr_2Ca_1Cu_2O_{3+x}$ no EPR signals can be detected either (Bente et al. /10/).

Summary Cu^{2+} detected in $YBa_2Cu_3O_{7-x}$ powder materials by EPR come from Cu^{2+}-containing impurity phases, which are correlated to the superconducting phase and/or form from it.

We thank Dr. U. Krebs for the Fe-containing powder material and Prof. Dr. K. Winzer for the twinned single crystals. The investigations are supported by the minister of science and technology (FK2 13N5493-0).

References

/1/ T. KOHARA, H. YAMAGATA, M. MATSUMURA, Y. YAMADA, I. NAKADA, E. SAKAGAMI, Y. ODA, and K. ASAYAMA, Physica 148B, 459 (1987).

/2/ F. MEHRAN, S.E. BARNES, T.R. MCGUIRE, W.J. GALLAGHER,

R.L. SANDSTROM, T.R. DINGER, and D.A. CHANCE, Phys. Rev. B 36, 740 (1987).

/3/ J. STANKOWSKI, P.K. KAHOL, N.S. DALAL, and J.S. MOODERA, Phys. Rev. B 36, 7126 (1987).

/4/ R. BARTUCCI, E. COLAVITA, L. SPARTELLI, G. BALESTRINO, and S. BARBANEN, Phys. Rev. B 37, 2313 (1988).

/5/ K. BENTE and D. SIEBERT, phys. stat. sol. (a) 106, K57 (1988).

/6/ J. ALBINO, O. DE AGUIAR, A.A. MENOVSKY, J. VAN DEN BERG, and H.B. BROM, J. Phys. C (1988).

/7/ O. BREMERT, C. MICHAELSEN, and H.U. KREBS, J. appl. Phys., in the press.

/8/ D.H.A. BLANK, H. KRUIDHOF, and J. FLOKSTRA, J. Phys. D 21, 226 (1988).

/9/ R. WÖRDENWEBER, K. HEINEMANN, and H.C. FREYHARDT, Cryogenics, in the press (1988).

/10/ K. BENTE, P. SCHMID, and D. SIEBERT, unpublished results (1988).

(Received September 22, 1988)

phys. stat. sol. (a) 111, K99 (1989)

Subject classification: 72.80 and 75.50, S8

Chernovtsy State University (a) and
Institute of Physics, Academy of Sciences of the Ukrainian SSR, Kiev[1] (b)

Peculiarities of Physical Properties of Lead and Tin Selenide Solid Solutions

By

V.I. GARASIM (a), D.M. ZAYACHUK (a), R.D. IVANCHUK (a), V.B. LOTOTSKII (a), P.M. STARIK (a), and V.A. SHENDEROVSKII (b)

Most recently growing interest is being attracted by the four-component solid solutions of $A^{IV}B^{VI}$ compounds. Among those solutions most investigated is the $Pb_{1-x}Sn_xTe_{1-y}Se_y$ system. The properties of the above solid solutions have been found to have no qualitative difference as compared with the properties of their ternary analogues or corresponding initial binary compounds.

We have found a qualitatively different situation when investigating another compound system based on $A^{IV}B^{VI}$ compounds, i.e. the $Pb_{1-x}Sn_xS_{1-y}Se_y$ quaternary solid solutions. As opposed to their ternary analogues (PbSnS- or PbSnSe-type solutions of the same crystal structure), we have observed a series of specific features in the behaviour of many physical properties of the above materials. It is the essentially non-monotonic temperature variation of the characteristics that we have observed. This fact is illustrated by the data concerning the temperature dependences of kinetic, optical, magnetic, and dynamic characteristics. Those data are given in Fig. 1 and 2 for $Pb_{0.92}Sn_{0.08}S_{0.80}Se_{0.20}$ single crystals grown by the Bridgman technique.

Temperature dependences of magnetic susceptibility of several samples are presented in Fig. 1. The magnetic susceptibility has been measured by the Faraday method with an accuracy of not less than 1% of the measured quantity. It is evident from Fig. 1 that in the temperature range from 77 to 300 K as a whole χ is increasing in magnitude with decreasing temperature, this fact is characteristic of crystals of the $A^{IV}B^{VI}$ system with low charge carrier concentration. Still in details the dependence of χ on T is essentially non-monotonic, in some sections sharp bends, steps occur that deviate from the general trend of its variation.

---

1) Prospekt Nauki 46, SU-252650 Kiev 28, USSR.

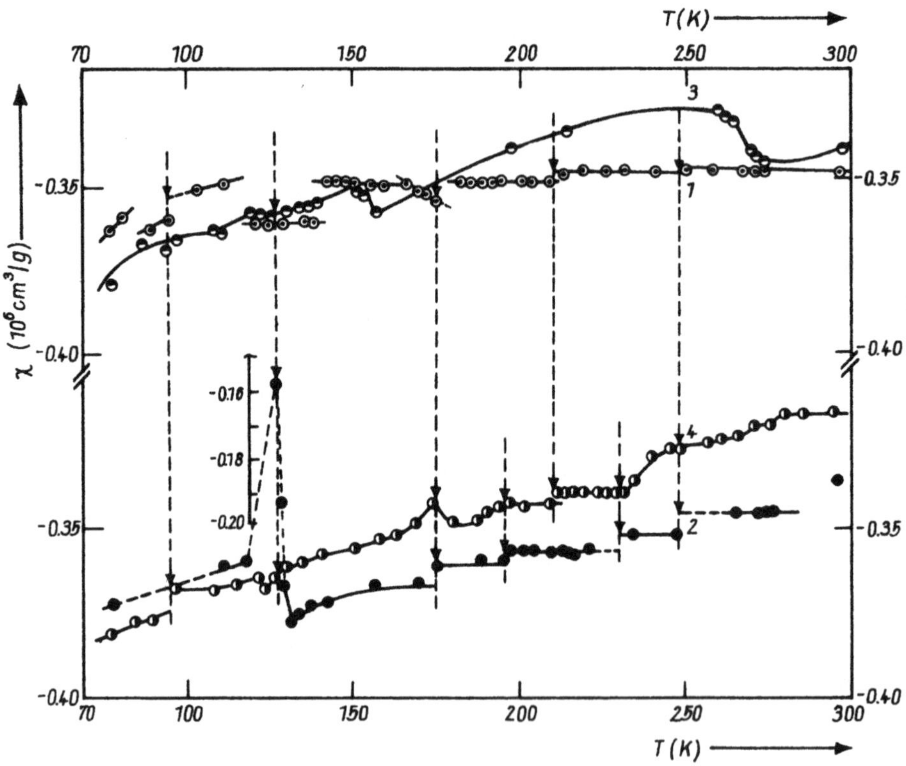

Fig. 1. Temperature dependence of the magnetic susceptibility of four samples of $Pb_{0.92}Sn_{0.08}S_{0.80}Se_{0.20}$ with hole concentration of the order of 10^{17} cm^{-3}, grown at different temperatures: (1) 1135 °C, (2) 1130 °C, (3) 1115 °C, (4) 1020 °C

In Fig. 2, as an example of one of the samples studied, we have presented the results of magnetic susceptibility (MS), electrical conductivity σ, Hall coefficient R, forbidden band gap E_g, and axial sound velocity v_e. The accuracy of E_g definition is ± 0.003 eV. The error in defining v_e, σ, R did not exceed 0.5, 2.0, and 3.0% of the measured quantity, respectively.

In the temperature region from 77 to 300 K, in the case of $A^{IV}B^{VI}$-type semiconductors crystallizing in the cubic phase, the $E_g(T)$ dependence is known

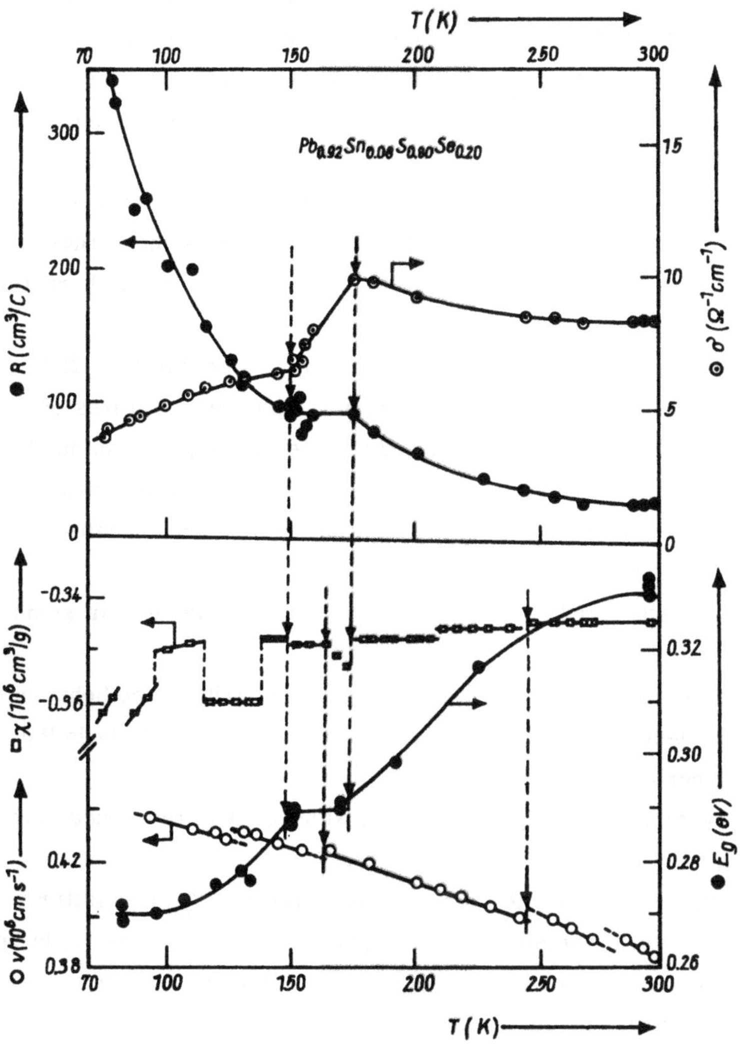

Fig. 2. Hall coefficient, electrical conductivity, magnetic susceptibility, width of the forbidden gap, and axial sound velocity as a function of temperature for $Pb_{0.92}Sn_{0.08}S_{0.80}Se_{0.20}$ grown at 1135 °C

to be (as a rule) a linear one, and the temperature coefficient of E_g is known to be constant. In the system under investigation dE_g/dT strongly depends on temperature, being essentially non-monotonic.

The sound velocity of semiconductors as a function of temperature in the above T region is also described by a linear dependence. As one can see in Fig. 2, for the system studied throughout the above temperature region, the experimental $v_e(T)$ dependence cannot be interpolated by a straight line. The deviations of the experimental data from a straight line (for quite a number of data) essentially exceed the experimental error. The resulting dependence can be interpolated by a set of linear sections, but then a number of discontinuities would occur in this dependence. The temperature of some of those sections agrees well with the temperatures of singularities of other characteristics.

The character of the temperature variation of σ (as shown in Fig. 2) is also typical for the system under investigation, i.e. up to temperatures of the order of 170 to 190 K, σ grows with T and then decreases. This is due to two competiting factors: the increase of the hole concentration with rising temperature (evidenced by the temperature dependence of the Hall coefficient) and the decrease of the hole mobility.

The temperature singularities in the characteristics studied exhibit some peculiarities:

- not at all those temperatures T_{oc} (at which a peculiarity in the behaviour of one of the coefficients studied has been observed) also a peculiarity in the behaviour of all other coefficients occurs;

- the temperature T_{oc} is sample dependent, although it is reproduced for the same sample;

- in the interval between two neighbouring temperatures T_{oc} (the latter being the same for several samples), the behaviour of any coefficient might be identical for all the samples, or might essentially differ from sample to sample both quantitatively and qualitatively.

Abrupt, stepwise changes of the temperature dependences of some physical characteristics or their breaks indicate the presence of a phase transition in the system. When a lot of thermal data (pertaining to such abrupt changes and breaks) are available, it might be evidence of more than a single phase transformation in the system. Although phase transitions, both single and multiple, have been observed in $A^{IV}B^{VI}$ systems by many workers /1 to 3/, their origin is not quite clear as yet. The physical nature of the above mentioned peculiarities should be further studied. It is only possible now to suppose that those peculiarities are connected with transformations in the "crystal matrix-defect" system. This supposition is corroborated by the magnetic susceptibility data. Since the hole concentration and foreign impurity concentration are low in the materials studied, and thus do not considerably contribute to χ, their magnetic susceptibility consists mainly of the lattice and defect contribution. If the above peculiarities were due only to the crystal matrix, then they must be identical for all the samples. If those peculiarities are caused by defects only, then T_{oc} should vary from samples to sample. As is well known, the presence of defects may cause both a random spread and a smearing of the local temperatures of phase transitions.

References

/1/ H. KAWAMURA, Lecture Notes Phys. 133, 470 (1980).
/2/ P.B. LITTLWOOD, Lecture Notes Phys. 152, 238 (1982).
/3/ A.G. CHANDOZHKO, E.I. SLYNKO, S.D. LETYUCHENKO, and K.D. TOVSTYUK, Ukr. fiz. Zh. 23, 1747 (1978).

(Received May 5, 1988; in revised form October 3, 1988)

phys. stat. sol. (a) 111, K105 (1989)
Subject classification: 72.80 and 75.50; S10.15

Matsusita Housing Products Co., Ltd.,
Housing Products Research Laboratory, Nara[1] (a) and
Osaka Municipal Technical Research Institute[2] (b)

Preparation and Magnetic and Electrical Properties of $La_{1-x}Ce_xCoO_3$ Compounds

By

K. TABATA (a) and H. KIDO[3] (b)

The perovskite-type oxides have been investigated for their magnetic, electrical, and catalytic properties /1 to 7/. In these compounds, the crystalline electric field and exchange energy are nearly equal in magnitude /2/. Thus, the electronic state of cobalt ions is very sensitive to its environment, which would be changed by introducing other ions with different ionic radius and valence state. In this study, we prepared $La_{1-x}Ce_xCoO_3$ ($0.010 \leq x \leq 0.030$) and mesured their magnetic and electrical properties.

$La_{1-x}Ce_xCoO_3$ ($0.010 \leq x \leq 0.030$) compounds were prepared from mixtures of metal acetates of each component /6/. First, the mixed acetate solution was evacuated to dryness in a rotary evaporator. The solid obtained was decomposed at 673 K for 1 h and calcinated in air at 1673 K for 5 h.

Phases of products were examined using X-ray powder diffractometry. The lattice constants were determined using silicon as an internal standard. Magnetic measurements were carried out by means of a magnetic torsion balance in the temperature range from 77 to 700 K. Electrical resistivities were measured on sintered pellets by a conventional four-probe method.

X-ray powder diffraction patterns of $La_{1-x}Ce_xCoO_3$ ($0.010 \leq x \leq 0.030$) were completely indexed on the basis of the rhombohedral perovskite-type structure. Lattice parameters, using hexagonal indexes, are shown in Fig. 1, together with those for $La_{1-x}Th_xCoO_3$ /7/. Lattice constants of $La_{1-x}Ce_xCoO_3$ were smaller than those of $La_{1-x}Th_xCoO_3$. Since the ionic radii of Th^{4+}, Ce^{3+}, and Ce^{4+} are 0.100, 0.101, and 0.080 nm /8/, this result shows that cerium ions in $La_{1-x}Ce_xCoO_3$ are tetravalent.

---

1) Yamato-Koriyama, Nara 639-11, Japan.

2) 1-6-50, Morinomiya, Joto-ku, Osaka 536, Japan.

3) To whom correspondence should be addressed.

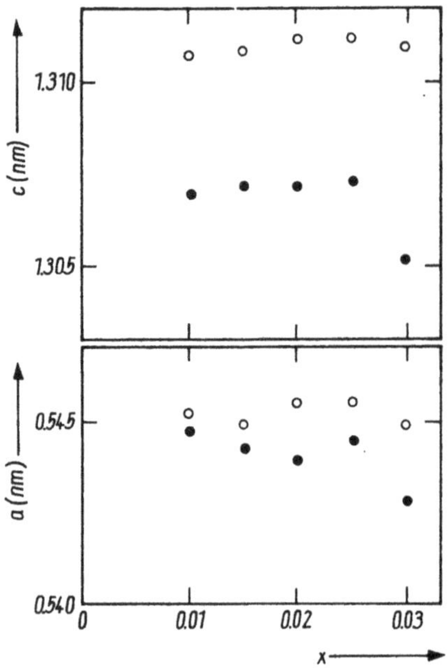

Fig. 1. Lattice constants of $La_{1-x}Ce_xCoO_3$ (●) together with that of $La_{1-x}Th_xCoO_3$ (o)

The magnetic susceptibility of these samples obeyed the Curie-Weiss law below 450 K, but had plateaus in the temperature range about from 450 to 600 K, as shown in Fig. 2 for x = 0.010. Effective magnetic moments, μ_{eff}, and paramagnetic Curie temperatures, Θ_a, are plotted against the compositional parameter x in Fig. 3. Values of both μ_{eff} and Θ_a showed a minimum at x = 0.020. The same results were obtained in the $La_{1-x}Th_xCoO_3$ system /7/. The μ_{eff} values were rather smaller than those expected from spin-only contributions of the high spin state Co^{3+} and divalent state cobalt ions. Thus, the cobalt ions in these compounds are believed to exist as tripositive high (Co^{3+}) and low spin (Co^{III}) states and divalent state. The electronic state of cerium

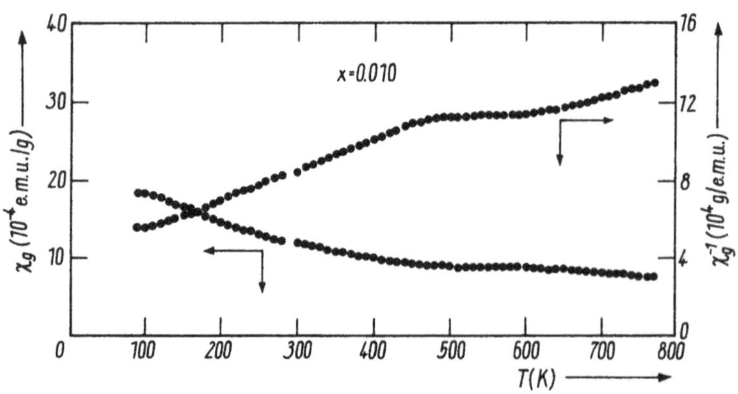

Fig. 2. Magnetic susceptibility and its inverse versus temperature for $La_{1-x}Ce_xCoO_3$ (x = 0.010)

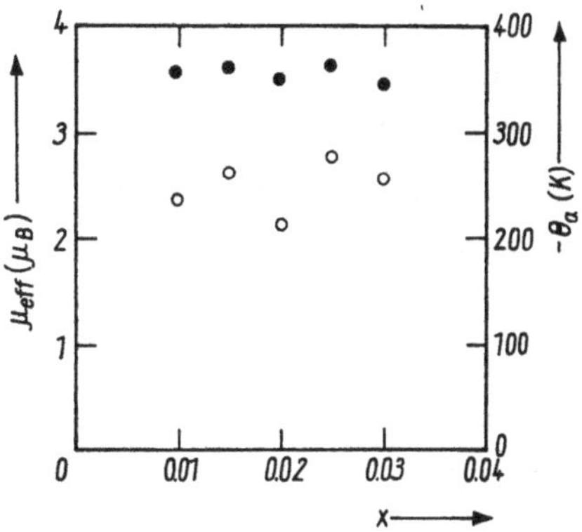

Fig. 3. Effective magnetic moment (●) and paramagnetic Curie temperature (○) of $La_{1-x}Ce_xCoO_3$

was thought to be tetravalent, because the observed values of μ_{eff} were smaller than those expected for cases including trivalent cerium ions. By assuming formulas of $La^{3+}_{1-x}Ce^{4+}_xCo^{3+}_{1-x-y}Co^{III}_yCo^{II}_xO_3$ or $La^{3+}_{1-x}Ce^{4+}_xCo^{3+}_{1-x-y}Co^{III}_yCo^{2+}_xO_3$ and using experimental values of μ_{eff}, y were calculated to be about 0.5, and had a maximum at x = 0.020. The increase of the non-magnetic low spin tripositive cobalt ion at x = 0.020 resulted in a decrease of μ_{eff} and Θ_a. The plateau in Fig. 2 indicates the change of the electronic state of tripositive cobalt ions from low spin to high spin state with increasing temperature /2, 7/.

$La_{1-x}Ce_xCoO_3$ ($0.010 \leqq x \leqq 0.030$) compounds are semiconducting in the temperature range from 150 to 300 K, as shown in Fig. 4 for x = 0.010. However, the electrical resistivity, ϱ, of these compounds did not obey the usual linear $\log_{10}\varrho$ versus 1/T plot, but obeyed the $\log_{10}\varrho$ versus $1/T^{1/4}$ plot. Thus, the electronic conduction in this range may be governed by the variable range hopping mechanism. A similar conduction behavior was observed for the related perovskite-type compounds /9, 10/.

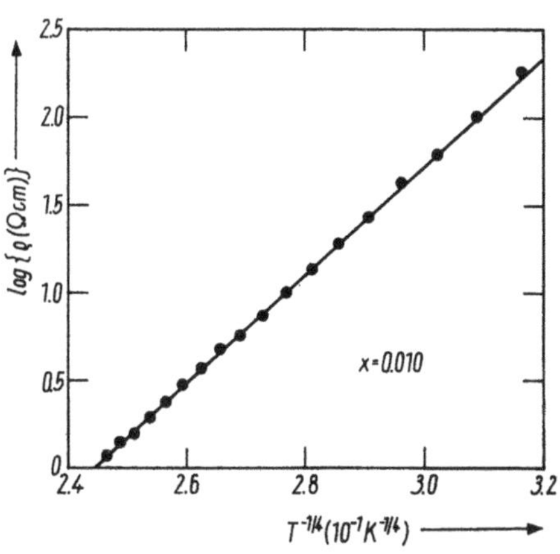

Fig. 4. Electrical resistivity versus temperature for $La_{1-x}Ce_xCoO_3$ (x = 0.010)

References

/1/ G.H. JONKER and J.H. VAN SANTEN, Physica 19, 120 (1957).

/2/ J.B. GOODENOUGH, J. Phys. Chem. Solids 6, 287 (1958).

/3/ P.M. RACCH and J.B. GOODENOUGH, Phys. Rev. 155, 932 (1967).

/4/ R.J.H. VOORHOECE, J.P. REMEIKA, P.E. FREELAND, and B.T. MATTIAS, Science 177, 353 (1972).

/5/ R.J.H. VOORHOECE, J.P. REMEIKA, and L.E. TRIMBLE, The Catalytic Chemistry of Nitrogen Oxides, Plenum Press, New York 1975 (p. 215).

/6/ K. TABATA, I. MATSUMOTO, and S. KOHIKI, J. Mater. Sci. 22, 3037 (1987).

/7/ K. TABATA and H. KIDO, phys. stat. sol. (a) 99, K121 (1987).

/8/ R.D. SHANON and C.T. PREWITT, Acta cryst. B25, 925 (1969); B26, 1046 (1970).

/9/ V. JOSHI, OM PARKASH, and G.N. RAO, J. Chem. Soc., Faraday Trans. II, 75, 1199 (1979).

/10/ Y. MIYAMOTO, M. KOIZUMI, S. KUME, J.P. DOUMERC, P. DORDOR, M. POUCHARD, and P. HAGENMULLER, Mater. Res. Soc. Symp. Proc. 22, 57 (1984).

(Received August 24, 1988)

phys. stat. sol. (a) 111, K109 (1989)
Subject classification: 78.40 and 78.60; S8.16

Institute of Solid State and Semiconductor Physics, Academy of Sciences of the Belorussian SSR, Minsk[1)]

Edge Absorption and Cathodoluminescence Study of $CuAl_xIn_{1-x}S_2$ Solid Solutions

By
I.A. AKSENOV, A.I. LUKOMSKII, and L.A. MAKOVETSKAYA

$CuAl_xIn_{1-x}S_2$ solid solutions aroused great scientific and practical interest since the $CuInS_2$ ternary compound (x = 0) is a promising material for fabricating solar cells /1/, and the $CuAlS_2$ ternary compound (x = 1) is a perspective candidate for blue to ultraviolet light emitting diode application /2/.

Single crystals of $CuAl_xIn_{1-x}S_2$ solid solutions have been obtained by the chemical vapour transport method /3/, and a phase diagram of the $CuInS_2$-$CuAlS_2$ system has been published /4/.

The aim of the present note is to study the absorption spectra of $CuAl_xIn_{1-x}S_2$ solid solutions in the spectral range immediately adjacent to the fundamental absorption edge as well as cathodoluminescence spectra of these solid solutions.

The absorption coefficient α was calculated from the transmission spectra $T = T(\lambda)$ making use of the reflection coefficient R calculated from the experimental data given in /5, 6/. The band gap E_g was determined from the dependence of $(\alpha h\nu)^2$ on $h\nu$ employing the intersection point of its linear part with the abscissae.

Fig. 1 shows the composition dependences of the band gap E_g for $CuAl_xIn_{1-x}S_2$ solid solutions at 300 and 77 K for the polarizations $\vec{E} \perp \vec{c}$ and $\vec{E} \parallel \vec{c}$ ($\vec{c}$ is the crystallographic [001] direction corresponding to the optical axis of the crystals). From Fig. 1 it is seen E_g changes nonlinearly with composition of the solid solutions in contrast with the preliminary data given in /7/. A similar character of $E_g(x)$ change is observed for solid solutions based on $A^{II}B^{VI}$ binary compounds and is well described by the relation

$$E_g(x) = a + bx + cx(1 - x) ,$$

where a and b are constants related to the values of the corresponding energy

---

1) Podlesnaya 17, SU-220726 Minsk, USSR.

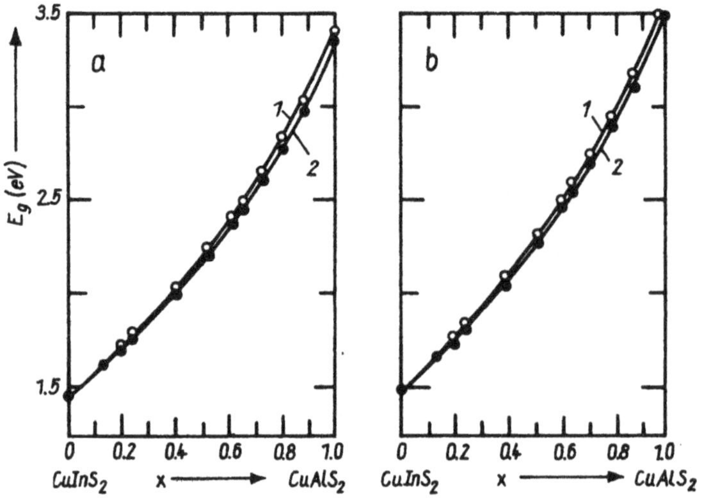

Fig. 1. The energy gap at a) 300 and b) 77 K versus composition of $CuAl_xIn_{1-x}S_2$ solid solutions; (1) $\vec{E} \perp \vec{c}$, (2) $\vec{E} \parallel \vec{c}$

gaps of the end members, c is the nonlinearity parameter.

We employed the above relation for describing the composition dependence of E_g of the investigated solid solutions. The analytical dependences of the band gap on solid solution composition (Fig. 1) at 77 and 300 K for the polarizations $\vec{E} \perp \vec{c}$ and $\vec{E} \parallel \vec{c}$ are described by

$$E_g^{300}(x) = 1.452 + 0.938x + 0.953x^2, \quad (\vec{E} \parallel \vec{c}),$$
$$E_g^{300}(x) = 1.452 + 1.011x + 0.942x^2, \quad (\vec{E} \perp \vec{c}),$$
$$E_g^{77}(x) = 1.475 + 0.962x + 1.018x^2, \quad (\vec{E} \parallel \vec{c}),$$
$$E_g^{77}(x) = 1.475 + 1.047x + 0.988x^2, \quad (\vec{E} \perp \vec{c}).$$

The cathodoluminescence spectra of $CuInS_2$, $CuAlS_2$, and of $CuAl_{0.75}In_{0.25}S_2$ solid solutions at 80 K are shown in Fig. 2. Spectrum 1 is typical of $CuInS_2$ single crystals and reveals one broad luminescence band with half-width 70 nm peaked at 890 nm. The same broad emiss on band has been found in the spectra of single crystals of all compositions, including $CuAlS_2$.

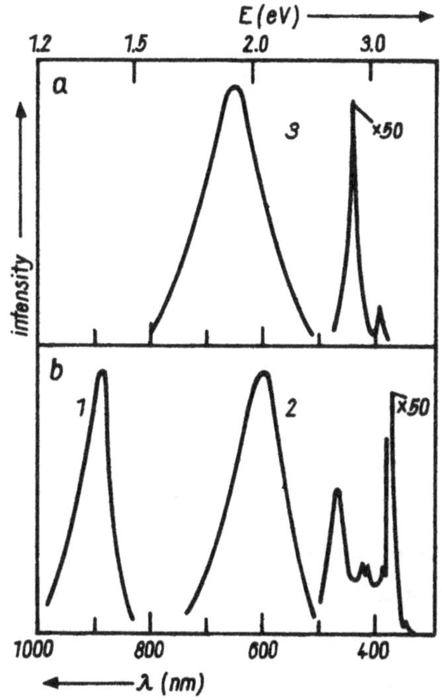

Fig. 2. The cathodoluminescence spectra of b) (1) $CuInS_2$, (2) $CuAlS_2$ ternary compounds, and that of a) (3) $CuAl_{0.75}In_{0.25}S_2$ solid solutions

We interpret the broad-band emission as donor-acceptor pair transition because this emission has shown a shift towards lower energies when the excitation intensity has been reduced.

Spectrum 2 is typical of $CuAlS_2$ single crystals and consists of seven emission bands at 353, 370, 374, 383.5, 414, 422.5, 467.5 nm with half-width up to 30 nm and the above mentioned broad band at 590 nm with half-width of 120 nm due to donor-acceptor transitions. The relative intensity of these emission bands was different depending on the sample and the electron beam excitation intensity.

The spectral position of the luminescence peak at 353 nm ($h\nu = 3.512$ eV) is in good agreement with that of the free A-exciton found by us from the λ-modulated reflection spectrum at $\vec{E} \parallel \vec{c}$. Therefore the band at 353 nm can be attributed to direct band-to-band recombination.

The relative intensity of the emission peaks at 370, 374, and 383.5 was different depending on sample and excitation intensity, but the spectral position of these luminescence bands was independent. Hence, these emission bands can be attributed to free-to-bound transitions.

The interpretation of the emission peaks at 414, 422.5, and 467.5 nm cannot be given at the present stage and further investigations are needed.

Spectrum 3 is typical of $CuAl_xIn_{1-x}S_2$ solid solutions with composition $x \geqq 0.4$. The spectra of solid solutions with $x < 0.4$ have one broad emission band which is analogous to the $CuInS_2$ spectrum. The spectra of solid solutions with $x \geqq 0.4$ contain three emission bands, that is one broad band with half-width of ≈ 120 nm due to donor-acceptor pair recombination and two compar-

Table 1

The spectral position of the emission peaks of $CuAl_xIn_{1-x}S_2$ solid solutions at 80 K under electron beam excitation

x	0.19	0.28	0.40	0.58	0.65	0.75	0.91
broad band (nm)	920	750	690	635	625	650	680
narrow bands (nm)	-	-	570 617	490 532	452 493	395 440	393.5 438

atively narrow bands with a half-width of ≈20 nm which can be interpreted as free-to-bound transitions.

The spectral positions of the emission peaks of the solid solutions investigated at 80 K under electron beam excitation are given in Table 1.

We should like to note that the shift of the spectral position of the emission bands with solid solution composition correlates but slightly with the composition dependence of the band gap. This discrepancy can be attributed to the non-stoichiometry of the investigated solid solutions.

References

/1/ S. WAGNER and P.M. BRIDENBAUGH, J. Crystal Growth 39, 151 (1977).

/2/ N. YAMAMOTO, Japan. J. appl. Phys. 19, Suppl. 3, 95 (1980).

/3/ I.A. AKSENOV, C.A. GRUTSO, L.A. MAKOVETSKAYA, G.P. POPELNYUK, and V.A. RUBTSOV, Izv. Akad. Nauk SSSR, Ser. neorg. Mater. 24, 560 (1988).

/4/ I.A. AKSENOV, L.A. MAKOVETSKAYA, G.P. POPELNYUK, and I.P. SHILOVICH, phys. stat. sol. (a) 105, K97 (1988).

/5/ W.N. HONEYMAN, J. Phys. Chem. Solids 30, 1935 (1969).

/6/ G.D. BOYD, H. KASPER, and J.H. MCFEE, IEEE J. Quantum Electronics 7, 563 (1971).

/7/ F. SUSUMU, Matsushita Electric Industrial Co., Ltd., Patent N56-32278 Japan, Int. cl.^{3} C30B 29/46, CO1G 15/00; 27.07.81, Appl. N 49-44077, Filed 18.04.74.

(Received July 13, 1988)

phys. stat. sol. (a) 111, K113 (1989)
Subject classification: 71.55 and 78.60; S8.12

Institute of Physics, N. Copernicus University, Toruń[1)]

Distribution of Traps in ZnSe:Cu Crystals Obtained by the Fractional Glow Technique

By
F. FIRSZT and H.L. OCZKOWSKI

Introduction Among the wide band gap II-VI semiconducting compounds zinc selenide appears to be one of the most promising materials for several opto-electronic applications. However, the growing interest in ZnSe for use in photodetectors or light emitting diodes emphasises the need for characterizing deep level parameters which may have important consequences for the performance of such devices.

The fractional glow technique is a useful method in the study of thermoluminescence phenomena which gives the possibility of estimating the values of trap depths E in semiconductors.

For some time past we have prepared single crystals of ZnSe non-intentionally doped and doped with Cu, Ag, and Al and studied their luminescent properties. We have observed rather strong thermoluminescence in these crystals. The purpose of this note is to examine some thermoluminescent properties of ZnSe doped with Cu. This work will be continued for zinc selenide crystals doped with other dopants.

Sample preparation The single crystals of ZnSe used in this experiment were grown from the melt by the high pressure Bridgman method from 6N electronic grade Koch-Light powder. Growth was performed in argon atmosphere (11 MPa). The crystals had cubic zincblende structure involving small amount of twins. The ZnSe crystals were cut into 1 mm thick plates parallel to the (111) crystallographic plane and annealed in molten zinc. The copper was then incorporated into ZnSe by thermal diffusion from a Zn+4%Cu alloy in an evacuated silica ampoule at 1270 K during 72 h. The crystal plates were then mechanically polished and chemically etched /1/.

Results and discussion Thermoluminescence (TL) was excited at 35 and 77 K with radiation from a high pressure 100 W mercury lamp. The ZnSe

---

1) Grudziądzka 5/7, PL-87-100 Toruń, Poland.

sample was illuminated during a period of 5 min without any filters and then with 366 nm (3.4 eV) filter for 10 min. It should be mentioned here that the TL can be also excited with photons whose energy is lower than the band gap of zinc selenide. In particular an effective excitation of TL with 2.27 eV photons (545 nm Hg line) was observed. In this experiment the normal glow curves (NGC) were recorded with linear heating for several heating rates w from 0.05 to 0.5 K/s.

Measurements of the energy distribution of traps have been performed by the Gobrecht-Hofmann method (fractional glow curve technique) in the temperature range from 85 to 250 K applying about 130 temperature cycles. The amplitude of temperature oscillations was changed from 3 to 5 K and was dependent on the intensity of fractional TL.

The thermoluminescence glow curves are presented in Fig. 1. When TL is excited at 35 K and NGC is measured in the range 35 to 300 K a strong peak at T = 70 K is observed and a relatively weak high temperature tail between 140 and 250 K (Fig. 1, curve a). The shape of the main peak suggests that it is probably not a single one. Curves b and c in Fig. 1 represent NGC for extreme heating rates of w = 0.05 K/s and w = 0.5 K/s. In this experiment the TL was excited at 77 K. Four peaks are resolved on NGC for the lowest

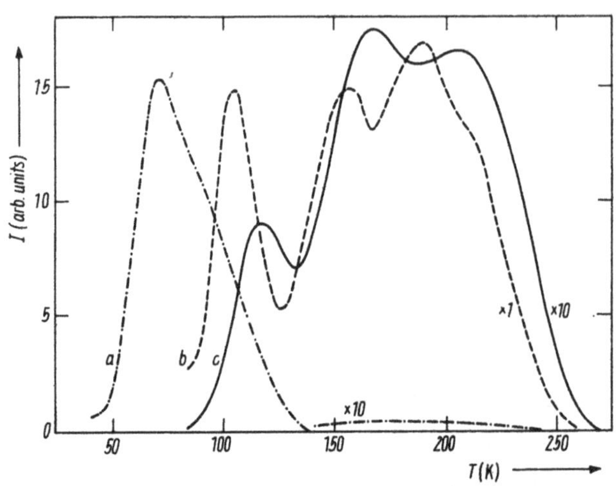

Fig. 1. Thermoluminescence glow curves for selected heating rates: (a) w = = 0.2, (b) w = 0.05, (c) w = 0.5 K/s

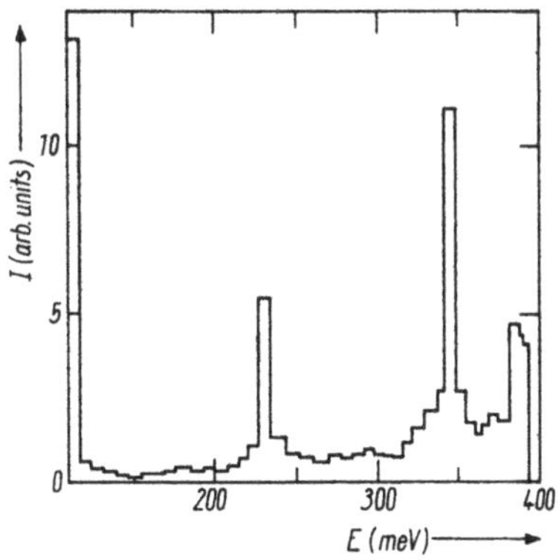

Fig. 2. Energetic distribution of traps obtained by the fractional glow technique

heating rate w = 0.05 K/s. These maxima correspond to the temperature 105, 160, 190, and 215 K.

The energetic distribution of traps obtained from FGC measurements in the temperature range 85 to 250 K is presented in Fig. 2. Unfortunately, the construction of our apparatus does not enable FGC measurements at temperatures lower than 85 K. The standard deviation energy determination of in the presented distribution is estimated to be about 5 meV and is much lower than the resolution of the FGC technique which is approximately equal to 2.5 kT /2 to 4/. In the distribution of traps one can see four main maxima corresponding to energies close to 120 meV (fraction A), 230 meV (B), 340 meV (C), and 380 meV (D). The traps with energy 340 meV are typical in ZnSe crystals and have been observed by several authors in TL, DLTS, and TSCAP /5 to 9/. Traps deeper than 0.3 eV are also known in literature /5 to 13/.

The traps with energy about 0.3 eV characterize a relatively wide energetic distribution (Fig. 2). In the case of our measurements there is no doubt that two narrow fractions of traps with energies close to 0.34 eV and 0.38 eV exist in the investigated ZnSe:Cu crystals. The existence of 0.38 eV traps is also revealed in the NGC graph for the lowest rate of heating w = 0.05 K/s.

It should be noted that the investigated crystals exhibit rather long-life phosphoresence when excited at 77 K as well as at 35 K. In the first case (excitation at 77 K) one can believe that the observed phosphorescence is due to the fraction of traps responsible for the strongest glow curve peak at about 70 K (Fig. 1, curve a). In this context the results of our measurements suggest that the shallowest fraction of traps (E = 0.12 eV) observed in the energetic distribution under excitation at T = 77 K may be a high-energy component of traps corresponding to the 70 K peak. In this case the estimated cross section for this fraction of traps (Table 1) may be too small.

Cross sections of particular fractions of traps have been calculated assuming monomolecular thermoluminescence kinetics. In this approximation the so-called

Table 1.

	E (meV)	s (s^{-1})	σ(m^2)
A	111 to 146	$1x10^4$	$1x10^{-25}$
B	146 to 278	$3x10^5$	$1x10^{-24}$
C	278 to 361	$6x10^6$	$1x10^{-22}$
D	361 to 393	$6x10^6$	$4x10^{-23}$

frequency factor s is given by the following formula:

$$s = \frac{E_m w}{kT_m^2} \exp\left(-\frac{E_m}{kT_m}\right) ,$$

where k is the Boltzmann constant, E_m energy of the maximum of distribution of a given fraction of traps, and T_m temperature of the corresponding maximum in the glow curve. There is some difficulty in estimating T_m for fraction D of traps (Fig. 1). The calculated values of the parameter s for particular fractions of traps are presented in Table 1. The last column of this table contains a rough estimation of the cross sections σ according to the Mott formula:

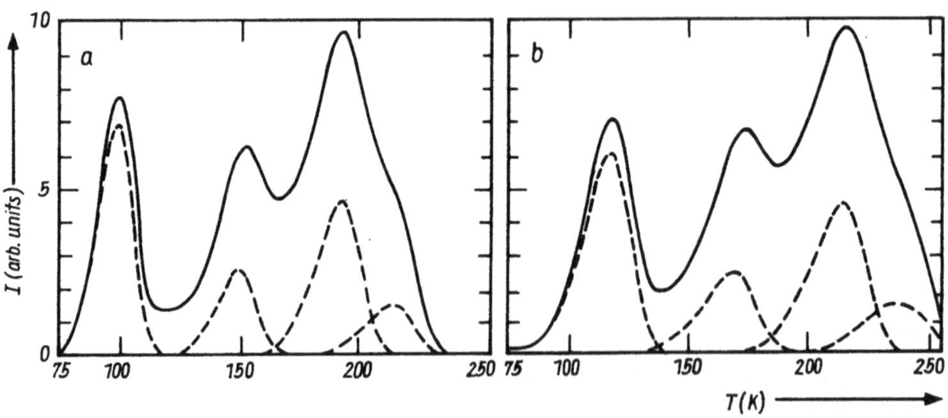

Fig. 3. Calculated shapes of glow curves for two different heating rates: a) w = 0.05 K/s, b) w = 0.5 K/s. In these calculations the experimental parameters (Table 1) are taken and monomolecular thermoluminescence kinetics is assumed. The final glow curves (solid line) are the superpositions of 45 elementary glow curves. The dashed lines represent the main four components of trap distribution (A, B, C, and D)

$$s = \frac{4\sqrt{6}\,\pi^{3/2}\, m^* k T_m^2}{h^3}\,\sigma ,$$

where $m^* = 0.17\ m_o$ is the effective electron mass in ZnSe.

Fig. 3. represents the theoretically calculated shapes of glow curves assuming monomolecular thermoluminescence kinetics, the experimental energetic distribution of traps (Fig. 2), and the parameters from Table 1. In these calculations only the s parameter for the fraction of traps 361 meV< E ≦ ≦ 393 meV was optimized. All other parameters are experimental.

Since the intensities of glow peaksks epend on details of the experimental procedure, it seems that the theoretical fitting is satisfactory.

References

/1/ W. BAŁA, F. FIRSZT, and Z. NOWAK, Proc. Conf. Physics, Jaszowiec 1983, Ed. J.J. GAŁAZKA, Vol. 6, 1983 (p. 306).

/2/ W. GOBRECHT and D. HOFMANN, J. Phys. Chem. Solids 27, 509 (1966).

/3/ G. RUDLOF, J. BECHERER, and M. GLAFEKE, phys. stat. sol. (a) 49, K121 (1978).

/4/ G. RUDLOF, J. BECHERER, and M. GLAFEKE, phys. stat. sol. (a) 52, K137 (1979).

/5/ H.L. OCZKOWSKI, phys. stat. sol. (a) 68, 199 (1981).

/6/ W. BAŁA, Acta phys. Polon. A69, 783 (1986).

/7/ D. VERITY, F.J. BRYANT, J.J. DAVIES, J.E. NICHOLLS, C.G. SCOTT, and D. SHAW, J. Phys. C 15, 5497 (1982).

/8/ K. KOSSAI, J. appl. Phys. 53, 1018 (1982).

/9/ D. VERITY, F.J. BRYANT, C.G. SCOTT, and D. SHAW, J. Crystal Growth 59, 234 (1982).

/10/ R.N. BHARGAWA, J. Crystal Growth 59, 15 (1982).

/11/ T. IDO and M. OKADA, J. Crystal Growth 72, 170 (1985).

/12/ P. BESSOMI and B.W. WESSELS, J. appl. Phys. 53, 3076 (1982).

/13/ A. HEURTEL, R. LEGROS, Y. MARFAING, and T. TRIBOULET, in: Defects in Semiconductors, Ed. H.J. VON BARDELEBEN, Mater. Sci. Forum 10/12, 833 (1986).

(Received August 17, 1988)

phys. stat. sol. (a) 111, K119 (1989)

Subject classification: 78.60; S12

National Environmental Engineering Research Institute, Nagpur (a)[1)] and Department of Physics, Nagpur University[2)] (b)

Radiation Induced Sensitization in Polyethylene

By

P. L. MUTHAL (a), S. V. MOHARIL (b), and B. T. DESHMUKH (b)

Thermoluminescence is now extensively used for dosimetry of ionizing radiations. It would be desirable to have polymers as phosphors in thermoluminescence dosimetry (TLD) as these materials are not affected much by humidity, exposure to air, etc., these can be made into tough pellets easy to handle and these are tissue equivalent. However, most polymers do not show any appreciable thermoluminescence above room temperature (RT) and for this reason have been virtually ruled out as possible dosimetry materials /1/. On the other hand, Hashimoto et al. have reported /2, 3/ glow peaks above RT in well oriented crystalline polymers. Recently, we have shown /4, 5/ that polyethylene crystallised in extended chain crystals (ECC) exhibits relatively intense TL near its melting point for irradiation at RT. Several problems have still to be overcome before this phosphor can be used in dosimetry. The intensity of luminescence is still weak and needs to be enhanced several folds. Impurities often increase the TL intensity. In some cases, pre-exposure to ionizing radiations followed by suitable thermal annealing results in increased response to the subsequent exposures. This is known as radiation induced sensitization /6/. In this note we report radiation induced sensitization for polyethylene crystallised in ECC.

The experimental setup used was the same as in our earlier work /4/. Extended chain crystals of polyethylene were obtained by crystallisation from the melt at 503 K under 600 MPa. A ^{60}Co source was used for irradiation. Glow curves were recorded with a linear heating rate of 50 K/min.

Fig. 1 shows glow curves for "virgin" sample exposed to 51.6 C/kg. The glow curve contains the well-known peaks P_1 to P_4. When the sample was stored for a week all TL decayed. This sample was then exposed again to 51.6 C/kg and the glow curve was recorded. It is seen (curve b) that the peak

---

1) Nehru Marg, Nagpur 440 020, India.

2) Nagpur 440 010, India.

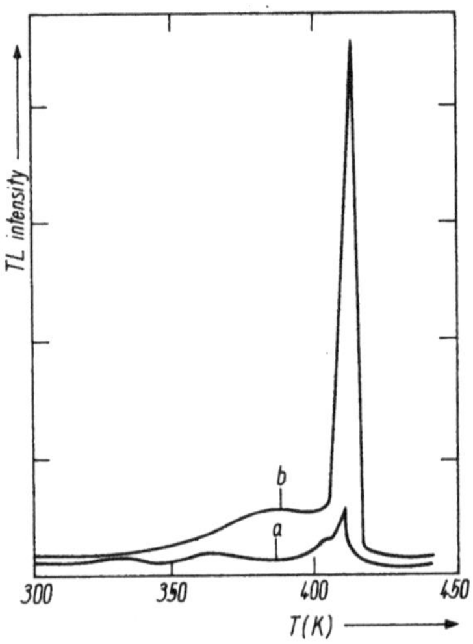

Fig. 1. Glow curves of extended chain crystals of polyethylene; (a) crystals exposed to 51.6 C/kg, (b) crystals in (a) stored for a week and then exposed again to 51.6 C/kg

height has increased nearly tenfold. Radiation induced sensitization with sensitization factor of nearly 10 is thus seen. More sensitization could not be achieved by repeating exposure/storing cycles. The sensitization was achieved for pre-exposures above 51.6 C/kg. More systematic studies on the variation of sensitization factor with pre-exposure could not be carried out due to low intensities. In order to accelerate the sensitization procedure, we attempted post-irradiation annealing at 400 K. However, this treatment destroyed sensitization. A more systematic study on the dependence of sensitization on the post-irradiation annealing temperature is necessary for reducing the duration of the sensitization treatment. All the same, the fact that radiation induced sensitization could be achieved itself is quite significant.

Having seen that the pre-exposure to ionizing radiations induces sensitization, it will be interesting to find out an explanation of the phenomenon. In inorganic phosphors such as LiF-TLD 100, radiation induced sensitization has been attributed to creation of traps /7/ or luminescence centres /8/ during the pre-exposure, saturation filling of traps during pre-exposure, and consequently removal of competition from these during the test exposure /9/, or to the increased detrapping /10/ or luminescence efficiency /11, 12/ in the sensitized sample. The last named is the most satisfactory explanation /12/. However, there is an important difference in the sensitization of LiF-TLD 100 and polyethylene. In the former case, sensitization is observed only for small test exposures, whereas for ECC polyethylene we observe that the sensitization and test exposures are of the same order. The basic mechanism thus can be different for these phosphors.

The traps responsible for the peak F_4 may originate in chain ends and pre-exposure may result in increasing concentration of the chain ends due to scission. In the sensitized sample there are thus more trapping sites available.

There is some evidence in favour of this hypothesis. In ECC polyoxymethylene irradiation resulted in such increase in the end group concentration /13/.

The above explanation is also consistent with the fact /14/ that in ECC polyethylene treated with fuming nitric acid (FNA) peak P_4 was found to be severely reduced. FNA treatment, though increases the ECC content of the sample by selective dissolution, results in substitution at end groups and destroys the traps responsible for P_4.

In conclusion, radiation induced sensitization results in about tenfold increase in sensitivity of ECC polyethylene. The sensitization most probably results from the increase in the concentration of chain ends.

The work was supported by the Council of Scientific and Industrial Research, New Delhi. P.L.M. thanks the Director of NEERI for giving permission to work on this problem.

References

/1/ R.J. FLEMING and J. HAGEKYRIAKOU, Radiat. Protect. Dosimetry 8, 99 (1984).

/2/ T. HASHIMOTO, T. SAKAI, and M. IGUCHI, J. Phys. D 12, 1567 (1979).

/3/ T. HASHIMOTO, K. OGITA, S. UMEMOTO, and T. SAKAI, J. Polymer. Sci., Phys. ed. 21, 1347 (1983).

/4/ P.L. MUTHAL, B.T. DESHMUKH, and S.V. MOHARIL, phys. stat. sol. (a) 101, 583 (1987).

/5/ S.V. MOHARIL, P.L. MUTHAL, and B.T. DESHMUK, J. Mater. Sci. Letters 6, 835 (1987).

/6/ A.R. LAKSHMANAN, R.C. BHATT, and K.G. VOHRA, phys. stat. sol. (a) 53, 617 (1979).

/7/ J.R. CAMERON and D.W. ZIMMERMAN, USAEC Rep. COO-1105-102 (1965).

/8/ J.R. CAMERON and D.W. ZIMMERMAN, USAEC Rep. COO-1105-113 (1966).

/9/ N. SUNTHERLINGAM and J.R. CAMERON, phys. stat. sol. 14, 397 (1969).

/10/ R. NINK and H.J. KOS, phys. stat. sol. (a) 35, 121 (1976).

/11/ S.V. MOHARIL, phys. stat. sol. (a) 77, 413 (1983).

/12/ S.V. MOHARIL, phys. stat. sol. (a) 79, 599 (1983).

/13/ M. JAFFE and B. WUNDERLICH, J. Polymer. Sci. A 6, 825 (1968).
/14/ P.L. MUTHAL, S.V. MOHARIL, and B.T. DESHMUKH, communicated to J. Mater. Sci. Letters.

(Received August 16, 1988)

phys. stat. sol. (a) 111, K123(1989)

Subject classification: 78.70; S11.1

Department of Physics, Nanjing University[1] (a)
and Nanjing Institute of Chemistry and Engineering, Nanjing[1] (b)

Positron Annihilation Lifetime Study of Positive Temperature Coefficient $BaTiO_3$ Samples

By

LING CHEN (a), MINGKANG TENG (a), GUANGHOU WANG (a), XIAOYUN LI (b), and TIANCHANG LU (b)

Introduction Semiconducting $BaTiO_3$ has been widely used to prepare ceramic capacitors and thermistors with positive temperature coefficient (PTC), etc. The mechanism of the doping effect in $BaTiO_3$ semiconductor has been attractive for quite a long time, but not very clear so far, especially because of the lack of knowledge of the location of doped ions. Furthermore, the details of the structure and chemistry of the grain boundaries cannot be identified by some given models /1, 2/. We used the positron annihilation lifetime spectra to study the doped vacancies in $BaTiO_3$ as well as their influence on PTC.

Experiments Two groups of $BaTiO_3$ samples with BaO excess were prepared by doping different concentrations of La_2O_3 and Nb_2O_5, respectively. Then they were measured by a positron annihilation lifetime spectrometer to find out the tendency and sensitivity of the positron parameters with the vacancy concentration.

Then, a third group of samples of two Sb-doped PTC $BaTiO_3$ semiconductors with excess BaO or TiO_2 were studied by the positron technique before and after being reduced in argon gas at 350 °C for 2 h.

All samples were home-made by ball-milling the $BaTiO(C_2O_4)_2 \cdot 4H_2O$ material (purity > 99.3%) with doped material, and dry pressed with polyvinyl alcohol adhesive. Then they were kept at a temperature of 1350 °C for 2 h and cooled at a rate of 100 K/h to 1050 °C, finally, down to room temperature uncontrolled.

The samples were also studied by a scanning electronic microscope (SEM), and it was found that the grain size of $BaTiO_3$ samples with BaO excess was about 5 μm and with TiO_2 excess was about 20 μm.

1) Nanjing, People's Republic of China.

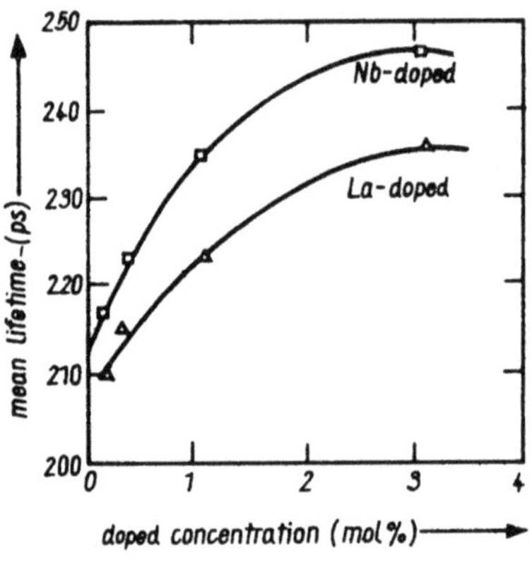

Fig.1

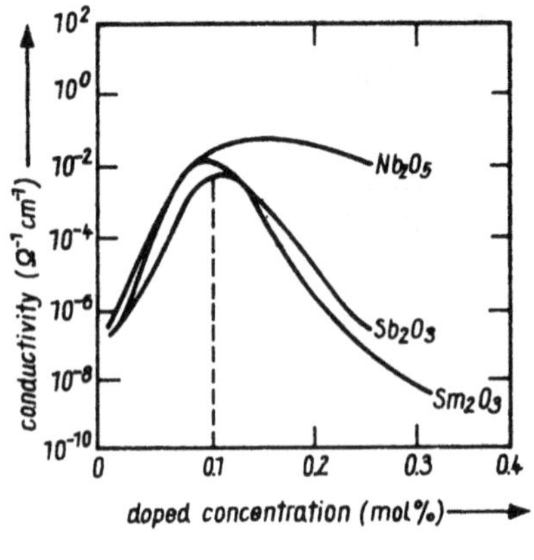

Fig.2

Fig. 1. Positron mean lifetime as a function of the dopant concentration

Fig. 2. Conductivity of $BaTiO_3$ with different dopants

The positron lifetime spectra were measured by a fast-fast coincidence positron lifetime spectrometer with system resolution 320 ps. A $^{22}NaCl$ positron source of 5 μCi was sandwiched between two pieces of samples 1 cm in diameter and 3 mm thick. The total counts of each spectrum were up to 10^6 with good statistics. The obtained spectra were analysed on the basis of the positron trapping theory and fitted by an IBM-PC/XT computer using the POSFIT-EXTENDED program /3 to 6/.

Results According to positron trapping theory, we can easily get the positron mean lifetime τ_m, the average annihilation rate R_a, and the capture rate of positrons K,

$$\tau_m = \sum(I_i * T_i) \;, \quad R_a = \sum(I_i/T_i) \;, \quad K = T_f^{-1}(\tau_m - T_f)/(T_d - \tau_m) \;,$$

where $T_f = 1/R_a$ is the positron free lifetime; $T_d = T_2$ is the trapping lifetime. Therefore, we fixed T_2 as 450, 470, 540, 580 ps for different kinds of vacancies during the data processing. Table 1 and 2 list the lifetime parameters of positron annihilation in the two kinds of doped vacancies of $BaTiO_3$. The vacancy concentrations were increased, but the kinds of vacancies were kept the same in the same group. Thus, in the process of fitting the positron spectra we

Table 1

Positron parameters of La-doped in $BaTiO_3$

No.	x(at%)	T_1 (ps)	T_2	I_1	I_2	τ_m	R_a(1/ns)	K(1/ns)
VB1	0.1	189±1	450	89.07±0.43	7.66±0.50	210±4	5.05±0.01	0.243
VB2	0.3	191±1	450	87.44±0.45	9.12±0.50	215±4	4.95±0.01	0.285
VB3	1.0	196±1	450	86.93±0.47	9.86±0.50	222±4	4.81±0.01	0.293
VB4	3.0	211±1	450	86.86±0.57	9.79±0.63	235±5	4.48±0.02	0.255

Table 2

Positron parameters of Nb-doped in $BaTiO_3$

No.	x(at%)	T_1 (ps)	T_2	I_1	I_2	τ_m	R_a(1/ns)	K(1/ns)
VT1	0.1	200±1	470	90.31±0.46	5.71±0.53	216±4	4.83±0.01	0.171
VT2	0.3	199±1	470	88.72±0.44	8.24±0.50	222±4	4.78±0.01	0.246
VT3	1.0	211±1	470	88.33±0.51	8.52±0.57	234±5	4.51±0.02	0.230
VT4	3.0	228±1	470	89.40±0.62	7.69±0.71	347±6	4.21±0.02	0.179

Table 3

Positron parameters of PTC $BaTiO_3$ before and after reduction

No.	process	T_1 (ps)	T_2	I_1	I_2	τ_m	R_a(1/ns)	K(1/ns)
one	before	192±1	540	89.72±0.14	8.14±0.18	220±1	4.95±0.01	0.281
	after	191±1	540	89.40±0.14	8.22±0.19	220±1	4.95±0.01	0.285
two	before	192±1	580	90.53±0.13	7.23±0.19	221±1	4.95±0.01	0.258
	after	192±1	580	91.27±0.13	6.57±0.18	218±1	4.97±0.01	0.234

fixed T_2 as 450 ps in Table 1 and 470 ps in Table 2. The data of Table 1 and 2 are plotted in Fig. 1, it can be clearly seen that τ_m increases with increasing vacancy concentration in the low concentration region and saturates in the high concentration region /7/. It shows that the positron lifetime parameters are sensitive to changes in the vacancy concentration in $BaTiO_3$ ceramics near the 0.1 mol% region. Our PTC $BaTiO_3$ samples were just doped by 0.1 mol% (Fig. 2).

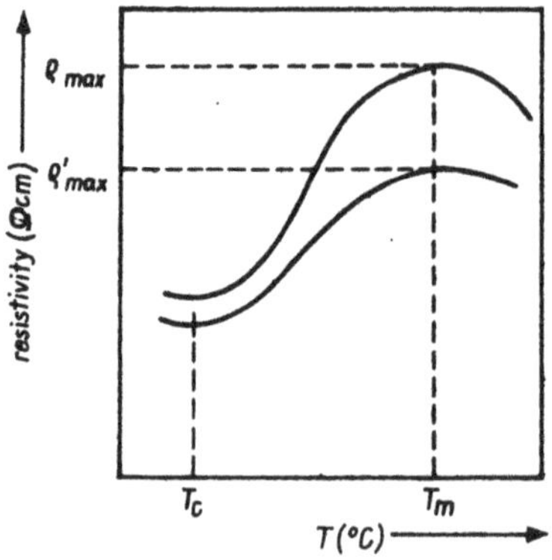

Fig. 3. Resistivity of PTC $BaTiO_3$ before (ϱ_{max}) and after (ϱ'_{max}) reduction ($\varrho_{max}/\varrho'_{max} \approx 10^{-2}$)

Otherwise, let us consider a radiation damage in the sample caused by the positron isotopic source which affects the experimental data significantly. The curves in Fig. 1 show that the vacancy concentration changes monotonously, this means that damage can be ignored.

Table 3 listed the positron parameters of two kinds of PTC $BaTiO_3$ ceramic samples with excess BaO and TiO_2, respectively, where T_2 is fixed before and after the reduction process because the kinds of vacancies were almost unchanged in argon gas at 350 °C for several hours.

The results from Table 3 show that the positron lifetime parameters are almost unchanged, whether the samples were reduced or not although the resistivity of the PTC samples changed by one to two orders of magnitude through the reduction processing (Fig. 3), indicating that both the type of vacancies and their concentration are not changed.

Discussion Pure single crystal $BaTiO_3$ does not exhibit the PTC anomaly, but the PTC effect can be achieved by doping. The PTC effect is related to the donor dopant concentration, thermal processing condition, as well as the gas during processing. For instance, if the oxygen pressure is low, the sample has no PTC effect. Existent experimental evidence has suggested that the mechanism of the PTC effect is a grain boundary rather than a bulk phenomenon /1/. A Schottky barrier layer model was proposed by Heywang /8/ and Jonker /9/ for n-type $BaTiO_3$ with semiconductor partially occupied donor states generated by chemisorption of oxygen. Thus electron depletion layers are formed at the grain boundary. An alternate view states that the traps are occupying a layer of variable thickness beneath the grain surface, which results from a gradient in the concentration of barium vacancies /10/.

From our experiments, we can conclude that the PTC effect is due to oxygen on the grain boundary rather than vacancies, and the experiment demonstrates that the Heywang-Jonker model is more reasonable.

References

/1/ L.M. LEZINSON, in: Grain Boundary Phenomenon in Electric Ceramics, Amer. Ceram. Soc. Inc.,1981 (p. 138).

/2/ Y.Y. SHEN, Electric Cells and Material (China) 12, 33 (1982).

/3/ G.H. WANG, D.X. SHEN, Y.Y. LU,et al., Nuclear Instrum. and Methods B14, 555 (1986).

/4/ W. BRANDT, Appl. Phys. 5, 1 (1974).

/5/ M. BERTOLACEINI et al., J. Phys. C 4, 734 (1971).

/6/ P. KIRKEGAARD, Comp. Phys. Commun. 7, 401 (1974).

/7/ Y.J. HE et al., J. Chin. Silicate Soc. 12, 3 (1984).

/8/ W. HEYWANG, J. Amer. Ceram. Soc. 47, 484 (1964).

/9/ G.H. JONKER, Solid State Electronics 7, 895 (1964).

/10/ J. DANIELS et al., Philips Tech. Rev. 38, 73 (1978/79).

(Received July 26, 1988)

phys. stat. sol. (a) 111, K129 (1989)

Subject classification: 78.60; S8.11

Sektion Physik der Humboldt-Universität zu Berlin,
Bereich Halbleiteroptik[1)] (a) and
Zentralinstitut für Elektronenphysik
der Akademie der Wissenschaften der DDR, Berlin (b)

Stability of Sm Centers in Electroluminescent $ZnS:SmF_3$ MISIM Structures during Ageing

By

K. LÖBE (a), R. BOYN (a), and R. REETZ (b)

The high-field ac electroluminescence of rare-earth (RE) doped ZnS films in MISIM type structures is of much interest in connection with the possible development of multicolor light-emitting displays /1/. One problem frequently encountered with such structures is the rather rapid degradation of luminescent efficiency found during operation.

It has been recently shown /2/ that for similar structures involving ZnS:Mn degradation is due to processes related to the insulators and/or the insulator - semiconductor interfaces. On the other hand, it is sometimes suspected that in the ZnS:RE case a decrease in the concentrations of the emitting RE centers occurring under electric load conditions (e.g. by outward drift of the RE ions or clustering due to supersaturation) could contribute to the ageing effect. In this note we present experimental results concerning this question, which were obtained on structures involving $ZnS:SmF_3$ active layers. Systems of this type are of particular interest in view of their relatively efficient red electroluminescence /3, 4/.

The structures consisted of layers deposited on glass substrates in the succession ITO-oxidic insulator-$ZnS:SmF_3$-oxidic insulator-Al. Electron-beam evaporation was employed for the electrode and insulator layers, and the active layer was produced by coevaporation of ZnS (electron-beam evaporation) and SmF_3 (thermal evaporation) from separate crucibles. The film thickness were some hundred nm. The structures were heat-treated at 350 °C. Electroluminescence was excited by sinusoidal voltage.

Fig. 1 shows a brightness versus voltage dependence measured after fabrication of the structure, which is similar to available data for $ZnS:SmF_3$ /3/

1) Invalidenstr. 110, DDR-1040 Berlin, GDR.

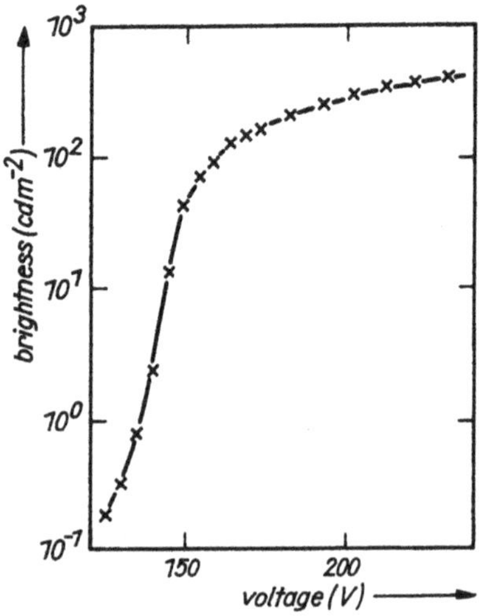

Fig. 1. Brightness-voltage dependence measured at 2 kHz and 300 K

including the values of the threshold voltage (U_{thr}) and of the maximum brightness attained (about 800 cd/m^2 at 5 kHz), and the voltage range of "saturation". When operating the structures for times up to 1000 h, we found a continuous increase of U_{thr} by about 20 or 30 V, and a continuous decrease of brightness in the voltage range of Fig. 1. (For most structures investigated the brightness dropped to unmeasurable values at times of some hundred hours.)

In Fig. 2 we present electroluminescence spectra of a long-lived structure in the range of the $^4G_{5/2} \rightarrow {}^6H_{9/2}$ Sm^{3+} transitions measured at various times during operation. The operating voltage was always about 10 V above U_{thr}; it was gradually increased because of the slow rise of U_{thr}. (In view of the degradation of luminescence, the 104 h, 500 h, and 995 h spectra were taken at a voltage somewhat higher than the operating voltage normally applied in this time range.)

The complex structure and the width of the spectra in Fig. 2 show that several centers contribute to the emission. (This situation is typical for RE doped II-VI compounds; see /5/ for a recent review.) The majority of these are different from the centers studied by us, with the aid of the dye-laser selective excitation technique, for ZnS films and bulk crystals doped with elemental Sm /6, 7/.

The important result appearing in Fig. 2 is that, within experimental accuracy, the emission spectrum does not change with time of operation. Thus we can conclude that, for the moderate electric loads applied, there is no change in the concentrations of the emitting centers (excluding the improbable case that all types of centers are eliminated at the same rate).

It should be noted that during the ageing procedure an increasing number of "black dots" (accompanied by the formation of pinholes in the structure)

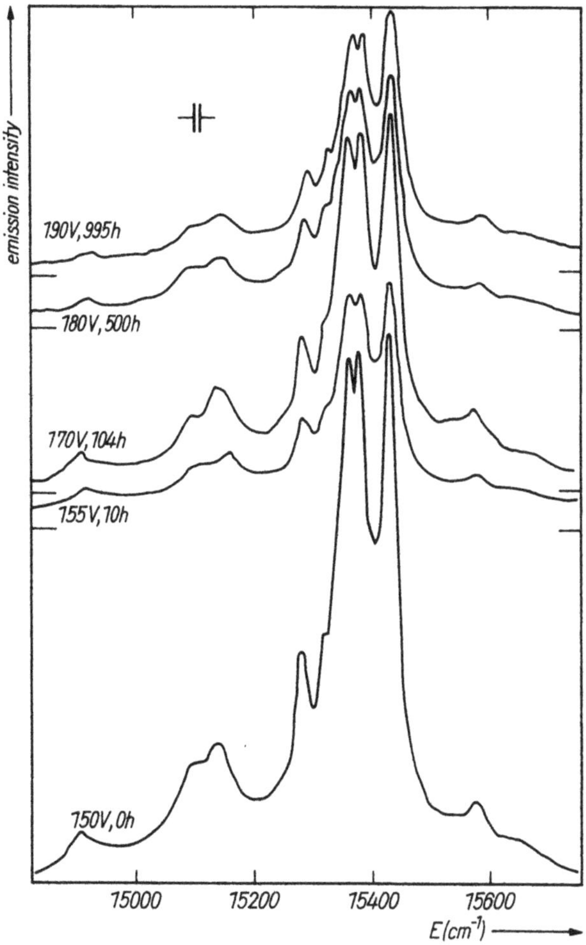

Fig. 2. Electroluminescence spectra due to $^4G_{5/2} \rightarrow {}^6H_{9/2}$ Sm^{3+} transitions measured at various times after the beginning of operation, at 2 kHz and 300 K. The measuring voltages are indicated for each spectrum. The emission intensity scales are different for various spectra. These data were obtained on a structure which had the same parameters, and was fabricated under the same conditions, as the structure used to measure the curve in Fig. 1.

were observed; their appearance is most probably initiated by local breakdown in the insulator films. As a consequence, degradation of brightness in the nondestructed areas should actually be considerably slower than appearing in the integral measurement.

The authors are grateful to Chr. Hottewitzsch (WFB) for measuring brightness-voltage characteristics.

References

/1/ R. MACH and G.O. MÜLLER, phys. stat. sol. (a) 69, 11 (1982).

/2/ G.O. MÜLLER, R. MACH, R. REETZ, and G.U. REINSPERGER, contributed paper to SID conference, Anaheim (USA) 1988.

/3/ J. OHWAKI, B. TSUJIYAMA, and H. KOZAWAGUCHI, Japan. J. appl. Phys. 23, 699 (1984).

/4/ T. TOHDA, Y. FUJITA, T. MATSUOKA, and A. ABE, Appl. Phys. Letters 48, 95 (1986).

/5/ R. BOYN, phys. stat. sol. (b) 148, 11 (1988).

/6/ K. LÖBE and R. BOYN, phys. stat. sol. (b) 140, K91 (1987).

/7/ K. LÖBE and R. BOYN, phys. stat. sol. (b) 148, 263 (1988).

(Received December 14, 1988)

Manuscripts and letters for physica status solidi (b) — basic research and physica status solidi (a) — applied research

should be sent to

Leipziger Straße 3—4, P. B. 1233, DDR-1086 Berlin

or

Kunst und Wissen, Erich Bieber OHG, Postfach 330104, D-1000 Berlin 33.

The object of "physica status solidi (b)" and "physica status solidi (a)" is to publish rapidly important and topical results concerning fundamental research in the field of solid state physics and aspects of application and experimental techniques of fundamental solid state research, respectively. The issues of "physica status solidi (b)" and "physica status solidi (a)" will be published on the 1st and the 16th day of each month, respectively. Either series will contain the following categories:

1. Review articles written by arrangement with the editors. They will be remunerated.
2. Original papers selected from those which have not been submitted for publication elsewhere, after being reviewed. (The latest date for acceptance of contributions is the 8th and the 23rd day of each month for "physica status solidi (b)" and "physica status solidi (a)", respectively, 50 days before the delivery of the journal.)
3. Short notes published in the off-set part of the journal. They are, if necessary after being reviewed, selected by editorial staff. (The latest date for acceptance of manuscripts for this part is the 12th and the 27th day of each month for "physica status solidi (b)" and "physica status solidi (a)", respectively, 16 days before the delivery of the journal.)
4. Pre-printed titles of papers to be published in the next issue of both series.

Editorial Information

Manuscripts should be submitted preferentially in English. Besides, papers in German, French, and Russian are admitted. The final decision on the incorporation of the papers in the series (b) or (a) is taken by the editorial staff after consultation of the referee. Papers on liquids are only published in physica status solidi if the problems treated are also relevant for the solid state. Authors are asked to observe the following rules:

1. Manuscripts have to be written as consisely as possible. At most, 20 and 4 pages (28 lines per page; abstracts, tables, references, figures, and captions of figures included, where 3 figures are counted as about one page) are admitted for original papers and short notes, respectively. All manuscripts should be typewritten on one side only. Two copies are required. Badly arranged manuscripts are returned.
2. The author's name, initials, and institution should be given.
3. One abstract of about ten lines in English and one in German, French, or Russian are required for original papers.
4. Because of lack of space conclusions (if any) must not have the character of a summary (i. e. substantially a repetition of the contents of the abstract).
5. Formulae should not be written too small and not with pencil. Separate lines for formulae are desirable. Indistinct writing will lead to a return of the manuscript.
6. Drawings should be enclosed on separate sheets of paper, not exceeding the maximum and minimum sizes of 20×30 and 10×15 cm^2, respectively. Photographs should be submitted, on white paper, only for review articles and original papers. Duplicates are required.
7. Captions for the figures should be submitted on separate sheets of paper, and special instructions with regard to the size of illustrations should be given on these.
8. References should be numbered in order of appearance and should be given, on a separate sheet, in the form used in physica status solidi.
9. SI-units should be used.
10. Each paper must contain a proposal for the subject classification according to the classification scheme appearing in the January issues of each annual set.

The authors will be sent galley proofs of review articles and original papers. Proof reading of short notes and pre-printed titles will be done by the editorial staff.

For each original paper and short note a total of 75 reprints is available free of charge. The number of reprints free of charge per review article is 25. Requests for additional reprints (against payment) should be given together with the submitted manuscripts. **The deadline for orders of additional reprints is the 15th and the 30th of each month** for "physica status solidi (b)" and "physica status solidi (a)", respectively, two weeks before the delivery of the journal.

Substance Classification*)

S1. Transition metals and their alloys and compounds (for oxides see S.10)
- S1.1. Fe
- S1.2. Other 3d metals
- S1.3. Noble metals
- S1.4. Lanthanides
- S1.5. Actinides
- S1.61. Carbides, silicides, germanides
- S1.62. Nitrides, phosphides, arsenides
- S1.63. Sulfides, selenides, tellurides

S2. Alkali metals and their alloys and compounds (for alkali halides see S9.11)

S3. Alkaline earth metals and their alloys and compounds (for alkaline earth halides, see S9.12)

S4. Other metals and their alloys

S5. Nonmetallic elements
- S5.11. Silicon
- S5.12. Germanium
- S5.2. Solidified Gases

S6. Group IV compounds (for transition metal compounds, see S1.61)

S7. Group V compounds (for transition metal compounds, see S1.62)
- S7.1. III–V compounds
- S7.11. Phosphides
- S7.12. Arsenides
- S7.13. Antimonides
- S7.15. Mixed crystals
- S7.16. Higher compounds

S8. Chalcogenides (for transition metal compounds, see S1.63)
- S8.1. II–VI compounds
- S8.11. Sulfides
- S8.12. Selenides
- S8.13. Tellurides
- S8.15. Mixed crystals
- S8.16. Higher compounds

S9. Halides
- S9.11. Alkali halides
- S9.12. Alkaline earth halides
- S9.15. Mixed crystals
- S9.16. Higher compounds

S10. Oxides
- S10.1. Simple oxides
- S10.15. Mixed crystals

S11. Other higher compounds
- S11.1. Salts of oxy-acids
- S11.2. Spinels, garnets, ferrites

S12. Organic compounds
- S12.1. Quasi-one-dimensional systems

Attention

The *Classification Scheme* on cover two is reproduced and used with permission of International Council for Scientific and Technical Information (ICSTI).
Because of the lack of space only the main points of the Classification Scheme are printed here. The *detailed subdivided scheme* is to be found in the January issues of each annual set and in the Index Volumes from 1986 on.
The *Subject Classification* given at the head of each paper begins with the main subject matter of the paper, and this is followed, if necessary, by figures referring to subsidiary subject matter and the type of solids investigated.
The papers published in this issue are listed in the order of their subject classification in the *Systematic List* at the end of the table of contents.

*) Outside the ICSTI Classification for Physics.

Bei Fragen zur Produktsicherheit wenden Sie sich bitte an:
If you have any questions regarding product safety,
please contact:

Walter de Gruyter GmbH
Genthiner Straße 13
10785 Berlin
productsafety@degruyterbrill.com